Communications in Computer and Information Science 2003

Rationale

The CCIS series is devoted to the publication of proceedings of computer science conferences. Its aim is to efficiently disseminate original research results in informatics in printed and electronic form. While the focus is on publication of peer-reviewed full papers presenting mature work, inclusion of reviewed short papers reporting on work in progress is welcome, too. Besides globally relevant meetings with internationally representative program committees guaranteeing a strict peer-reviewing and paper selection process, conferences run by societies or of high regional or national relevance are also considered for publication.

Topics

The topical scope of CCIS spans the entire spectrum of informatics ranging from foundational topics in the theory of computing to information and communications science and technology and a broad variety of interdisciplinary application fields.

Information for Volume Editors and Authors

Publication in CCIS is free of charge. No royalties are paid, however, we offer registered conference participants temporary free access to the online version of the conference proceedings on SpringerLink (http://link.springer.com) by means of an http referrer from the conference website and/or a number of complimentary printed copies, as specified in the official acceptance email of the event.

CCIS proceedings can be published in time for distribution at conferences or as post-proceedings, and delivered in the form of printed books and/or electronically as USBs and/or e-content licenses for accessing proceedings at SpringerLink. Furthermore, CCIS proceedings are included in the CCIS electronic book series hosted in the SpringerLink digital library at http://link.springer.com/bookseries/7899. Conferences publishing in CCIS are allowed to use Online Conference Service (OCS) for managing the whole proceedings lifecycle (from submission and reviewing to preparing for publication) free of charge.

Publication process

The language of publication is exclusively English. Authors publishing in CCIS have to sign the Springer CCIS copyright transfer form, however, they are free to use their material published in CCIS for substantially changed, more elaborate subsequent publications elsewhere. For the preparation of the camera-ready papers/files, authors have to strictly adhere to the Springer CCIS Authors' Instructions and are strongly encouraged to use the CCIS LaTeX style files or templates.

Abstracting/Indexing

CCIS is abstracted/indexed in DBLP, Google Scholar, EI-Compendex, Mathematical Reviews, SCImago, Scopus. CCIS volumes are also submitted for the inclusion in ISI Proceedings.

How to start

To start the evaluation of your proposal for inclusion in the CCIS series, please send an e-mail to ccis@springer.com.

Preface

The objective of the second International Conference on Biomedical Engineering Science and Technology (ICBEST): Roadway from Laboratory to Market was to establish itself as a leading international forum for researchers and industry practitioners to discuss the most recent fundamental advancements in the state of the art and to put cutting-edge technologies in biomedical engineering into practice. The conference was centred around the theme "Computing in Biomedical Research". Researchers, as well as professionals from both the business world and academic institutions, provided contributions in order to facilitate the exchange of ideas and information regarding the most recent developments in the field of biomedical engineering science and technology.

The previous ICBEST event was held in 2018, with the themes: biomedical signal processing, medical image processing, biomedical instrumentation, etc. The second ICBEST 2023 was held at the National Institute of Technology Raipur (NIT Raipur), Raipur (India), on February 10–11, 2023. ICBEST 2023 was organised in hybrid mode, i.e., the research papers were presented online and offline in both modes. Two eminent academicians, Guoan Zheng from the University of Connecticut, USA and Sanjeev Kumar Mahto from IIT-BHU, India, delivered invited talks on recent trends in biomedical research as keynote speakers. Research articles focused on artificial intelligence in healthcare, computational mechanics in healthcare, and health informatics. This volume includes the role of computing in advancing biomedical research and its potential to bring new biomedical technologies to market. The 32 full papers included in this book underwent a thorough review process, resulting in the selection of the best submissions from a pool of 60. The type of review was double blind, with at least two reviewers for each paper.

We would like to thank all those who submitted papers, the members of the programme committee and the external reviewers for their valuable reviews, the chairs of the special track, as well as the members of the organising committee, for all their support and contributions to the success of the conference.

February 2023

Bikesh Kumar Singh
G.R. Sinha
Rishikesh Pandey

Organization

Chief Patron

A. B. Soni	National Institute of Technology Raipur, India

Patron

P. Diwan	National Institute of Technology Raipur, India

Program Committee Chairs

Bikesh Kumar Singh	National Institute of Technology Raipur, India
G. R. Sinha	International Institute of Information Technology, Bangalore, India
Rishikesh Pandey	University of Connecticut, USA

Secretaries

Saurabh Gupta	National Institute of Technology Raipur, India
Nishant Kumar Singh	National Institute of Technology Raipur, India
M. Marieswaran	National Institute of Technology Raipur, India

Technical Coordinators

Arindam Bit	National Institute of Technology Raipur, India
Neelamshobha Nirala	National Institute of Technology Raipur, India
Sumit Kumar Banchhor	National Institute of Technology Raipur, India
Sharda Gupta	National Institute of Technology Raipur, India
Raj Kumar Sahu	National Institute of Technology Raipur, India

Student Coordinators

N. P. Guhan Seshadri	National Institute of Technology Raipur, India
Kushangi Atrey	National Institute of Technology Raipur, India
Yogesh Sharma	National Institute of Technology Raipur, India
Pankaj Jain	National Institute of Technology Raipur, India
Divya Singh	National Institute of Technology Raipur, India
Bhanupriya Mishra	National Institute of Technology Raipur, India
Resham Raj Shivwanshi	National Institute of Technology Raipur, India
Padmini Sahu	National Institute of Technology Raipur, India
Nitesh Kumar Singh	National Institute of Technology Raipur, India
Hardik Thakkar	National Institute of Technology Raipur, India
Sumit Kumar Roy	National Institute of Technology Raipur, India
Ashish Kumar Dewangan	National Institute of Technology Raipur, India
Vaibhav Koshta	National Institute of Technology Raipur, India
Soumya Jain	National Institute of Technology Raipur, India

Advisory and Technical Committee

Shrish Verma	National Institute of Technology Raipur, India
R. K. Tripathi	National Institute of Technology Raipur, India
P. Y. Dhekne	National Institute of Technology Raipur, India
Arif Khan	National Institute of Technology Raipur, India
S. K. Mahto	IIT BHU, India
Neeraj Sharma	IIT BHU, India
O. P. Vyas	IIIT Allahabad, India
G. R. Sinha	IIIT Bangalore, India
Anuj Jain,	MNNIT Allahabad, India
Rakesh Kumar Sinha	BIT Mesra, India
Aparajita Ojha	IIITDM Jabalpur, India
Ashutosh Kumar Singh	NIT Kurukshetra, India
Kavita Thakur	Pt. Ravishakar Shukla University, India
Deepak Garg	Bennett University, India
Arnab Bhattacharya	IIT Kanpur, India
Ravi Prakash Tewari	MNNIT Allahabad, India
B. Geethanjali	SSN College of Engineering, India
Deepak Joshi	IIT Delhi, India
J. Sivaraman	NIT Rourkela, India
Abhishek K. Tiwari	MNNIT Allahabad, India
Mitul K. Ahirwar	MANIT Bhopal, India
Xiaohong Gao	Middlesex University London, UK

Contents

An Enhancement in K-means Algorithm for Automatic Ultrasound Image Segmentation

Lipismita Panigrahi[1](✉) and Raghab Ranjan Panigrahi[2]

[1] Department of Computer Science, University of Texas at San Antonio, San Antonio, Texas, USA
lipismita.panigrahi@utsa.edu

[2] Department of Computer Science and Engineering, SOA University, Bhubaneswar, India

Abstract. Breast malignancy is a relatively frequent disease that affects people all over the world. When interpreting the lesion component of medical images, inter- and intra-observer errors frequently happen, leading to considerable diversity in result interpretations. To combat this variability, computer-aided diagnosis (CAD) systems are essential. Automatic segmentation is an essential and critical step in CAD systems toward boundary detection, feature extraction, and classification. The aim of this study is to incorporate an Ant colony optimization (ACO) to initialize the cluster center and replace the Euclidean distance (ED) with the Manhattan distance (MD), in the traditional K-means algorithm to segment the BUS images with maximal area preservation. The Jaccard index (JI), Dice similarity (DS), and Area difference (AD) are the cluster validation measures used to compare the efficiency of the proposed method with other state-of-the-art methods. A total of 1293 BUS images are used in this study. According to the quantitative experimental findings, the suggested method can successfully segment the BUS images with an accuracy of 91.66%. Compared to existing methods, the proposed approach accomplishes segmentation more quickly and accurately.

Keywords: BUS images · segmentation · ACO · Vector field Convolution · K-Means algorithm

1 Introduction

Breast malignancy is a most prevalent cause of death for many adolescent women worldwide [1]. It needs early discovery to reduced early mortality [2]. Combining breast ultrasound (BUS) imaging with machine learning can lead to better breast cancer detection, segmentation, and classification results [3–6]. During the exploratory phase of microarray data analysis, biologists often use clustering algorithms to identify related biological groups. Users may wish to consider the various clustering algorithms available in the machine learning literature and choose the clustering method that best suits their data set or application. From last decades there are different clustering algorithms used for BUS image segmentation, but K-means algorithm gets popularity among all [4, 7–10]. The following section describes a few of the clustering algorithms utilised for medical image segmentation.

B. K. Singh et al. (Eds.): ICBEST 2023, CCIS 2003, pp. 1–8, 2024.
https://doi.org/10.1007/978-3-031-54547-4_1

2 Related Work

Gumaei, A. *et. al* [7] proposed a model by utilising Gamma distribution to simulate the k-means approach. Later on, Panigrahi *et. al* [4] proposed a methodological approach that employs textural cues to automatically identify a seed point in a BUS image. Subsequently, the seed point is used to determine the segmentation methods like Otsu, Active contour, and K-means and presents a relative analysis between them. Finally, the quantitative experimental results conclude that the K-means algorithm outperforms the others. Subsequently, Samundeeswari, E. S. *et. al* [11] proposed a clustering-based segmentation method in which an Ant Colony Optimization (ACO) is incorporated in a traditional K-Means algorithm to segment the region with maximal boundary preservation. Further, Chowdhary *et. al* [9] proposed an effective segmentation and categorization system for medical images. In this study, the author introduced intuitionistic possibilistic fuzzy c-mean (IPFCM) and Fuzzy Support Vector Machine (SVM) for achieving an optimal categorization result. In this study, the author hybridized Intuitionist fuzzy c-mean (IFCM) and possibilistic fuzzy c-mean (PFCM) algorithms to overcome the shortcoming of FCM. Further, Hassan *et. al* [10] present a comparative analysis of many clustering methods for identifying breast malignancy in terms of their degree of accuracy and also analyze the pros and cons of the employed methods for producing mammography and ultrasound medical images in addition to describing the imaging procedure. Later on Chandra *et. al* [11] proposed super pixel-based disease localization and severity assessment (DLSA) framework for medical images.

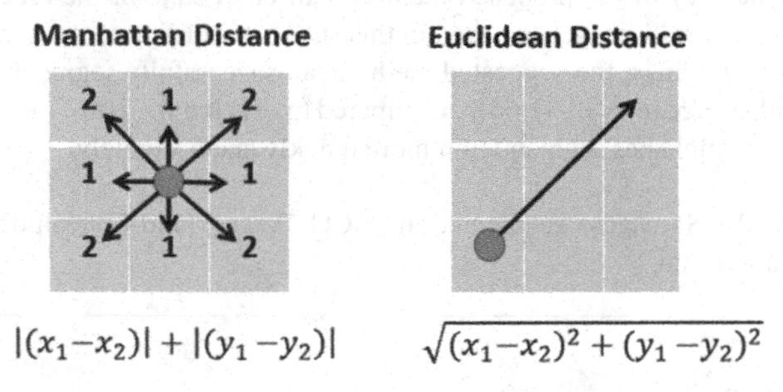

Fig. 1. Difference between ED and Manhattan distance

Despite the K-means method has been frequently utilized for BUS image segmentation, there are certain limitations.

- The random initialization of the cluster centers produces unstable cluster outcomes [12].
- Due to its sensitivity to outliers and noise, it might not converge to the global optimum [12]
- Images with minimal noise can be segmented using Euclidean distance (ED), while images with outliers and other imaging errors cannot be segmented [13]. Figure 1 presents the difference between ED and Manhattan distance (MD).
- ED is slower than MD [14].

- ED is not scaled invariant and it is less useful when the dimensionality of the data sets is increased [15].

Therefore, this study suggested an automatic segmentation method for BUS images with maximum boundary preservation by incorporating an ACO method for automatically initializing the cluster center and replacing the ED with the MD, in a traditional K-means algorithm to overcome these limitations.

3 Proposed Method

3.1 Image Acquisition

A total of 1293 breast US images were used in this investigation. From which 620 images were collected from the Dr. Bhimrao Ambedkar Memorial Hospital in Raipur, Chhattisgarh. The remaining images were collected from online sources [16]. Breast US images have low resolution, poor contrast, and indistinct edges because they are sensitive to noise and speckle, a crucial artefact. As a result, speckle reducing anisotropic diffusion (SRAD) filter is employed in this study to create a multi-scale image series and minimise noise and speckle at different sizes [2].

3.2 Proposed Segmentation Method

The proposed method is shown in Fig. 2. The objective function J_m of the proposed method is derived from the objective function of the conventional K-means clustering algorithm. However, the standard K-means clustering algorithm has a number of drawbacks that are discussed in Sect. 2. The following can be done to overcome the constraints indicated using the suggested method:

1. In an effort to address the first issue, the suggested approach used an ACO for automatic initialize the cluster center. The ACO method for automatically initializing the cluster center was proposed by Samundeeswari, E. S. *et. al* [12]. Insects called ants reside in a group known as a colony. In its pursuit of nourishment, each ant develops a unique solution that benefits the entire colony. The greatest pheromone deposition, a chemical signal that affects how colony members behave, helps to pinpoint the quickest route between a food source and an ant nest. Each unladen ant starts off carrying weight and looks for a local minima until the best solution is identified to drop the weight or it takes K steps in accordance with the neighbourhood search function as described in Eq. (1).

$$f(n) = \max\left(0, \frac{1}{\sigma^2}\sum_{j\in k}\left(1 - \frac{d(i,j)}{\alpha}\right)\right) \tag{1}$$

where, $d(i,j) \in [0, 1]$ is the difference between the data points i and j, $\alpha(i,j) \in [0, 1]$ [0, 1] is an element of scaling, $\sigma^2 = ((K * 2) + 1)^2$ is the scope of the nearby neighbourhood search, and 'n' might range between 1 to σ^2. The ACO algorithm for initialize cluster center and the experimental result can be found in [12].

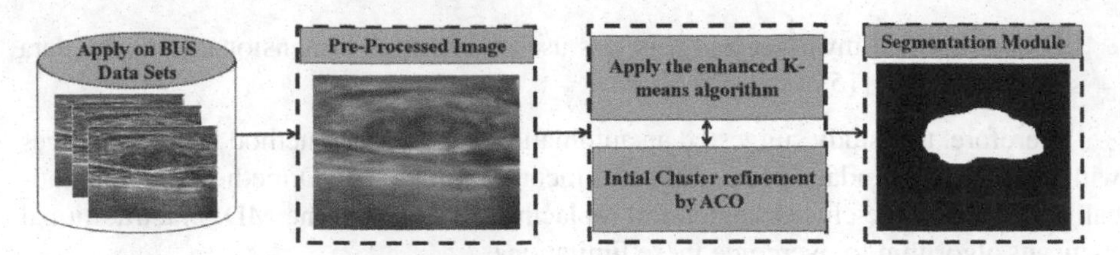

Fig. 2. Overview of the Proposed method.

2. In order to overcome the 2nd–4th shortcoming, the proposed method incorporated an MD in place of an ED in the traditional K-means algorithm. A cyclic process for automatic segmentation is described in Algorithm 1.

Algorithm 1: Enhanced K-Means clustering Algorithm

Input:

(1) $I = \{i_1, i_2, \ldots i_n\}, i_i \in R^N$, // the BUS data set

(2) Initialize C, $2 \le C \le n$, // the number of clusters

Output: Segmented BUS images.

Algorithm:

Step 1: Insert a input image I_i.

Step 2: Generate a de-noise image Di.

Step 3: Define the number of clusters C.

Step 4: Initialize the cluster centroids 'a_j' using ACO algorithm where j=1..C.

Step 5: Measure the difference between the cluster centroids 'a_j' and each data point by Eq. (2).

$$J = \sum_{i=1}^{n} \sum_{j=1}^{C} \left(\left| x_i - a_j \right| \right) \tag{2}$$

Step 6: Determine the new cluster centroid a_j by averaging the pixel values of the cluster it belongs to.

Step 7: If there is a relocation, proceed to step 5, repeat this process until the convergence is found,

Step 8: End

4 Experimental Evaluation and Result Discussion

This experiment is executed using an i5 processor, 8 GB of RAM, and Windows 10 with MATLAB 2022a. 1293 BUS images are used in this study, of which 528 were malignant and 765 were benign masses. Initially, the BUS images are preprocessed with an SRAD filter to produce de-noised images. Further, to solve the cluster initialization problem, the ACO method is incorporated. Finally, to segment, the BUS images with maximal area preservation, we replaced the ED with the MD in the traditional K-means algorithm and proposed an enhanced K-means algorithm. The segmentation result of the proposed method is presented in Fig. 3. Figure 3. (a), (b), and (c) represent the original image, the ground truth image given by the radiologist, and the preprocessed image, respectively. The segmented image by K-means, Fuzzy C-means, Ostu, and the proposed method are presented in Fig. 3 (d-f) respectively.

To prove the efficiency of the proposed method with other cutting-edge methods, the Jaccard index (JI), Dice similarity (DS), and Area difference (AD) are the cluster validation measures used, and the mathematical equations are described in Table 1. The performance measures [1], such as false positive (FP), true positive (TP), true negative (TN), and false negative (FN), that are used to calculate the quantitative and qualitative measures are also described in Eq. (3–6). Where I_{GT} and I_{Auto} are the resultant images of the ground truth and automated model, respectively. Tables 2 and 3 describe the quantitative and qualitative performance results of the clustering-based segmentation method. Experimental findings imply that the proposed approach can be used as an expert system to assist medical professionals in detecting breast lesions by providing objective evidence. The results obtained so far are promising and effective compared to state-of-the-art algorithms. Finally, the computational time discussed in Table 4 proves the computational efficiency of the proposed method. The proposed enhanced K-means clustering method achieves 91.66% accuracy, which is better than the other cutting-edge method.

$$TP = |I_{GT} \cap I_{Auto}| \quad (3)$$

$$FP = |I_{GT} \cup I_{Auto} - I_{GT}| \quad (4)$$

$$FP = |I_{GT} \cup I_{Auto} - I_{Auto}| \quad (5)$$

$$TN = T - |I_{GT} \cup I_{Auto}| \quad (6)$$

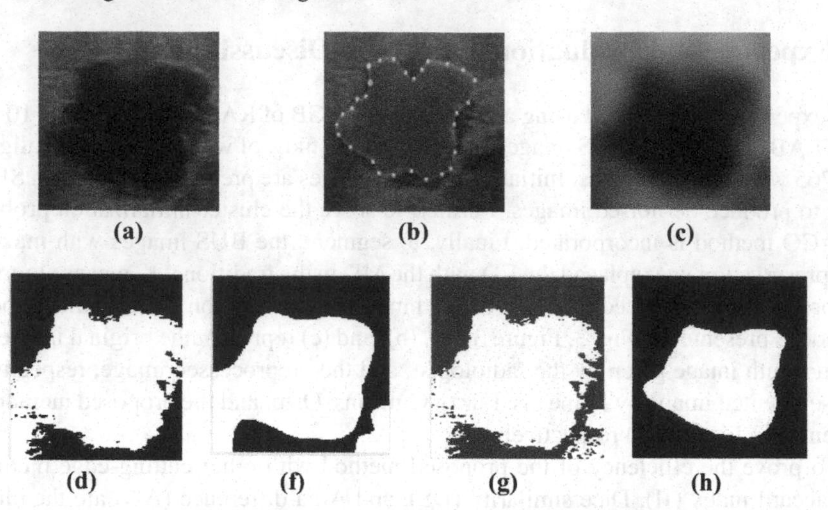

Fig. 3. (a) Original image, (b) Ground truth image given by radiologist, (c) preprocessed image, segmented image by (d) K-means (ED), (e) Fuzzy C-means, (f) Ostu and (g) proposed method

Table 1. Quantitative and Qualitative measures used to proof the efficiency of the proposed method [17]

Performance measures (%)	Mathematical equation
ACC	$\frac{TP+TN}{TP+TN+FP+FN}$
Recall	$\frac{TP}{TP+FN}$
Specificity	$\frac{TN}{TN+FP}$
Precision	$\frac{TP}{TP+FP}$
F-Score	$2 \times \frac{\Pr ecision \times \text{Recall}}{\Pr ecision + \text{Recall}}$
JI	$\frac{\lvert I_{GT} \cap I_{Auto} \rvert}{\lvert I_{Auto} \rvert + \lvert I_{GT} \rvert - \lvert I_{GT} \cap I_{Auto} \rvert}$
DS	$2 \times \frac{\lvert I_{GT} \cap I_{Auto} \rvert}{\lvert I_{Auto} \rvert + \lvert I_{GT} \rvert}$
AD	$\frac{A_{GT} - A_{Auto}}{A_{GT}} \times 100$

Table 2. Quantitative performance (%) result of clustering based segmentation method

Methods	ACC	F-score	Specificity	Precision	Recall
K-means (with ED)	75.02	74.11	14.76	73.20	74.88
Fuzzy C-means	88.33	87.21	13.55	87.56	83.23
Ostu	69.26	68.43	16.78	66.36	65.78
Proposed	**91.66**	**90.23**	**12.34**	**90.99**	**91.23**

Table 3. Qualitative performance result of clustering based segmentation method

Methods	JI	DS	AD
K-means (with ED)	0.621	0.632	3.663
Fuzzy C-means	0.734	0.724	1.712
Ostu	0.582	0.578	4.113
Proposed	**0.911**	**0.893**	**0.878**

Table 4. Computational time of the methods (sec)

Proposed	2.112
Methods	Times
K-means (with ED)	3.773
Fuzzy C-means	2.887
Ostu	4.334

5 Conclusions

Due to the variability in tissue texture, segmenting the tumor region in the BUS image is an important step. In order to characterize the lesion, a decent segmentation method must be created to segment the lesion region with an appropriate border. This work developed an enhanced K-Means by replacing the ED with MD and the initial cluster center initialized by an ACO method. It is proved that the proposed algorithm results in good segmentation with 91.66% which is better than the traditional K-means clustering. The proposed system is intended to function as a component of the CAD system to facilitate the detection of the lesion section for subsequent breast cancer diagnosis.

Acknowledgements. We are thankful to Dr. Satyabhuwan Singh Netam (Associate Professor and Head of Department) and Dr. Bhagyashri, Dr. Bhimrao Ambedkar Memorial Government Medical Hospital, Raipur, Chhattisgarh, India, for providing us the data. We are also thankful to ethical committee NIT Raipur for providing ethical permission.

References

1. Panigrahi, L., Verma, K., Singh, B.K.: Ultrasound image segmentation using a novel multi-scale Gaussian kernel fuzzy clustering and multi-scale vector field convolution. Expert Systems with Applications **115**, 486–498 (2019)
2. Panigrahi, L., Verma, K., Singh, B.K.: Hybrid segmentation method based on multi-scale Gaussian kernel fuzzy clustering with spatial bias correction and region-scalable fitting for breast US images. IET Comput. Vis. **12**, 1067–1077 (2018)
3. Singh, B.K., Verma, K., Panigrahi, L., Thoke, A.S.: Integrating radiologist feedback with computer aided diagnostic systems for breast cancer risk prediction in ultrasonic images: An experimental investigation in machine learning paradigm. Expert Syst. Appl. **90**, 209–223 (2017)
4. Panigrahi, L., Verma, K., Singh, B.K.: An enhancement in automatic seed selection in breast cancer ultrasound images using texture features: In: 2016 Int. Conf. Adv. Comput. Commun. Informatics, IEEE, pp. 1096–1102 (2016)
5. Panigrahi, L., Verma, K., Singh, B.K.: Evaluation of Image Features Within and Surrounding Lesion Region for Risk Stratification in Breast Ultrasound Images. IETE J. Res. 1–12 (2019)
6. Bafna, Y., Verma, K., Panigrahi, L., Sahu, S.P.: Automated boundary detection of breast cancer in ultrasound images using watershed algorithm, pp. 729–738. In Ambient communications and computer systems, Springer, Singapore (2018)
7. Gumaei, A., El-Zaart, A., Hussien, M., Berbar, M.: Breast segmentation using k-means algorithm with a mixture of gamma distributions. In: 2012 symposium on broadband networks and fast internet (RELABIRA), pp. 97–102 (2012)
8. Moftah, H.M., Azar, A.T., Al-Shammari, E.T., Ghali, N.I., Hassanien, A.E., Shoman, M.: Adaptive k-means clustering algorithm for MR breast image segmentation. Neural Computing and Applications. **24**(7), 1917–1928 (2014)
9. Chowdhary, C.L., Mittal, M., Pattanaik, P.A., Marszalek, Z.: An efficient segmentation and classification system in medical images using intuitionist possibilistic fuzzy C-mean clustering and fuzzy SVM algorithm. Sensors **20**(14), 3903 (2020)
10. Hassan, N.S., Abdulazeez, A.M., Zeebaree, D.Q., Hasan, D.A.: Medical images breast cancer segmentation based on K-Means clustering algorithm: a review. Ultrasound **27**, 28 (2021)
11. Chandra, T.B., Singh, B.K., Jain, D.: Disease Localization and Severity Assessment in Chest X-Ray Images using Multi-Stage Superpixels Classification. Computer Methods and Programs in Biomedicine, 106947 (2022)
12. Samundeeswari, E.S., Saranya, P.K., Manavalan, R.: Segmentation of breast ultrasound image using regularized K-means (ReKM) clustering. In: 2016 international conference on wireless communications, signal processing and networking (WiSPNET), pp. 1379–1383 (2016)
13. Chen, S., Zhang, D.: Robust image segmentation using fcm with spatial constrains based on new kernel-induced distance measure. IEEE Trans Systems Man Cybernet **34**, 1907–1916 (2004)
14. Sharma, S.K., Kumar, S.: Comparative analysis of Manhattan and Euclidean distance metrics using A* algorithm. J. Res. Eng. Appl. Sci **1**(4), 196–198 (2016)
15. Distance Measures in Data Science (linkedin.com) accessed on 13.12.2022.
16. Faust, O., Acharya, U.R., Meiburger, K.M., Molinari, F., Koh, J.E., Yeong, C.H., Ng, K.H.: Comparative assessment of texture features for the identification of cancer in ultrasound images: a review. Biocybernetics and Biomedical Engineering. **38**(2), 275–296 (2018)
17. Chandra, T.B., Verma, K., Jain, D., Netam, S.S.: Segmented Lung Boundary Correction in Chest Radiograph Using Context-Aware Adaptive Scan Algorithm, pp. 263–275. Advances in Biomedical Engineering and Technology. Lecture Notes in Bioengineering. Springer, Singapore (2021)

Silent Speech Interface Using Lip-Reading Methods

Raghupathy Jothibalaji(✉), S. Siva Adithya, N. V. Saravanan, and M. Dhanalakshmi

Department of Biomedical Engineering, Sri Sivasubramaniya Nadar College of Engineering, Chennai 603110, India
raghupathy1940@bme.ssn.edu.in

Abstract. A silent speech interface (SSI) maps articulatory movement data to forecast the speech of mute people. SSIs have vast potential for improving oral communication in people who have had laryngectomy surgery or those with severe voice abnormalities. In this paper, we present a lip-reading recognition algorithm to recognize English words from the lips when speaking. The lip contour and English words were distinguished using a variety of lip metric metrics, including width, height, contour points, area, and ratio (width/height). In this test, fifteen talkers mouthed (silently articulated) sentences to finish a phrase-reading assignment in an experimental setting. Among the fifteen participants, 89% and 81% of the mouthed phrases were correctly recognized for two different classifier cases respectively. The automated ROI detection, elimination of lip shape effects, and operation without prior data training are strengths of the proposed method. The experimental outcomes showed the viability and potential of lip-reading-based silent speech interfaces for future clinical applications.

Keywords: Silent Speech Interface · Total Laryngectomy · Image processing

1 Introduction

One of the most promising areas of study in human-computer interaction for helping persons with voice impairments is the Silent Speech Interface (SSI) [1]. Due to its distinctive characteristics and broad usage, The lip-reading domain of study is becoming more and more popular nowadays. Lip reading's primary objective is to recognize lips and keep track of changes in their position and orientation throughout a speech [2]. There are several uses for lipreading, all of which are heavily incorporated to improve the daily system. Lip-reading is employed as a security measure for people who suffer from hearing loss or other hearing problems. Lip-reading is also used in forensic investigations to extract spoken words from video and detect keywords from a person's facial expression. Lip-reading methods can also be employed as a security measure in establishments with restricted access, such as banks, colleges, and museums. There are several methods for locating and recognising lips using digital imagery, according to the Current Lip Detection Trends. They fall into one of two categories [7]:

B. K. Singh et al. (Eds.): ICBEST 2023, CCIS 2003, pp. 9–23, 2024.
https://doi.org/10.1007/978-3-031-54547-4_2

i. Lip detection methods based on models. Active Appearance Models (AAM), Active Shape Models (ASM), Shapes, and deformable templates, are examples of such models.
ii. Methods for detecting lips based on images. Spatial information, pixel color, intensity, corners, lines, edges, and motion are all examples.

Lips move in different ways for different people. The opening of the mouth varies from person to person. The area of the mouth either lengthens or contracts depending on the letter when a letter is spoken. Only the movements of the lips may be determined by their outward appearance. It is impossible to duplicate the texture of lip movement. Even identical twins cannot imitate lip movements. The vast majority of the research done on lip-reading makes use of 2D picture lip detection. When reading lips, the tongue and teeth are also taken into account as features. The primary target for this treatment is those who have had total laryngectomy and are suffering from speech difficulties such dysarthria, apraxia, and aphasia. Patients with advanced laryngeal cancer may benefit from total laryngectomy therapy. This procedure also results in the loss of one's natural voice, which has a detrimental impact on daily communication activities [3]. Patients are given the Lip Recognition System to help them speak normally after undergoing total laryngectomy surgery.

Audio-visual speech recognition (AVSR), visual-only speech recognition (VSR), speaker identification, intelligent human-computer interaction, human expression recognition, lip segmentation for VSR or AVSR, Automated speech recognition (ASR), and techniques for human lip recognition have been widely developed. But in a noisy environment, these techniques can fail [4].

In recent studies, face detection, mouth detection, the location of the region of interest (ROI), and the extraction of visual information are the three phases in the shared processing mechanism of the proposed algorithms for detecting human lips. Accurately identifying human lips from continuous image sequences still poses a difficulty due to variations in lip forms and hues. The methods mentioned above might be successful at lip contour detection or preliminary speech recognition, however, if you speak in English, they won't be able to grasp anything you say. They are affected by changes in the background situation as well as initial accuracy. Additionally, the numerous electrodes used and their placement on the subject's face in the current silent speech interface techniques are uncomfortable. These modern procedures use sophisticated machine learning algorithms to produce the desired outcomes, which can be uncomfortable for persons who may have undergone a laryngectomy [5].

The aforementioned problem was resolved by developing a lip recognition system that can distinguish each English word in real-time while speaking. When converting video data into frames, many lip properties, including height, breadth, area, and ratio of the lips, among others, were employed as features for lip-reading detection [6]. By establishing a connection between these extracted lip traits and the lip parameters of each English vowel, it would be feasible to properly identify every English word.

Our suggested method leverages lip pictures to get lip points from the video's extracted frames, giving us more accurate findings than the existing approach. The study presented here differs significantly from the researcher's earlier work in that it adds to the body of knowledge in the field by considering five unique common terms used on a daily basis for each subject and by including more data, whereas the earlier study complicated the process by including other languages and more complex words and phrases. The present method is more accurate than the previous works and correctly recognizes the terms given that English is the only language being utilized.

We executed the above procedures using two variants of the same model, the first with one thick layer and the second with two dense layers, assessing the correctness of the results from both methods, deciding which is superior, and providing justifications.

2 Methodology

The procedure follows input extraction in the form of short recordings of different speakers voicing out five common words. The workflow of our study involves:

1. Acquiring data set (audio and video format).
2. MATLAB is used to transform video data into frame-by-frame images.
3. Feature extraction and selection
4. Classification using different ResNet-50 classifiers
5. Comparing the results obtained from the classifiers

2.1 Dataset/Data Acquisition

The evaluation of classifiers is an essential component of any built system. As a result, a new database has been gathered, and evaluations are computed using the findings of previously read research papers. One issue that has to be fixed during testing is the system's resilience for lip recognition. It is necessary to collect data from a large number of people. Background information, such as the surrounding environment, is used to help it extract features more effectively.

The subject was requested to sit down in front of a camera that was roughly one meter distant. A colleague uses their phone to illustrate to the participant how to utilize the term. Throughout the recording, subjects were instructed to keep their heads still and display neutral facial expressions. Naturally, the video recordings show erratic head and torso movements. The recording was done in a room with less background noise and lamps that were pointed directly at the subjects' faces to improve extraction in MATLAB. The audio and video were captured with Canon EOS 200d cameras.

In total, 70 input films are obtained from 14 people, 6 of whom are female, and 8 of whom are male. (Audio, front video). Each participant was required to utter five frequently used words in daily life. The words "hi," "bye," "morning," "night," and "stop" were pronounced. The project only used the graphic component as the required input. All of our data contains an approximate 1-second video, and the number of frames calculated is 24 frames/second (per video). The sampling rate computed was 14400 Hz. A sample frame extracted from the video dataset of two different speakers speaking the word 'hello' is illustrated in Fig. 1. a and b.

2.2 Workflow

A workflow is described as a series of tasks that process data along a predetermined path from beginning to end. The data acquisition is in the form of short video clippings.

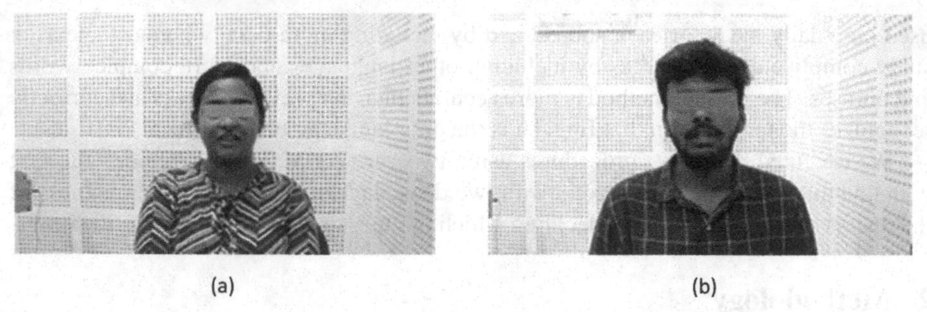

Fig. 1. (a) A Sample frame obtained from the video dataset of a single speaker and (b) One of the many frames obtained from the video dataset of a single speaker

The video is fed into a MATLAB program that creates a frames counter. A frame rate of 24 frames per second is achieved. As a consequence, 3304 total frames are produced. To identify the uttered word, specific mouth and lip traits are carefully selected and eliminated. Examples of features concerning lip detection are geometric-based such as the mouth's height and width and the corresponding area of the mouth. Many more features are taken into consideration to compute the word. The classifier model also receives the characteristics. The Extracted frames are augmented using an Image Data Generator and split into training, testing, and validation datasets. The Global Average Pooling layer and Dense layer are added to the ResNet50 model using a tensor flow library, and the model gets trained using the training dataset. Additionally, testing is implemented using a dataset that predicts the words from the frames (Fig. 2).

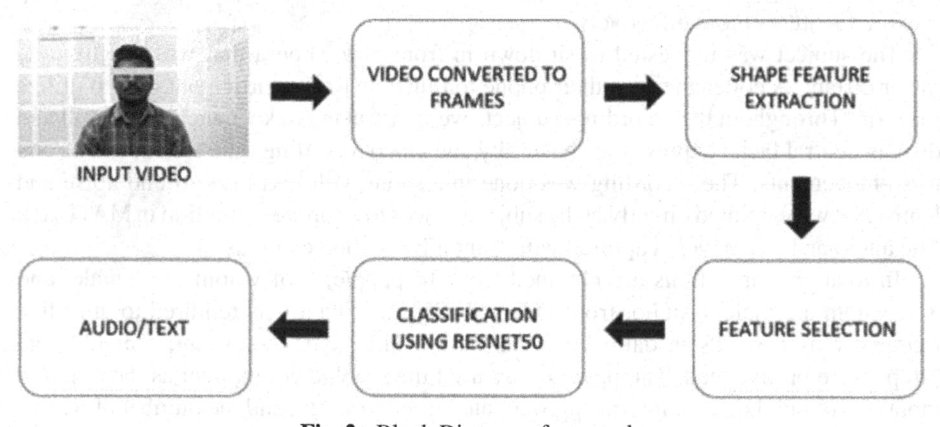

Fig. 2. Block Diagram of our study

2.3 Data Pre-processing

The received input video data is subjected to the frame converter algorithm built in MATLAB. With the help of this tool, frames are created from videos (pictures). Our dataset consists of around one-second-long video fragments. The entire number of frames for one speaker saying "bye" is shown in Fig. 3 below.

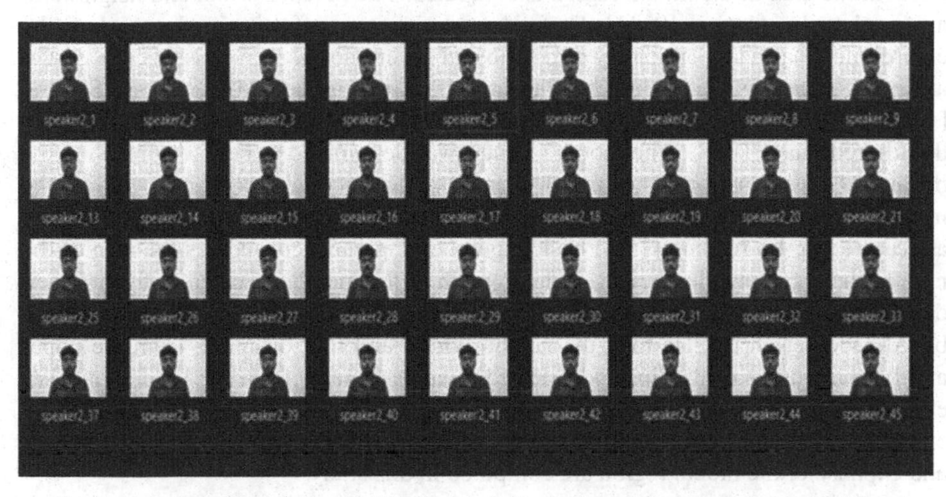

Fig. 3. Frames extracted from a video dataset of a single speaker uttering the word 'bye'

2.4 Feature Extraction

The enormous number of variables involved in complicated data analysis is one of the main challenges. A classification algorithm may overfit training examples and perform poorly on fresh samples [7].

Feature extraction is the process of transforming raw data into numerical attributes while possibly retaining the information from the original data set. The purpose of feature extraction is to reduce the amount of resources required to explain a large amount of data. It yields better results when compared to machine learning performed directly on the raw data. This section focuses on the many traits and qualities ingested to create a classification model to ultimately identify the words spoken using the SSI approach [7]. Using the methodologies described above, the two classifier variants that have been suggested could recognise the mouth ROI automatically, reduce the influence of individual differences, and outperform earlier techniques.

In the first instance, we gave a quick overview of both the human capacity for lip-reading and VSR. Lip-reading involves lip detection and minute changes in the lip position while the words are spoken [7]. The majority of the work in VSR was performed as part of the development of the AVSR structure since the visual signal correlates and complements the auditory signal and thus boosts the system's performance. With the visual indication, little was accomplished. The extraction of visual speech characteristics and feature detection are the two important phases in the bulk of suggested lip-reading techniques. The following are some of the existing feature extraction methods [8]:

2.4.1 Methods Based on Geometric Features

The lips are indeed the region of interest in lip geometry-based feature extraction. As a result, the area of the mouth region is computed. The mouth's width and height ratio are employed as features [9–11]. These characteristics are effective in places where the environment has a higher density of people and is very noisy. As a consequence, visual attributes are necessary. The lip's height and width ratios are adopted by Sahu and Sharma [12]. After extracting the Region of Interest, they examined the bivariate distribution of the filtered image (ROI). The mouth's center was determined to be a feature using the mean of that distribution. Thabet et al. normalized the picture's height and width of the mouth's region of interest before concatenating all the height and width of the images as features for lip reading [15]. Physical geometric parameters such as the area of the mouth, the corresponding height, and the width of the mouth are taken as features and depicted by Mathulaprangsan et al. [13]. The six features for lip reading are proposed by Alizadeh et al. [9]. He took the height, width, and four distances from the upper and bottom curves of the lips and selected them as features from the lip contour. Wu et al. analyze the significant words that the visual system recognizes. [14]. For feature extraction, the lips were curved less. Geometric parameters like width, shape, height, and expanse of the mouth region are computed in this study.

2.4.2 Appearance-Based Parameters

The location of the teeth and tongue are two appearance-based factors that are taken into account for lip movement detection. The geometry-based features have some difficulties, such as accurately recognizing the mouth feature, and lighting circumstances, and to address these issues, research is performed on appearance-based characteristics. A different technique for obtaining attributes uses pixel data as features [15]. An improved local binary pattern (ILBP) derived from three orthogonal planes was scrutinized for the movement and spatial distribution of the mouth region. A feature that is also used is the binary representation of the lip [16]. The mouth opening, which consisted of the tongue and teeth, was a characteristic used by Leung et al. [17]. The area surrounding the teeth was chosen as the ROI, and its contour served as a feature in later operations [11]. To define the contour and shape of the lips and features, the Active Shape Model (ASM) was used. The approach of lip segmentation with a weak level was challenging to use to detect the color contrast of a picture. The ASM model was thus employed as a feature and for the contour detection of the lips [18].

2.4.3 Image-Based Approaches

The work of (Lucey & Sridharan, 2008) was intended to be posing invariant. Their AVSR method began with head pose approximation and face recognition and was planned to identify speech, nevertheless of head or face [21]. They employed an approach of pose estimation provided by (Viola & Jones, 2001). Pose estimation determines the visual feature extraction technique used on the front face, right, and left face profiles. HMM was used to identify the feature vectors, and then linear discriminative analysis was used to reduce them (LDA). Methods bring out the visual features and information by modifying the mouth resemblance to a field of characteristics using various transform techniques such as Discrete Cosine transform (DCT), and Discrete Wavelet (DWT). The above transforms are crucial in reducing dimensionality and eliminating redundant data.

2.5 Feature Selection

By employing the pertinent data and the elimination of irrelevant data, feature selection is a technique for lowering the input variable for your model. A detailed overview is shown below:

2.5.1 Height and Width Features of the Mouth

Lip contour points were used as form features to recognize speech. We believe employing all or part of the lip outline and points to decide the lip's outer shape is superfluous (redundant). The width and height of the mouth, backed up by a bounding ellipse, are sufficient to estimate the true outline of the lips. Two signals (W and H) are produced by changes in lip height and width during the utterance of a specific syllable.

2.5.2 Mutual Information (M) Feature

When a new/different phoneme is spoken, the aspect of ROI changes in time. The mouth will change appearance when it transitions from phoneme to phoneme. Using mutual information to track elements of the change in the mouth region between ROIs is a smart idea.

2.5.3 Ratio of Vertical Edges to Horizontal Edges (ER)

ROI's vertical edge to horizontal edge ratio is calculated using the Sobel edge detector (ER). ER provides useful data on the appearance of the mouth at a specific period. The number of horizontal edges increases as the mouth is horizontally stretched, resulting in a decrease in ER.

2.5.4 Ratio of Vertical to Horizontal Features, Denoted by 'R'

The coefficients in each of the three other sub-bands have a Laplacian distribution with 0 means. The further a coefficient deviates from the mean in each non-LL-sub-band, the more likely it is associated with a significant visual characteristic. We utilize the approach mentioned above to detect feature-related pixels in the non-LL sub-bands as significant coefficients. The number of vertical features acquired from wavelet sub-band HL divided by the number of horizontal features obtained from LH yields the following ratio (R).

2.5.5 Appearance of the Tongue (RC)

Some phonemes, such as [th], need the appearance of the tongue to help with utterance. The tongue's red color is difficult to depict, so the ratio of the red color to the size of the ROI is taken into account. All features are scaled to the range [0, 1] above and below the critical mass (M) threshold.

2.5.6 Appearance of Teeth (T)

Some phonemes, such as[s], include the appearance and aspect of the teeth. As a result, detecting white matter (teeth) in the relevant ROI acts as a visual cue for pronouncing them. The major traits that distinguish teeth from those in other sections of the ROI are low saturation and high-intensity readings.

The features selected and incorporated for the classification are tabulated in Table 1.

Table 1. .

Features	Equations/Algorithms
Height and Width features of the mouth	W – width H - height
Mutual information (M) feature	$M(A;B) = \sum_{A}\sum_{B} p(A,B) log \frac{p(A,B)}{p(A)p(B)}$
The ratio of vertical edges to horizontal edges (ER)	$ER = \frac{\sum_{x=1}^{W}\sum_{y=1}^{H}\sum_{i=1}^{1}\sum_{j=1}^{1} \lvert ROI(x+i,y+j)(Sv(i+1,j+1)\rvert}{\sum_{x=1}^{W}\sum_{y=1}^{H}\sum_{i=1}^{1}\sum_{j=1}^{1} \lvert ROI(x+i,y+j)(Sh(i+1,j+1)\rvert}$
Ratio of vertical to horizontal features	$R = \frac{\sum_{x}\sum_{y}\{1(HLmedian-\sigma HL)\leq HL(x,y)\leq(HLmedian+\sigma HL),0\ otherwise}{\sum_{x}\sum_{y}\{1(HLmedian-\sigma HL)\leq HL(x,y)\leq(HLmedian+\sigma HL),0\ otherwise}$
Appearance of tongue (RC)	$RC = \frac{\sum_{x=1}^{W}\sum_{y=1}^{H} red(ROI(x,y))}{W.H}$
Appearance of teeth (T)	$T = \sum_{x=1}^{W}\sum_{y=1}^{H} t(x,\ the\ y)$

The features and their mathematical algorithms are derived appearance above features is taken, selected and the features are acted upon and incorporated into the classification process using classifier mode and l ResNet 50 CNN.

3 Classification

The ResNet (Residual Network) is a Convolutional Neural Network (CNN) that is designed to overcome the problem of "Vanishing Gradient", allowing for the creation of networks with thousands of convolutional layers that surpass shallower networks. A vanishing gradient occurs during backpropagation. When there are too many layers, the gradient gets smaller and smaller until it eventually vanishes. Because of this, optimization is stopped there.

In this section, we develop two ResNet models that are employed to identify the words voiced by the individual with the features and the dataset loaded to the classifier model.

These two models differ by the number of layers present, Model - I consists of one dense layer, and Model - II has 2 dense layers. And each of them has one Global Averaging Pool layer and a Max pool layer.

4 Residual Block

The fundamental building components of ResNet networks are residual blocks. ResNet adds intermediate inputs to the output of a series of convolution blocks to enable dense convolutional structures. Skip connections are utilized to ensure a smoother gradient flow and the preservation of important features until the final layers. They do not add to the network's computational load. A residual block is depicted in the image below: The input to the ResNet block is x, which is the output from the preceding layers. F(x) is a convolutional neural network with many blocks (Fig. 4).

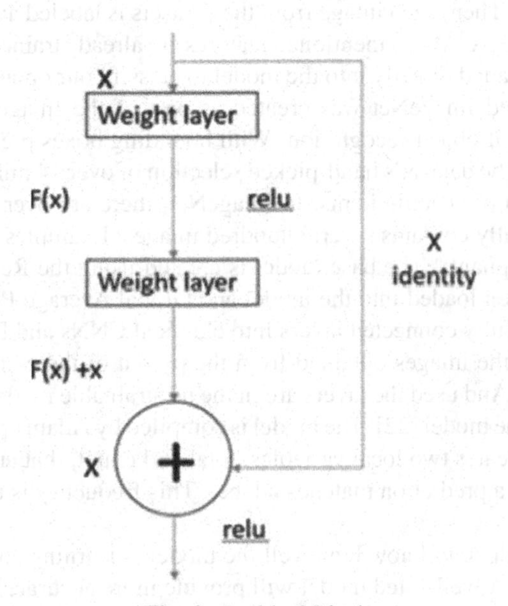

Fig. 4. Residual Block

Where, x is the input features loaded into the ResNet50 model and the F(x) is the output obtained from the weight layer. In this model, there is a special feature in the model known as "Skip connection" as we can infer from the block diagram, the calculation of weights is determined by the input "x" is loaded as the output by skipping all the weight layers [23].

Firstly, we split the database into training data, testing data and validation data using split folder API, 80% of data is for training the ResNet50 model, 10% is for testing, and

10% is for validation. After Data augmentation, the e feed the classifier 5 classes, each with 728 frames in the training dataset, 241 frames in the test dataset, and 243 frames in the validation dataset. Regarding the features, extracted from the dataset, the train test ratio of the features is 80:20, i.e., 80% for training and remaining 20% for testing.

4.1 Model Characteristics

ResNet50 is a type of RNN model that is preferred for the proposed model in which we created 2 different models of the same classifier model. Model - I consists of one dense layer which consists of 32 convolutional layers and a global average pool layer and max pool layer, whereas Model - II, which contains Two dense layers, 64 convolutional layers, a global average pool layer, and a max pool layer.

4.2 Training

We loaded the pre-processed data from MATLAB and loaded it into the classifier and Data Augmentation if done using Image Data Generator from Tensorflow which will increase our dataset. Then, each image from the datasets is labeled and those images are resized to 122x122x3 px. Above mentioned features are already trained in the "Imagenet" classifier which is loaded directly into the model to classify our obtained dataset. A vast visual database called ImageNet was created as part of the ImageNet project to aid in the advancement of object recognition. With bounding boxes present in at least one million of the shots, the dataset's hand-picked selection of over 14 million images allows for the identification of specific items. In ImageNet, there are over 20,000 categories, each of which typically contains several hundred images. Examples of these categories are "apple" and "elephant." The base model is created using the Resnet50 API, which obtained features been loaded into the model and Global Average Pooling layer which designed to replace fully connected layers into classical CNNs and Dense layers which are used to classify the images obtained from the output of the convolution layers are added to our model. And used the layers are in the no -trainable mode which will reduce the learning rate of the model [22]. The model is compiled by Adam optimization, and the accuracy function creates two local variables, total and count, that are used to calculate the number of times a prediction matches a label. This frequency is ultimately returned as accuracy.

Model fitting is done to know how well the model is learning and is generalized to similar trained data. A well-fitted model will provide more accurate results. Overfitting models fit the data too closely [24].

4.3 Testing

Testing the learned model is done by loading the trained model and then using the testing dataset which is loaded to the model, and the model is executed to predict training data to check the accuracy of the trained model using the validation dataset (Fig. 5 and Fig. 6).

The accuracy of the ResNet50 classifier was 0.89775 or 89.7% in Model - I and 0.8189 or 81.89% in Model - II. By observing the confusion matrix we can compare

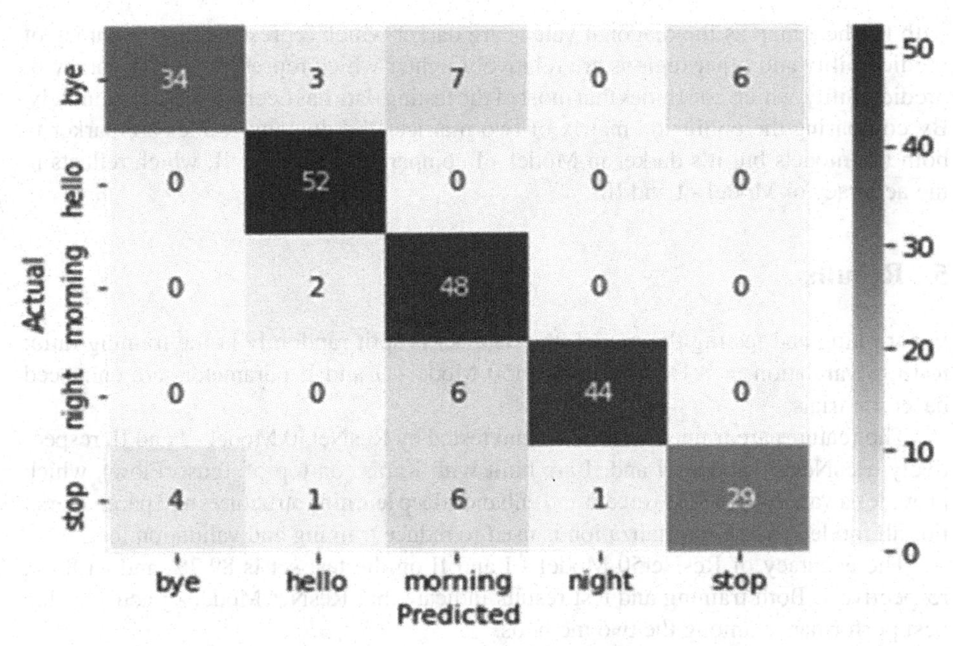

Fig. 5. Confusion Matrix for ResNet50 model - I

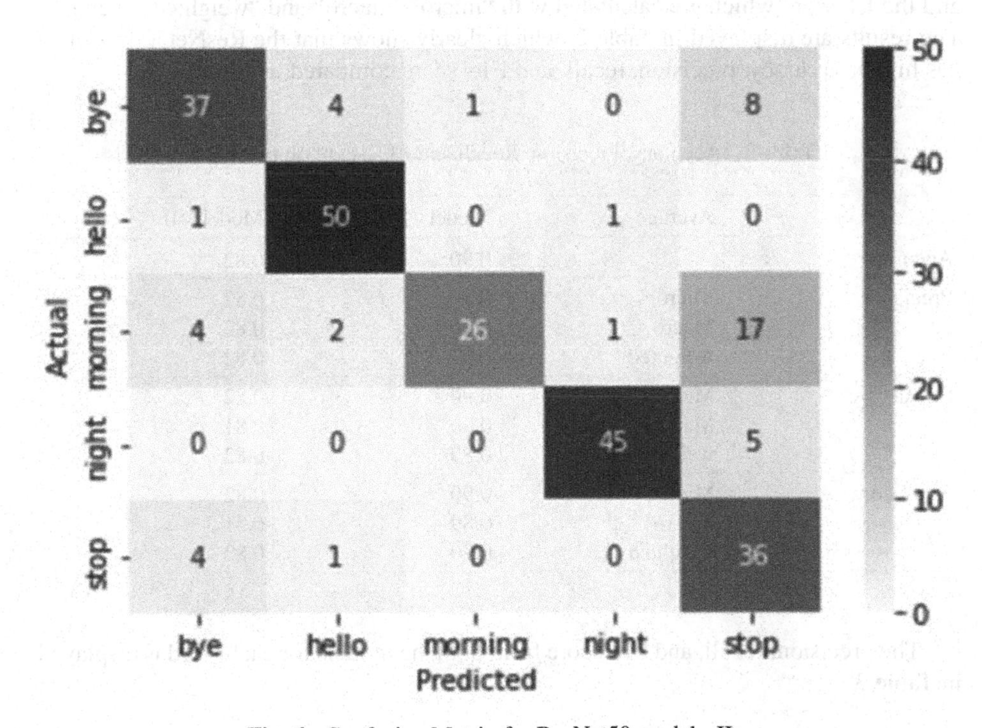

Fig. 6. Confusion Matrix for ResNet50 model - II

with the heat map as the diagonal values are darker which represents high accuracy of predictability and other regions are relatively lighter which represents less accuracy of predictability, which concludes that most of the testing data has been predicted accurately. By comparing the confusion matrix of two matrices the diagonal values are darker in both the models but it's darker in Model - I compared to Model - II, which reflects in the accuracy of Model - I and II.

5 Results

For training and testing the model, the data set is split randomly at the training ratio: testing: validation = 8:1:1. The ResNet50 Model - I and II parameters are enhanced based on trials.

The features are trained, validated, and tested by ResNet50 Model - I and II, respectively. ResNet50 Model - I and II are built with Keras (on top of TensorFlow), which provides a variety of APIs to create and enhance deep learning structures and parameters.. For all models, batch normalization is used to reduce training and validation loss.

The accuracy of ResNet50 Model - I and II on the test set is 89.7% and 81.89%, respectively. Both training and test results indicate that ResNet Model - I achieves the best performance among the two methods.

The evaluation metrics for the ResNet50 models considered are Precision, Recall, and the F1-score which are calculated with "micro","macro" and "weighted" averages. The results are displayed in Table 2, which clearly shows that the ResNet50 model - I has higher accuracy, precision, recall, and F1 - score compared to model - II.

Table 2. Accuracy, Precision, Recall and F1 - score on the test set

	Average	Model - I	Model – II
Accuracy		0.90	0.82
Precision	Micro	0.90	0.82
	Macro	0.90	0.82
	weighted	0.90	0.82
Recall	Micro	0.90	0.82
	Macro	0.90	0.81
	weighted	0.89	0.82
F1 - Score	Micro	0.90	0.82
	Macro	0.89	0.81
	weighted	0.90	0.82

The precision, recall, and F1 - score from both the models for each word is displayed in Table 3.

Table 3. Precision, Recall, and F1 - score of each word predicted by ResNet50 models

	Precision		Recall		F1 - Score	
	Model - I	Model - II	Model - I	Model - II	Model - I	Model - II
Bye	0.82	0.80	0.92	0.90	0.87	0.85
hello	0.94	0.94	0.96	0.88	0.95	0.91
morning	0.91	0.77	0.78	0.86	0.84	0.81
night	0.96	0.98	0.96	0.80	0.96	0.88
stop	0.85	0.61	0.85	0.61	0.85	0.61

6 Discussion

The SSI was put into place with a primary focus on persons who had total laryngectomy and speech difficulties. SSI comes off as a suitable means to aid muted persons. The ideology was to extract a set of trained common words from subjects and visualize and work on the image data collected without their audio. Due to the devoid of subjects who suffer from speech disorders, we utilized this methodology which entitles the future scope. The image processing paradigm follows a simple yet effective path to derive an effective outcome. The initial goal was to break down the video data into frames and supplement them for greater accuracy. Based on frames/second conversion, the sequence follows the basis for feature extraction. The lips and mouth play a crucial role in recognizing the words from the images. A thorough literature review was performed to study the existing feature extraction techniques. Consequently, features were selected to suit our study and were briefly summarized. The results instilled from the confusion matrices for two Resnet-50 classifier different models were to show the variation in the accuracy of detection of the words spoken based on the difference in the number of convolution layers. Model I with lesser convolutional overlaps can be observed to produce higher efficiency and accuracy to phase out the words. The scope to bridge the research gap between more efficient machine learning algorithms and easy identification of words for these speech-impaired people is still under development.

References

1. Zhang, X., et al.: Understanding pictograph with facial features: end-to-end sentence-level lip reading of Chinese. In: Proceedings of the AAAI Conference on Artificial Intelligence, Vol. 33, No. 01, pp. 9211–9218 (2019)
2. Zhang, C., Zhao, H.: Lip reading using local-adjacent feature extractor and multi-level feature fusion. In: Journal of Physics: Conference Series, Vol. 1883, No. 1, p. 012083. IOP Publishing (2021)
3. Lin, B., Yao, Y., Liu, C., Lien, C., Lin, B.: Development of Novel Lip-Reading Recognition Algorithm. IEEE Access **5**, 794–801 (2017). https://doi.org/10.1109/ACCESS.2017.2649838
4. Abdullah, A., Chemmangat, K., A Computationally Efficient sEMG based Silent Speech Interface using Channel Reduction and Decision Tree based Classification, Procedia Computer Science (2020)

5. Chawla, M.: Review and analysis of various lip reading system techniques. Int. J. Res. Appl. Sci. Eng. Technol. **6**, 4094–4098 (2018). https://doi.org/10.22214/ijraset.2018.4675
6. Jain, A., Rathna, G.N.: Lip reading using simple dynamic features and a novel ROI for feature extraction. In: Proceedings of the 2018 International Conference on Signal Processing and Machine Learning (SPML '18), pp. 73–77. Association for Computing Machinery, New York, NY, USA (2018). https://doi.org/10.1145/3297067.3297083
7. Kapkar, P.P., Bharkad, S.D.: Lip Feature Extraction And Movement Recognition Methods: A Review
8. Hassanat, A.B., Jassim, S.: Visual words for lip-reading. In Mobile Multimedia/Image Processing, Security, and Applications 2010, Vol. 7708, pp. 86–97. SPIE (2010)
9. Alizadeh, S., Boostani, R., Asadpour, V.: Lip feature extraction and reduction for HMM-based visual speech recognition systems||. In: 2008 9th International Conference on Signal Processing, pp. 561–564. Beijing (2008)
10. Egorov, E., Kostyumov, V., Konyk, M., Kolesnikov, S.: LRWR: Large-Scale Benchmark for Lip Reading in Russian language (2021). arXiv preprint arXiv:2109.06692
11. Li, M., Cheung, Y.: A novel motion based lip feature extraction for lip-reading||. In: 2008 International Conference on Computational Intelligence and Security, pp. 361–365. Suzhou (2008)
12. Nair, A., Shashikumar, N., Vidhya, S., Senthil Kumar, K.: Design of a Silent Speech Interface using Facial Gesture Recognition and Electromyography (2017). https://doi.org/10.1007/978-981-10-4220-1_22
13. Mathulaprangsan, S., Wang, C., Kusum, A.Z., Tai, T., Wang, J.: A survey of visual lip reading and lip-password verification. In: International Conference on Orange Technologies (ICOT), pp. 22–25. Hong Kong (2015)
14. Sahu, V., Sharma, M.: Result based analysis of various lip tracking systems. In: International Conference on Green High Performance Computing (ICGHPC), pp. 1–7 (2013)
15. Thabet, Z., et al.: Lipreading using a comparative machine learning approach. First International Workshop on Deep and Representation Learning (IWDRL), pp. 19–25. Cairo (2018)
16. Wu, P., Liu, H., Li, X., Fan, T., Zhang, X.: A novel lip descriptor for audio-visual keyword spotting based on adaptive decision fusion||. IEEE Trans. Multimedia **18**(3), 326–338 (2016)
17. Rathee, N., Ganotra, D.: Analysis of human lip features: a review||. Int. J. Applied Systemic Studies **6**(2), 137–184 (2015)
18. Liu, H., Fan, T., Wu, P.: Audio-visual keyword spotting based on adaptive decision fusion under noisy conditions for human-robot interaction. In: IEEE International Conference on Robotics and Automation (ICRA), pp. 6644–6651. Hong Kong (2014)
19. Li, Y., Takashima, Y., Takiguchi, T., Ariki, Y.: Lip reading using a dynamic feature of lip images and convolutional neural networks. In: IEEE/ACIS 15th International Conference on Computer and Information Science (ICIS), pp. 1–6. Okayama (2016)
20. Gomez, E., Travieso, C.M., Briceno, J.C., Ferrer, M.A.: Biometric identification system by lip shape. In: Proceedings. 36th Annual 2002 International Carnahan Conference on Security Technology, pp. 39–42. Atlantic City, USA (2002)
21. Agrawal, S., Omprakash, V.R., Ranvijay: Lip reading techniques: A survey. In: 2nd International Conference on Applied and Theoretical Computing and Communication Technology (iCATccT), pp. 753–757. Bangalore (2016)
22. Leung, S.H., et al.: Automatic Lipreading with Limited Training Data. In: 18th International Conference on Pattern recognition (ICPR'06), pp. 881–88. Hong Kong (2006)
23. Lee, D., Myung, K.: Read my lips, login to the virtual world. In: IEEE International Conference on Consumer Electronics (ICCE), pp. 434- 435. Las Vegas (2017)
24. Matthews, I., Cootes, T.F., Bangham, J.A., Cox, S., Harvey, R.: Extraction of visual features for lipreading||. IEEE Trans. Pattern Anal. Mach. Intell. **24**(2), 198–213 (2002)

25. Chen, J., Tiddeman, B., Zhao, G.: Real-time lip contour extraction and tracking using an improved active contour model, In: Bebis, G., et al. (eds.) Advances in Visual Computing. ISVC, vol. 5359, pp. 236–245 (2008)
26. Sangve, S., Mule, N.: Lip Recognition for Authentication and Security. IOSR Journal of Computer Engineering (IOSR-JCE) **16**(3), 18–23 (2014)
27. Lucey, P., Sridharan, S.: A visual front-end for a continuous pose-invariant lipreading system. In: Proceedings of the 2nd International Conference on Signal Processing and Communication Systems, 15–17 December 2008. Australia, Queensland, Gold Coast (2008)
28. Hassanat, A.: Visual Passwords Using Automatic Lip Reading. Int. J. Sci. Basic and Applied Res. (IJSBAR) **13**, 218–231 (2014)
29. Gomez, E., Travieso, C.M., Briceno, J.C., Ferrer, M.A.: Biometric identification system by lip shape. In: Proceedings. 36th Annual 2002 International Carnahan Conference on Security Technology, pp. 39–42. Atlantic City, USA (2002)
30. Wu, Z., Shen, C., Van Den Hengel, A.: Wider or deeper: revisiting the resnet model for visual recognition. Pattern Recogn. **90**, 119–133 (2019)
31. Afouras, T., Chung, J.S., Zisserman, A.: Deep lip reading: a comparison of models and an online application (2018). arXiv preprint arXiv:1806.06053
32. Wang, J., Samal, A., Green, J.R.: Preliminary test of a real-time, interactive silent speech interface based on electromagnetic articulograph. ACL/ISCA Workshop on Speech and Language Processing for Assistive Technologies, pp. 38–45. Baltimore, MD (2014)

Performance Analysis of Recent Algorithms for Compression of Various Medical Images

Apeksha Negi, Nidhi Garg(✉), and Sunil Agrawal

ECE Department, UIET, Panjab University, Chandigarh, India
{nidhi_garg,s.agrawal}@pu.ac.in

Abstract. Data compression refers to a method that converts or transforms data from a characterization to a novel compressed characterization, where the distributed information remains the same by reducing the number of bits. With the increasing demand for storing and transmitting clinical scans, the shortage of sufficient storage space and communication bandwidth occurs. The terms "lossy" and "lossless compression" refer to two larger types of image compression methods. The lossy techniques loss some information at the time of compression on the other hand lossless may not loss information. The medical images have two important segments which are ROI (Region of Interest) and non-ROI. The ROI segment will be compressed using lossless technique whereas the lossy techniques will be applied on the non-ROI segment. At first one lossless and one lossy technique Lempel-Ziv-Welch (LZW) and Singular Value Decomposition (SVD) is implemented respectively. Then three combination of hybrid technique i.e. lossless and lossy image compression technique Huffman + Singular Value Decomposition (SVD), Lempel-Ziv-Welch (LZW) + Discrete Cosine Transform (DCT) and Set Partitioning in Hierarchical Trees (SPHIT) + Discrete Wavelet Transform (DWT) is implemented respectively. The image compression techniques are implemented in MATLAB and results is analysed in terms of Peak Signal-to-Noise Ratio (PSNR), Mean Squared Error(MSE), Structural Similarity Index Measure (SSIM) and Compression Ratio (CR). In this SPHIT + DWT shows better result in terms of PSNR and CR while compared to other techniques.

Keywords: Lossy · Lossless · Huffman · SVD · LZW · DCT · SPHIT · DWT

1 Introduction

The advancement of technologies these days has driven the production, transmission and compression of huge amounts of data on a daily basis. As a result, data compression is required to reduce the basic dimensions of the data such as text, image, audio, video or any form of digital data so as to enable smaller memory usage and transfer the data in accelerated mode. Data compression can reduce storage usage and speed up the transfer of data [1]. Lossy and lossless are the two main kinds of image compression. Both of these methods are elaborated below:

B. K. Singh et al. (Eds.): ICBEST 2023, CCIS 2003, pp. 24–34, 2024.
https://doi.org/10.1007/978-3-031-54547-4_3

i. Lossless compression: Lossless compression compresses data without losing information. This method is typically the benchmark in clinical imaging because it reduces the load concerning the storage and transferring of such data, while simultaneously preventing the loss of critical medical data that can cause serious medical and legal consequences. Lossless compression keeps the image quality as it is. Compressed data will be just as competent as organic data [2].
ii. Lossy compression: Lossy compression methods have potential to obtain a high-level compression performance through the permanent removal of some of the information from the organic data. Lossy compression represents a technique that removes irrelevant information. The file decompressed by this technique is not the same as the original data. The biggest gain of lossy compression is that it greatly lessens the size of the original data significantly, which also implies that it will yield lower quality than the organic data [3].

1.1 Medical Image Compression

Image processing is an important branch of clinical research/medical practice. In the past years, with the advances in digital imaging techniques, medical image analysis has been developed substantially [4]. The concept of digital image processing relies on the processing of digital images by digital computers. Actually, a digital image is a special arrangement of finite number of constituents. Each constituent has its own location and value and is designated as a pixel, picture or picture element. Medical imaging for diagnostic purposes is commonly used in medical field.

Advances in scanning technology over the past decades, have made such systems increasingly important, producing images of greater precision and better quality due to higher bit resolutions [5]. Such developments led to an increase in the amount of information that requires processing, transmission, and storage. It is primarily suitable with the use of volumetric scanning techniques such as computed tomography (CT), magnetic resonance imaging (MRI), and X-rays that produce a large number of photographic portions. These actualities, coupled with the growing number of medical images archival databases established by medical individuals, have prompted research into effective compression techniques for these indications. Therefore, implementing compression algorithms for preventing the inception of errors in the output data of the scanning system is of great worth.

The decorrelation algorithm, the primary compression engine, and the format method are the three basic parts of the majority of medical image compression algorithms [6]. These three elements are shown in Fig. 1.

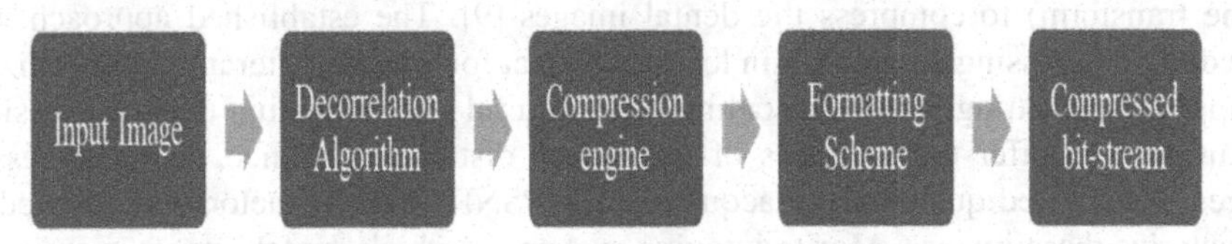

Fig. 1. Block diagram of a Medical Image Compression System

Algorithmic decorrelation approaches exploit data redundancy in clinical images using prediction models. Predictive models reduces the variance between successive instances, chunks or volumes and uses motion compensation and estimation or DPCM (differential pulse code modulation) schemes to generate residual data. In contrast, multi-resolution models use a transforms, such as the discrete wavelet transforms (DWT) or discrete cosine transforms (DCT) to decorrelate the image data. Decorrelated photographs and remaining data are compressed in lossless fashion using entropy encoders including Huffman, arithmetic, and run-length coding. Before entropy coding, decorrelated photographs are generally quantized to make compression better. The lossless coding approaches do not apply quantization to the decorrelated data. At last, the compressed image goes through a formatting process and ensures proper use of the encoded data.

2 Literature Review

In this module, we are going to discuss about various lossless and lossy techniques and the way in which they used by different researchers to get favorable results.

Q. Min, et.al (2022) introduced a compression method that simplifies the transmission and storage of medical image data by integrating anatomical data with a unique deep neural network model [7]. Using anatomical parameters, such as image density, organ size and relative position, this technique initially segmented the medical imaging data into distinct regions. The best predictors for each region were then created through training a deep neural network.These predictors can be adaptively exchanged depending on the characteristics of the compressed region. An entropy coding approach was used to finally condense the residuals. A review of this compression method reveals that the proposed "divide and conquer" strategy indeed achieves high compression performance and prediction accuracy that outperformed Joint Photographic Experts Group (JPEG) 2000 by 38%.

B. P.V., et.al (2021) intended an ROI-based method for compressing the MRI (magnetic resonance imaging) images to deal with the issue of frequency elements of the processed medical image [8]. The ROI (region of interest) was separated from the non-ROI using FCM (Fuzzy C-Means) technique. The non-ROI was compressed on the basis of CAE (Capsule Autoencoder). The implementation of Discrete Cosine Transform (DCT) was done with Huffman Run-Length Encoding (RLE) in order to compress the ROI. The Compression Ratio (CR) and Peak Signal to Noise Ratio (PSNR) metrics were considered to analyze the intended method. The results confirmed that the intended method performed more successfully in comparison with other techniques.

O. Krylova, et.al (2021) established an AGU-M coder on the basis of DCT (discrete cosine transform) to compress the dental images [9]. The established approach was helped in compressing the images in lossless format for which no iteration was utilized. An appropriate setting of a fixed scaling factor, useful for controlling the compression, was utilized to offer the faintness of introduced distortions. Hence, the compressed images of expected quality were acquired. The PSNR-HVS-M factor was adopted to characterize these images. Morita imaging system, in which dental images were comprised, executed in the experimentation. A team of dentistry experts analyzed the quality of visual image. Experimental results show the superiority of the established approach over the traditional approach and the CR (compression ratio) of this approach was found higher.

P. T. Akkasaligar, et.al (2020) proposed a cryptosystem with a new encryption mechanism and lossless compression method [10]. The security of clinical images was enhanced using the chaos-based DNA cryptography. The lossless discrete haar wavelet transform (DHWT) method was used to compress the original clinical image. This method emphasized on restoring the compressed picture into binary form and splitting it into 4 sub-images. The chaotic sequences obtained from 4D Lornez chaotic map were utilized to shuffle the sub image pixels. The DNA coding rules were useful to develop 4 unique DNA structures. The results proved the robustness of the projected mechanism against various attacks with regard to security. Moreover, the projected mechanism was applicable on telemedicine and e-health applications.

X. Song, et.al (2020) developed a hybrid spatial prediction algorithm for enhancing the process to compress the medicinal image in lossless format [11]. This algorithm aimed to partition the input image into the objective and the background areas. Therefore, a block-based spatial prediction method was proposed by exploiting the structural similarity and symmetry of target area. The pixel-based predictive model was implemented to achieve the background area. Both methods were integrated in the developed algorithm such that the images were compressed. The results of experiments indicated that the developed algorithm performed well in contrast to other techniques.

O. H. Nagoor, et.al (2020) recommended a technique of compressing an image in which a RNN (Recurrent Neural Network) was combined with a 3D sequence predictive model [12]. This technique focused on learning voxel long neighbour dependencies in 3D (three-dimensional). For this, LSTM (Long Short-Term Memory) was implemented. Later, the arithmetic coding was exploited to compress the residual error. The experiential outcomes proved that the recommended technique offered the efficacy to compress the image up to 15% in comparison with the conventional techniques. Moreover, this technique was capable of generalization for hidden modalities of computerized tomography (CT) and magnetic resonance imaging (MRI).

A. P. Mukhopadhyay, et.al (2019) presented lossless scalable RBC for Digital Imaging and Communications in Medicine (DICOM) images in which Discrete Wavelet Transform (DWT) and distortion limiting compression method were exploited for the rest of parts of the image [13]. This approach removed the noisy data in the background and restored parts of the image losslessly. The ROI (Region of Interest) method, to compress the image, deployed DWT and Set Partitioning in Hierarchical Trees (SPHIT) algorithm for constructing an algorithm. The presented approach was analyzed with regard to metrics. The results validated that the presented ROI method performed more effectively in augmenting the quality of ROI at least Mean Square Error (MSE) and Peak Signal to Noise Ratio (PSNR).

S. Gao, et.al (2019) designed a DL (deep learning) model in order to improve the compressed brain images [14]. An effectual post-processing network was deployed with the purpose of mitigating the compression artifacts. The new network elements were presented to develop the designed model for 3D volumetric brain image. The adaptability of this model was also proved for other kinds of biomedical images. The experimental results exhibited that the designed model was more efficient as compared to the traditional techniques. Furthermore, the designed model was assisted in enhancing the quality of compressed brain images up to 0.8 dB and offering the CR (compression rate) of 160.

3 Research Methodology

The magnetic resonance imaging (MRI) images are the medical images that can be compressed for the transmission. In the MRI image important part is called region of Interest and other part is called Non Region of Interest. In this research work, the lossless and lossy image compression techniques are applied on ROI part and Non ROI part respectively. The Block diagram is represented in Fig. 2.

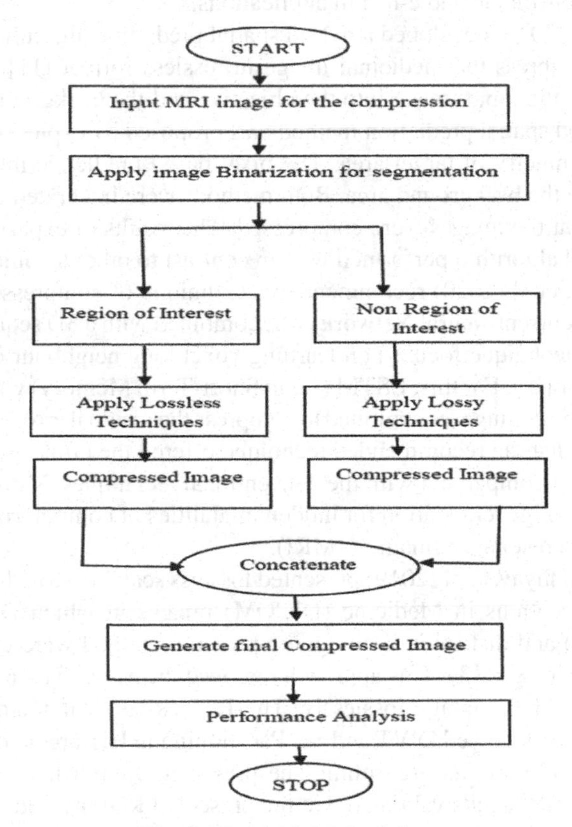

Fig. 2. Research Methodology

3.1 Lossless Image Compression Techniques

The various lossless techniques are applied on the Region of Interest Part. The various lossless techniques are given below: -

3.1.1 The Lempel-Ziv-Welch (LZW) Algorithm

The non-statistical lossless compression algorithm LZW is well-known and has been used for both text and binary data. Due to its non-statistical nature, it is not required to understand the same statistical information about the data being compressed as Huffman, Fano, and Arithmetic coding. LZW compresses data using an infinite-sized dictionary, which means it employs a dictionary to store symbol sequences from the source data stream. The fundamental set of symbols are then allocated a set of indices (codeword locations) in the dictionary. To match a sequence that already exists in the dictionary, the longest sequence imaginable will be used. In case of failure, the following shorter (and already matched) sequence is compressed. The new longer sequence will be added to the dictionary in the hopes that it will help with the compression of subsequent sequences. The decoder must already have all of the previous sequences from the dictionary in order to decode an entry. At both the encoder and decoder sides, the dictionary is merely locally maintained; no actual transfer or synchronisation is required. Actually, this limits the dictionary to brief local use. LZW compression uses simply simple searches rather than many computations or even sorting.

3.1.2 Huffman Coding

A lossless picture compression approach is the Huffman coding. Because it employs "variable length coding," symbols in the data are translated to binary symbols dependent on how frequently they appear. The binary sign for the character with the highest probability is shorter. Huffman coding is a greedy algorithm because it always chooses a local optimal solution and hopes that it will also be the global optimal solution when presented with multiple possibilities for each phase of the problem-solving process. Depending on the qualities of the data, Huffman coding can typically save 20% to 90% of the space when compressing data. Prior to encoding, Huffman coding must be constructed. This involves assigning a code value of "0" or "1" from bottom to top based on the likelihood, and then encoding the data from top to bottom. The following are the precise steps to create Huffman coding: The data are encoded in accordance with the designed Huffman tree after (1) computing the probability of each pixel's data and (2) setting the two least probability data as "0" or "1" and combining them till the probability is normalised. Since each Huffman-encoded code word is unique and cannot be the front part of another, the code words can be communicated together without the use of extra isolation symbols. If there are no errors in the broadcast, the receiver can accurately decipher them.

3.1.3 Set Partitioning in Hierarchical Trees (SPHIT) Technique

In the field of wavelet transformation, an image is compressed using the SPIHT compression algorithm. Wavelet transformed images can be used to create a SOT (spatial orientation tree) that depicts the relationship between a parent and its children. Each node in the tree represents a coefficient, often referred to as a pixel, in the changed image. The top left node of size (m/2L)*(n/2L) is referred to as the root node of the SOT if an image of size mn is transformed by an L-level DWT. Considering that a 1616 image is transformed by a 3-level DWT. The square is denoted by R, which stands for the root, while the 2x2 pixels with the numerals 0, 1, 2, and 3 correspond to the root.

SPIHT defines a significance function for a given set T that determines whether any pixels in the set T exceed a specific threshold. Equation (1) illustrates the meaning of a set T at the nth bit plane, S n (T):

$$S_n(T) = \{1, \max\{|ci, j|\} \geq 2n \\ 0, \textit{otherwise} \tag{1}$$

when $S_n(T)$, is '0', T is called a non-significant set. Otherwise, T is called a significant set.

T is referred to as a non-significant set when S n (T), is '0'. T is said to as a significant set if not. Significant amounts must be broken into subgroups and retested for significance; insignificant numbers can be represented as one bit of '0'. A given set T and its descendants [denoted by D(T)] combined are computed by SPIHT based on the zero-tree hypothesis [2] by evaluating the importance of T ∪ D(T) (the union of T and D(T)) and by expressing T ∪ D(T) as a single symbol "zero" if T ∪ D(T) is insignificant. On the other hand, T is divided into subsets and each of them is individually evaluated if T ∪ D(T) is significant. The entire image is divided into 4*4 sets (sets made up of 4*4 pixels), and the relevance of the union of each 4*4 set and its descendants is assessed in order to lessen the complexity of SPIHT. From the most significant bit-plane to the least significant bit-plane, the SPIHT method encrypts wavelet coefficients bit-plane by bit-plane [15].

3.2 Lossy Image Compression Techniques

Below are the various Lossy Image Compression techniques: -

3.2.1 Singular Vector Decomposition (SVD) Algorithm

SVD (Singular Vector Decomposition) is one of the most popular image compression approaches. This approach is different from 'neighbourhood-based' algorithms. Generally speaking, the main purpose of using SVD is to decrease the number of features of a data set. The most interesting part is the matrix factorization where one holds the same dimensions. Broadly stated, matrix factorization is the method to find matrices whose output is the rating matrix. The given formula is shown in equation number (2):

$$A = USV^T \tag{2}$$

where A is the specified $n \times m$ matrix, U is the $n \times n$ matrix that includes the eigenvectors of AA^T, S represents the $n \times m$ matrix holding the square root of the eigenvalues related to AA^T on its diagonal. Finally, V is the $m \times m$ matrix containing the AA^T's eigenvectors.

3.2.2 Discrete Cosine Transform (DCT)

The Discrete Cosine Transform (DCT), which aims to decorrelate the image data, is an orthogonal transform. After decorrelation, it is possible to independently encode each transform coefficient without sacrificing compression effectiveness. A signal is converted from a spatial representation to a frequency representation using the DCT. The DCT sums sinusoids of different sizes and frequency to represent an image. The main visually important information about a typical image is contained in just a few DCT coefficients, which is a feature of DCT. Once the DCT coefficients have been calculated, they are normalised using a quantization table with various scales specified by the JPEG standard and calculated using psychovisual evidence. Entropy and compression ratio are impacted by the choice of quantization table. Greater mean square error and better compression ratio are negatively correlated with the value of quantization, as well as with the quality of the reconstructed image. The less significant frequencies are eliminated during the Quantization stage of a lossy compression approach, and the frequencies that are left are then employed to extract the image during the decomposition stage. An effective lossy coding scheme is used to further compress the quantized coefficients after they have been rearranged in a zigzag pattern.

3.2.3 Discrete Wavelet Transform (DWT)

The discrete wavelet transform (DWT), widely used as important tool for signal and image processing, has the following salient properties. 1) The basis generated by the orthogonal (or bi-orthogonal) mother wavelet (MW) is a fully orthogonal basis, thus allowing error free reconstruction of the analytic signal. 2) octave analysis is done and MW has the bandpass filter characteristic with octave width in the frequency domain, so the data is highly compressed. Mattal's fast approach involves a down-sampling step, so the DWT has no shift invariance. MW affects efficiency, and furthermore, octave analysis may not be the best criterion for adapting the analytic signal to obtain octave band characteristic such as data compression and noise cancellation. Images are applied individually to each image component using the DWT. The DWT decorrelates the image to different scale sizes while largely preserving its spatial connectivity. In the one-dimensional (1-D) DWT, the high pass filter (H) and low pass filter (L) divide the line of pixels into two equal-length lines. Four sub-bands are created when a two-dimensional (2-D) image is filtered in the LL, LH, HL, and HH directions in both the vertical and horizontal planes. Other sub-bands are used to filter out missing features, leaving the LL sub-band with a low resolution copy of the original image. At the scale size specified in the wavelet, the sub-bands contain horizontal (LH), vertical (HL), and diagonal (HH) edges.

4 Result and Discussion

This research work is based on the medical image compression. The ROI is compressed by the lossless image compression; the non-region of interest part will be encrypted by the lossy image compression technique. The certain combinations are created which include Huffman coding which is lossless compression with SVD image compression with lossy image compression. The second combination is LZW with DCT. The third combination is SPHIT with DWT. The single image compression techniques are close implemented like LZW and SVD. PSNR, MSE, SSIM and Compression ratio are performance metrics used to compare various image compression techniques. The lossless and lossy schemes which is discussed in paper, has major use in the clinical applications. The MRI image is large in size and at the time of sharing with the entities it can be compressed using proposed framework. The proposed framework results will lead to less loss of useful information and maintain quality of the image (Fig. 3).

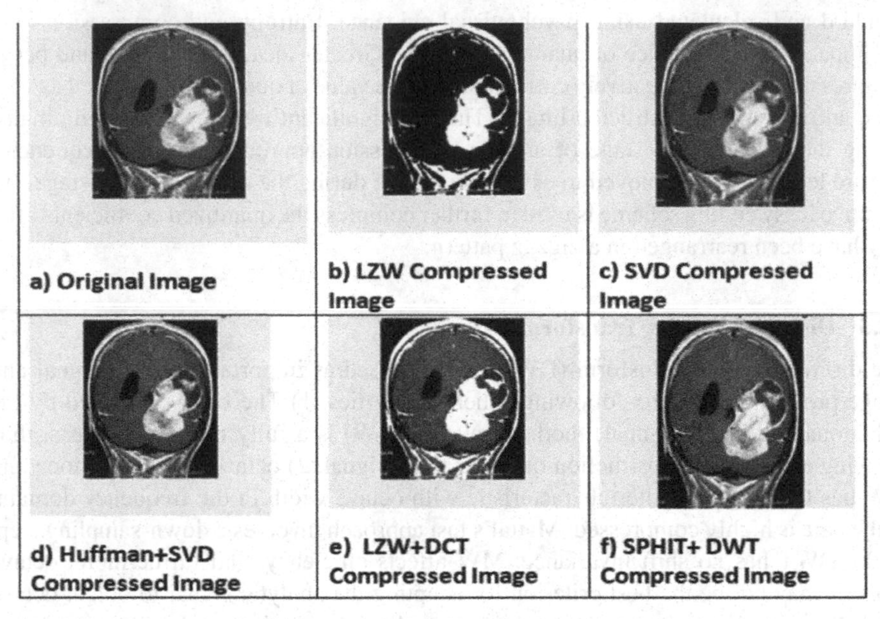

Fig. 3. Simulation results of techniques used.

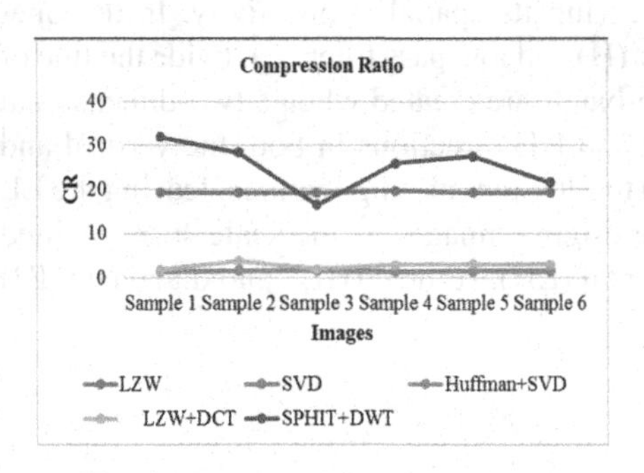

Fig. 4. Compression Ratio Analysis

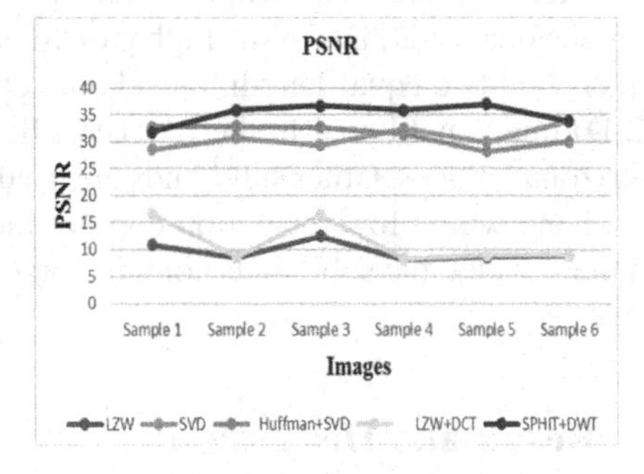

Fig. 5. PSNR Analysis

As shown in Fig. 4, it is analysed from the results that compression ratio of LZW technique is high as compared to other technique. The LZW is the lossless technique but when it is combined with the lossy technique which is DCT it gave least compression ratio. As shown in Fig. 5, the PSNR value of the SPHIT + DWT is high as compared to other compression techniques. As shown in Fig. 6, the MSE value of the SPHIT + DWT is the least as compared to the other image compression algorithms. As shown in

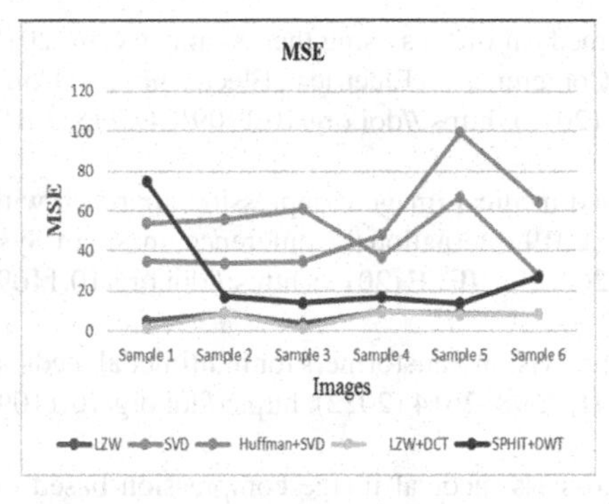

Fig. 6. MSE Analysis

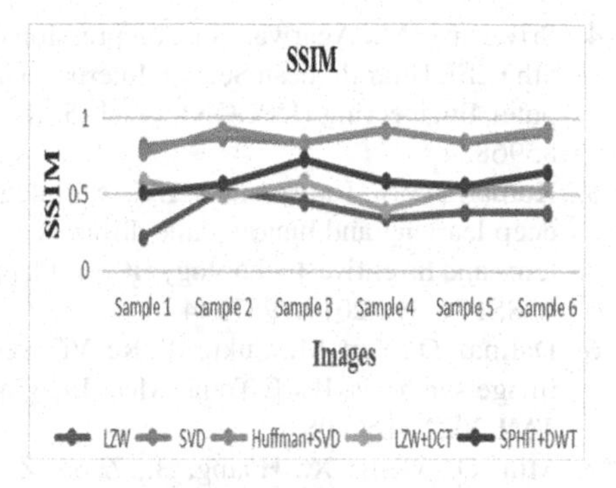

Fig. 7. SSIM Analysis

Fig. 7, it is analysed that Huffman + SVD has maximum SSIM as compared to other techniques.

5 Conclusion

Lossless and lossy are the two classes of image compression adopted by the research community for medical image processing. Lossy compression needs an exact rebuilt photograph of the original photograph from the duplication. In this research work hybrid technique of lossy and lossless compression is used. The medical image is segmented into ROI and non-ROI segments. The ROI part will be segmented using lossless techniques and non-ROI part will be segmented using lossy techniques. The three combinations of lossy and lossless techniques is created which are Huffman + SVD, LZW + DCT and SPHIT + DWT. PSNR, MSE, SSIM and Compression ratio are used to compare performance of technique. The individually one lossless and one lossy technique LZW and SVD is implemented respectively. The proposed model can be further extended using deep learning techniques to maintain parametric values.

References

1. Chen, Z., Gu, S., Lu, G., Xu, D.: Exploiting Intra-Slice and Inter-Slice Redundancy for Learning-Based Lossless Volumetric Image Compress. In: Budhewar, A.S., Doye, D.D. (eds.) Performance Analysis of Compression Techniques for Multimedia Data. In: Studies in Fuzziness and Soft Computing, pp. 243–256. IEEE (2019). https://doi.org/10.1007/978-3-030-03131-2_12.ion. IEEE Trans. Image Process. 31, 1697–1707 (2022). https://doi.org/10.1109/TIP.2022.3140608
2. Krishnaswamy, R., NirmalaDevi, S.: Efficient medical image compression based on integer wavelet transform. In: 2020 Sixth International Conference on Bio Signals, Images, and Instrumentation (ICBSII), pp. 1–5. IEEE (2020). https://doi.org/10.1109/ICBSII49132.2020.9167597
3. Budhewar, A.S., Doye, D.D.: Performance analysis of compression techniques for multimedia data. In: Studies in Fuzziness and Soft Computing, pp. 243–256. IEEE (2019). https://doi.org/10.1007/978-3-030-03131-2_12

4. Srivastava, M., Agarwal, S.: Compression of medical images using their symmetry. In: 2018 5th IEEE Uttar Pradesh Section International Conference on Electrical, Electronics and Computer Engineering (UPCON), pp. 1–5. IEEE (2018). https://doi.org/10.1109/UPCON.2018.8596824
5. Kumar, B.P.S., Ramanaiah, K.V.: An integrated medical image compression approach with deep learning and binary plane difference. In: 2019 International Conference on Smart Systems and Inventive Technology (ICSSIT), pp. 202–205. IEEE (2019). https://doi.org/10.1109/ICSSIT46314.2019.8987944
6. Dalmaz, O., Yurt, M., Cukur, T.: ResViT: residual vision transformers for multimodal medical image synthesis. IEEE Trans. Med. Imaging **41**, 2598–2614 (2022). https://doi.org/10.1109/TMI.2022.3167808
7. Min, Q., Wang, X., Huang, B., Zhou, Z.: Lossless medical image compression based on anatomical information and deep neural networks. Biomed. Signal Process. Control **74**, 103499 (2022). https://doi.org/10.1016/j.bspc.2022.103499
8. Bindu, P.V., Afthab, J.: Region of interest based medical image compression using DCT and capsule autoencoder for telemedicine applications. In: 2021 Fourth International Conference on Electrical, Computer and Communication Technologies (ICECCT), pp. 1–7. IEEE (2021)
9. Kryvenko, L., Krylova, O., Krivenko, S., Lukin, V.: A fast noniterative visually lossless compression of dental images using BPG coder. In: 2022 11th Mediterranean Conference on Embedded Computing (MECO), pp. 1–6. IEEE (2022). https://doi.org/10.1109/MECO55406.2022.9797114
10. Akkasaligar, P.T., Biradar, S.: Medical image compression and encryption using chaos based DNA cryptography. In: 2020 IEEE Bangalore Humanitarian Technology Conference (B-HTC), pp. 1–5. IEEE (2020). https://doi.org/10.1109/B-HTC50970.2020.9297928
11. Song, X.: A hybrid spatial prediction algorithm for lossless compression of CT and MRI medical images. In: 2021 40th Chinese Control Conference (CCC), pp. 2918–2922. IEEE (2021). https://doi.org/10.23919/CCC52363.2021.9549895
12. Nagoor, O.H., Whittle, J., Deng, J., Mora, B., Jones, M.W.: MedZip: 3D medical images lossless compressor using recurrent neural network (LSTM). In: 2020 25th International Conference on Pattern Recognition (ICPR), pp. 2874–2881. IEEE (2021). https://doi.org/10.1109/ICPR48806.2021.9413341
13. Mukhopadhyay, A.P., Mohapatra, S., Bhattacharya, B.: ROI based medical image compression using DWT and SPIHT algorithm. In: 2019 International Conference on Vision Towards Emerging Trends in Communication and Networking (ViTECoN), pp. 1–5. IEEE (2019). https://doi.org/10.1109/ViTECoN.2019.8899545
14. Gao, S., Xiong, Z.: Deep Enhancement for 3D HDR brain image compression. In: 2019 IEEE International Conference on Image Processing (ICIP), pp. 714–718. IEEE (2019). https://doi.org/10.1109/ICIP.2019.8803781
15. Jin, Y., Lee, H.-J.: A block-based pass-parallel SPIHT algorithm. IEEE Trans. Circuits Syst. Video Technol. **22**, 1064–1075 (2012). https://doi.org/10.1109/TCSVT.2012.2189793

Smart Gaming System for Hand Rehabilitation

J. Jane Elona(✉) and T. Jayasree

College of Engineering, Anna University, Chennai 600025, India
janeelonaremy@gmail.com, jai_t@annauniv.edu

Abstract. The human body part that is most often represented through icons is the hand. Shaking hands is indeed a sign of greeting and camaraderie, and it can symbolize strength, power, and protection as well as compassion, kindness, and consistency. The hand is an incredible feat of anatomical engineering, both in terms of its structure and function. So, any damage to the hand's supporting structures has the possibility of causing severe impairment. Even the tiniest hand injuries require a good medical evaluation in order to lower this danger. Many injuries and diseases affecting the upper extremities are treated with hand rehabilitation. This covers sudden trauma, Carpel Tunnel syndrome, wrist drop, overuse disorders, and various problems. In order to motivate patients with these conditions a virtual gaming is designed using UNITY-Game Engine for the betterment of these syndromic patients. The proposed system gives the most efficient strategy of hand tracking that uses 2D visual information in 3D to make the patient focus on rehabilitation to increase their hand movements, the patient receives treatment while manipulating a physical object that is monitored to assess how well they are doing without the direct observation of an occupational therapist. The gaming module, its fundamental structure and the supporting subsystems are designed in informal way that make it easier for patients and therapists to use. In order to create more gaming frame works that would be used in all rehabilitation clinics for these pushing activities, a preliminary usability test is instigated.

Keywords: Hand Rehabilitation · UNITY Game Engine · Virtual-Reality (VR)

1 Introduction

Nowadays, many countries are being through a serious society aging problem including India. The Elder people are susceptible to many diseases such as stroke, rheumatic and other syndromes like carpel tunnel, wrist- drop, etc. that have a high impact on those patients' life. The rehabilitation method for such patients are mainly focused on hand-functional abilities that includes pinching, grabbing, and catching. Currently the available rehabilitation training methods are very taxing due to boring exercises. For stroke patients with hand impairment, the One-On-One passive training method is the only one rehabilitation training technique widely used in India. In this approach, the patient spends a lot of time performing FNMs (Functional Neuromuscular Stimulation) in a hospital rehabilitation unit while being watched by physiotherapists for rehabilitation. This process takes a long time, and often tedious for the patients. The time taken by the doctors and the patients are massively wasted.

B. K. Singh et al. (Eds.): ICBEST 2023, CCIS 2003, pp. 35–46, 2024.
https://doi.org/10.1007/978-3-031-54547-4_4

However, traditional (non-computerized) stroke therapy is mostly performed in the hospitals and other healthcare centers [1]. These centers are mostly located in the urban areas. Patients from rural areas need to travel for their treatment. Therefore, stroke patients with limited mobility facing some difficulties in getting into the centers and most of them may not be able to get the treatment. In addition, traditional rehabilitation methods need a lot of facilities and human resources because every exercise needs a different equipment and a therapist can treat only one stroke patient at a time. Several studies proved that duration, capacity and intensity of exercise sessions are the most important factors for effective motor rehabilitation. Only steady and intense exercises alone will be helpful for the patients to regain their motor functions. If the patients cannot have enough access to get treatment due to lack of therapists and facilities, the efficient treatment is not guaranteed.

In order to overcome these problems, a computer-based virtual-reality (VR) rehabilitation system is introduced to help the stroke patients regaining their basic motor functions. VR technologies have been deployed since early 1990s to provide the patients a natural engagement mode and the ability to strengthen their motor skills. It can be achieved by using haptic technologies integrated with virtual environments to let the patients feel and touch virtual things. The size of haptic gadgets, such as the Cyber Glove [2], have limited usage of the technology, particularly for stroke patients [3].

This paper presents the Virtual Reality (VR)-based hand Rehabilitation technique that uses a unity software to develop VR based games such as Pushing the Tower and sorting the Cubes. Here, the VR based game will provide artificial environment which will make them to interact with the reel tangible objects by pushing and touching them. This set-up requires the patient to use tangible objects to accomplish their objectives. Using VR-REHAB, patients are able to experience the real force while they perform their exercises. The ultimate goal of the project is to enable the patients to get involved in the daily activities that will be helpful for them to regain the motor skills. Furthermore, the studies have been evaluated for three subjects in the interval of five days. Their efficiencies and improvement have been noted according to their performances.

1.1 Need for Study

To overcome the insufficient accessibility of treatment due to lack of therapists and facilities, the gaming system is introduced for effective treatment. The smart gaming system is cost effective when compare with other traditional therapy. The VR- based therapy help the syndromic patients to concentrate more on rehabilitation exercises. In addition, displaying fully immersive virtual objects in real-world scene promotes engagement of patients in their treatment sessions.

1.2 Medical Aspects of a Gaming System

The healthcare system faces additional challenges as the population continues to grow. More people than ever are living with illness and disability, which puts more stress on the healthcare system. Theoretically, therapists should be able to handle heavier caseloads and more patients by integrating more telemedicine into conventional rehabilitation for those with chronic diseases. An economically sound way to give patients more practice chances during non-treatment time is to use a gaming-based system as a therapy supplement. Earlier, the rehabilitation procedures could take months and even years for the patients enabling to get their full potential abilities to do the task as normal people. In olden days, if the people suffering with chronic illness, the therapy would be required for their entire live time. Nowadays, it is possible to make these patients to get back to their normal life sooner than earlier period. This can be achieved by the emerging technologies such as VR based therapies and their advancement in medicine. It is frequently difficult to treat neurological problems with therapy that is both intense enough to restore muscular strength and function in a limb or joint and transferable enough to allow for skill recovery. The hand activation requirements vary depending on the game. Children with hemiplegic cerebral palsy have shown improved upper limb function in those exercises that call for bilateral tasks. Patients must not only enjoy participating in many areas of the rehabilitation process, but also feel secure while having fun.

2 Literature Review

A Virtual Reality based Training and Assessment System was developed, that contains two rehabilitation training games and one assessment system [4]. The virtual system might draw patients' attention and lessen how tedious rehabilitation training and evaluation are for them. Compared with the existing rehabilitation assessment methods, the proposed virtual assessment system can give the assessment results similar to Fugl-Meyer Assessment, which is more quantitative, interesting and convenient. The process of assessment needs the coordination between doctor and patient, which is more complicated. This research presented a virtual reality-based training and assessment system for hand rehabilitation based on the current scenario. Considering patients' needs, two rehabilitation games are designed to increase the interests and curiosities of patients. Meanwhile, an iterative learning control algorithm is also added to increase the effectiveness of hand rehabilitation training.

Utilizing 2D visual information, such as color and shape, is one of the most efficient hand tracking approaches. Hence a depth-space Kalman filter is used for 3D hand tracking [5]. But when tracking is done in a variety of lighting situations, techniques based on visual sensors are particularly sensitive. Additionally, because hand motions are made in 3D space, the performance of hand gesture identification utilizing 2D information is necessarily constrained. Using a 3D depth sensor and a Kalman filter, this research proposes a revolutionary real-time 3D hand tracking approach in depth space. We identify potential hands using motion clusters and predetermined wave motion, and we monitor hand locations using a Kalman filter. We contrast the performance of the suggested method with the existing ones to confirm its efficacy.

A Home-Based Functional Electrical Stimulation-Assisted Hand Therapy were developed in three case reports of a home-based intervention for children with hand hemiplegia that integrates custom video games with contralaterally controlled functional electrical stimulation (CCFES) [6]. With CCFES (controlled functional electrical stimulation), the less-affected hand's voluntary opening regulates the more-affected hand's stimulated opening. Though they have showed potential in the treatment of hemiplegia, video games that encourage goal-oriented, skill-requiring mobility have not yet been paired with electrical stimulation in young patients. The feasibility of this intervention for home use was assessed by device logs, end of-treatment interviews, and motor function/impairment assessments. They found preliminary indications that CCFES-integrated video game therapy can provide a high dose of hand motor control therapy at home and in the lab. Improvements in motor outcomes were also observed.

Here a study based on the topic of tracking the development of virtual reality applications for rehabilitation [7]. It is based on the relatively recent therapeutic use of virtual reality (VR) for rehabilitation, which is rapidly increasing in response to the development of science and technology. Since implementation happens so quickly, determining research goals and proving the effectiveness of interventions are frequently more reactive than proactive. This study employed information science analytical tools to investigate if VR rehabilitation applications have developed as a separate area of study or simply a methodology in core fields like biomedical engineering, medicine, and psychology.

In study of Hand Tracking-based Motion Control for Robot Arm Using Stereo Camera [8], an application that put in use for controlling a robot arm. The hand motion was recorded by the camera, which also tracked it and logged the coordinate location or 3D hand motion position. The system was made to monitor the hand's pitch, yaw, and roll motion. In addition, the device was able to track the forward and backward motion of the hands. After that, our system saved and analyzed the hand tracking data. Using TCP/IP connectivity, these data were sent to the arm robot and converted into movement by the arm robot. They focus on developing robot arms that can be manipulated by hand.

The simulated tasks for poststroke rehabilitation have been developed recently, where some movement patterns have been recognized based on their functional goals [9]. Then, haptic-based simulated tasks for these movements are developed. While a patient is practicing these simulated exercises, the system collects different types of data that can be used later for movement evaluation. Exercises created for the PHANToM and Cybergrasp haptic interfaces are currently part of the system's implementation.

The framework of therapy for hand-motor rehabilitation that comprises of a variety of activities [10]. Unlike other attempts, this framework was built based on well-established exercises that are used in the Jebsen Test of Hand Function and Box and Block tests [11]. The framework comprises four components: 1) the sensory component that communicates with the application programming interface for haptic devices (API); 2) the haptic-simulation component that is responsible for haptic/graphic scene synchronization; 3) an application component that hosts the exercises; and 4) the haptic data component that acts as a haptic data repository [12, 13].

Several systems have been developed that incorporate haptic technologies in rehabilitation. The effectiveness of haptics in the field of rehabilitation has been demonstrated by these systems. One of the first attempts was the Burdea et al. investigation. In addition, Rydmark et al. [14] have developed some haptic-based exercises for brain-injured patient. Results proved that using such system could promote rehabilitation [15]. Haptic-based rehabilitation inherits the advantages of VR based rehabilitation because haptic systems usually incorporate VR environment as a main component of its system setup. In addition, the patient is able to touch and manipulate the virtual objects using such systems [16].

Here the combining LMC (Leap Motion Control) with 2D and 3D VR settings had a similar positive impact on functional and clinical results. Additionally, individuals with Parkinson's disease reported higher levels of motivation in 3D VR settings compared to 2D environments, but also more stress and strain [17]. LMC (Leap Motion Control) was also used for testing the feasibility of free hand movement for stroke rehabilitation [18], recognition of American and Greek Static Signs [19], and development of a videogame-based therapy for elderly people rehabilitation [20].

Additionally, due to the convenience and availability of LMC with the rapid development of VR software, hardware and simulation games, the technology contributed to variety of applications at homes with the assistance of a Virtual Therapist (VT) [21]. Similar in-home rehabilitation therapy using a wearable wristband device to detect hand movements and transfer sensory data to companion videogame has proven to give successful results [22]. The rehabilitation process must be engaging and entertaining using VR, provide real-time feedback, and monitor response via stored activity data. The overall setup must be noncomplicated to allow use not only in a controlled environment but also at patient's home after a few demonstrating minutes at the lab/clinic [23].

Researchers in [24] developed an AR system to train a stroke patient for grasp-and-release tasks. In order to enable the patient to grip and manipulate virtual items, the system displays the patient's actual hand inside a virtual environment on a head-mounted display (HMD). In addition, this system is equipped with an assistive glove to help the patient extend their fingers during the grasp-and-release task. Using a joystick, a therapist can modify positions, orientations, and sizes of virtual objects in the virtual environment to best fit the patient's needs [25].

From our literature review, we found that most of the existing systems for rehabilitation of hand-motor function use hand-wearable haptic devices that support multipoint haptic interaction. Mostly, these devices tend to be bulky and require a significant effort to wear even by healthy people. This is a serious problem to stroke patients or any syndromic patient like Carpel Tunnel and wrist drop patient have difficulties in moving their fingers freely. In addition, hand-wearable haptic devices are not only costly but also are attached to special supportive machines that prohibit its associated system from being portable. This is not only the problem with such devices; their continuous usage results in tiredness and some fatigue in the arm and shoulder of patients. Thus, some patients may start to lose their interest in these systems, particularly those who are in senior ages.

Additionally, such haptic devices require a calibration process that takes time and require patients to make difficult gestures with their affected hands. Other than haptic devices there are many 3D exergames developed in rehabilitation field by using leap motion controller (LMC), but these are very expense for a normal people to use it. Our proposed system was developed for the purpose of cost-effectiveness in rehabilitation field. So that the patient can freely use this system without any oculus for VR, or any LMC (Leap Motion Control). Therefore, the VR rehabilitation systems can change the traditional boring rehabilitation exercises into entertaining ones, carrying out visual and auditory motivational component which is cheaper. It does not require huge setup for making it in practical world. In addition, these components are preprogrammed to encourage patients by producing motivational messages (visual or audio) in response to events a patient may trigger. Events are classified based on how much a patient may progress in her/his correspondent exercise.

3 Architecture of Hand Tracking System

In the host computer, the software of rehabilitation training and assessment system is designed by Unity Engine. Unity3d developed by Unity Technologies is a multi-platform and comprehensive game development tool, which can create three dimensional objects used to interact with people. Meanwhile, it supports building the corresponding executable file.

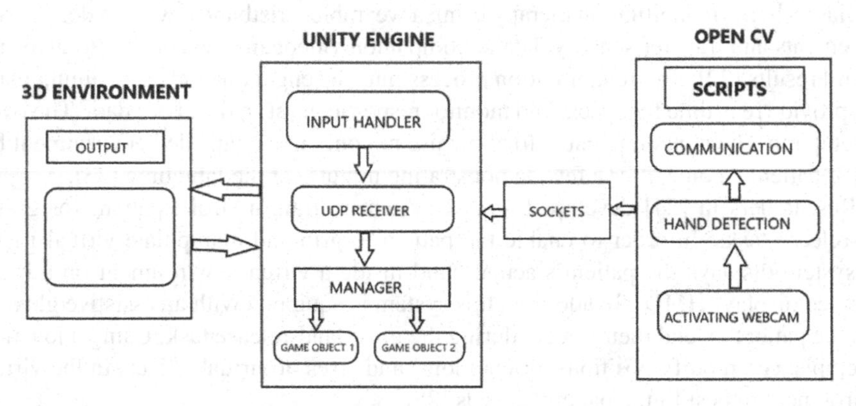

Fig. 1. Block diagram of assessment system

From the above Fig. 1, PyCharm software is used to interlink with the Unity Engine. Where the face, hand and its postures are detected and these data are transferred to Unity Engine through UDP Sockets. The Face and hand detection are done with the help of a web-camera. Activating webcam by installing CV-ZONE. Some other libraries are installed for reading data and for communication. Here the libraries are available in packages in the PyCharm interpreter. The PyCharm interpreter should be in the VR environment for accessing data in 3D environment. Depending on the system and the build options, various socket families are supported by this module. The address family given when the socket object was established determines the address format that is required by that specific socket object.

The socket () function creates a socket object whose methods perform the various socket system calls, making it simple to convert the socket library interface and Unix system call to Python's object-oriented paradigm. Parameter types are at a marginally higher level than the C interface. For instance, much like with read () and write () activities on Python files, buffer generation on receiving operations is automatic, and buffer length is implicit in transmit actions.

3.1 Design of 3D Hand Tracking System

According to the suggestions of rehabilitation therapist, the fingers separation motion has a vital significance for hand rehabilitation. Therefore, combined with the uniqueness of the proposed mechanical structure, freely floating hand points is chosen to be one of the rehabilitation training tasks. Its function and the design of scene are worth to be considered. In view of the situation that patient's part fingers lost movement function, this paper divides the game into the task mode. The total setup is in Desktop model. Hence it will be easily for hand rehabilitation patient to sit and interact in large screen. To design any 3D tracking system first we need to analyze the patient's rehabilitation status. This will help the developers to construct the outline of game for the particular patient but in this case, we have taken for a common people who are suffering from Carpel Tunnel and Wrist drop syndromic conditions.

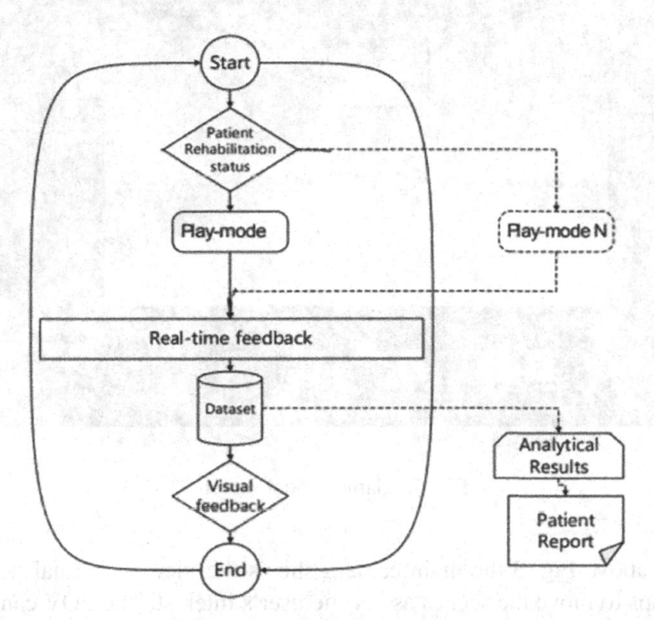

Fig. 2. Game operating flowchart

The above Fig. 2 gives the clear-cut idea of designing any game. Here the patient rehabilitation status was identified and modified in such a way that the carpel-tunnel patients are not able to move their hand functionalities. This game was developed as a 3D environment using Unity Game Engine (2019.2.16f1 "64-bit") with its Core Unity and Interaction Engine packages. Computer Vision has the capability of detecting movements, orientation and gestures of UL which is the main pillar for the rehabilitation exercises sought from the developed game. The developed game operating flow chart, allows the patient to interact with the game on different play-modes. Play-mode N resembles future modes that can be developed and added to the game. Each play-mode is aimed to motivate the patient to make certain UL movements and functions, explained as follows,

Play Mode 1: Represented in Fig. 5, designed for exercises based on pushing and Sorting of objects with physical simulation properties. The pushing exercise suggests that low accuracy movements is required to accomplish the game level. While the Sorting exercises is possible on static or moving platforms, which requires high accuracy in hand movements.

Once all the primary setup is done the following VR game can be implemented by assigning a cube or any gamified object as per the user's interest in the manager.

Under the situation of task mode, some environmental factors are designed and the cubes are laid as a floor in VR environment and the camera is kept behind the scene. Here the object is given with a rigidity weight oh 10kg for the gravitational pulling effect. When the cubes are assembled as a tower with a mass. When the Game starts, it gives an animation effect on the screen.

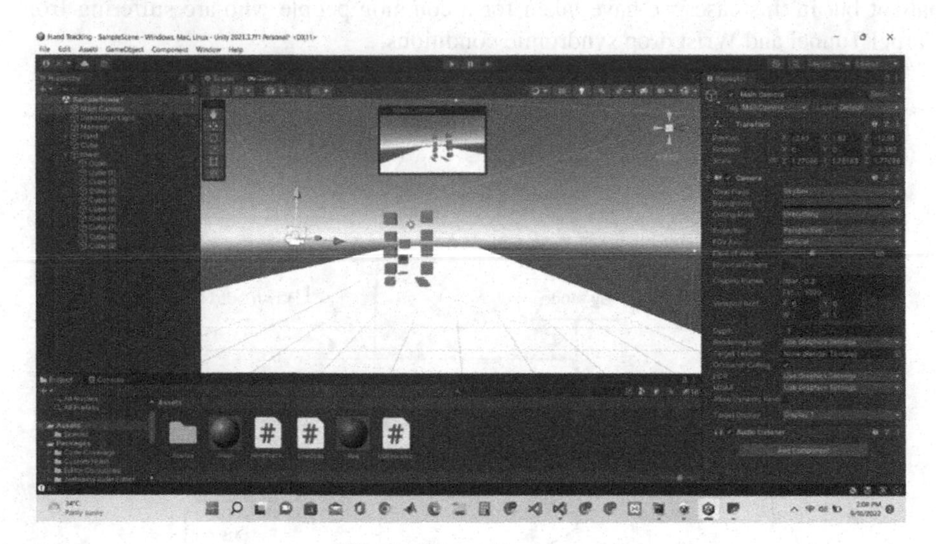

Fig. 3. Game virtual screen

From the above Fig. 3 the main camera shows the view of Virtual Scene and the XYZ axis helps to move the scenes as per the user's interest. The FOV can be adjusted for better visualization and here it is set as 60%. Thus, the design and organization of asserts for motion tracking system. If there is any misalignment while playing a game, it can be sorted out in the main camera scene.

As UNITY- software is user-friendly it takes less than few mins to alter the scenes. It is possible to make many scenes with tangible objects. As this game is developed for understanding and analyzing the game engine, it does not require much objects and scene to incorporate in it. And also, this take less time to create a scene which is very much useful for any rehabilitation center who are with less capital. In over populated country like India require more rehabilitation centers and there are more than 172 centers are running freely for the welfare of the people. In such cases these short time scenery game will definitely helpful for all kinds of people.

4 Result and Discussion

The 3D motion tracking system for hand rehabilitation using Unity software is designed. The 3D tracking of hands in purely 2D images is an extremely challenging problem for using in real-time. Here the assessment is based on the Desktop mode using webcam.

At first the detection of hand points using PyCharm Software is implemented. And the 21 points are marked using Media Pipe packages. A palm detection model that produces an orientated hand bounding box from the complete picture and the points are sent to the UNITY software. Where two of the models used by the MediaPipe pipeline to precisely localize the key points of 21 3D palm coordinates in the identified hand region. The coordinates for X, Y, and Z will be displayed in square braces in the console for designing game.

There will be 21 points showing different angles, this information's are very much useful in in getting data of a particular region in hand. For basic level games, like pushing etc. doesn't need all the coordinate values but for the advanced gaming system which includes exercises requires 21 coordinates location.

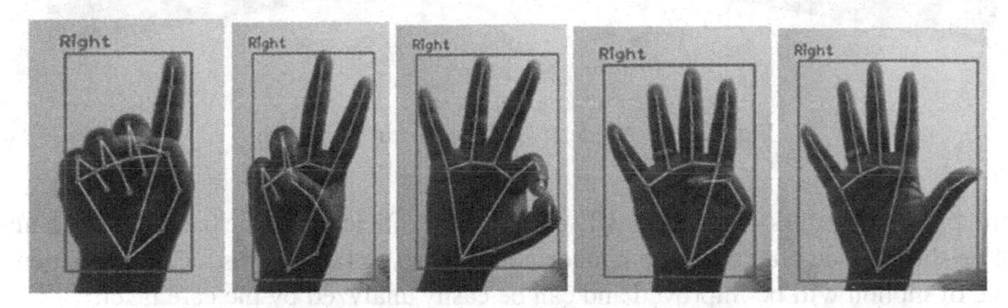

Fig. 4. Testing Real Time Hand Tracking in PyCharm

Different signs are tested for checking the speed of motion tracking, which is shown in Fig. 4. Testing the performance is very important for the patient involvement. The latency rate is minimum and gives the real time data without any lag. It is necessary to check the latency rate and to make sure to keep it minimum. If incase the rate is high, it is very difficult to make the syndromic patients to concentrate on the virtual game. They feel really bored because of the poor refreshing rate.

Hence testing the latency rate is very much important in developing any sort of a game. After checking the latency rate in PyCharm. The data are made to transfer through the sockets which will communicate to the client, by plotting the 21 coordinates in VR environment it will resembles the hand structure in the main camera scene in UNITY. The Pushing task exercise don't have time limit. As far as the patient can able to play the game it will continue. Here we have taken 3 subjects and all perceived the environment as virtual environment not the real one. This is the ultimate task of our project to make them involved in virtual environment and fortunately the game made them to involve in it.

The blocks were created and they are arranged in the form of a tower which is shown in the Fig. 5. Each block is given some mass for the animation effect. Here the task begins for the stroke patient. The patients are guided in such a way to move their hand to push the block. Their task is to move their hand forward or backward motion for making the cube to fall off. The score board detects the number of towers that the patient made to push. It will automatically calculate and give the patients efficiency. Here the patient's efficiency can be calculated by encountering their daily activity. That is the system generated the daily data and regulate their efficiency by taking in the consideration of time with the

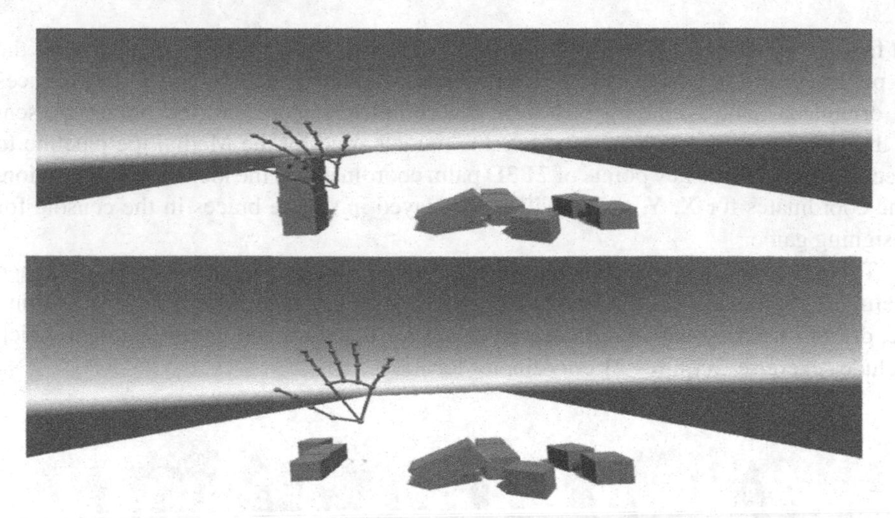

Fig. 5. Play Mode 1: Pushing the cube exercise

score, by this way the system can easily calculate the efficiency of the patient. The results will be shown once after the patient made all the cubes to fall, by this means the patients range of motion will be improved and can be easily analyzed by the care taker.

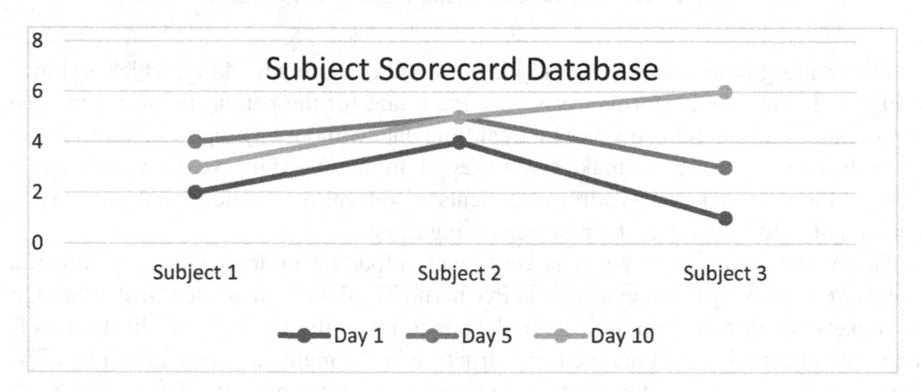

Fig. 6. Subject Scorecard Database

The above Fig. 6 represents the player scorecard for 10 days. During the testing process the subjects are instructed to play the game in a standard position. This is because change in position may lead errors when finding the efficiency. Some initial validation tests were performed on the developed game to verify its operation and accuracy in detecting hand movements. Images of some tests are shown in Fig. 4 and Fig. 5. These images were taken during tests of Collecting, Sorting and Hand Gestures detection exercises, respectively, which showed promising results. The developed game allowed for an immersive and engaging user experience, physical simulation of motion, real-time feedback and visualization of results. Thus far, results indicate that the developed smart VR game is adequate and useful for in-home UL rehabilitation exercises.

5 Future and Conclusion

As for future developments on the reported assessment, additional play modes and exercises will be added for advanced and accurately studied rehabilitation movements. Parameters, such as moving distance, accuracy of movement and time duration of the exercise will be used to increase the difficulty level. Furthermore, a scoring system will be created to determine the progress of the patient as a result of the rehabilitation program. Results can be represented as reports generated from the dataset in Unity Game Engine. Accessing and navigation of the UI(User Interface), as part of rehabilitation process, is planned using finger gestures combined with UL(Upper Limb) movement. Finally, VR capabilities, visual and sound effects will be added to improve the 3D environment and user experience.

Patients with UL (Upper Limb) disabilities can find a new way to recover using convenient and low-cost hardware and software applications. The field of VR could largely contribute to the disability and rehabilitation technology sector through real-time motion detection, data analysis and the promising results of the widely reported LMC (Leap Motion Control) technology. As the rehabilitation technologies continue to develop, it is expected to see more game developments aiming at improving abilities and functions of many parts of the human body. This is proven by the above evaluation as the complexity of the game is very minimum and the time duration to create the game is very less when compare to other 3D motion tracking games. Therefore, this game can be used in all developing countries due to its user-friendly nature.

References

1. Rijken, M., Dekker, J.: Clinical experience of rehabilitation therapists with chronic diseases: A quantitative approach: Clin. Rehabil. **12**(2), 143–150 (1998)
2. Immersion Corporation, Cyberglove. [Online]. Available: www. immersion.com
3. Kayyali, R., et al.: Occupational therapists' evaluation of haptic motor rehabilitation. In: Proc. 29th Annu. Int. Conf. IEEE EMBS (2007)
4. Park, S., Yu, S., Kim, J., Kim, S., Lee, S.: 3D hand tracking using Kalman filter in depth space. EURASIP J. Adv. Sign. Proc. (2012)
5. Michael, J.F., Curby, A., Suder, R., Katholi, B., Knutson, J.S.: Home-Based Functional Electrical Stimulation-Assisted Hand Therapy Video Games for Children With Hemiplegia: Development and Proof-of-Concept. IEEE Trans. Neural Syst. Rehabil Eng. (2020)
6. Liu, S., Meng, D., Cheng, L., Huang, F.: A virtual reality based training and assessment system for hand rehabilitation. In: Ninth International Conference on Intelligent Control and Information Processing (ICICIP) (2018)
7. Keshne, E.A., Weiss, P.T., Geifman, D., Raban, D.: Tracking the evolution of virtual reality applications to rehabilitation as a field of study. J. Neuro Eng. Rehabilit. volume Published: 21 June (2019)
8. Sugandi, B., Toar, H., Alfianto, A.: Hand Tracking-based Motion Control for Robot Arm Using Stereo Camera. In: International Conference on Applied Engineering (ICAE) (2018)
9. Yeh, S.-C., et al.: In: Dainoff, M.J. (ed.) Ergonomics and Health Aspects, HCII 2007, LNCS 4566, pp. 378–387. © Springer-Verlag, Berlin Heidelberg (2007)
10. Alamri, A., Cha, J., El Saddik, A.: AR-REHAB: an augmented reality framework for poststroke-patient rehabilitation. IEEE Transactions on Instrumentation and Measurement **59**(10), (2010)

11. Stanislaus, A.A., Winters, J.M.: Use of remote functional assessment tools to enhance rehabilitation research and practice. In: Proceedings of the Second Joint 24th Annual Conference and the Annual Fall Meeting of the Biomedical Engineering Society (2002)
12. Alamri, A., et al.: Haptic exercises for measuring improvement of poststroke rehabilitation patients. In Proc. Int. Workshop Med. Meas. Appl. (2007)
13. Barghout, A., Alamri, A., Eid, M., El Saddik, A,: Haptic rehabilitation exercises performance evaluation using automated inference systems. Int. J. Adv. Media Commun. **3**(1/2), 197–214 (2009)
14. Broeren, J., Björkdahl, A., Pascher, R., Rydmark, M.: Virtual reality and haptics as an assessment device in the postacute phase after stroke: CyberPsychol. Behavior **5**(3), 207–211 (2002)
15. Broeren, J., Georgsson, M., Rydmark, M., Sunnerhagen, K.S.: Virtual reality in stroke rehabilitation with the assistance of haptics and telemedicine. In: Proc. 4th Int. Conf. Disability, Virtual Reality Assoc. Technol. (2002)
16. Burdea, G.: Review paper: Virtual rehabilitation Benefits and challenges. In: Proc. Int. Med. Inf. Assoc. Yearbook Med, pp. 170–176. Inf., Heidelberg, Germany (2003)
17. Cikajlo, I., Potisk, P.: Advantages of using 3D virtual reality based training in persons with Parkinson's disease: a parallel study. J. Neuro Eng. Rehabilit. **16**(119), 1–14 (2019)
18. Khademi, M., et al.: Free-hand interaction with leap motion controller for stroke rehabilitation. In: CHI '14 Extended Abstracts on Human Factors in Computing Systems (2014)
19. Mapari, R.B., Kharat, G.: American static signs recognition using leap motion sensor. In: Proceedings of the Second International Conference on Information and Communication Technology for Competitive Strategies (2016)
20. Iosa, M., et al.: Leap motion controlled videogame-based therapy for rehabilitation of elderly patients with subacute stroke: a feasibility pilot study. Topics in Stroke Rehabilitation **22**(4), 306–316 (2015)
21. Sourial, M., Elnaggar, A., Reichardt, D.: Development of a virtual coach scenario for hand therapy using LEAP motion. In: 2016 Future Technologies Conference (FTC) (2016)
22. Mazlan, A., et al.: Rehabilitation Hand Exercise System with Video Games. Int. J. Adv. Trends Comp. Sci. Eng. **9**(1), 545–551 (2020)
23. Fotopoulos, D., et al.: Gamifying motion control assessments using leap motion controller. In: Hasman, A., et al., (eds.) Data, Informatics and Technology: An Inspiration for Improved Healthcare, Volume 251, Series: Studies in Health Technology and Informatics. IOS Press Ebooks (2018)
24. Luo, X., Kenyon, R., Kline, T., Waldinger, H., Kamper, D.: An augmented reality training environment for post-stroke finger extension rehabilitation. In: Proc. 9th ICORR, pp. 329–332 (2005)
25. Luo, X., et al.: Integration of augmented reality and assistive devices for post-stroke hand opening rehabilitation. In: Proc. Conf. IEEE Eng. Med. Biol. Soc., pp. 6855–6858 (2005)

An Ample Review of Various Deep Learning Skills for Identifying the Stages of Sleep

P. K. Jayalakshmi(✉), P. Manimegalai, and J. Sree Sankar

Karunya Institute of Technology and Sciences, Coimbatore, India
jayalakshmip@karunya.edu.in

Abstract. Sleep is an important part of everyone's life. Sleep disorders are very common nowadays, when neglected may cause many neurological problems. Common problems like sleep interruptions, snoring, etc. can be detected by using sleep stage analysis. However, the conventional methods used for sleep analysis are very time-consuming. This limitation can be avoided by using an automatic tool for diagnosis based on artificial intelligence. Different artificial intelligence technologies like deep learning ensure the full utilization of data with very less information loss. A detailed study of different models from the years is provided here. The studies here use deep learning model techniques to analyze the sleep stages using polysonogram (PSG) signals. Additionally, this investigation demonstrates the use of electrocardiogram (ECG) signals with convolutional neural networks for the analysis of the various stages of sleep. It demonstrates the excellent categorization performance of CNN networks, especially 1D CNN. An examination of the studies reveals that EOG and EMG signals may also be utilised in future automated detection systems, in addition to EEG signals. This led us to the conclusion that PSG signals can also be used in conjunction with deep learning algorithms in addition to EEG signals. This study examines various techniques using PSG signals and deep learning skills that have been published in latest years.

Keywords: Sleep stage analysis · PSG signals · 1D CNN · Deep learning models

1 Introduction

Life is not possible without sleep. The regulation of numerous bodily processes as well as human wellbeing depend on getting enough sleep [1, 2]. A person's behaviour while sleeping and the corresponding physiological changes that happen to the electrical patterns of the waking brain during sleep are two ways that scientists define sleep [107]. The physiological criteria are determined based on the results of EEG, electrooculography (EOG), and electromyography (EMG). It's important to compare the sleepiness and lethargy and has to find out which process is happening. But weariness might be a secondary effect of sleepiness. When sleep starts to take hold, many behavioural and physiological characteristics progressively change.

B. K. Singh et al. (Eds.): ICBEST 2023, CCIS 2003, pp. 47–65, 2024.
https://doi.org/10.1007/978-3-031-54547-4_5

The sleep can be catogarized in to two types. They are Non-rapid eye movement (NREM) and Rapid eye movement [REM]. The absence of this leads to sleep disorders [3]. Poor quality of sleep also leads to many sleep disorders. In one paper the author Strangers et al. highlighted the sleep related issues globally [4]. The paper says that almost 18.6% of the adult population out of 200 million have the sleep-related issues. REM sleep predominates in the final third of adult human sleep while slow-wave sleep predominates in the first third. According to the conventional Rechtschaffen and Kales (R-K) [24] grading manual, NREM sleep, which makes up between 75 and 80 percent of adult humans' sleep time, is split into four stages (NREM stages 1 through 4). However, the most recent American Academy of Sleep Medicine (AASM) scoring guideline [21] divides this into 3 stages mostly based on brain signal analysis criteria. For the total time of sleep one –fourth is occupied by rapid eye moment stage. of which may have a characteristic saw-tooth appearance—dominate the EEG tracings.

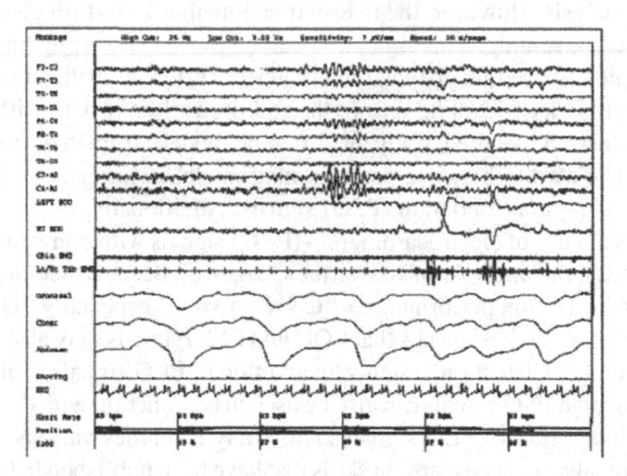

Fig. 1. Recordings from polysomnography demonstrating rapid eye movement (REM) sleep [107]

Characteristic changes are noticed in blood pressure, breathing, oxygen saturation and pulse rate with the rhythmic change in the eye movements rapidly. The normal adult has an orderly transition from alertness to the start of sleep, Non - rem sleep sleep, and finally REM sleep. Non - rem sleep is characterised by slow eye movements, a gradually diminished reaction to external stimulation. Deep sleep is characterised by reduction in sensory response, the absence of muscle tone, low voltage and swift brain activity. Many conventional methods are there for the manual sleep stage scoring [5, 6]. But in many cases human interventions can cause errors in sleep stage analysis [7]. The convenience of the patient and the cost of the diagnostics also to be considered. All the sleep labs required a dedicated facility for the measurement. Also most of the sleep labs are in the hospitals and diagnostic centres only and the patient has to travel and spend the whole night for the analysis. Obviously the cost of the diagnosis is very high considering with less patient convenience. Portable monitoring devices will serve the purpose to some extent, but the diagnostic limitations, failure of the device, reliability concerns will come in to picture [8]. To coverup all the disadvantages mentioned above the only thing to be done is the change of method used for sleep analysis. Deep learning technology can be incorporated for the classification. This research focuses more on the benefits and constraints of deep learning models for classifying sleep stages. The primary goal of this endeavour is to give the professionals working on this platform in-depth information. This includes the research over many years along with their classification of sleep stage performance. Deep learning is a promising technology which can be used effectively for this classification process. Because in future all the studies will be based on deep leraning and its related technologies.

2 History and Background

The development of EEG and sleep starts from the embryo state itself and develops in all stages of line with the nervous system development [107]. These morphogenetic changes will be pointedly influenced by co-occurring medical or neurological conditions, as well as by neurological, environmental, and genetic factors. Sleep requirements drastically differ from childhood to old age. Babies who are just born have a irregular sleep pattern, requiring a total of 16 h of sleep every day. A child's daily sleep demand decreases to approximately 10 h between the ages of 3 and 5, and to approximately 11 h between the ages of 9 and 10. With age, sleep develops a biphasic cycle.. Adults typically sleep for 7.5 to 8 h each night in a monophasic pattern; but, as people age, the rhythm shifts back to being biphasic. In About 50% of newborn infants' sleep time is spent in the REM stage, but by the time they are 6 years old, this percentage has dropped to 25%, which is more in line with the usual adult pattern. The adult sleep cycle of NREM/REM cycling is formed by the age of three months. It was in 1965 marked a golden era in the sleep analysis with the discovery of Obstructive sleep apnea [9]. Recent studies, it shows sleep apnea can cause cardiovascular diseases and can be a reason for death if no treatment options are provided [10, 11]. It is also mentioned that poor sleep quality may lead to many sleep disorders like sleep apnea, insomnia, etc. [12, 13]. The various symptoms that lead to insomnia are snoring, extreme daytime fatigue etc. They may be due to some blockage in the respiratory tract [15]. It may lead to cardiovascular complications which may cause diseases like stroke, heart failure, etc. [14, 16]. Also, the most important thing is that the lack of sleep may affect the performance and productivity of the individual during the daytime [17, 18]. Currently, overnight polysomnography is a useful technique for assessing sleep stages [6]. Electroencephalogram (EEG), electrooculogram (EOG), electromyography (EMG), electrocardiogram (ECG), airflow, and blood oxygenation are among the data captured by PSG [5]. The sleep should be appraised by dividing these PSG recordings into units termed epochs [21]. This Epoch is graded and divided into various sections. It is easier to use a programmed diagnostic tool (PDT) for the purpose as the technology is increasing. It is proved that these technologies are far better than the conventional methods [22–24].

3 Polysomnogram(PSG) Analysis with Programmed Diagnostic Tools

A thorough examination performed to identify sleep abnormalities is termed polysomnography, commonly known as a sleep study. Your pulse rate, breath, oxygen saturation, ocular and muscle activity of legs, and brain waves are all monitored during a polysomnography scan [108]. With the help of polysomnography, it can determine whether or not your sleep patterns are disrupted, when they do, and reasons. Home sleep apnea testing equipment comes in a variety of designs and configurations. They often keep track of the heart rate, oxygen levels, breathing rate, and airflow. One technique additionally includes data on blood vessel tone. Both traditional machine learning and deep learning may be used as the foundation for the diagnostic tool for the analysis. However regular machine learning often faces the problem of overfitting samples [26]. Some data reduction steps like feature extraction should be done to reduce the overfitting problem [27]. This can be done manually with human experts. The categorization machine learning models can then use this as input. However, the conversion of PSG recordings to a low-dimensional vector in this method can cause information loss. So thus can be concluded that machine learning is not suited for handling high-dimensional data [27]. So it is limited to classifying PSG signals. With deep learning, there is no information loss when using it. They develop the models so that they can automatically detect PSG signal features. It can be utilised while handling enormous volumes of data and makes accurate predictions [29]. This suggests that DL approaches are ideal for categorizing high-dimensional PSG signals.

4 Deep Learning Models

Python programming is commonly used in developing deep learning models. The normal python tools used are Tensor flow, Keras, Pytorch, Lasagne. A small comparison between these pyton tools are given below in table [109] (Table 1).

Table 1. Comparison between the different python tools [109]

	Keras	Pytorch	Lasagne
Description	The most sophisticated and user-friendly library of the group is definitely Keras. Francis Chollet, another member of the Google Brain team, is responsible for its writing and upkeep. It gives users the option of using TensorFlow's or Theano's symbolic graph to run the models they create. If you have prior expertise with machine learning in Lua, Keras is probably worth a look because of its Torch-inspired user interface. The Keras community is fairly extensive and quite active, in part because of its good documentation and relative simplicity of usage. The TensorFlow team recently revealed plans to include Keras support in all future releases, making Keras a subset of the TensorFlow project	PyTorch, the most recent addition to our collection of deep learning frameworks for Python. It's a loose translation of the Torch library from Lua to Python, and it stands out for having the support of the Facebook Artificial Intelligence Research team (FAIR) and being able to handle dynamic computation graphs, which is something that competitors like Theano, TensorFlow, and derivatives lack. Although PyTorch's place in the Python deep learning environment is still up for debate, all indications are that it will be a very credible alternative to the other frameworks on our list	Lasagne provides abstractions on top of Theano to improve its suitability for deep learning because Theano's primary goal is to be a library for symbolic mathematics. Lasagne enables users to conceive at the Layer level when designing network models rather than in terms of function relationships between symbolic variables. Lasagne offers a multitude of common components to assist with layer definition, layer initialization, model regularisation, model monitoring, and model training while requiring little sacrifice in terms of flexibility
Advantages	Support of Theano or tensor flow will be there. It is a high level interface with easier learning curve	The support for dynamic graphs is high. It is a blend of high level and low level APIs	Highly flexible
Disadvantages	Flexibility is less	References are limited	Community is small

The ratio of each deep learning tool used to develop a model for automatic sleep stage analysis is shown in the chart below (Fig. 2).

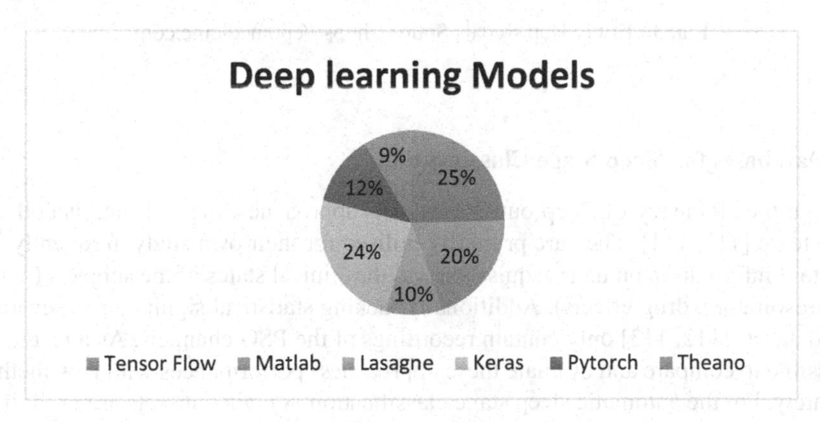

Fig. 2. Deep learning Models

The analysis shows Keras is the high-level application programming interface for python. It can be used with either tensor flow and theano.

5 Classification of the Various Stages of Sleep Using Deep Learning Methods

5.1 Sleep Stages

Humans can experience six discreate stages of sleep. It is explained in detail by Rechts Chaffen & Kales [51] in their paper. The six stages are classified as (1)Wakefulness(W) (2)Rapid eye movement sleep (REM) (3) Four stages of non REM sleep(NREM)-slots 4. As per the EEG characteristics W stage happens when the brain is fully active [52]. Generally it is shown by alpha rhythms of high frequency. In NREM stages the theta rhythms are dominating when S1 stage occurs. K complex waveform occurs in S2 stage. It lasts for 1–2 s. In S3 low frequency delta rhythms appear and will continue to dominate in S4 also. S4 is followed by REM stage. The stages of sleep classification based on EEG study is shown in (Fig. 3) Fig. 6

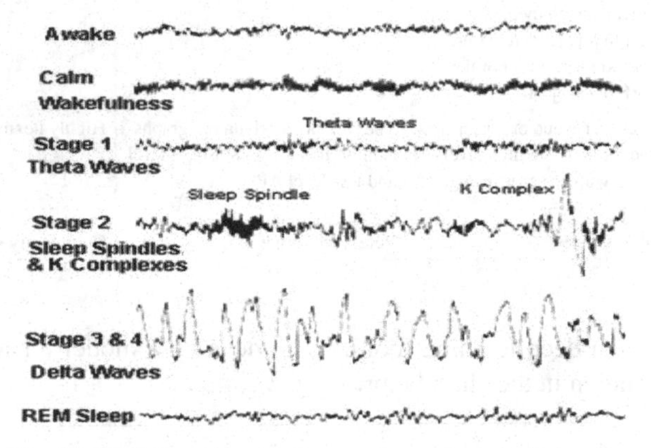

Fig. 3. EEG Sleep signals Source: https://epomedicine.com

5.2 Databases for Sleep Stage Classification

To judge the efficiency of sleep outline analysis approaches, the self data is collected by the team [110, 111]. They are primarily utilised for their own study, frequently lack important information on their acquisition and the clinical states of the subjects (neural, cardiorespiratory, drug effects). Additionally lacking statistical significance, several of these datasets [112, 113] only contain recordings of the PSG channels. As a result, it is impossible to compare and evaluate these approaches' performances with new methods accurately. For the automatic sleep stage classification, 8 main data sets are used. It can be downloaded directly from Physionet [53]. They are

1. Sleep-EDF [54]

2. The expanded Sleep-EDF [54]
3. University college of Dublin sleep apnea database [53]
4. The sleep heart health study [55, 56]
5. Massacusettsets Institute of technology data set [57]

The data set called ISRUC sleep data set [58] can be downloaded from the official website. But for the dataset from the Montreal Archive of sleep studies(MASS) [59] requires permission to download. The table below shows the sleep stage classification summary approaches with deep learning different sleep data bases (Tables 2, 3 and 4).

Table 2. Sleep-EDF data set expanded

Author	Signal	Sample	Approach	Programming tool used	Accuracy
Zhu et al. [63] 2020	EEG	15,188	CNN	-	93.7%
Qureshi et al. [64] 2019	EEG	41,900	CNN	-	92.5%
Yildirim et al. [65] 2019	EEG	15,188	CNN	KERAS	90.8%
Hsu et al. [66] 2013	EEG	2880	RNN	-	87.2%
Michielli et al. [67] 2019	EEG	10,280	RNN-LSTM	MATLAB	86.7%
Wei et al. [68] 2017	EEG	42,308	RNN	PYTORCH	83.9%
Mousavi et al. [69] 2019	EEG	42,308	CNN-RNN	TENSORFLOW	84.3%
Seo et al. [70] 2020	EEG	42,308	RNN	PYTORCH	83.9%
Zhang et al. [71] 2020	EEG	-	CNN	-	83.6%
Supratak et al. [72] 2017	EEG	41,950	CNN-LSTM	TENSORFLOW	82.0%
Phan et al. [73] 2019	EEG	-	CNN	TENSORFLOW	81.9%
Vilamala et al. [74] 2017	EEG	-	CNN	-	81.3%
Phan et al. [75, 76] 2018	EEG		CNN		79.8%
Xu et al. [77] 2020	PSG SIGNALS	-	DNN	-	86.1%

Table 3. Sleep –EDF dataset

Author	Signal	Sample	Approach	Programming tool used	Accuracy
Wang et al. [78] 2018	EEG	-	CNN	-	-
Fernandez-Blanco et al. [79] 2020	EEG		CNN	-	92.7
Jadhav et al. [80] 2020	EEG	62,177	CNN	-	83.3
Mousavi et al. [69] 2019	EEG	222,479	CNN	TENSOR FLOW	80.0
Tsinalis et al. [81] 2016	EEG		CNN		74.0
Yildirim et al. [65] 2019	EOG	127,512	1D-CNN	KERAS	88.8
Yildirim et al. [65] 2019	EEG + EOG	127,512	1D-CNN	KERAS	91.0

Table 4. Mix sleep databases

Author	Signal	Sample	Approach	Programming tool used	Accuracy
Seo et al. [70] 2020	EEG	57,395	RNN	PyTorch	86.5
Supratak et al. [72] 2017	EEG	58,600	CNN- BiLSTM	TENSORFLOW	86.2
Phan et al. [73] 2019	EEG		CNN	TENSORFLOW	78.6
Dong et al. [83] 2018	EOG		RNN-LSTM	THEANO	85.9
Chambon et al. [84] 2018	EEG/EOG + EMG		CNN	KERAS	
Phan et al. [73] 2019	EEG/EOG + EMG		CNN	TENSORFLOW	83.6

5.3 Input Signals Used for Sleep Stage Classifications

The sleep stage classification uses PSG signals the most frequently. The physiological signals that are gathered throughout a sleep cycle are called polysonogram (PSG) recordings. It is a multivariate system made up of records from the EEG, ECG, EMG, and EOG. PSG recordings are the most common input type here. PSG systems, which contain data from the ECG, EEG, EOG, and EMG, are used in a variety of research.

EEG is the most widely used in all of these research. It's because EEG signals serve as the basis for distinctive waves like alpha, beta, theta, and delta. Below is a statistical breakdown of various signal inputs (Fig. 4).

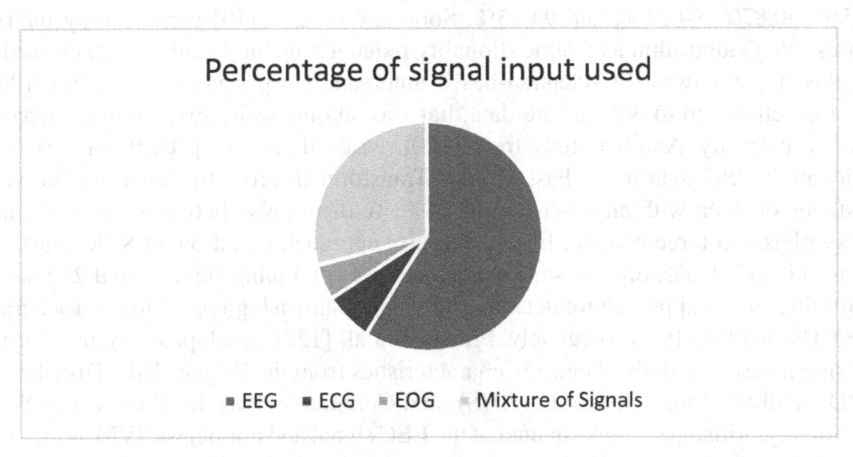

Fig. 4. Percentage of different input signals

EOG & EMG is also important while analysing REM sleep stages. In some cases the combination of the signals are also used. ECG [96] is also an important parameter for the analysis. The ECG signals can be given as direct input to deeplearning models. For the analysis purpose using ECG the use cardiorespiratory spectrogram. In this method cardiovascular [97], respiratory [98] or blood flow in brain [99] changes are noticed and analysed.

5.4 Classification and Feature Extraction Methods Used for Sleep Stage Classification

Numerous 1 channel and more channels automatic sleep stage grading techniques have been documented here. Two-channel (Fp1 and Fp2) EEG signals were divided into quasi-stationary components in the work by [115], characteristics were extracted using Short Time Fast Fourier, dimension reduction was completed using the Fuzzy C-Means technique, and a multiclass SVM was utilised to create an ASSC system. The accuracy rate of this project's output was 71.27%. Mustfa et al. [116] employed six different Brain activity along with a range of signal processing components, such as time domain, frequency domain, and quasi characteristics. As classifiers for the five phases of sleep, SVM and Random Forests (RF) were also considered. The results showed that frontal EEG data with spectral linear properties and an RF produced the best results and outperformed SVM. [117] used different statistical and spectral data collected from a single EEG channel to characterize various sleep phases using a Bootstrap Aggregating (Bagging) technique. The accuracy scores were 85.58%, 86.25%, 87.58%, 89.65%, and 95.06%, respectively, for the six-stage, five-stage, four-stage, three-stage, and two-stage techniques, according to the authors. Similar to this, [118] proposed a study that employed a 1 channel EEG-based method for sleep stage analysis called Complete Ensemble Empirical Mode Decomposition. The various steps of sleep were categorised using bagging. For the 6-Class, 90.89% for the 5-Class, 92.74% for the 4-Class, 94.25% for the 3-Class, and 99.54% for the 2-Class, the accuracy of this task was, in that order, 86.63%, 90.87%, 94.21%, and 94.23%. Sotlo and others [119] used entropy metrics features, the Q-algorithm as a dimensionality reduction method, and J-mean clustering as a classifier for two EEG-channel-based automatic sleep stage scoring. According to best practises, up to 80% of the data that was automatically discriminated may be classified correctly. Another study from [120] retrieved several spectral features from multichannel PSG data using Fast Fourier Transform in order to categorise the various stages of sleep with an efficiency of 84%. To distinguish between the awake and drowsy phases, a three-channel EEG waveform approach based on an SVM classifier was used in [121]. The drowsy state was defined as combining stages 1 and 2 of sleep. The findings of the approach for detecting sleepiness shown high precision and accuracy of 98.01% and 97.91%, respectively. Fraiwa and all [122] developed a system for the assessment based on derived entropy characteristics from the Wigner-Ville Distribution (WVD), Hilbert-Hough Spectrum (HHS), and Continuous Wavelet Transform(CWT). Additionally, utilising a single channel of the EEG signal and multiclass SVM as the classifier, the study from [123] conducted sleep stage classification based on extracting nine graph domain features from a Visibility Graph (VG) and a Horizontal VG (HVG) with 85% accuracy.

6 Different Approaches Used for the Sleep Stage Analysis

6.1 CNN Based Models

CNN is the most common approach used in many cases of analysis.1D,2D and 3D approaches are used in all the cases.1D to 2D transformations are highlighted [101–103] in many papers and carried out with caution due to the useful information loss. Twenty-five papers were reviewed and the statistical data is given below (Fig. 5)

Commonly used ones are 1D CNN which is used to process 1D signals [104]. 2D models are used for 2D signals which requires deeper architectures. So comparing with 2D,1D architectures are much simpler, faster, and more efficient, and it is mainly used for real-time deployments. So it is mainly used to analyze sleep patterns [104].

6.2 RNN/LSTM-Based Models

The sleep stage classifications using RNN/LSTM are very few in number compared with the study using CNN models. The 4-layer RNN model [66] has the best performance. The overall accuracy obtained then is 87.2%. Various sleep datasets are used in all the studies of RNN/LSTM models. The introduction of the Elman network increases the accuracy of all the studies. An accuracy of 86.7% is also obtained by using 2 LSTM units [67]. The mixed models are also used for the studies [83]. An end-to-end RNN model called SeqSleepNet was proposed in which short-term Fourier –Transform was used for converting PSG signals into power spectra signals. The accuracy of that model is 87.4 which is the highest among the RNN/LSTM models (Table 5).

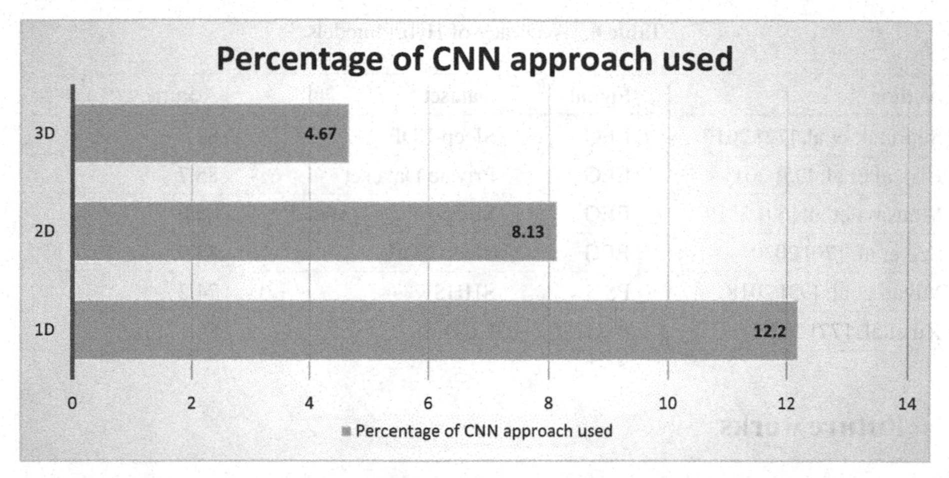

Fig. 5. Different approaches used for sleep stage analysis

Table 5. Accuracy of RNN/LSTM models

Author	Signal	Dataset	Accuracy (%)
Hsu et al. [66] 2013	EEG	Sleep EDF	87.2
Phan et al. [76] 2018	EEG	Sleep EDF	79.1
Dong et al. [83] 2018	EEG + EOG	MASS	83.4
Dong et al. [83] 2018	EEG + EOG	MASS	85.9
Phan et al. [85] 2019	EEG + EOG + EMG	MASS	87.4

6.3 Hybrid Models

There have been a few studies on the use of hybrid models. Two bidirectional LSTM layers and CNN layers made up the hybrid model that performed the best for EEG signals. In order to extract representative characteristics, CNN layers were used, while LSTM layers were used to collect data on the temporal stages of sleep [72]. The architecture of the bidirectional RNN units was different between the prior model [70] and this model. The table below shows the accuracy found for each model (Table 6).

Table 6. Accuracy of Hybridmodels

Author	Signal	Dataset	Accuracy (%)
Supratak et al. [72] 2017	EEG	Sleep-EDF	82
Biswal et al. [95] 2017	EEG	Private Data set	85.7
Mousavi et al. [69] 2019	EEG	Sleep-EDF expanded	80.0
Seo et al. [70] 2020	EEG	Sleep EDF	83.9
Yuvan et al. [93] 2018	PSG	SHHS	74.2
Xu et al. [77] 2020	PSG	UCD	86.1

7 Future works

The automatic discovery of sleep stages using machine knowledge is an effective method for detection. Thinking about the future of this technology gives light to many possibilities. The most comprising technology that must have developed in the future is the web-based sleep stage classification using EEG signals. The EEG signal acquired is sent to a cloud database for storing. That can be accessed directly by a doctor or clinician for the diagnosis purpose and the detailed analysis can be sent to the mobile phone of the patient. As we know everything is online nowadays, and this technology will be a boon to doctors and patients with sleep problems.

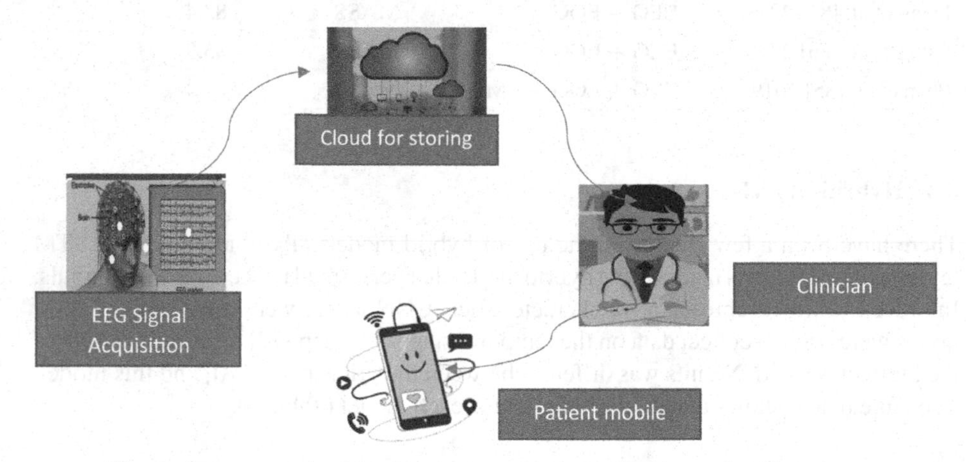

Fig. 6. Sleep stage classification system using EEG signals: A cloud based one

8 Conclusions

Nowadays the sleep disorders are one of the most common problem faced by many people. In that sleep apnea is a most dangerous sleep order which can cause many cardiovascular diseases. So the need of an efficient diagnostic tool is necessary in all cases.

In this paper from the analysis we can infer the tool for the diagnosis requires deep learning techniques. Here in most cases PSG signals are used as inputs. The CNN model can perform very well using the EEG signals and also it offers the highest performance. In most of the cases 1D CNN model is used along with the EEG signals. It is noted that other than EEG signals the mixture of signals are also used. So the future models can concentrate more on the mixture of signals. Also apart from the CNN models RNN/LSTM and Hybrid models can be explored more to get more accuracy & efficiency. In the coming years more research in emerging efficient prototypes for the sleep stages analysis has to be developed as sleep is an important factor in everyone's life.

References

1. Laposky, A., Bass, J., Kohsaka, A., Turek, F.W.: Sleep and circadian rhythms: Key components in the regulation of energy metabolism. FEBS Lett. **582**, 142–151 (2007)
2. Cho, J.W., Duffy, J.F.: Sleep, sleep disorders, and sexual dysfunction. World J. Men's Health **37**, 261–275 (2019)
3. Institute of Medicine (US): Committee on Sleep Medicine and Research. In: Colten, H.R., Altevogt, B.M. (eds.) Sleep Disorders and Sleep Deprivation: An Unmet Public Health Problem. National Academies Press, Washington, DC, USA (2006)
4. Stranges, S., Tigbe, W., Gómez-Olivé, F.X., Thorogood, M., Kandala, N.-B.: Sleep problems: an emerging global epidemic? Findings from the INDEPTH WHO-SAGE study among more than 40,000 older adults from 8 countries across Africa and Asia. Sleep **35**, 1173–1181 (2012).
5. Schulz, H.: Rethinking sleep analysis. J. Clin. Sleep Med. **4**, 99–103 (2008)
6. Spriggs, W.H.: Essentials of Polysomnography; Jones & Bartlett Learning: Burlington. MA, USA (2014)
7. Silber, M.H., et al.: The visual scoring of sleep in adults. J. Clin. Sleep Med. **3**, 121–131 (2007)
8. Corral, J., Pepin, J.-L., Barbé, F.: Ambulatory monitoring in the diagnosis and management of obstructive sleep apnoea syndrome. Eur. Respir. Rev. **22**, 312–324 (2013)
9. Jung, R., Kuhlo, W.: Neurophysiological studies of abnormal night sleep and the Pickwickian syndrome. Prog. Brain Res. **18**, 140–159 (1965)
10. Bahammam, A.: Obstructive sleep apnea: from simple upper airway obstruction to systemic inflammation. Ann. Saudi Med. **31**, 1–2 (2011)
11. Marshall, N.S., Wong, K.K.H., Liu, P.Y., Cullen, S.R.J., Knuiman, M., Grunstein, R.R.: Sleep apnea as an independent risk factor for all-cause mortality: The Busselton health study. Sleep **31**, 1079–1085 (2008)
12. Hirotsu, C., Tufik, S., Andersen, M.L.: Interactions between sleep, stress, and metabolism: from physiological to pathological conditions. Sleep Sci. **8**, 143–152 (2015)
13. Schilling, C., Schredl, M., Strobl, P., Deuschle, M.: Restless legs syndrome: evidence for nocturnal hypothalamic-pituitary-adrenal system activation. Mov. Disord. **25**, 1047–1052 (2010)
14. Hungin, A.P.S., Close, H.: Sleep disturbances and health problems: Sleep matters. Br. J. Gen. Pract. **60**, 319–320 (2010)
15. Hudgel, D.W.: The role of upper airway anatomy and physiology in obstructive sleep. Clin. Chest Med. **13**, 383–398 (1992)
16. Shahar, E., et al.: Sleep-disordered breathing and cardiovascular disease. Am. J. Respir. Crit. Care Med. **163**, 19–25 (2001)

17. Williamson, A., Lombardi, D.A., Folkard, S., Stutts, J., Courtney, T.K., Connor, J.L.: The link between fatigue and safety. Accid. Anal. Prev. **43**, 498–515 (2011)
18. Léger, D., Guilleminault, C., Bader, G., Lévy, E., Paillard, M.: Medical and socio-professional impact of insomnia. Sleep **25**, 625–629 (2002)
19. Mohsenin, V.: Obstructive sleep apnea and hypertension: A critical review. Curr. Hypertens. Rep. **16**, 482 (2014)
20. Balachandran, J.S., Patel, S.R.: Obstructive sleep apnea. Ann. Intern. Med. 161 (2014)
21. Iber, C., Ancoli-Israel, S., Chesson, A.L., Quan, S.F.: The AASM Manual for the Scoring of Sleep and Associated Events: Rules, Terminology and Technical Specification; American Academy of Sleep Medicine: Darien. IL, USA (2007)
22. Svetnik, V., et al.: Evaluation of automated and semi-automated scoring of polysomnographic recordings from a clinical trial using zolpidem in the treatment of insomnia. Sleep **30**, 1562–1574 (2007)
23. Pittman, M.S.D., et al.: Assessment of automated scoring of polysomnographic recordings in a population with suspected sleep-disordered breathing. Sleep **27**, 1394–1403 (2004)
24. Anderer, P., et al.: An E-health solution for automatic sleep classification according to Rechtschaffen and Kales: Validation study of the Somnolyzer 24 $\times$ 7 utilizing the siesta database. Neuropsychobiology **51**, 115–133 (2005)
25. Acharya, U.R., et al.: Nonlinear dynamics measures for automated EEG-based sleep stage detection. Eur. Neurol. **74**, 268–287 (2015)
26. Mirza, B., Wang, W., Wang, J., Choi, H., Chung, N.C., Ping, P.: Machine learning and integrative analysis of biomedical big data. Genes **10**, 87 (2019)
27. Faust, O., Razaghi, H., Barika, R., Ciaccio, E.J., Acharya, U.R.: A review of automated sleep stage scoring based on physiological signals for the new millennia. Comput. Methods Programs Biomed. **176**, 81–91 (2019)
28. Shoeibi, A., et al.: Epileptic Seizure Detection Using Deep Learning Techniques: A Review (2020). arXiv, arXiv:2007.01276
29. Faust, O., Hagiwara, Y., Hong, T.J., Lih, O.S., Acharya, U.R.: Deep learning for healthcare applications based on physiological signals: A review. Comput. Methods Programs Biomed. **161**, 1–13 (2018)
30. Silva, D.B., Cruz, P.P., Molina, A., Molina, A.M.: Are the long–short term memory and convolution neural networks really based on biological systems? ICT Express **4**, 100–106 (2018)
31. Hubel, D.H., Wiesel, T.N.: Receptive fields, binocular interaction and functional architecture in the cat's visual cortex. J. Physiol. **160**, 106–154 (1962)
32. Yamashita, R., Nishio, M., Do, R.K.G., Togashi, K.: Convolutional neural networks: An overview and application in radiology. Insights Imaging **9**, 611–629 (2018)
33. Tabian, I., Fu, H., Khodaei, Z.S.: A convolutional neural network for impact detection and characterization of complex composite structures. Sensors **19**, 4933 (2019)
34. Krizhevsky, A., Sutskever, I., Hinton, G.E.: Imagenet classification with deep convolutional neural networks. Commun. ACM **60**, 84–90 (2017)
35. Goehring, T., Keshavarzi, M., Carlyon, R.P., Moore, B.C.J.: Using recurrent neural networks to improve the perception of speech in non-stationary noise by people with cochlear implants. J. Acoust. Soc. Am. **146**, 705 (2019)
36. Coto-Jiménez, M.: Improving post-filtering of artificial speech using pre-trained LSTM neural networks. Biomimetics **4**, 39 (2019)
37. Lyu, C., Chen, B., Ren, Y., Ji, D.: Long short-term memory RNN for biomedical named entity recognition. BMC Bioinform. **18**, 462 (2017)
38. Graves, A., Liwicki, M., Fernández, S., Bertolami, R., Bunke, H., Schmidhuber, J.: A Novel connectionist system for unconstrained handwriting recognition. IEEE Trans. Pattern Anal. Mach. Intell. **31**, 855–868 (2008)

39. Kumar, S., Sharma, A., Tsunoda, T.: Brain wave classification using long short-term memory network based OPTICAL predictor. Sci. Rep. **9**, 1–13 (2019)
40. Kim, B.-H., Pyun, J.-Y.: ECG identification for personal authentication using LSTM-based deep recurrent neural networks. Sensors **20**, 3069 (2020)
41. Yu, Y., Si, X., Hu, C., Zhang, J.-X.: A review of recurrent neural networks: LSTM cells and network architectures. Neural Comput. **31**, 1235–1270 (2019)
42. Hochreiter, S., Schmidhuber, J.: Long short-term memory. Neural Comput. **9**, 1735–1780 (1997)
43. Gers, F.A., Schmidhuber, J., Cummins, F.: Learning to forget: Continual prediction with LSTM. In: Proceedings of the 9th International Conference on Artificial Neural Networks—ICANN'99. Edinburgh, UK (1999)
44. Masuko, T.: Computational cost reduction of long short-term memory based on simultaneous compression of input and hidden state. In: Proceedings of the 2017 IEEE Automatic Speech Recognition and Understanding Workshop, pp. 126–133. Okinawa, Japan (2017)
45. Dash, S., Acharya, B.R., Mittal, M., Abraham, A., Kelemen, A. (eds.): Deep Learning Techniques for Biomedical and Health Informatics. Springer International Publishing, Cham, Switzerland (2020)
46. Rumelhart, D.E., McClelland, J.L.: Learning internal representations by error propagation. In: Parallel Distributed Processing: Explorations in the Microstructure of Cognition: Foundations, pp. 318–362. MIT Press, Cambridge, MA, USA (1987)
47. Voulodimos, A., Doulamis, N., Doulamis, A., Protopapadakis, E.: Deep learning for computer vision: a brief review. Comput. Intell. Neurosci. **2018**, 1–13 (2018)
48. Testolin, A., Diamant, R.: Combining denoising autoencoders and dynamic programming for acoustic detection and tracking of underwater moving targets. Sensors **20**, 2945 (2020)
49. Trabelsi, A., Chaabane, M., Ben-Hur, A.: Comprehensive evaluation of deep learning architectures for prediction of DNA/RNA sequence binding specificities. Bioinformatics **35**, i269–i277 (2019)
50. Long, H., Liao, B., Xu, X., Yang, J.: A hybrid deep learning model for predicting protein hydroxylation sites. Int. J. Mol. Sci. **19**, 2817 (2018)
51. Hori, T., et al.: Proposed sments and amendments to 'A Manual of Standardized Terminology, Techniques and Scoring System for Sleep Stages of Human Subjects', the Rechtschaffen & Kales (1968) standard. Psychiatry Clin. Neurosci. **55**, 305–310 (2001)
52. Carley, D.W., Farabi, S.S.: Physiology of sleep. Diabetes Spectr. **29**, 5–9 (2016)
53. Goldberger, A.L., et al.: PhysioBank, PhysioToolkit, and PhysioNet. Circulation **101**, e215–e220 (2000)
54. Kemp, B., Zwinderman, A., Tuk, B., Kamphuisen, H., Oberye, J.: Analysis of a sleep-dependent neuronal feedback loop: The slow-wave microcontinuity of the EEG. IEEE Trans. Biomed. Eng. **47**, 1185–1194 (2000)
55. Zhang, G.-Q., et al.: The national sleep research resource: Towards a sleep data commons. J. Am. Med. Inform. Assoc. **25**, 1351–1358 (2018)
56. Quan, S.F., et al.: The sleep heart health study: Design, rationale, and methods. Sleep **20**, 1077–1085 (1997)
57. Ichimaru, Y., Moody, G.: Development of the polysomnographic database on CD-ROM. Psychiatry Clin. Neurosci. **53**, 175–177 (1999)
58. Khalighi, S., Sousa, T., Santos, J.M., Nunes, U.: ISRUC-Sleep: A comprehensive public dataset for sleep researchers. Comput. Methods Programs Biomed. **124**, 180–192 (2016)
59. O'Reilly, C., Gosselin, N., Carrier, J., Nielsen, T.: Montreal archive of sleep studies: An open-access resource for instrument benchmarking and exploratory research. J. Sleep Res. **23**, 628–635 (2014)

60. Li, Q., Li, Q.C., Liu, C., Shashikumar, S.P., Nemati, S., Clifford, G.D.: Deep learning in the cross-time frequency domain for sleep staging from a single-lead electrocardiogram. Physiol. Meas. **39**, 124005 (2018)
61. Tripathy, R., Acharya, U.R.: Use of features from RR-time series and EEG signals for automated classification of sleep stages in deep neural network framework. Biocybern. Biomed. Eng. **38**, 890–902 (2018)
62. Radha, M., et al.: Sleep stage classification from heart-rate variability using long short-term memory neural networks. Sci. Rep. **9**, 1–11 (2019)
63. Zhu, T., Luo, W., Yu, F.: Convolution-and attention-based neural network for automated sleep stage classification. Int. J. Environ. Res. Public Health **17**, 4152 (2020)
64. Qureshi, S., Karrila, S., Vanichayobon, S.: GACNN SleepTuneNet: A genetic algorithm designing the convolutional neuralnetwork architecture for optimal classification of sleep stages from a single EEG channel. Turk. J. Electr. Eng. Comput. Sci. **27**, 4203–4219 (2019)
65. Yıldırım, Ö., Baloglu, U.B., Acharya, U.R.: A deep learning model for automated sleep stages classification using PSG signals. Int. J. Environ. Res. Public Health **16**, 599 (2019)
66. Hsu, Y.-L., Yang, Y.-T., Wang, J.-S., Hsu, C.-Y.: Automatic sleep stage recurrent neural classifier using energy features of EEG signals. Neurocomputing **104**, 105–114 (2013)
67. Michielli, N., Acharya, U.R., Molinari, F.: Cascaded LSTM recurrent neural network for automated sleep stage classification using single-channel EEG signals. Comput. Biol. Med. **106**, 71–81 (2019)
68. Wei, L., Lin, Y., Wang, J., Ma, Y.: Time-frequency convolutional neural network for automatic sleep stage classification based on single-channel EEG. In: Proceedings of the 2017 IEEE 29th International Conference on Tools with Artificial Intelligence, pp. 88–95. Boston, MA, USA, IEEE, Piscataway, NJ, USA (2017)
69. Mousavi, S., Afghah, F., Acharya, U.R.: SleepEEGNet: Automated sleep stage scoring with sequence to sequence deep learning approach. PLoS ONE **14**, e0216456 (2019)
70. Seo, H., Back, S., Lee, S., Park, D., Kim, T., Lee, K.: Intra- and inter-epoch temporal context network (IITNet) using sub-epoch features for automatic sleep scoring on raw single-channel EEG. Biomed. Signal Process. Control **61**, 102037 (2020)
71. Zhang, X., et al.: Automated multi-model deep neural network for sleep stage scoring with unfiltered clinical data. Sleep Breath. **24**, 581–590 (2020)
72. Supratak, A., Dong, H., Wu, C., Guo, Y.: DeepSleepNet: A model for automatic sleep stage scoring based on raw single-channel EEG. IEEE Trans. Neural Syst. Rehabil. Eng. **25**, 1998–2008 (2017)
73. Phan, H., Andreotti, F., Cooray, N., Chén, O.Y., de Vos, M.: Joint classification and prediction CNN framework for automatic sleep stage classification. IEEE Trans. Biomed. Eng. **66**, 1285–1296 (2019)
74. Vilamala, A., Madsen, K.H., Hansen, L.K.: Deep convolutional neural networks for interpretable analysis of EEG sleep stage scoring. In: Proceedings of the 2017 IEEE 27th International Workshop on Machine Learning for Signal Processing (MLSP), pp. 1–6. Tokyo, Japan (2017)
75. Phan, H., Andreotti, F., Cooray, N., Chen, O.Y., de Vos, M.: DNN filter bank improves 1-max pooling CNN for single-channel EEG automatic sleep stage classification. In: Proceedings of the 2018 40th Annual International Conference of the IEEE Engineering in Medicine and Biology Society (EMBC), pp. 453–456. Honolulu, HI, USA (2018)
76. Phan, H., Andreotti, F., Cooray, N., Chen, O.Y., De Vos, M.: Automatic sleep stage classification using single-channel EEG: learning sequential features with attention-based recurrent neural networks. In: Proceedings of the 2018 40th Annual International Conference of the IEEE Engineering in Medicine and Biology Society (EMBC). Honolulu, HI, USA (2018)

77. Xu, M., Wang, X., Zhangt, X., Bin, G., Jia, Z., Chen, K.: Computation-efficient multi-model deep neural network for sleep stage classification. In: Proceedings of the ASSE '20: 2020 Asia Service Sciences and Software Engineering Conference, pp. 1–8. Association for Computing Machinery, Nagoya, Japan, New York, NY, USA (2020)
78. Wang, Y., Wu, D.: Deep learning for sleep stage classification. In: Proceedings of the 2018 Chinese Automation Congress (CAC), pp. 3833–3838. IEEE, Xi'an, China, Piscataway, NJ, USA (2018)
79. Fernandez-Blanco, E., Rivero, D., Pazos, A.: Convolutional neural networks for sleep stage scoring on a two-channel EEG signal. Soft. Comput. **24**, 4067–4079 (2019)
80. Jadhav, P., Rajguru, G., Datta, D., Mukhopadhyay, S.: Automatic sleep stage classification using time-frequency images of CWT and transfer learning using convolution neural network. Biocybern. Biomed. Eng. **40**, 494–504 (2020)
81. Tsinalis, O., Matthews, P.M., Guo, Y., Zafeiriou, S.: Automatic Sleep Stage Scoring with Single-Channel EEG Using Convolutional Neural Networks. Imperial College London, London, UK (2016)
82. Sokolovsky, M., Guerrero, F., Paisarnsrisomsuk, S., Ruiz, C., Alvarez, S.A.: Deep learning for automated feature discovery and classification of sleep stages. IEEE/ACM Trans. Comput. Biol. Bioinform. 17 (2019)
83. Dong, H., Supratak, A., Pan, W., Wu, C., Matthews, P.M., Guo, Y.: Mixed neural network approach for temporal sleep stage classification. IEEE Trans. Neural Syst. Rehabil. Eng. **26**, 324–333 (2018)
84. Chambon, S., Galtier, M.N., Arnal, P.J., Wainrib, G., Gramfort, A.: A deep learning architecture for temporal sleep stage classification using multivariate and multimodal time series. IEEE Trans. Neural Syst. Rehabil. Eng. **26**, 758–769 (2018)
85. Phan, H., Andreotti, F., Cooray, N., Chén, O.Y., de Vos, M.: SeqSleepNet: End-to-end hierarchical recurrent neural network for sequence-to-sequence automatic sleep staging. IEEE Trans. Neural Syst. Rehabil. Eng. **27**, 400–410 (2019)
86. Zhang, J., Yao, R., Ge, W., Gao, J.: Orthogonal convolutional neural networks for automatic sleep stage classification based on single-channel EEG. Comput. Methods Programs Biomed. **183**, 105089 (2020)
87. Zhang, J., Wu, Y.: Complex-valued unsupervised convolutional neural networks for sleep stage classification. Comput. Methods Programs Biomed. **164**, 181–191 (2018)
88. Sors, A., Bonnet, S., Mirek, S., Vercueil, L., Payen, J.-F.: A convolutional neural network for sleep stage scoring from raw single-channel EEG. Biomed. Signal Process. Control **42**, 107–114 (2018)
89. Fernández-Varela, I., Hernández-Pereira, E., Alvarez-Estevez, D., Moret-Bonillo, V.: A Convolutional Network for Sleep Stages Classification (2019). arXiv, arXiv:1902.05748v1
90. Zhang, L., Fabbri, D., Upender, R., Kent, D.T.: Automated sleep stage scoring of the Sleep Heart Health Study using deep neural networks. Sleep 42 (2019)
91. Cui, Z., Zheng, X., Shao, X., Cui, L.: Automatic sleep stage classification based on convolutional neural network and fine-grained segments. Complexity **2018**, 1–13 (2018)
92. Yang, Y., Zheng, X., Yuan, F.: A study on automatic sleep stage classification based on CNN-LSTM. In: Proceedings of the ICCSE' 18: The 3rd International Conference on Crowd Science and Engineering, pp. 1–5. Association for Computing Machinery, Singapore, New York, NY, USA (2018)
93. Yuan, Y., et al.: A hybrid self-attention deep learning framework for multivariate sleep stage classification. BMC Bioinform. **20**, 1–10 (2019)
94. Biswal, S., Sun, H., Goparaju, B., Westover, M.B., Sun, J., Bianchi, M.T.: Expert-level sleep scoring with deep neural networks. J. Am. Med. Inform. Assoc. **25**, 1643–1650 (2018)
95. Biswal, S., et al.: SLEEPNET: Automated Sleep Staging System via Deep Learning (2017). arXiv, arXiv:1707.08262

96. Hoshide, S., Kario, K.: Sleep Duration as a risk factor for cardiovascular disease—A review of the recent literature. Curr. Cardiol. Rev. **6**, 54–61 (2010)
97. Woods, S.L., Froelicher, E.S.S., Motzer, S.U., Bridges, S.J.: Cardiac Nursing, 5th edn. Lippincott Williams and Wilkins, London, UK (2005)
98. Krieger, J.: Breathing during sleep in normal subjects. Clin. Chest Med. **6**, 577–594 (1985)
99. Madsen, P.L., et al.: Cerebral O2 metabolism and cerebral blood flow in humans during deep and rapid-eye-movement sleep. J. Appl. Physiol. **70**, 2597–2601 (1991)
100. Klosh, G., et al.: The SIESTA project polygraphic and clinical database. IEEE Eng. Med. Boil. Mag. **20**, 51–57 (2001)
101. Yıldırım, Ö., Talo, M., Ay, B., Baloglu, U.B., Aydin, G., Acharya, U.R.: Automated detection of diabetic subject using pre-trained 2D-CNN models with frequency spectrum images extracted from heart rate signals. Comput. Biol. Med. **113**, 103387 (2019)
102. Pham, T.-H., et al.: Autism spectrum disorder diagnostic system using HOS bispectrum with EEG Signals. Int. J. Environ. Res. Public Health **17**, 971 (2020)
103. Khan, S.A., Kim, J.-M.: Automated bearing fault diagnosis using 2D analysis of vibration acceleration signals under variable speed conditions. Shock. Vib. **2016**, 1–11 (2016)
104. Kiranyaz, S., et al.: 1D convolutional neural networks and applications: A survey. Mech. Syst. Signal Process, 151 (2021)
105. Patanaik, A., Ong, J.L., Gooley, J.J., Ancoli-Israel, S., Chee, M.W.L.: An end-to-end framework for real-time automatic sleep stage classification. Sleep, 41 (2018)
106. Terzano, M.G., et al.: Atlas, rules, and recording techniques for the scoring of cyclic alternating pattern (CAP) in human sleep. Sleep Med. **2**, 537–553 (2001)
107. Chokroverty, S.: An overview of normal sleep. In: Chokroverty, S. (ed.) Sleep disorders medicine: basic science, technical considerations and clinical aspects, 3rd ed. Elsevier, Butterworth, Philadelphia (2009)
108. https://www.mayoclinic.org/tests-procedures/polysomnography/about/pac-20394877
109. https://www.kdnuggets.com/2017/02/python-deep-learning-frameworks-overview.html
110. Krakovska, A., Mezeiova, K.: Automatic sleep scoring: a search for an optimal combination of measures. Artif. Intell. Med. **53**(1), 25–33 (2011)
111. Koch, H., et al.: Automatic sleep classification using a data-driven topic model reveals latent sleep states. J. Neurosci. Methods **235**, 130–137 (2014)
112. Koupparis, A.M., Kokkinos, V., Kostopoulos, G.K.: Semi-automatic sleep EEG scoring based on the hypnospectrogram. J. Neurosci. Methods **221**, 189–195 (2014)
113. Brignol, A., Al-Ani, T., Drouot, X.: Phase space and power spectral approaches for EEG-based automatic sleep-wake classification in humans: a comparative study using short
114. Aboalayon, K., Faezipour, M., Almuhammadi, W., Moslehpour, S.: Sleep Stage Classification Using EEG Signal Analysis: A Comprehensive Survey and New Investigation. Entropy 18. MDPI AG, 272 (2016). https://doi.org/10.3390/e18090272
115. Huang, C.-S., et al.: Applying the fuzzy c-means based dimension reduction to improve the sleep classification system. In: Proceedings of the IEEE International Conference on Fuzzy Systems (FUZZ), pp. 1–5. Hyderabad, India (2013)
116. Radha, M., Garcia-Molina, G., Poel, M., Tononi, G.: Comparison of feature and classifier algorithms for online automatic sleep staging based on a single EEG signal. In: Proceedings of the 36th IEEE Annual International Conference of Engineering in Medicine and Biology Society, pp. 1876–1880. Chicago, IL, USA (2014)
117. Hassan, A.R., Bashar, S.K., Bhuiyan, M.I.H.: On the classification of sleep states by means of statistical and spectral features from single channel electroencephalogram. In: Proceedings of the IEEE International Conference on Advances in Computing, Communications and Informatics (ICACCI), pp. 2238–2243. Kochi, India (2015)

118. Hassan, A.R., Bhuiyan, M.I.H.: Computer-aided sleep staging using complete ensemble empirical mode decomposition with adaptive noise and bootstrap aggregating. Biomed. Signal Process. Control **24**, 1–10 (2016)
119. Rodríguez-Sotelo, J.L., et al.: Automatic sleep stages classification using EEG entropy features and unsupervised pattern analysis techniques
120. Lan, K.-C., et al.: Using off-the-shelf lossy compression for wireless home sleep staging. J. Neurosci. Methods **246**, 142–152 (2015)
121. Yu, S., et al.: Support vector machine based detection of drowsiness using minimum EEG features. In: Proceedings of the IEEE International Conference on Social Computing (SocialCom), pp. 827–835. Alexandria, VA, USA (2013)
122. Fraiwan, L., Lweesy, K., Khasawneh, N., Fraiwan, M., Wenz, H., Dickhaus, H.: Time frequency analysis for automated sleep stage identification in fullterm and preterm neonates. J. Med. Syst. **35**, 693–702 (2011)
123. Zhu, G., Li, Y., Wen, P.P.: Analysis and classification of sleep stages based on difference visibility graphs from a single-channel EEG signal. IEEE J. Biomed. Health Inform. **18**, 1813–1821 (2014)
124. Shuyuan, X., et al.: An improved k-means clustering algorithm for sleep stages classification. In: Proceedings of the 54th IEEE Annual Conference on Society of Instrument and Control Engineers of Japan (SICE), pp. 1222–1227. Hangzhou, China (2015)

Deep Transfer Learning for Schizophrenia Detection Using Brain MRI

Siddhant Mudholkar, Amitesh Agrawal, Dilip Singh Sisodia(✉), and Rikhi Ram Jagat

Department of Computer Science and Engineering, National Institute of Technology Raipur, Raipur, India
{dssisodia.cs,rrjagat.phd2019.cse}@nitrr.ac.in

Abstract. Schizophrenia causes hallucinations, delusions, and excessive disorganization. Early diagnosis and treatment lessen family issues and social expenses. It needs multiple psychiatrist consultations and brain MRIs to diagnose. Schizophrenia has no objective medical index. Machine learning and deep learning algorithms simplify disease diagnosis. Handcrafted features require domain specialists to develop the feature set, which involves time and experience in machine learning models. Deep Learning (DL) models employ brain MRI scans to predict schizophrenia; however, they require a large dataset to train, which increases computing time. But, only a few schizophrenia-detecting MRI datasets are openly available. Previously, transfer learning-based DL models have been trained on sagittal, coronal, and axial 3D MRI images. But in this model, unnecessary information and noisy features reduce performance. Therefore, we employ the axial view of brain scans as it contains the subcortical region and ventricular areas which contribute most to the prediction of schizophrenia. Axial view images are used to train transfer learning-based VGG19 model for schizophrenia identification. This study uses a COBRE-published brain MRI dataset. The collection includes 146 patients' brain MRIs. We analyzed 3D MRI images to produce axial brain slices. Thresholding improves intensity differentiation in axial view images and data augmentation reduces overfitting and reduces data; both of them are being used in preprocessing. The suggested model utilizes a VGG-19 pre-trained network with fully connected layers. Adam optimizer for optimization, ReLU for hidden layers, and sigmoid for last layer activation functions. Our model was evaluated using accuracy, precision, recall, and F1 score. The dataset model was 90.9% accurate. It is compared using standard metrics to different categorization models and existing models. We improved accuracy by a maximum of 3% and a minimum of 1%.

Keywords: Schizophrenia · Deep Learning · Classification · Transfer Learning · VGG19 · Axial view

1 Introduction

Schizophrenia is a highly significant medical condition that has an impact on a person's behavior and thought process. Schizophrenia symptoms can range from psychotic symptoms, which alter a person's thinking and include hallucinations and delusions,

B. K. Singh et al. (Eds.): ICBEST 2023, CCIS 2003, pp. 66–82, 2024.
https://doi.org/10.1007/978-3-031-54547-4_6

to negative symptoms, such as a loss of motivation or enjoyment in daily activities, difficulty expressing emotions, and withdrawal from social interactions, to cognitive symptoms, which cause issues like poor memory and lack of concentration. One in every 300 people on the planet has it [1]. There is an increase in the number of patients with schizophrenia-related issues each year.

One of the top 10 global causes of disability in developed countries is schizophrenia. The fundamental problem is that schizophrenia has a mortality rate of roughly 2.5, which results in a drop in life expectancy of between 15 and 20 years [2]. Many people end their lives because of terrible delusions and hallucinations. It is an extremely serious, long-lasting mental condition. It causes a variety of different psychological problems. Patients find it increasingly challenging to distinguish between reality and their hallucinations. One of the most expensive disorders to treat, schizophrenia can be identified early on to save both time and money. Women are more likely to exhibit signs of schizophrenia than men do after the age of 30, with symptoms often appearing between the ages of 16 and 25 in women [3]. The average age of onset for men is 18, whereas it is 25 for women. Schizophrenia is exceedingly rare in those under the age of 10 and people over the age of 40. In young people, treatment and diagnosis should be started as soon as possible because they still have a full life ahead of them. It takes a long time to diagnose schizophrenia, which involves frequent clinic visits and doctor interviews based on patient behavior. There may be instances where a case is ignored or misdiagnosed because of the subjective nature of a doctor's diagnosis. As a result, numerous studies are being conducted to identify schizophrenia as quickly as possible, which can help patients save money on medical expenses and time, as well as lessen the financial strain on their families. As many additional diseases are currently detected using computer-aided diagnosis methods, such as the segmentation of brain tumors [4], the detection of Parkinson's disease [5], and the detection of Alzheimer's disease [6]. Since the amount of data available from Magnetic Resonance Imaging (MRI) is restricted, many machine learning models rely on genetic data to function. However, as more data becomes accessible, we can now directly use MRI data to construct a deep learning model that can assist in more quickly and accurately diagnosing schizophrenia. By combining our deep neural network model and brain MRI scans to detect this condition early and rapidly, the likelihood of extending the patient's life expectancy and expediting diagnosis and treatment will rise. Even after utilizing the model, we still require the doctor's help because it is not entirely correct, but it will still provide the patient with some suggestions concerning schizophrenia. We took this as a challenge and experimented with several methods to achieve decent results with a very little dataset of brain MRI data from schizophrenia patients. The dataset of available data is really small and it is very difficult to train any kind of model from that data. We have developed a deep learning model that can assist us in recognizing schizophrenia using simply an MRI scan to fully resolve this issue. With very little data, we were nevertheless able to train our model and provide excellent results. The major goal is to use the axial view, which displays the subcortical and ventricular regions that are most significant for detection [7, 8], to only extract the most crucial information. Reduce the number of parameters to speed up and improve the accuracy of the computation. Transfer learning with VGG19 was used to increase the accuracy of schizophrenia identification, and some fully connected layers were added after the transfer learning model to allow it to learn from and forecast using medical pictures. Comparative analysis of the most recent, state-of-the-art image classification models used for transfer learning and hyperparameter fine tuning.

2 Related Work

Many studies have already been conducted in this area, but they all used machine learning; there aren't many deep learning models available for this problem statement [9, 10]. More than 55% of studies involving neuroimaging employed support vector machines [11]. Shi et al. [12] presented a model to distinguish between healthy controls (HC) and schizophrenics (SZ). Utilizing fMRI and sMRI data, their study employed a cutting-edge technique termed multimodal imaging and multilevel characterization using multiclassifier (M3), which can also pinpoint the area that is most crucial for the detection of schizophrenia. AUC was 0.8491, sensitivity was 66.69%, specificity was 93.75%, and classification accuracy was 83.49% for this approach. There are many papers available which summarized the research done in field of neuroimaging using deep learning [13–15]. For the purpose of detecting schizophrenia, Zheng et al. [16] suggested a model using fMRI data and a 2D CNN network that uses transfer learning to improve the VGG-16 network. The proposed model used a VGG-16 model which was pretrained on the imageNet Dataset which was used to analyze the axial view of the brain MRI images. Other models including AlexNet, VGG16, and ResNet were outperformed by this model using transfer learning, which had accuracy ratings of 78.36%, 85.27%, and 83.09%, respectively. Based on VGG-16, fMRI has a classification accuracy of up to 84.3%. While the proposed model in this study outperformed the existing VGG-16 algorithm with an accuracy of 87.85% in the COBRE dataset. A new preprocessing method was proposed by Oh J et al. [17] by turning the images into videos, where each slice of the transverse section was turned into a frame, and the frames were then stitched together to create a video. The structural MRI data sets were utilized to categorize patients with schizophrenia and healthy individuals using a three-dimensional convolutional neural network (3DCNN) architecture. The overall accuracy of the 3D CNN models was 97%, which is among the highest. Results from this model also demonstrated the brain's regional analysis. Latha et al. [18] the proposed method to use deep belief networks (DBNs) to analyze the schizophrenic subjects based on the ventricle region. They assessed the effectiveness of their model using the COBRE dataset. The least number of errors are made by DBN on the segmented ventricle area. The recommended method yielded a segmented ventricle image with a high area under the curve (0.899), adequately demonstrating its efficacy. A model using 3D MRI T1 weighted images was put forth by Zhang et al. [19]. They used a variety of preprocessing techniques. First, they used the Brain Extractor Tool to obtain an image of the entire brain, which they then transferred onto the 3D VGG-11 model. The suggested approach involved modifying the VGG-11 model and using a squeeze excitation block (SE-VGG-11BN Model). This model's accuracy was 92.1% when tested using independent data obtained from full brain scans. On the COBRE dataset, the accuracy for T1W (whole head) was 0.867, with 0.897 sensitivity, 0.861 specificity, and 0.938 AUC. Since 3D CNN takes into account the entire brain and extracts the key features from brain MRI images, we can conclude from all the studies conducted to date that it offers overall better accuracy than any 2D model. However, running a 3DCNN model requires a lot of time, money, and resources. Additionally, the studies mentioned above show that 2DCNN can be quite effective at predicting disease if we can identify the regions of interest or those that contribute the most. The increase in the dataset can also help us achieve higher and better accuracy with less computing power and in a shorter amount of time. According to [19], the ventricular and subcortical regions are the most affected by schizophrenia. These regions are also the most exposed in an axial view of the brain, which aids in more precise illness prediction. Table 1 shows the summary of related works mentioned above, which contains the dataset used, preprocessing pipeline, and methodology.

Table 1. Summary of related work

S. No	Author	Dataset Used	Preprocessing	Methods
1	Yassin et al. [20]	Private Dataset of 131 patients	FreeSurfer image analysis suite v.6.0	6 different classifiers, SVM and LR best results
2	Shi et al. [12]	Center for Biomedical Research Excellence (COBRE)	Data Processing & Analysis of Brain Imaging and Data Processing Assistant for Resting-State fMRI	Multimodal imaging and Multilevel characterization using Multiclassifier(M3)
3	Liu et al. [21]	Private data of 76 patients	Freesurfer image analysis suite v 5.3.0	Multiple kernel framework
4	Zheng et al. [16]	COBRE	Thresholding, Binarization	2DCNN using Transfer learning
5	Oh J et al. [17]	COBRE, MCICShare, NMorphCH, NUSDAST, BrainGluSchi	Converting the image slices into a video	3DCNN architecture
6	Latha et al. [18]	COBRE	Skull stripping to remove non brain tissues	Deep Belief Networks (DBNs)
7	Zhang et al. [19]	COBRE, NMorphCH, BrainGluSchi	Brain affine registration with skull stripping	3D SE-VGG-11BN Model

3 Proposed Methodology

In this section we will propose an extended VGG19 model, brain MRI images are only required for detection. Figure 1 depicts the complete workflow of the proposed extended VGG19 model. We first did data analysis on each MRI scan to find the correct intensity of the axial view of the brain image. The data is then preprocessed using threshold and smoothening. The data is then split into training (70%), validation (15%), and testing set (15%). Then this preprocessed data is used to train our Deep Transfer Neural Network model where the validation dataset helps us to increase the overall performance. Then this trained model is tested on testing data. We have used 5 different deep learning models on different optimizers with different learning rates and used different activation functions and evaluating performance metrics on all of them. We have also used different preprocessing techniques in order to achieve quality data.

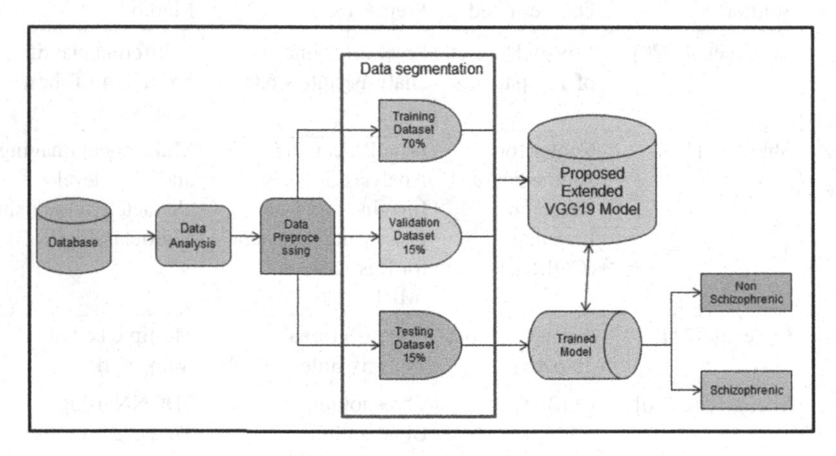

Fig. 1. System workflow

3.1 Dataset

The Center for Biomedical Research Excellence (COBRE) dataset, which is a public dataset available at http://fcon1000.projects.nitrc.org/indi/retro/cobre.html, is the one used in the model. In the dataset, there are 148 patients (ages ranged from 18 to 65 in each group), 74 of whom are healthy, 72 of whom have schizophrenia, and 2 of whom are unidentified. For each participant, the dataset includes the following information. Anatomical fMRI, resting fMRI, and observable participant variables like gender, age, handedness, and diagnostic data are all provided. This information, where fMRI stands for function Magnetic Resonance Imaging, is crucial to determining further observations on the patients, which handedness patient are frequently diagnosed with schizophrenia. Table 2 shows the sample characteristics of COBRE dataset.

Table 2. Sample characteristics of COBRE dataset

Characteristics		Values
No. of images		146
No. of normal images		72
No. of schizophrenia images		74
Patient demographics		
	Age, mean (SD), years	36.97 (12.78)
	Age, range (min/max)	18/65
	Female, No./total (%)	37/146 (25.34)
Image quality		
	Acquisition year	2009 to 2013
	Scanner field strength	3T
	No, with excessive noise	1
	No, with image error	-
Psychiatric diagnosis		
	Schizophrenia (broad)	-
	Schizophrenia (strict)	74
	Schizoaffective disorder	-

3.2 Data Analysis

The COBRE dataset included weighted scans of the brain and 3D T1 weighted images (the "spin-lattice" relaxation time) with dimensions of (192*256*256). Brain MRI scans are required for the successful prediction of schizophrenia. These images can be obtained in a variety of views, including coronal, axial, and sagittal views, but we specifically need the axial view. By carefully analyzing the frames of pictures throughout time, we were able to extract the 2D axial image from the brain MRI that is depicted in Fig. 2. We then preprocessed this data into a more significant format for the model and trained our model using it. Therefore, in order to obtain a 2D axial representation of the brain before using the data, we preprocessed the 3D images. Figure 3 depicts the subcortical region and ventricular area, which are most visible in an axial view and contribute most to the detection of schizophrenia [19]. As MRI images are stored in the NIfTI (Neuroimaging Informatics Technology Initiative) file format, which contains different slices of the brain at various points in time, we must adjust the Z axis of the image over time to obtain the correct view of the brain, which can aid in disease prediction. We then tested 36 different time frames for the 256-dimensional Z axis of the image, forming the images at a 7-frame interval, and examined the resulting data. After examining the data, we determined the optimal time period during which the axial picture of the brain was exact, distinct, and intensified.

3.3 Data Preprocessing

In the preprocessing stage, we added binary threshold to the image to help the model distinguish between various brain properties. We also explored various types of smoothing, such as gaussian blur, median blur, bilateral filter, blur, with a 5 x 5 matrix, to amplify the

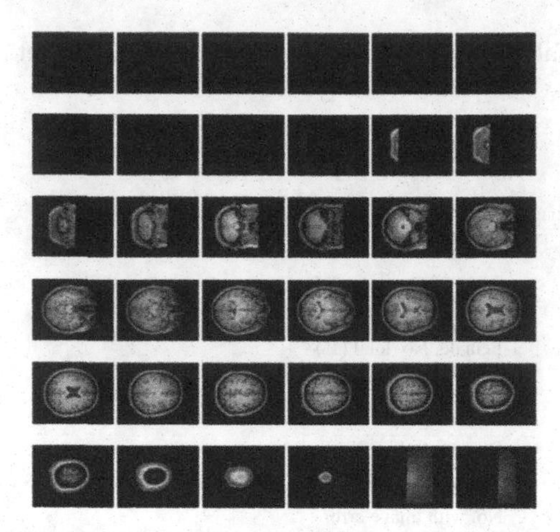

Fig. 2. Images of brain at different time intervals

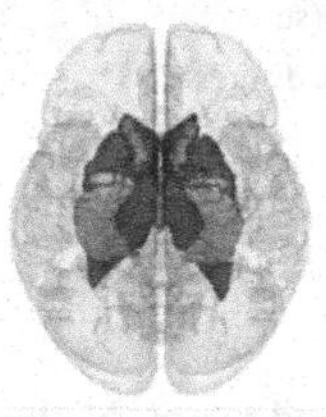

Fig. 3. Subcortical region of brain [19]

data and strengthen the model. As a result, we increased our dataset by using 7 images per patient for training and validation purposes instead of just 1 image per patient for testing, as our dataset was too limited for the model to be trained well to predict on fresh data. In order to generate more data, we also explored data augmentation by using various filters on the same data. This increased the size of our training and validation data by seven times, which improved the model's accuracy and robustness. Figure 4 shows the preprocessed image including thresholding and filtering of the image.

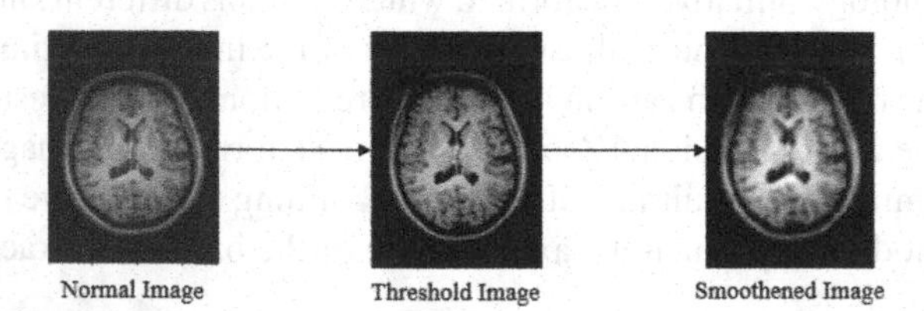

Fig. 4. Data preprocessing pipeline

3.4 Proposed Extended VGG19 Model

We have proposed the extended VGG19 model, a sequential model that builds on the VGG19 model for transfer learning and adds four additional deep neural layers for improved learning and prediction.

VGG-19: The VGG19 model is a variation of the VGG model that, in essence, has 19 layers (16 convolution layers, 3 Fully connected layer, 5 Max Pool layers and 1 SoftMax layer). The convolution filter collects the input information from each layer's many feature maps, which each have many neurons. The convolutional layer's parameter sharing approach significantly reduces the number of parameters. The input feature dimension is down sampled using Max Pooling. To optimize and enhance the network, VGG19 is utilized as transfer learning with a model of ImageNet weights that has already been trained.

The convolutional layer is expressed as follows:

$$y_j^s = f\left(\sum_{i \in M_j} y_i^{s-1} * t_{ij}^s + a_j^s\right). \tag{1}$$

In Eq. (1), let's assume that input layer or pooling layer is represented by $s-1$ where s is the convolutional layer, then y_i^s is the s convolutional layer's j^{th} feature map; the right-hand side of the Eq. (1) shows the $s-1$ layer's feature map. Convolutional operation is performed with the jth convolution kernel t_{ij}^s of the s layer and sum where a represents the bias and $f(\cdot)$ represents the ReLU activation function.

The pooling layer, which fulfils the job of scaling dimensionality, immediately follows the convolutional layer. The following is the calculation equation:

$$y_i^s = f\left(\alpha_j^s down\left(y_i^{s-1}\right) + a_j^s\right). \tag{2}$$

In Eq. (2), The pooling function, $down(\cdot)$, finds the highest value possible for a particular area of the feature map; α_j^s and a_j^s, reflect the bias and weight of the pooling, respectively.

Extended VGG-19: After processing the feature in all of the layers of VGG19, we omit the final layer and add an additional layer to flatten the output of the model (VGG19 was pretrained on the ImageNet dataset, but its weights cannot be trained). This layer is then followed by three fully connected hidden layers with 128 neurons each and a dropout layer with 30% of closed neurons, using ReLU (Rectified Linear Unit) as the activation function. The output layer, which utilizes a sigmoid activation function to forecast the model's ultimate output, was added after the hidden layers. Figure 5 depicts the entire extended VGG19 architecture with additional layers. We use the sequential model of the keras library to fine-tune our hyperparameters so that we may assign a different set of weights to each predictor. Modified VGG19 has 3 fully connected neural network layers with 3,156,225 trainable parameters and 20,024,384 non-trainable parameters. Table 3 displays the sequential model details including layer size and parameter information. Following the pretrained VGG19 model, we added 3 dropout layers, 3 fully connected layers, and 1 output layer. Our loss function was Binary Cross entropy, and we used the

Adam function as our optimizer to minimize the loss using Learning Rate $\alpha = 10^{-3}$. The model was trained over 100 epochs with the help of Early Stopping callback functionality, which prevents overfitting and ends model training once the loss has converged.

Transfer learning: Deep learning's need for numerous training models can be overcome through transfer learning. It is possible to train a model with fewer data sets by employing a pretrained model that has been trained on a sizable amount of data. The feature extraction layer of the model is kept while the model's classification layer is retrained using a training method termed fine-tune. The pretraining model's fixed feature extraction layer employs the VGG19 network, which was pretrained on the ImageNet data set with no weights allowed to train. The model's knowledge is then passed to our extended layers, which use both the knowledge and image inputs. Figure 6 depicts the transfer learning model that is being employed. We use a four-layered DNN to predict the final output after training the schizophrenia classification model on VGG19.

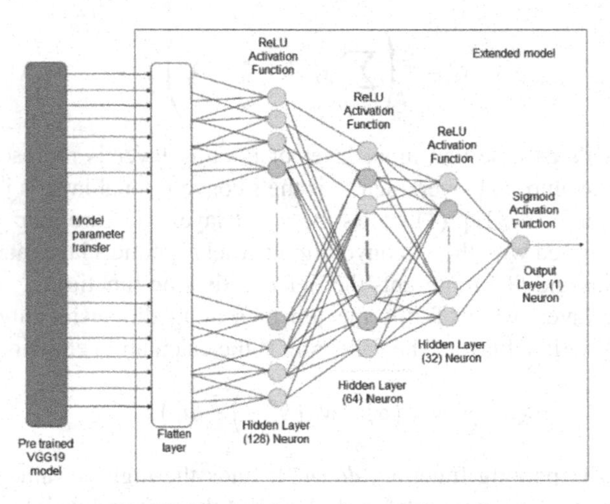

Fig. 5. Proposed extended vgg19 mode

Evaluation measures: We use metrics that are routinely used in classification tasks, such as precision, recall, accuracy, and F1 Score. True positive (TP) indicates that the positive class is anticipated to be a positive class and that the model accurately predicted the number of sample positive classes. False negative (FN) denotes the projection of the positive class as a negative class and indicates that the model correctly anticipated the presence of the sample's negative classes. False positives (FP) occur when the model incorrectly predicts the number of positive classes of samples while the negative class is projected to be a positive class.

$$Recall = \frac{TP}{TP + FN} \tag{5}$$

Table 3. Model summary of proposed extended VGG19 model

Layer (type)	Output Shape	Param #
vgg19 (Functional)	(None, 6, 8, 512)	20024384
flatten_2 (Flatten)	(None, 24576)	0
dropout_6 (Dropout)	(None, 24576)	0
dense_16 (Dense)	(None,128)	3145856
dropout_7 (Dropout)	(None,128)	0
dense_17 (Dense)	(None, 64)	8256
dropout_8 (Dropout)	(None, 64)	0
dense_18 (Dense)	(None, 32)	2080
dense_19 (Dense)	(None, 1)	33
Total params: 23,180,609		
Trainable params: 3,156,225		
Non-trainable params: 20,024,384		

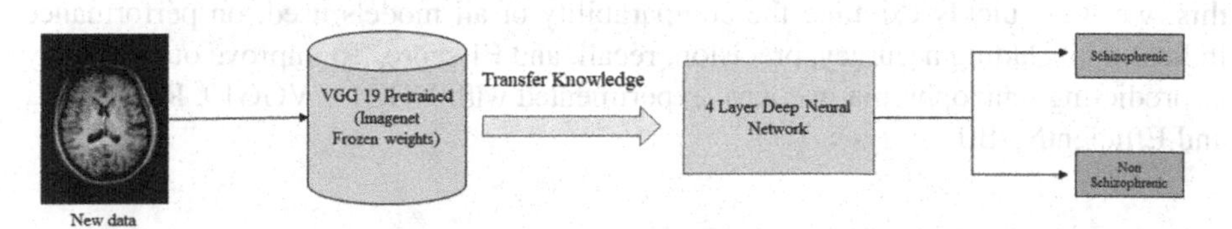

Fig. 6. Transfer learning process

Equation (5) shows the calculation of recall which indicates how many actual positive was correctly identified.

$$Precision = \frac{TP}{TP + FP} \tag{6}$$

Equation (6) shows the calculation of precision which indicates how many positive outcomes were predicted correctly.

$$Accuracy = \frac{TP + TN}{TP + TN + FP + FN} \tag{7}$$

Equation (7) shows the calculation of accuracy which indicates the fraction of outcomes model predicted correctly.

$$F1Score = \frac{TP}{TP + \frac{1}{2}(FP + FN)} \tag{8}$$

Equation (8) shows F1 score which combines the recall and precision metric into one for easier evaluation.

4 Experimental Result

The proposed extended VGG-19 model was trained and tested on Google colab [22]. The virtual machine is powered by a 2.30 GHz Intel(R) Xeon(R) CPU with two cores and two threads each. There was 12 GB of RAM and 25 GB of storage space. Nvidia T4 GPU with 16 GB of RAM and 8.1 TFLOPs (10–12 floating-point operations per second) is provided by Colab. Tensorflow, Keras, Matplotlib, ScikitLearn, nibabel, and OpenCV are among the packages used to train the model. The COBRE dataset, which consists of 3D MRI imaging data, is being used. The data set is broken down into three sections: training, validation, and testing. Training and validation were used to train the model, while testing will be used here to assess the model. The testing dataset consists of 21 sets of brain MRI pictures, 10 of which are of patients with schizophrenia and 11 of which are of patients without schizophrenia. Similar analysis and preprocessing techniques are used for test data, but only 1 slice of the axial view of the brain MRI image is used for evaluation. In this experiment, we'll test various cutting-edge deep learning categorization models for transfer learning. The proposed extended VGG19 model will be further tested utilising various hyperparameters, including the activation function and preprocessing method, to compare them. The findings will then be compared to the most recent state-of-the-art model which uses the same dataset. Following this, we will quickly examine the comparability of all models used, on performance indicators including accuracy, precision, recall, and F1 score. To improve our accuracy in predicting schizophrenia, we have experimented with VGG16, VGG19, ResNet101, and EfficientNetB0.

4.1 Transfer Learning Using VGG16

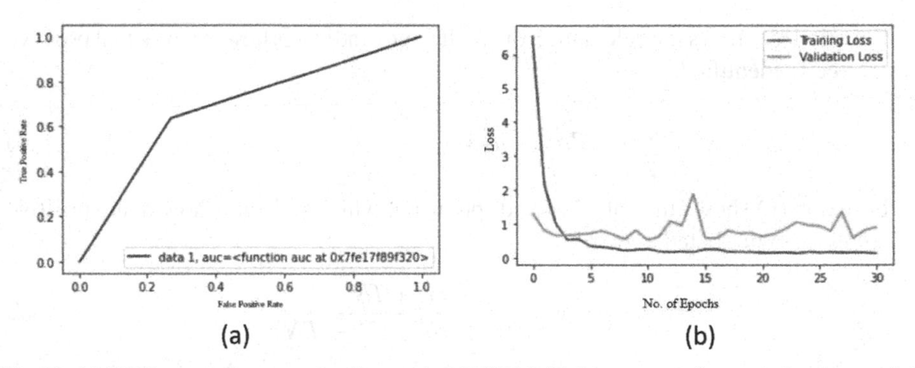

Fig. 7. (a) ROC curve for pretrained extended VGG16 (b) Loss curve for pretrained extended VGG16

Transfer learning was utilised with the VGG16 model in the previous State-of-the-art model; therefore, we trained our dataset using this model as our pre trained model. The ROC curve of the VGG16 model is shown in Fig. 7 (a), with the x axis indicating the false positive rate and the y axis representing the true positive rate. With a test loss of

0.86, we were able to attain an accuracy of 0.71. Recall value, F1 score, and Precision were each 0.63, 0.66, and 0.63, respectively. The graph of training loss and validation loss with respect to the number of epochs is shown in Fig. 7 (b).

4.2 Transfer Learning Using ResNet101

ResNet50 has done a great job classifying images, while ResNet101, an updated version of ResNet50, has done a much better job in almost every situation. So, we used ResNet101 to train our dataset. With a test loss of 0.36, we were able to attain an accuracy of 0.77. Recall value, Precision, and F1 score were each 0.76, 0.72, and 0.72, respectively. The ROC curve for the ResNet101 model is displayed in Fig. 8 (a). The graph of training loss and validation loss with respect to the number of epochs is shown in Fig. 8 (b).

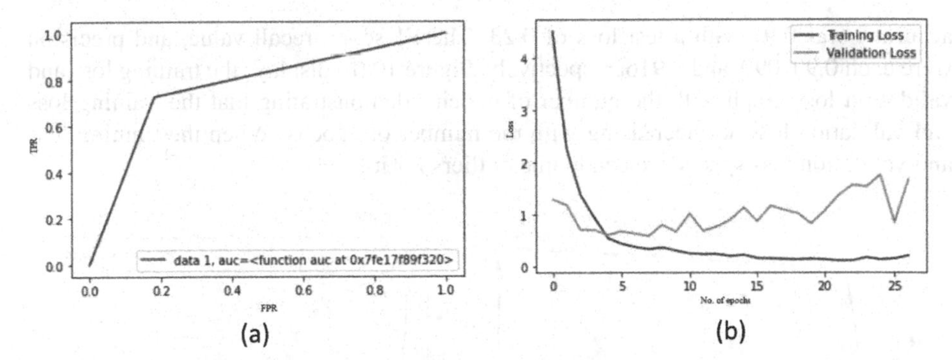

Fig. 8. (a) ROC curve for pretrained extended ResNet101 (b) Loss curve for pretrained extended ResNet101

4.3 Transfer Learning Using EfficientNetB0

EfficientNet exhibits strong transfer performance on the CIFAR-100 dataset and achieves cutting-edge accuracy. Therefore, we used EfficientNetB0 to train our dataset, which allowed us to obtain a maximum accuracy of 0.81, surpassing that of VGG16 and ResNet101 with a test loss of 0.47. Precision, recall value, and F1 score were each 0.81, 0.81, and 0.78, respectively. The EfficientNetB0's ROC curve is displayed in Fig. 9 (a). The graph of training loss and validation loss with respect to the number of epochs is shown in Fig. 9 (b).

4.4 Proposed Extended VGG19 Model

VGG19 is an updated version of VGG16 that uses 19 layers to train the model as opposed to 16 layers in VGG16. To attain greater accuracy and to make our modern learn more than the model provided in the publication, we trained our dataset using the VGG19 model. The ROC curve for the VGG19 is depicted in Fig. 10 (a), with the x axis showing the false positive rate and the y axis showing the true positive rate. The highest

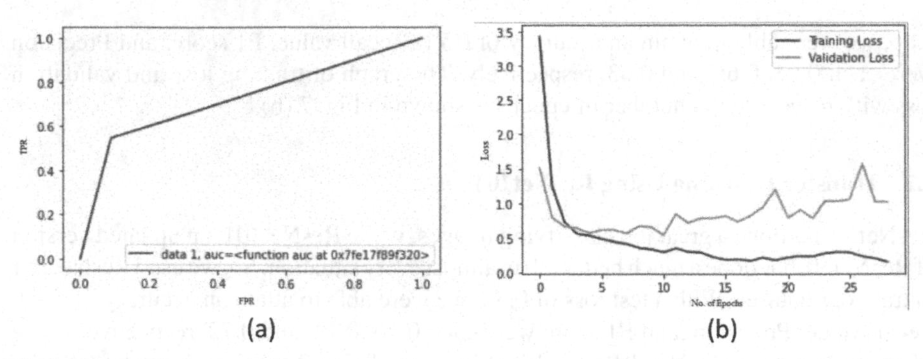

Fig. 9. (a) ROC curve for pretrained extended EfficientNetB0 (b) Loss curve for pretrained extended EfficientNetB0

accuracy was 0.91 with a test loss of 0.23. The F1 score, recall value, and precision were each 0.90, 090, and 0.916, respectively. Figure 10 (b) displays the training loss and validation loss graph with the number of epochs, demonstrating that the training loss and validation loss are decreasing with the number of epochs. When the training loss and validation loss spikes, we can blame outliers for it.

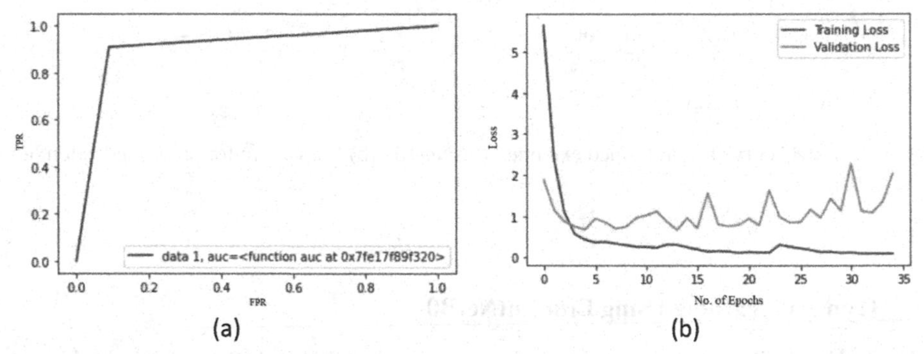

Fig. 10. (a) ROC curve for proposed extended VGG19 (b) Loss curve for proposed extended VGG19 model

Table 4 compares several hyperparameters utilized in training to achieve the best outcomes. The first row displays the various activation functions that were applied to the model's hidden layers and their corresponding outcomes. The second row displays the various filtering techniques that are used to reduce overfitting. For the activation function, we utilized the Tanh function, which gave us an accuracy with precision of 0.77, an F1 score of 0.78, and a recall of 0.81. Next, we tried the sigmoid activation function, which gave us an accuracy with precision of 0.71, an F1 score of 0.63, and a recall of 0.63. The ReLU activation function gave us the best results, with an accuracy of 0.91, a precision of 0.90, an F1 score of 0.90, and a recall of 0.92. In order to get the best results, we tested filtering techniques besides activation functions. Gaussian blur was utilized, and it produced accuracy of 0.72, precision of 0.81, F1 score of 0.75, and recall of 0.81. We chose an average blur, which produced the best results for precision (0.90), F1 score (0.90), and recall (0.92).

Table 4. Comparison of different hyperparameters in the proposed extended VGG19 model

Activation Function	Hyperparameter	Accuracy	Precision	Recall	F1-score
	ReLU	**0.91**	**0.90**	**0.91**	**0.90**
	Tanh	0.77	0.81	0.81	0.78
	Sigmoid	0.71	0.66	0.63	0.63
Filter for image processing	Gaussian Blur	0.72	0.81	0.81	0.75
	Blur	**0.91**	**0.90**	**0.91**	**0.90**

4.5 Comparison with Existing Models

We will contrast our model with the one presented in the publication, which used VGG16 as transfer learning and adjusted the VGG16 model by reducing layers, while our model employed VGG19, followed by a 4-layer model, employing ReLU activation function in hidden layers and sigmoid at the output layer. Table 5 displays the performance metrics of the proposed model in comparison to the models presented by Zheng et al. [16] and Latha et al. [18]. It is evident from the performance metrics that our model performs better than the existing model.

The accuracy, precision, recall, and F1 scores of all the Deep Neural Network (DNN) models outlined above will now be compared.

Figure 11 compares the accuracy, precision, recall, and F1 scores of all the tested models, and it is clear from this comparison that the suggested model outperforms all other models, including the current one.

Table 5. Comparison of proposed extended VGG19 model with existing models

	Accuracy	Precision	Recall	F1-score
Zheng et al. [16]	0.87	0.87	0.890	0.85
Latha et al. [18]	0.90	0.93	0.875	0.89
Proposed extended VGG19 model	0.91	0.90	0.916	0.90

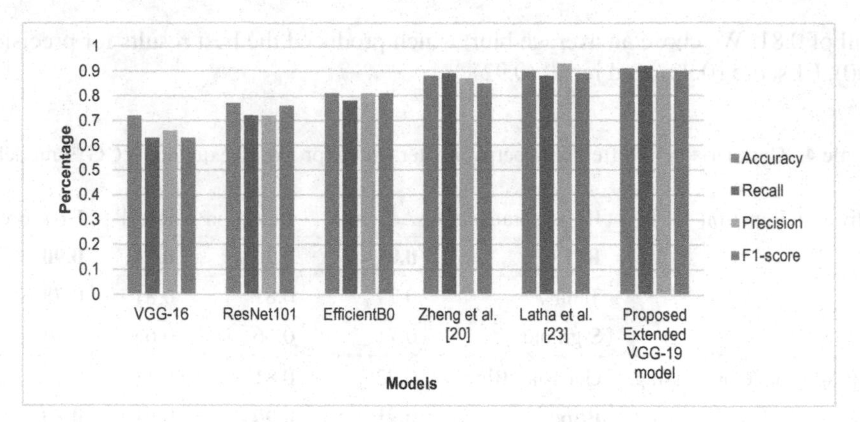

Fig. 11. Comparison of different pretrained models and state-of-the-art models on standard evaluation metrics

5 Conclusion

Previous models that used deep learning, which requires a lot of data, used all the slices of the brain MRI images including all perspectives of viewing, which caused noise in the data as some superfluous information was entering the model, which decreased the model's performance. In order to solve the problem of extraneous data or noisy data, we proposed a model that used unprocessed 3D scans of the brain and only obtained the axial view of the image after data analysis. This is because the subcortical region and ventricular areas are most exposed in these scans and they are the main contributing factors in cases of schizophrenia. Then the data were processed through the preprocessing pipeline after analysis in order to threshold the image and apply image augmentation. To address the massive data challenge, we constructed the model based on transfer learning using the VGG-19 pretrained model and training parameters for our dataset. Then, using fully connected layers, we transfer the knowledge to make predictions about whether the person has schizophrenia or not. Additionally, this research compares the outcomes of various convolution models that have already been pre-trained using the ImageNet dataset. The most recent 2D CNN model currently in use utilizes VGG-16 to obtain parameters before using the data to train an improved classifier. When we compared this model to the best 2D CNN model currently available, we found that accuracy had increased by a maximum of 3% and a minimum of 1%. We used our model to execute more quickly than 3D CNN-based models while suffering little accuracy loss. Over the COBRE dataset, we evaluated the performance of our suggested model. We used evaluation criteria like accuracy, precision, recall, and F1-Score to compare various models. According to evaluation results, our suggested model works quicker and better than the current models for detecting schizophrenia using MRI. The accuracy and evaluation metrics of all developed models can be enhanced by expanding the data due to the extremely low amount of publicly available MRI scan data. The VGG-19 model may be paired with another model for increased accuracy. Utilizing other phenotypic data along with MRI scans offers yet another opportunity for development. This classification technique can also be used to other brain conditions whose detection requires brain MRI images.

References

1. "Institute of health metrics and evaluation (IHME). Global health data exchange (GHDX)." https://vizhub.healthdata.org/gbd-results?params=gbd-api-2019-permalink/af5cd3f71b82a07b7823ae8d7ccc7c23, accessed 01 Dec. 2022
2. Ringen, P.A., Engh, J.A., Birkenaes, A.B., Dieset, I., Andreassen, O.A.: Increased mortality in schizophrenia due to cardiovascular disease – a non-systematic review of epidemiology, possible causes, and interventions. Front Psychiatry **5**(SEP), 1 (2014). https://doi.org/10.3389/FPSYT.2014.00137
3. Sham, P.C., MacLean, C.J., Kendler, K.S.: A typological model of schizophrenia based on age at onset, sex and familial morbidity. Acta Psychiatr. Scand. **89**(2), 135–141 (1994). https://doi.org/10.1111/j.1600-0447.1994.tb01501.x
4. Zhao, X., Wu, Y., Song, G., Li, Z., Zhang, Y., Fan, Y.: A deep learning model integrating FCNNs and CRFs for brain tumor segmentation. Med. Image Anal. **43**, 98–111 (2018). https://doi.org/10.1016/J.MEDIA.2017.10.002
5. Alissa, M., Lones, M., Petrick, R., Vallejo, M.: Parkinson's disease diagnosis using deep learning (2018)
6. Gao, S., Lima, D.: A review of the application of deep learning in the detection of Alzheimer's disease. Int. J. Cognitive Comp. Eng. **3**, 1–8 (2022). https://doi.org/10.1016/J.IJCCE.2021.12.002
7. Brugger, S.P., Howes, O.D.: Heterogeneity and homogeneity of regional brain structure in schizophrenia: a meta-analysis. JAMA Psychiat. **74**(11), 1104–1111 (2017). https://doi.org/10.1001/JAMAPSYCHIATRY.2017.2663
8. Gutman, B.A., et al.: A meta-analysis of deep brain structural shape and asymmetry abnormalities in 2,833 individuals with schizophrenia compared with 3,929 healthy volunteers via the ENIGMA Consortium. Hum. Brain Mapp. **43**(1), 352–372 (2022). https://doi.org/10.1002/HBM.25625
9. Lai, J.W., Ang, C.K.E., Rajendra Acharya, U., Cheong, K.H.: Schizophrenia: a survey of artificial intelligence techniques applied to detection and classification. Int. J. Environ Res. Public Health **18**(11), 6099 (2021). https://doi.org/10.3390/IJERPH18116099/S1
10. Veronese, E., Castellani, U., Peruzzo, D., Bellani, M., Brambilla, P.: Machine learning approaches: From theory to application in schizophrenia. Comput. Math. Methods Med. **2013** (2013). https://doi.org/10.1155/2013/867924
11. Sakai, K., Yamada, K.: Machine learning studies on major brain diseases: 5-year trends of 2014–2018. Jpn. J. Radiol. **37**(1), 34–72 (2019). https://doi.org/10.1007/S11604-018-0794-4
12. Shi, D., et al.: Machine learning of schizophrenia detection with structural and functional neuroimaging. Dis Markers **2021** (2021). https://doi.org/10.1155/2021/9963824
13. Krizhevsky, A., Sutskever, I., Hinton, G.E.: ImageNet classification with deep convolutional neural networks. Accessed: 30 Nov. 2022. [Online]. Available: https://code.google.com/archive/p/cuda-convnet/
14. Simonyan, K., Zisserman, A.: Very deep convolutional networks for large-scale image recognition. In: 3rd International Conference on Learning Representations, ICLR 2015 - Conference Track Proceedings (2014). https://doi.org/10.48550/arxiv.1409.1556
15. He, K., Zhang, X., Ren, S., Sun, J.: Deep residual learning for image recognition. Proceedings of the IEEE Computer Society Conference on Computer Vision and Pattern Recognition, vol. 2016-December, pp. 770–778 (2016). https://doi.org/10.1109/CVPR.2016.90

16. Zheng, J., et al.: Diagnosis of schizophrenia based on deep learning using fMRI. Comput Math Methods Med **2021** (2021). https://doi.org/10.1155/2021/8437260
17. Oh, J., Oh, B.L., Lee, K.U., Chae, J.H., Yun, K.: Identifying schizophrenia using structural MRI with a deep learning algorithm. Front Psychiatry **11** (2020). https://doi.org/10.3389/fpsyt.2020.00016
18. Latha, M., Kavitha, G.: Detection of schizophrenia in brain MRI images based on segmented ventricle region and deep belief networks. Neural Comput. Appl. **31**(9), 5195–5206 (2019). https://doi.org/10.1007/s00521-018-3360-1
19. Zhang, J., et al.: Detecting schizophrenia with 3d structural brain MRI using deep learning (2022). [Online]. Available: http://schizconnect.org/
20. Yassin, W., et al.: Machine-learning classification using neuroimaging data in schizophrenia, autism, ultra-high risk and first-episode psychosis. Transl Psychiatry **10**(1) (2020). https://doi.org/10.1038/s41398-020-00965-5
21. Liu, J., Li, M., Pan, Y., Wu, F.X., Chen, X., Wang, J.: Classification of schizophrenia based on individual hierarchical brain networks constructed from structural MRI images. IEEE Trans. Nanobioscience **16**(7), 600–608 (2017). https://doi.org/10.1109/TNB.2017.2751074
22. "Google Colab." https://colab.research.google.com/, accessed 01 Dec. 2022

An Efficient Approach for Early Prediction of Sudden Cardiac Death Using Two-Stage Feature Selection and Gradient Boosting Classification

Shaik Karimulla(✉) and Dipti Patra

IPCV Laboratory, Department of Electrical Engineering, National Institute of Technology Rourkela, Rourkela, Odisha, India
{520EE1006,Dpatra}@nitrkl.ac.in

Abstract. Sudden cardiac death (SCD) is one of the leading causes of death worldwide, resulting in unpredicted loss of heart function. This complex problem occurs in people with or without a history of cardiac illness. The symptoms of SCD start 1 h prior to its onset. The early detection of SCD may save many lives around the world. Hence it is vital to develop an accurate and precise method for identifying individuals at risk of developing SCD. This paper presents an efficient methodology for the early prediction of SCD using heart rate variability (HRV) and wavelet transform analysis by comparing diseased and non-diseased subjects. To accomplish this, the ECG signals of Normal sinus rhythm (NSR), Sudden cardiac death (SCD), and coronary artery disease (CAD) subjects were collected and pre-processed. HRV signals were derived from the ECG signal to extract various time domain, frequency domain, and non-linear method-based features. These features along with wavelet features and statistical features were considered for the selection of significant features. In this work, a two-stage feature selection method is proposed based on mutual information (MI) and recursive feature elimination (RFE) along with gradient boosting (GB) classification for accurately detecting SCD. Using the proposed MI-RFE-GB scheme, we achieved SCD detection 1 h before its onset with accuracy, sensitivity, specificity, and precision at 97.60%, 97.54%, 98.80%, and 97.59% respectively. The experimental results of the proposed scheme demonstrate the superiority over state-of-the-art methods. However, the current study can be extended using various cardiac disease datasets that cause for the development of SCD.

Keywords: Artifact correction · heart Rate Variability · wavelet transform · gradient boosting classifier · feature selection

1 Introduction

According to World health organization (WHO) statistics, cardiovascular diseases (CVDs) are the leading cause of mortality worldwide, taking 17.9 million lives annually [1]. Electrocardiogram (ECG) is a non-invasive golden standard to identify the heart's

B. K. Singh et al. (Eds.): ICBEST 2023, CCIS 2003, pp. 83–97, 2024.
https://doi.org/10.1007/978-3-031-54547-4_7

electrical activity. Sudden cardiac death (SCD) is a severe cardiac disorder caused for unconsciousness and death within a few minutes due to the development of irregularities in the electrical conduction system of the heart. SCD is responsible for 25% of deaths among CVDs [2]. SCD is associated with patients with or without a history of cardiac diseases. The common symptoms of SCD start one hour prior to its onset [3]. The survival rate after the incidence of SCD is about 10% due to a lack of early prediction techniques. The main cardiac abnormalities which are responsible for the onset of SCD are coronary artery disease (CAD), ventricular fibrillation (VF), ventricular flutter (VFL), ventricular tachycardia (VT) and bradyarrhythmia etc. [4]. Individuals with a history of CAD are more susceptible to developing SCD due to the thinning of the coronary arteries caused by the obstruction of blood flow to the heart muscle due to the build-up of plaque in the arteries. Delaying the CAD diagnosis leads to the body receiving inadequate oxygen, which can result in cardiac arrest and heart failure [5]. Many authors proposed different strategies for the early prediction of SCD using ECG and HRV signals.

Heart rate variability (HRV) provides the most significant electrophysiological marker used to identify cardiovascular problems. HRV refers to the fluctuation in the time interval between successive heartbeats. The autonomic nervous system (ANS) is significantly influenced by the transitory signal known as HRV, which is acquired from the electrocardiogram's RR intervals [6]. A healthy heart requires higher HRV to react to environmental changes and balance the two ANS branches.

The wavelet transform is a strong time–frequency analysis and signal coding method for complicated nonstationary data. Wavelets record both short-term, high-frequency, and longer-term, low-frequency information. The approach is excellent for analyzing transients, aperiodicity, and other non-stationary signal properties because tiny variations in signal shape may be emphasized over the scales of interest [7]. The multiplicity of wavelet functions allows the most suited one to be picked for the signal under study. The main contributions of this study are (i) the development of an accurate strategy for early prediction of SCD by classifying NSR SCD and CAD subjects. (ii) Multistage feature selection algorithm is proposed to identify the best subset of features using mutual information- recursive feature elimination along with gradient boosting classification algorithm. (iv) The proposed method includes the fusion of HRV, wavelet, and statistical features that provides enhanced classification results which could be helpful in the field of early prediction of SCD.

2 Methods and Materials

2.1 Data

In this study, three datasets were examined for classification, a total of 58 ECG signals acquired from the physioBank database. To maintain the balance among the dataset sampling rate of the NSR dataset is up-sampled to 250 Hz. The symptoms of SCD start 1 h before its onset, so a 1-h signal is acquired for each subject randomly chosen from the 24-h data for NSR and CAD [8]. For SCD, data is collected one hour prior to the commencement of VF. The collected 1-h ECG signal is further segmented into 12 non-overlapping segments for short-term analysis of HRV. The complete description of the dataset is mentioned in Table 1.

Table 1. The complete description of the dataset used in the study of SCD prediction

Dataset	No. of subjects	Sampling frequency	No. of segments	Gender and Age (range)
NSR	18	128 Hz (Up-sampled to 250 Hz)	216	5 males (26–45) 13 females (20–50)
SCD	20	250 Hz	240	10 males (34–80) 8 females (30–89) 2unknown
CAD	20	250 Hz	240	16 males (44–73), 4 females (62–67)

2.2 Pre-processing

In this stage, the ECG signal of a 5-min duration is denoised using the sym6 discrete wavelet transform (DWT) [9]. The denoised signal is used to identify the QRS peak detection using wavelet transform [10]. The HRV signal is derived from the ECG signal with the help of Wavelet-based R peak detection algorithm. Further, it is processed for artifact correction to remove ectopic beats, extra beats, and missing beats. The artifact correction algorithm is developed using median filtering and the threshold value. The steps involved in the correction algorithm followed as (i) The average RR interval value is calculated from the median filtered HRV signal. The 20% of the average value is set as the threshold limit. (ii) Any RR interval greater than the average RR interval + threshold value is considered an artifact. (iii) Any RR interval less than the average RR interval - threshold value is considered an artifact. The identified artifact is corrected by using cubic spline interpolation. The effectiveness of artifact correction of a 5-min HRV signal is represented in Fig. 1.

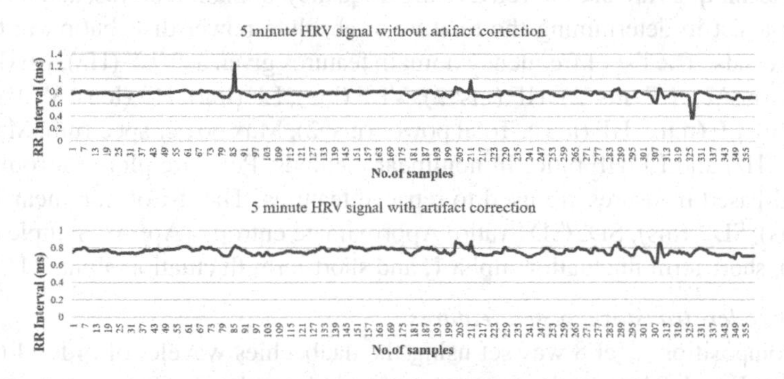

Fig. 1. Comparison of 5-min HRV signal with and without artifact correction.

Block diagram representation of the proposed method for early prediction of sudden cardiac using two-stage feature selection algorithm is mentioned in Fig. 2

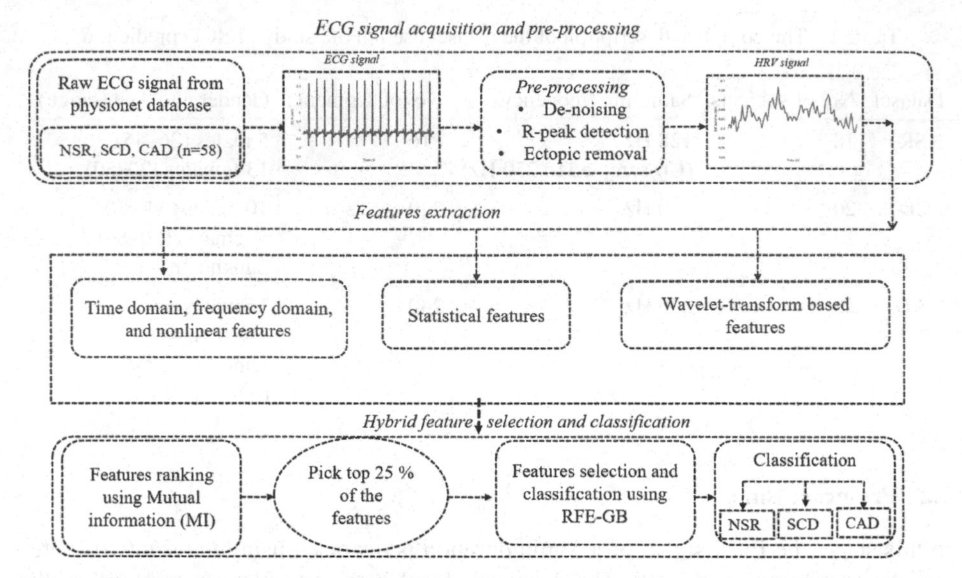

Fig. 2. Block diagram representation of proposed scheme for early prediction of SCD using HRV features and MI-RFE-GB based feature selection.

2.3 Features Extraction

Heart rate variability features:
In this study, 35 features extracted from the HRV signal can be divided into three methods: time domain, frequency domain, and nonlinear methods. The statistical and geometric methods applied in time domain analysis to retrieve the features and the list is Mean RR (ms), SDNN (ms), Mean HR (beats/min), SD HR (beats/min), Min HR (beats/min), Max HR (beats/min), RMSSD (ms), NN50 (beats), pNN50 (%), RR triangular index, Stress index, TINN (ms) Max RR (s), Min RR(s), and Mean absolute deviation. Fourier transform techniques are used to retrieve the frequency-domain features, and these measurements assist in determining the relative or absolute power distribution in the four frequency bands. The list of frequency domain features given as VLF (Hz), LF (Hz), HF (Hz), VLF (ms^2), LF (ms^2), HF (ms^2), VLF (log), LF (log), HF (log), VLF (%), LF (%), HF (%), LF (n.u.), HF (n.u.), Total power (ms^2), Max power spectrum, Maximum frequency (Hz) and LF/HF ratio. In nonlinear methods, Poincare plots, entropy-based, and fractal-based measures are used to retrieve features. The list of nonlinear features is SD1 (ms), SD2 (ms), SD2/SD1 ratio, Approximate entropy (ApEn), Sample entropy (SampEn), short-term fluctuation alpha 1, and short-term fluctuation alpha 2 [11].

Discrete wavelet- transform based features:
DWT decomposition level 8 was set using the daubechies wavelet of order 4 (db4) as illustrated in Fig. 4. The wavelet features extracted from the DWT decomposition level 8 are absolute mean, Energy, standard deviation and variance. The approximation and detail coefficients shown in Fig. 3.

Other statistical features:

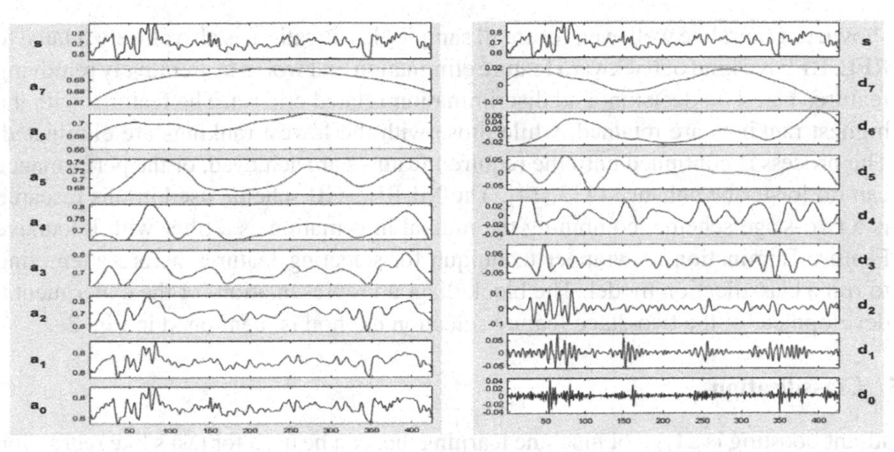

Fig. 3. DWT applied on HRV signal up to level 8 using daubechies4 for 3 classes

The derived HRV signals are used to extract the extra statistical features such as Skewness, Kurtosis, Interquartile range, and Histogram.

2.4 Feature Selection

This study proposes a two-stage feature selection scheme for selecting best significant features using Mutual information (MI) and Recursive feature elimination (RFE) methods.

a. Mutual Information based FS

The concept of feature selection is extensively used in statistics and machine learning, Previous to developing a model, it includes identifying a subset of pertinent features. Since the 1970s, the selection of features has been a growing and developing area of research. It has successfully removed redundant and irrelevant features, increased the performance of learning tasks, and improved abilities to understand learning methods' outcomes. According to information theory, the degree of uncertainty in X caused by knowing anything about Y is called mutual information [12].

$$I(X;Y) = \sum_{X,y} P(x,y) \log \frac{P(x,y)}{P(x)P(y)} \quad (1)$$

where the marginal probability distribution functions for X and Y are p(x) and p(y), respectively, and P (x, y) is the joint probability distribution function of X and Y.

b. Recursive Feature Elimination (RFE)

A recently created feature selection technique for problems with limited sample classification is called recursive feature elimination (RFE). It is a successful method for choosing features for limited samples. By eliminating the least significant features whose removal would have the least impact on the training mistakes, RFE tries to increase the generalization performance. Random forest algorithm, which has been shown to generalize well even for small sample classification, is also closely related to RFE. RFE is a kind of backward feature elimination that works by iteratively removing features based on decision- and discrimination-related criteria. The features with the highest rankings are retained, while those with the lowest rankings are eliminated. The process is continued until the required features are achieved, or the performance can no longer be enhanced [13, 14]. The MI-RFE-GB scheme used in this research is a two-stage scheme, combining the mutual information as a filter with Recursive Feature Elimination as wrapper technique for selecting features before attempting to run a classification model. The block diagram representation for the experimental development of the two-stage feature selection method is mentioned in Fig. 2.

2.5 Classification

Gradient boosting is a type of machine learning that can be used for tasks like regression and classification. It gives a prediction model in the form of a group of weak prediction models, usually decision trees. Gradient-boosted trees are the name of the algorithm that is made when a decision tree is a weak learner [15]. It usually does better than a random forest. A gradient-boosted trees model is built in stages, like other boosting methods. However, it is more general than the other methods because it can be used to optimize any loss function that can be differentiated. In this study, Gradient boosting and a group of classifiers used for classification.

3 Results

This study developed a novel methodology for implementation for early prediction of SCD by classifying NSR, SCD, and CAD subjects. 35 HRV features were extracted from 696 HRV signals of 3 classes with a duration of 5 min for each signal. The feature selection analysis has been implanted on these 87 features for SCD detection analysis. The mean and standard deviation values of each feature are mentioned in Table 2.

Table 2. Mean and SD values of the time domain, frequency domain, and nonlinear methods of three classes.

Time domain features			
HRV feature	NSR	SCD	CAD
Mean RR (ms)	744.14 ± 127.80	849.77 ± 222.31	856.21 ± 136.50
SDNN (ms)	35.17 ± 18.65	85.60 ± 57.34	31.79 ± 14.97
Mean HR (beats/min)	82.99 ± 14.31	75.42 ± 19.39	71.89 ± 11.711
Min HR (beats/min)	69.77 ± 12.31	63.58 ± 16.41	64.35 ± 8.94
Max HR (beats/min)	99.55 ± 16.14	104.36 ± 41.98	82.65 ± 16.13

(continued)

Table 2. (*continued*)

Time domain features			
HRV feature	NSR	SCD	CAD
RMSSD (ms)	**32.59 ± 21.38**	**114.77 ± 78.24**	**23.86 ± 14.11**
NN50 (beats)	**39.27 ± 36.78**	**133.96 ± 107.89**	**8.10 ± 9.27**
pNN50 (%)	10.60 ± 11.57	39.55 ± 29.41	5.01 ± 5.77
SD HR (beats/min)	3.79 ± 1.49	9.84 ± 8.05	2.89 ± 1.90
RR triangular index	8.54 ± 3.44	12.57 ± 9.80	6.81 ± 2.42
TINN (ms)	182.66 ± 85.21	451.25 ± 234.66	151.32 ± 84.19
Stress Index	14.32 ± 6.08	8.15 ± 5.42	15.59 ± 6.083
Max RR (s)	0.981 ± 0.14	0.97 ± 0.14	0.97 ± 0.13
Min RR(s)	0.72 ± 0.14	0.72 ± 0.14	0.71 ± 0.14
Mean absolute deviation	0.04 ± 0.02	0.04 ± 0.02	0.04 ± 0.02
Frequency domain features			
VLF (Hz)	0.035 ± 0.00	0.03 ± 0.00	0.03 ± 0.00
LF (Hz)	0.09 ± 0.02	0.08 ± 0.03	0.07 ± 0.02
HF (Hz)	0.21 ± 0.07	0.25 ± 0.03	0.17 ± 0.03
VLF (ms^2)	88.02 ± 129.81	330.42 ± 677.83	112.00 ± 150.31
LF (ms^2)	831.75 ± 910.35	3093.9 ± 5716.17	754.73 ± 938.77
HF (ms^2)	619.5 ± 1754	4561.61 ± 7705.95	249.97 ± 416.29
VLF (log)	2.76 ± 2.31	4.36 ± 1.91	4.15 ± 1.12
LF (log)	6.20 ± 1.11	6.69 ± 1.82	6.12 ± 1.04
HF (log)	5.50 ± 1.22	7.12 ± 1.83	4.79 ± 1.17
VLF (%)	5.61 ± 6.41	5.18 ± 4.82	12.05 ± 7.65
LF (%)	60.91 ± 15.22	37.73 ± 13.50	67.51 ± 12.16
HF (%)	33.41 ± 17.18	56.84 ± 15.94	20.40 ± 11.71
LF (n.u.)	**64.97 ± 16.94**	**40.21 ± 15.31**	**76.91 ± 12.61**
HF (n.u.)	34.96 ± 16.93	59.54 ± 15.21	23.06 ± 12.611
Total power (ms^2)	1539.94 ± 2395.62	7995.35 ± 13,585	1116.94 ± 1387.53
LF/HF ratio	2.75 ± 2.34	0.83 ± 0.71	4.80 ± 3.66
Max power spectrum	0.38 ± 0.24	0.39 ± 0.25	0.39 ± 0.25
Maximum frequency	**16.34 ± 6.74**	**16.06 ± 7.15**	**16.64 ± 8.01**
Nonlinear methods features			
SD1 (ms)	**23.07 ± 15.14**	**81.29 ± 55.44**	**16.92 ± 10.01**

(*continued*)

Table 2. *(continued)*

Time domain features			
HRV feature	NSR	SCD	CAD
SD2 (ms)	43.63 ± 22.44	87.99 ± 61.74	41.28 ± 19.27
SD2/SD1	1.97 ± 0.58	1.12 ± 0.36	2.60 ± 0.67
Approximate entropy (ApEn)	**1.17 ± 0.08**	**0.98 ± 0.21**	**0.80 ± 0.07**
Sample entropy (SampEn)	1.64 ± 0.26	1.30 ± 0.58	1.20 ± 0.26
Short-term fluctuations, alpha 1	**1.13 ± 0.24**	**0.69 ± 0.22**	**1.32 ± 0.22**
Long-term fluctuations, alpha 2	0.33 ± 0.2	0.34 ± 0.16	0.42 ± 0.16

From the Table 2, it was clear that 7 out of 40 features were selected using MI-RFE-GB feature selection scheme. The selected features RMSSD (ms), NN50(beats), SD1 (ms) have highest mean and SD values in case of SCD compared to NSR and SCD. The features LF (n. u), Short-term fluctuations alpha1and Maximum frequency have highest values of mean and SD in case of CAD compared to NSR and SCD.The remaining selected feature approximate entropy has highest values of mean and SD in case of NSR compared to SCD and CAD.

Table 3. Mean and SD values of other statistical features.

Statistical features	NSR	SCD	CAD
Skewness	-0.26 ± 0.86	-0.20 ± 0.84	
Kurtosis	0.93 ± 3.53	0.85 ± 3.36	0.93 ± 3.42
Interquartile range	**0.07 ± 0.04**	**0.07 ± 0.04**	**0.06 ± 0.04**
Histogram_7	4.03 ± 14.07	4.96 ± 17.94	9.11 ± 34.91
Histogram_8	67.26 ± 81.85	69.86 ± 82.56	69.40 ± 82.23
Histogram_9	80.33 ± 62.22	81.26 ± 63.56	79.65 ± 64.21

From the Table 3, it was clear that 1 out of 6 other statistical features was selected using MI-RFE-GB feature selection scheme. The selected feature Interquartile range has highest mean and SD values in case of NSR and SCD compared to CAD.

Table 4. Mean and SD values of absolute mean, energy, standard deviation and variance of detail coefficients d0-d7.

Class	Feature	D0	D1	D2	D3	D4	D5	D6	D7
NSR	Absolute mean	**0.0031 ± 0.001**	0.009 ± 0.003	0.017 ± 0.006	0.026 ± 0.009	0.03 ± 0.01	0.04 ± 0.01	**0.06 ± 0.02**	0.07 ± 0.02
	Energy	0.04 ± 0.01	0.07 ± 0.02	0.115 ± 0.03	0.14 ± 0.03	0.17 ± 0.04	0.20 ± 0.05	0.24 ± 0.06	0.27 ± 0.07
	Standard deviation	0.04 ± 0.01	0.07 ± 0.02	0.113 ± 0.03	0.14 ± 0.03	0.17 ± 0.04	0.20 ± 0.05	0.23 ± 0.06	0.26 ± 0.07
	Variance	0.001 ± 0.001	0.006 ± 0.004	0.013 ± 0.007	0.022 ± 0.01	0.032 ± 0.01	0.04 ± 0.02	0.05 ± 0.03	0.07 ± 0.04
SCD	Absolute mean	**0.003 ± 0.002**	0.001 ± 0.008	0.026 ± 0.015	0.04 ± 0.02	0.05 ± 0.03	0.07 ± 0.04	**0.09 ± 0.05**	0.11 ± 0.06
	Energy	0.07 ± 0.05	0.12 ± 0.08	0.16 ± 0.09	0.20 ± 0.10	0.24 ± 0.12	0.28 ± 0.14	0.32 ± 0.15	0.36 ± 0.17
	Standard deviation	0.07 ± 0.05	0.12 ± 0.07	0.16 ± 0.09	0.20 ± 0.10	0.23 ± 0.11	0.27 ± 0.13	0.31 ± 0.14	0.34 ± 0.16
	Variance	0.008 ± 0.01	0.02 ± 0.03	0.03 ± 0.04	0.05 ± 0.05	0.07 ± 0.07	0.09 ± 0.09	0.11 ± 0.11	0.14 ± 0.14
CAD	Absolute mean	**0.008 ± 0.002**	0.02 ± 0.007	0.04 ± 0.01	0.06 ± 0.02	0.09 ± 0.03	0.12 ± 0.04	**0.16 ± 0.05**	0.19 ± 0.06
	Energy	0.06 ± 0.01	0.11 ± 0.02	0.17 ± 0.04	0.22 ± 0.05	0.27 ± 0.06	0.33 ± 0.07	0.38 ± 0.09	0.43 ± 0.10
	Standard deviation	0.06 ± 0.01	0.11 ± 0.02	0.16 ± 0.03	0.21 ± 0.04	0.26 ± 0.05	0.30 ± 0.06	0.35 ± 0.07	0.39 ± 0.08
	Variance	0.004 ± 0.002	0.01 ± 0.007	0.02 ± 0.01	0.04 ± 0.02	0.07 ± 0.03	0.09 ± 0.04	0.12 ± 0.05	0.16 ± 0.07

From the Table 4, it was clear that 2 out of 36 features were selected using MI-RFE-GB two-stage feature selection. The selected features Wavelet absolute mean_0 and Wavelet absolute mean_6 have highest mean and SD values in case of CAD compared to NSR and SCD. Box plot representation of all selected features using the MI-RFE-GB method is shown in Fig. 6. The selected features using the proposed two-stage FS algorithm are highlighted in Tables 2, 3, and 4 respectively.

The current study proposes MI-RFE-GB procedures to evaluate significant features for accurate SCD prediction results. The steps involved in the development of two-stage feature selection are followed as, Step 1: Initially, a mutual information-based technique is used to rank all 86 features. Figure 3 indicates the descending ranking order of the features. Step 2: Pick the top 25% of the features for further feature selection and classification using the RFE-GB technique. Step 4: The MI-RFE-GB method was used to choose find the best combination of features along with a classification accuracy of 97.60%. Step 5: The list of selected features are Short-term fluctuations, alpha 1,

Approximate entropy (ApEn), SD1 (ms), NN50 (beats), RMSSD (ms), Interquartile range, Wavelet absolute mean_0, Maximum frequency, LF (n.u.) and Wavelet absolute mean_6.

The performance of MI-RFE-GB is compared with various feature selection schemes such as RFE-RF, RFE-GB, MI-RFE-GB and MI-RFE-RF based feature selection is mentioned in Fig. 4.

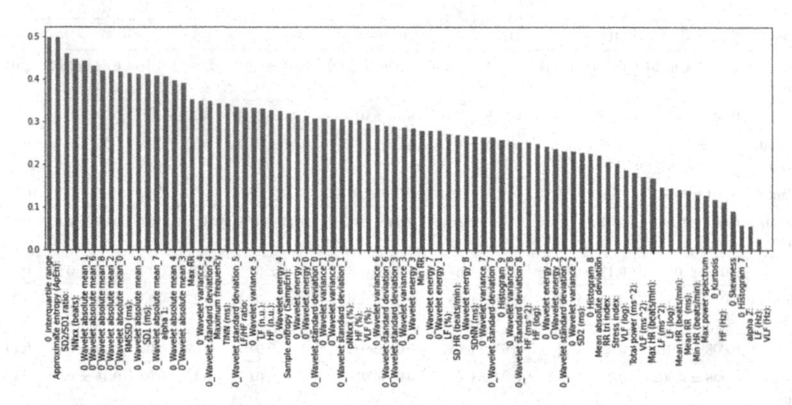

Fig. 4. Features ranking in descending using Mutual information (MI)

The comparison of performance analysis RFE-GB and RFE-RF with and without MI is mentioned in Fig. 5.

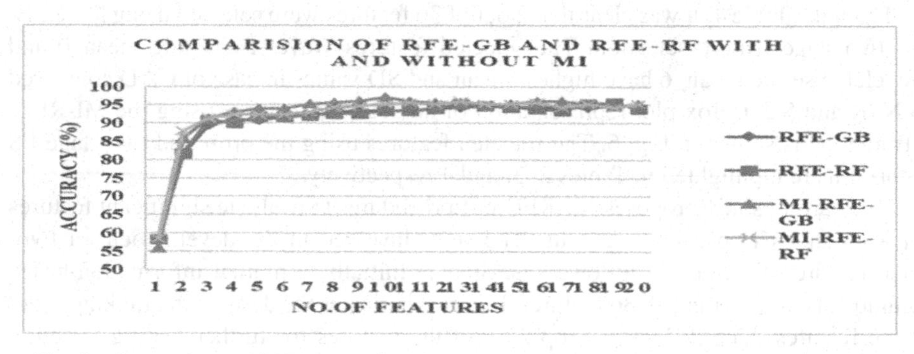

Fig. 5. Comparison of RFE-GB and RFE-RF with and without Mutual information stage

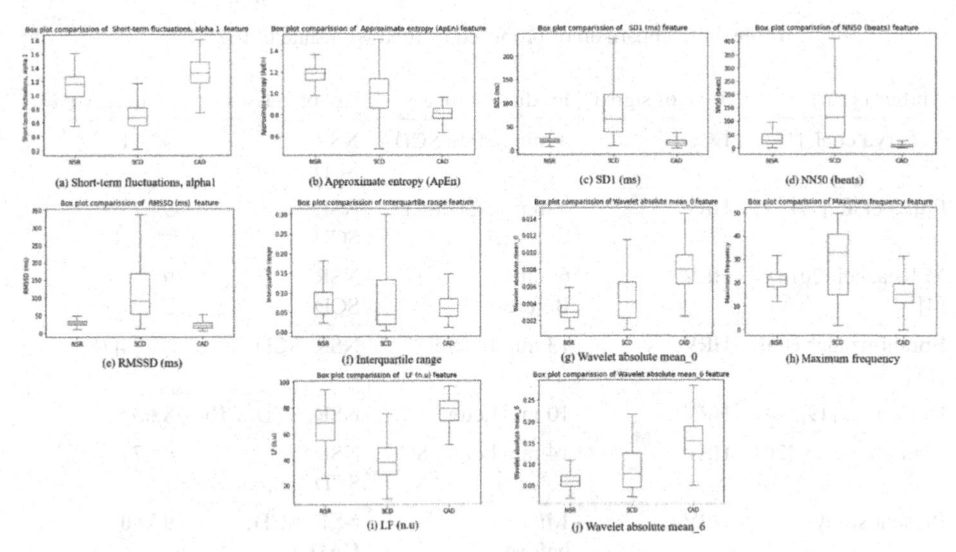

Fig. 6. Box plot representation of MI-RE-GB based selected features (a) Short-term fluctuations, alpha 1, (b) Approximate entropy (ApEn), (c) SD1 (ms), (d) NN50 (beats), (e) RMSSD (ms), (f) Interquartile range, (g) Wavelet absolute mean_0, (h) Maximum frequency, (i) LF (n.u.) and (j) Wavelet absolute mean_6.

Table 5. Comparison of feature selection models in terms of number of features vs accuracy.

Feature selection method	Number of features	Accuracy (%)
HRV features	40	94.74
Wavelet transform features	36	85.86
RFE-RF	28	95.65
RFE-GB	20	95.25
MI-RFE-GB	10	97.13
MI-RFE-RF	13	96.1

4 Discussion

This research proposes a new methodology for the early detection of SCD using HRV features. Initially, ECG signals of classes were taken from the physioBank database. Wavelet transform is used to denoise the ECG signal, i.e., to remove the baseline wandering and powerline interface, etc. Wavelet based R peak detection using Hilbert transform algorithm is used to derive QRS peak from the given signal. Heart rate variability (HRV) signal is derived from a denoised ECG signal may contain different artifacts such as extra beat, missing beat, ectopic peak, etc., which were corrected using a artifact correction algorithm. The artifact correction algorithm has two steps for determination of the artifact i.e., (i) Median filtering is implemented on the HRV signal (ii) Take the average of the HRV signal after median filtering (iii) Setting a threshold value for RR interval gives

Table 6. Comparison of proposed method with state of the art

Author (year)	Type of signal	Prediction time	No. of classes	Accuracy (%)
Acharya et al. [16]	ECG	4 min before SCD	NSR SCD	92.11
Fujita et al. [17]	HRV	4 min before SCD	NSR SCD	94.7
M Khazaei (2018) [4]	HRV	6 min Before	NSR SCD	95
Ebrahimzadeh et al. [18]	HRV	13 min before	NSR, SCD	84.24[18]
Devi et al. [19]	HRV	10 min before	NSR, SCD, CHF	83.33
Manhong et al. [20]	HRV	14 min before SCD	NSR SCD	94.7
Present study	HRV	**1 h before**	**NSR, SCD, CAD**	**97.60**

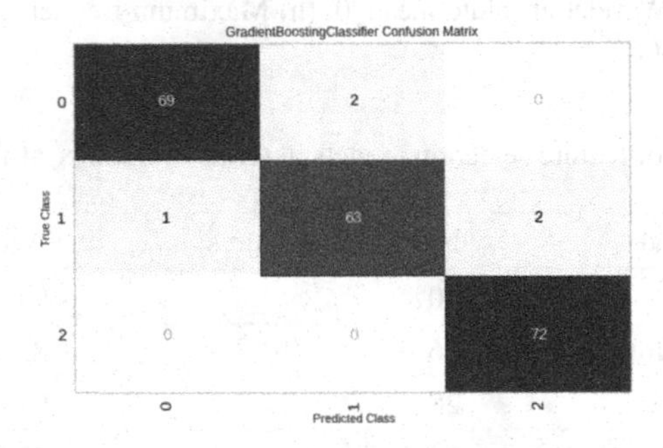

Fig. 7. Confusion matrix for the classification using selected features (MI-RFE-GB).

artifact confirmation whether the signal is having or not. The complete explanation of the artifact correction algorithm is mentioned in Sect. 2. The pre-processed HRV signal is used to extract the features using the time domain in which 15 features were extracted using statistical and geometric methods, the frequency domain in which 18 features were derived using a Fourier transform-based tool, and nonlinear methods in which 7 features were extracted from HRV signal using Poincare plot analysis, entropy and detrended based fluctuations. Wavelet transform based features such as absolute mean, energy, standard deviation and variance of 8 level decomposition signals were calculated. The extracted features were further used for the classification, and the performance evaluation of different classification models was tested using all features without feature selection, which caused more computational time and deduction in accuracy. Feature selection is an important task used to find the optimal set of features that produces good classification accuracy with less computational time. In this study, the proposed two-step feature selection technique using mutual information recursive feature elimination with

Gradient boosting classifier (MI-RFE-GB), which produced a classification accuracy of 97.60% with ten selected features.

This study uses two popular machine learning algorithms i.e. Random forest (RF) and Gradient boosting classifiers. Table 5 indicates the comparison of performance analysis of ensemble of features and various combinations of subset of features. The individual performance of HRV features, Wavelet decomposition features and statistical features has produced a classification accuracy of 94.74%, and 85.86% and 96.17% with 40, 36 and 82 features respectively. The proposed method of ensemble of features along with feature selection has produced superior results than individual performance of different feature combinations.

Most of the previous studies for early detection of SCD have taken two classes such as NSR and SCD. The detailed description of comparison of proposed method with state-of the art as followed as Acharya et al. have developed a SCD detection methodology by extracting nonlinear features from the ECG signal. Using decision tree (DT) and Support vector machine (SVM), authours have predicted SCD 4 min before its onset with an accuracy of 86.8% [16]. Fujitha et al. have developed a SCD detection methodology by extracting nonlinear features from the HRV signal. Using Support vector machine (SVM, authours have predicted SCD 4 min before its onset with an accuracy of 94.7% [17]. M khazaei et al. have predicted SCD 6 min before with an accuracy of 95% using HRV signal and RQA analysis [4]. Compared to prior studies, the current study has gained importance by increasing the SCD prediction time, no, of classes (three), and new feature selection algorithm. The present study of early detection of SCD is compared with state-of-the-art techniques.

5 Conclusion

The present study suggests a unique method for assessing SCD risk by analyzing HRV and wavelet transform in healthy and unhealthy cardiac conditions. According to available data, the experiment's findings showed a significant difference in extracted features between the SCD and non-SCD groups. This illustrates that even though other cardiac abnormalities including CAD may be ruled out, SCD risk can be identified. Linear (time and frequency domain) and nonlinear features were retrieved from HRV signals, and wavelet coefficient features (D0-D7) and statistical features. The most effective set of features for classification was found using the developed MI-RFE-GB hybrid feature selection method. The accuracy, sensitivity, specificity, and precision of the proposed technique are 97.60%, 97.54%, 98.80%, and 97.59% respectively for classifying normal and abnormal states. The suggested technique may be included in automated diagnostic tools and continuous monitoring systems to increase survival rates by accurately predicting SCD at an early stage. The present study has limitations on number of classes considered for the early detection of SCD. The present study can be extended in future by adding more number of classes, prediction time and various methods such as time frequency analysis, deep learning etc.

References

1. Kaptoge, S., et al.: World health organization cardiovascular disease risk charts: revised models to estimate risk in 21 global regions. Lancet Glob. Heal. **7**(10), e1332–e1345 (2019). https://doi.org/10.1016/S2214-109X(19)30318-3
2. Moran, A.E., Roth, G.A., Narula, J., Mensah, G.A.: 1990-2010 Global cardiovascular disease atlas. Glob. Heart **9**(1), 3–16 (2014). https://doi.org/10.1016/j.gheart.2014.03.1220
3. Haqqani, H.M., Chan, K.H., Kumar, S., Denniss, A.R., Gregory, A.T.: The contemporary era of sudden cardiac death and ventricular arrhythmias: basic concepts, recent developments and future directions. Hear. Lung Circ. **28**(1), 1–5 (2019). https://doi.org/10.1016/S1443-9506(18)31972-3
4. Khazaei, M., Raeisi, K., Goshvarpour, A., Ahmadzadeh, M.: Early detection of sudden cardiac death using nonlinear analysis of heart rate variability. Biocybern. Biomed. Eng. **38**(4), 931–940 (2018). https://doi.org/10.1016/j.bbe.2018.06.003
5. Rohila, A., Sharma, A.: Detection of sudden cardiac death by a comparative study of heart rate variability in normal and abnormal heart conditions. Biocybern. Biomed. Eng. **40**(3), 1140–1154 (2020). https://doi.org/10.1016/j.bbe.2020.06.003
6. Robinson, B.F., Epstein, S.E., Beiser, G.D., Braunwald, E.: Control of heart rate by the autonomic nervous system. Studies in man on the interrelation between baroreceptor mechanisms and exercise. Circ. Res. **19**(2), 400–411 (1966). https://doi.org/10.1161/01.RES.19.2.400
7. Addison, P.S.: Wavelet transforms and the ECG: A review. Physiol. Meas. **26**(5) (2005). https://doi.org/10.1088/0967-3334/26/5/R01
8. Holstila, E., Vallittu, A., Ranto, S., Lahti, T., Manninen, A.: Helsinki. Cities as Engines Sustain. Compet. Eur. Urban Policy Pract. 175–189 (2016). https://doi.org/10.4324/9781315572093-15.
9. Lin, H.Y., Liang, S.Y., Ho, Y.L., Lin, Y.H., Ma, H.P.: Discrete-wavelet-transform-based noise removal and feature extraction for ECG signals. Irbm **35**(6), 351–361 (2014). https://doi.org/10.1016/j.irbm.2014.10.004
10. Rakshit, M., Das, S.: An efficient wavelet-based automated R-peaks detection method using Hilbert transform. Biocybern. Biomed. Eng. **37**(3), 566–577 (2017). https://doi.org/10.1016/j.bbe.2017.02.002
11. Shaffer, F., Ginsberg, J.P.: An overview of heart rate variability metrics and norms. Front. Public Heal. **5**(September), 1–17 (2017). https://doi.org/10.3389/fpubh.2017.00258
12. Hoque, N., Bhattacharyya, D.K., Kalita, J.K.: MIFS-ND: A mutual information-based feature selection method. Expert Syst. Appl. **41**(14), 6371–6385 (2014). https://doi.org/10.1016/j.eswa.2014.04.019
13. Senan, E.M., et al.: Diagnosis of chronic kidney disease using effective classification algorithms and recursive feature elimination techniques. J. Healthc. Eng. **2021** (2021). https://doi.org/10.1155/2021/1004767
14. Theerthagiri, P.: Predictive analysis of cardiovascular disease using gradient boosting based learning and recursive feature elimination technique. Intell. Syst. with Appl. **16**(September), 200121 (2022). https://doi.org/10.1016/j.iswa.2022.200121
15. Shi, H., Wang, H., Huang, Y., Zhao, L., Qin, C., Liu, C.: A hierarchical method based on weighted extreme gradient boosting in ECG heartbeat classification. Comput. Methods Programs Biomed. **171**, 1–10 (2019). https://doi.org/10.1016/j.cmpb.2019.02.005
16. Acharya, U.R., et al.: An integrated index for detection of sudden cardiac death using discrete wavelet transform and nonlinear features. Knowledge-Based Syst. **83**(1), 149–158 (2015). https://doi.org/10.1016/j.knosys.2015.03.015
17. Fujita, H., et al.: Sudden cardiac death (SCD) prediction based on nonlinear heart rate variability features and SCD index. Appl. Soft Comput. J. **43**(2016), 510–519 (2016). https://doi.org/10.1016/j.asoc.2016.02.049

18. Ebrahimzadeh, E., et al.: An optimal strategy for prediction of sudden cardiac death through a pioneering feature-selection approach from HRV signal. Comput. Methods Programs Biomed. **169**, 19–36 (2019). https://doi.org/10.1016/j.cmpb.2018.12.001
19. Devi, R., Tyagi, H.K., Kumar, D.: A novel multi-class approach for early-stage prediction of sudden cardiac death. Biocybern. Biomed. Eng. **39**(3), 586–598 (2019). https://doi.org/10.1016/j.bbe.2019.05.011
20. Shi, M., et al.: Early detection of sudden cardiac death by using ensemble empirical mode decomposition-based entropy and classical linear features from heart rate variability signals. Front. Physiol. **11**(February), 1–16 (2020). https://doi.org/10.3389/fphys.2020.00118

Malware Detection Framework Based on Iterative Neighborhood Component Analysis for Internet of Medical Things

Santosh K. Smmarwar(✉), Govind P. Gupta, and Sanjay Kumar

Department of Information Technology, National Institute of Technology, Raipur, India
{sksmmarwar.phd2019.it,gpgupta.it,skumar.it}@nitrr.ac.in

Abstract. The advancement of the medical equipment and its integration with the Internet of Medical Things *(IoMT)* has facilitated the remote monitoring of the health-related information of the patient and its analytics by the expert. *IoMT* network generates the large amount of medical data that are remotely accessed by the smart hospital, smart ambulance, and doctors for better observation and monitoring of the patient. In recent years, the smart connectivity of *IoMT* devices faces various known and unknown malware threats. To protect the *IoMT* system, there is a need to device an intelligent malware detection system for IoMT network. Thus, this paper proposes a hybrid Deep learning model by integrating the Convolutional Neural Network (*CNN*) with Iterative Neighbourhood Component Analysis (*INCA*), for malware detection and identification task, named as *(CNN-INCA-MD)*. The proposed model is evaluated on *IoT* malware dataset and achieved an accuracy of 96.98% with precision of 96.68%, recall of 96.67% and f1score of 96.66%. The proposed model is robust in detecting malware and outperforms from existing models.

Keywords: Malware Detection · Cyber-attacks · *IoMT* · *CNN*

1 Introduction

The use of *IoT* enabled technological advancement in medical fields such as internet of health devices (*IoHD*) which equipped the hospital with smart connectivity and smart health services in treatment of patients via *IoMT* devices. The *IoMT* environment is collection small things like smart devices, smart watch, smart ambulance, smart application, smart city and smart user connected to the internet for communicate among each devices [1]. The smart connectivity between devices having security issues, weak password and vulnerable infrastructure that causes to malware attacks by cyber criminals while communication between doctors and patients for smart treatment [2]. The *IoMT* devices consists of number of sensors and actuators which are used to collect and store sensitive data of patients for further effective and accurate analysis about the disease by medical expert. Different types of sensor devices like motion sensor, breathing sensor, temperature, blood pressure sensors are used to monitor health of patients [3]. In the

B. K. Singh et al. (Eds.): ICBEST 2023, CCIS 2003, pp. 98–106, 2024.
https://doi.org/10.1007/978-3-031-54547-4_8

year 2022 more than 13% of US population's healthcare data have been breached by cyber attackers and observed different types of malware attack. The malware attacker developing various types of new variants of malware to invade the healthcare systems without detecting by antivirus [4]. The conventional method of malware detection like signature, heuristic and sandboxing approaches are incapable of detecting hidden pattern of malware and resulted into poor accuracy, low detection rate and high false rate [5]. Nowadays, the research community uses the artificial intelligence (*AI*) based method like machine learning(*ML*) and deep learning(*DL*) to detect the new variants or zero-day malware with high accuracy and low false rate. *ML* based methods requires feature engineering of malware binaries which is computationally expensive and takes much time [6]. To overcome the need of feature engineering deep learning play the important role in malware detection instead of *ML* based approaches [7]. In order to sort out these problems and challenges of malware detection. The author concentrated to detect the new variants of malware family using the proposed deep convolutional enabled malware detection framework. The prime contribution of the proposed work are:

- Proposed a three phase *CNN-INCA-MD* enabled malware detection framework for *IoMT* environment.
- The proposed framework includes three phase which are feature extraction by *CNN*, feature selection by *INCA* and finally, malware classification, respectively.
- *CNN* model is utilized to extract significant features from malware samples and *INCA* is used to select the optimal feature to improve the accuracy of malware classification.

2 Related Work

This section gives the brief about the recent work of malware detection in *IoMT* environment based on *ML* and *DL* techniques. For instance, wazid et al. [8] detail about the different types of *IoMT* malware detection techniques and associated research issues and problems of cyber-attacks and their types existed for *IoT* devices. Alsubaei et al. [9] presented the security assessment framework for *IoMT* malware detection and detailed the number of security and privacy issues having in medical domain like awareness among the patients, medical staff and doctors of using the internet of health devices securely. Jeon et al. [10] introduced the *CNN* enabled *IoT* malware detection framework using dynamic analysis approach to minimize the damage of devices by early detection of malware for cloud environment. However, this framework lacks the scalability and huge amount of malware dataset to perform better. Liaqat et al. [11] proposed the cyber threat detection model for *SDN* (software defined Network) enabled *IoMT* environment using the hybrid *CNN* and *LSTM* model. Saheed et al. [12] demonstrated the cyber-attack detection architecture for *IoMT* environment by utilizing the Deep *RNN* and *ML* techniques for efficient malicious threat detection. However, the proposed work needs huge amount of malware sample and lacks scalability. Kumar et al. [3] presented the ensemble learning based fog-cloud enabled cyber-attack detection for *IoMT* environment and discuss the limitation of early dataset which suffers from poor detection rate and the lacks latest *IoMT* dataset. Khan & Akhunzada [13] designed the hybrid *SDN* based malware detection architecture. However this method is ineffective for any type of new malware detection. Ravi et al. [7] introduced the *DL* system based on multi-view attention malware detection in healthcare system. The proposed model used the windows and android malware samples for model assessment and achieved the accuracy more than 97%. Ravi et al. [14] proposed the attention based mechanism based on multidimensional *DL* model in *IoMT* environment for cross-platform malware detection with more than 94% accuracy.

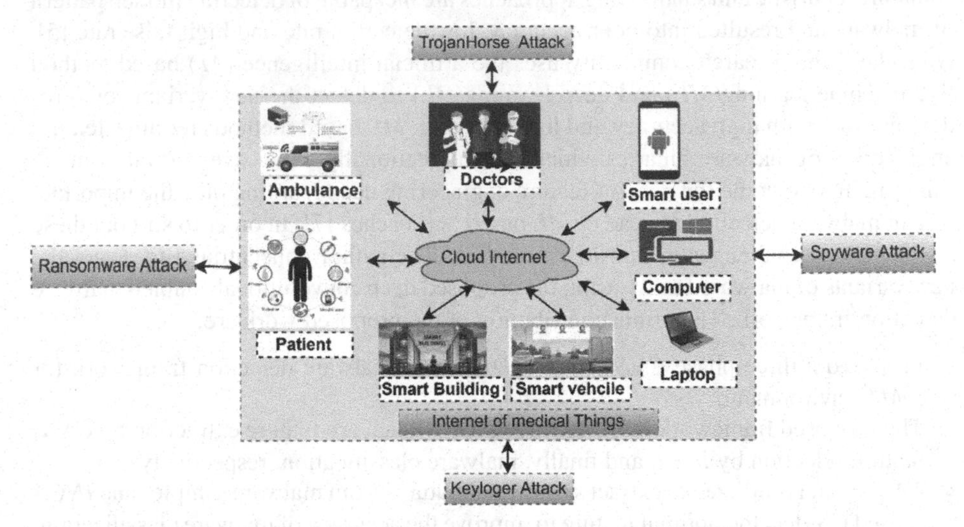

Fig. 1. A scenario of different types of malware attack on *IoMT* environment

3 Proposed Work

The proposed model mentioned in Fig. 2 is designed to malware detection for *IoMT* environment. This model has three phases: phase 1 includes feature extraction using *CNN*, phase 2 applied the *INCA* feature selection to select most optimal feature to improve the accuracy and phase 3 for automatic malware classification.

3.1 Convolutional Neural Network (*CNN*)

It is widely used deep learning model utilized in various areas such as computer vision, image classification, speech recognition, and fraud detection for excellent results. The outstanding popularity of *CNN* is due to its capability of automatically extract the important features from the input images. It is a collection of three layers such as convolutional layer, pooling layer, fully connected layer. It is effective for a huge dimensional data such as images [13]. Feature extraction reduces the dimensions and complexity of data. This increases the convolutional network training efficiency, reduces input data dimensions and parameter sizes [15]. This work employed two dimensional *CNN* model consisting of four convolution layers, four pooling layers and one fully connected layer (flatten, dropout and dense) [16]. The extracted features is then passes to *INCA* feature selection module to select the optimal features. Finally the fully connected layer generates flatten vector of the all features into single feature vector which is used by softmax function for malware classification.

3.2 Iterative Neighborhood Component Analysis *(INCA)*

The proposed model applied the *INCA* method which is a modified version of neighborhood component analysis *(NCA)* feature selection algorithm. It is used to select the most discriminative or optimal feature selection to increase the predictive accuracy [17]. *INCA* work on the basis of weight calculation of each feature attributes. Initially it assign the fixed weight (value 1) to each feature attributes. The weight is updated by using the adam optimizer and stochastic gradient descent function with fitness function based on distant correlation between features. In this method two types weights are created that is informative and redundant weight. Informative weights are indicated by larger value and redundant weight by smaller weights value. The main motive of using *INCA* feature selection is to get the optimal features, for this an iterative error calculation is used in *INCA*. In this feature vector with the minimum error value is selected as the optimum feature vector [18].

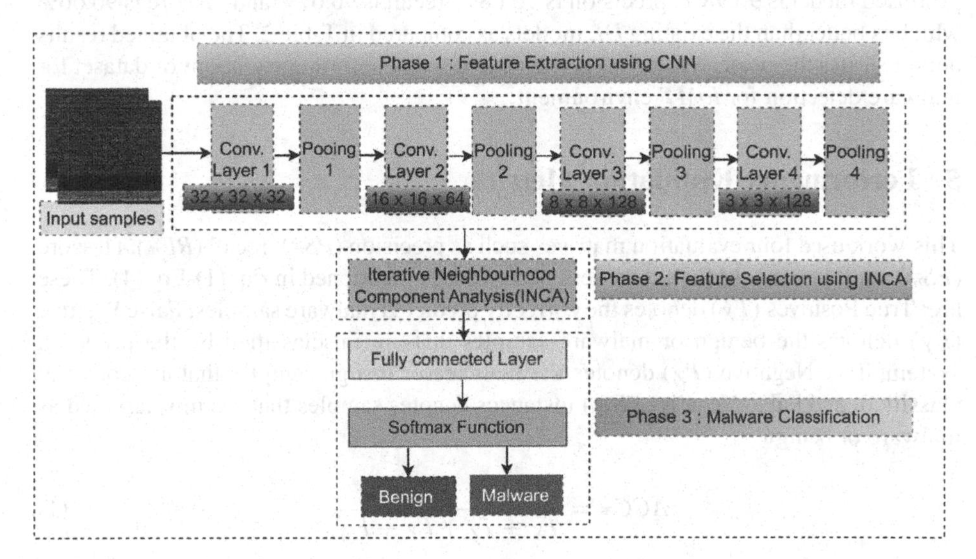

Fig. 2. Proposed malware detection framework based on *CNN-INCA*

4 Results Analysis

This section demonstrates the experimental analysis of the proposed *CNN-INCA-MD* framework for malware detection. Experimental work used python programming language on Intel core i3-2330M, processor 2.20GHz and 8 GB RAM. The proposed work used *IoT* malware dataset collection of benign samples (2486) and malware samples (14733) collected from VirusTotal and VirusShare [19] for performance evaluation and comparison of presented malware detection framework.

The dataset is divided into train and test ratio of 70: 30. The model is trained using train data which contains the total 12049 malware and benign samples, and tested using test samples of 5170 of benign and malware samples. The proposed model classify malware with higher accuracy, precision and recall. The model is trained with four convolutional layer, four pooling layer, four batch normalization and leaky Relu with different number of filter size like 32, 64, 128, pool size is 2×2, stride size 2×2, kernel size 3×3. After fourth layer of convolution the extracted feature is passed through *INCA* module and fully connected layers. The fully connected layer consists of flatten layer, dense layer, dropout and softmax function which is used for malware classification. The model is compiled with adam optimizer to reduce the overfitting and minimize the loss of model. The model is run by taking number of different hyper parameter as mentioned Table 1 and epoch 30 with Adam optimizer, a batch size of 64 and learning rate of 0.0001. The training and testing performance of the model in terms of confusion matrix, model accuracy and model loss are mentioned in Figs. 3, 4 and 5. The accuracy of proposed model is 96.98%, precision is 96.68%, recall is 96.67% and f1score is 96.66% which is better than the existing *DL* models as compared in Table 2. The obtained results demonstrates the presented model is robust and scalable for huge amount of dataset for malware detection for *IoMT* environment.

5 Performance Evaluation Metrics

This work used four evaluation matrices such as precision (*PS#*), recall (*RE#*), f1-score (*FS#*), and accuracy (*ACC**) to assess performance mentioned in Eq. (1)-Eq. (4). These are: True Positives (T_X) denotes the correctly predicted malware samples, False Positive (F_Y) denotes the benign or malware samples that are misclassified by the proposed system, True Negative (T_Z) denotes the malware or benign samples that are correctly classified, and False Negative (F_W) instances denotes samples that are misclassified as malware or benign.

$$ACC* = \frac{T_X + T_Z}{T_X + T_Z + F_Y + F_W} \tag{1}$$

$$PS\# = \frac{T_X}{T_X + F_Y} \tag{2}$$

$$RE\# = \frac{T_X}{T_X + F_W} \tag{3}$$

$$FS\# = \frac{2 \times (RE\# \times PS\#)}{RE\# + PS\#} \tag{4}$$

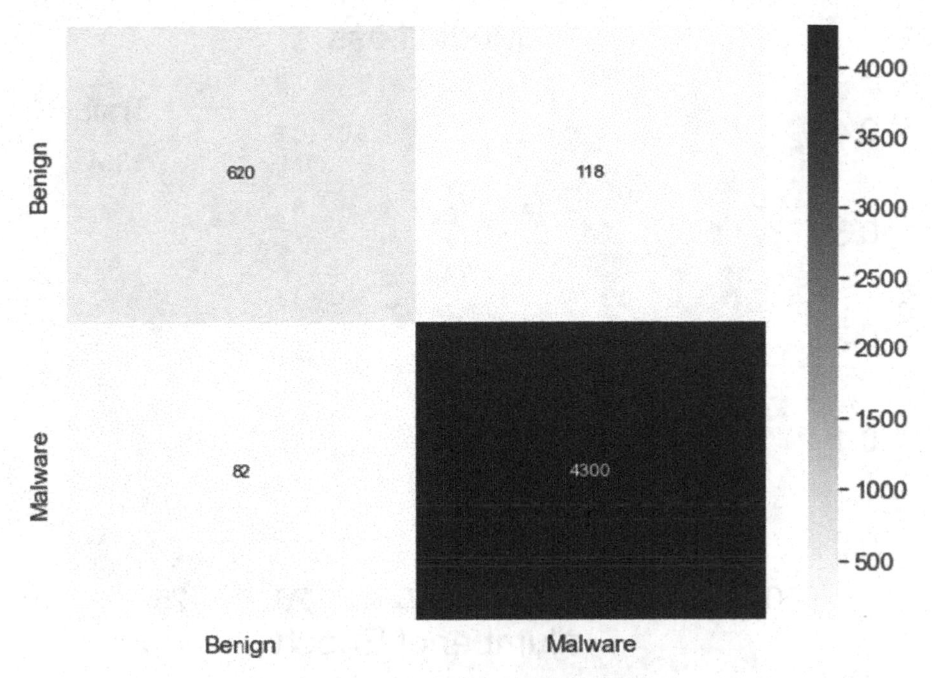

Fig. 3. Confusion matrix of proposed

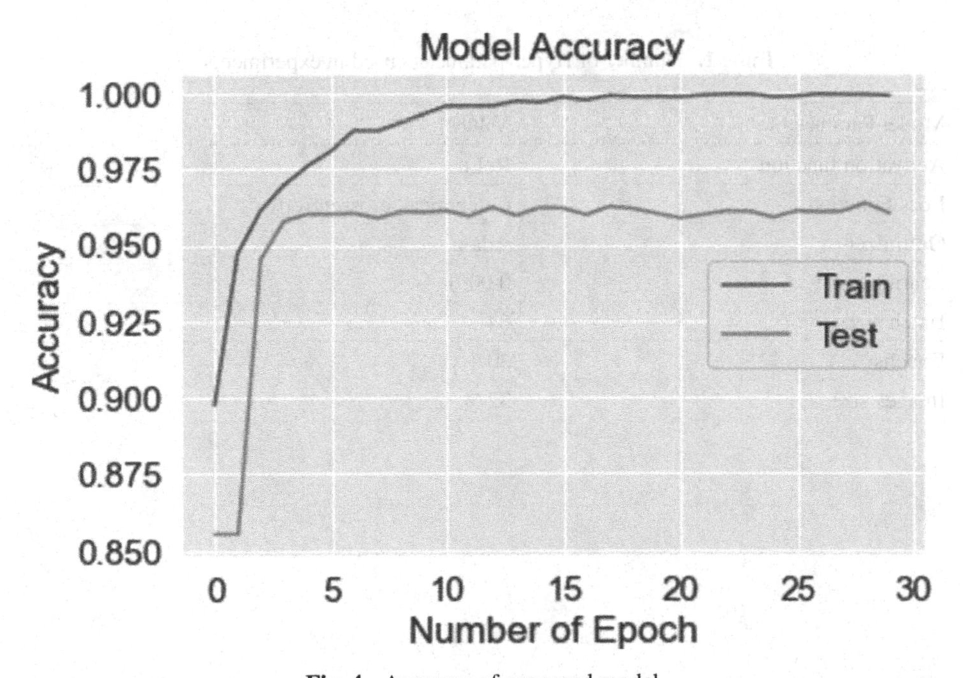

Fig. 4. Accuracy of proposed model

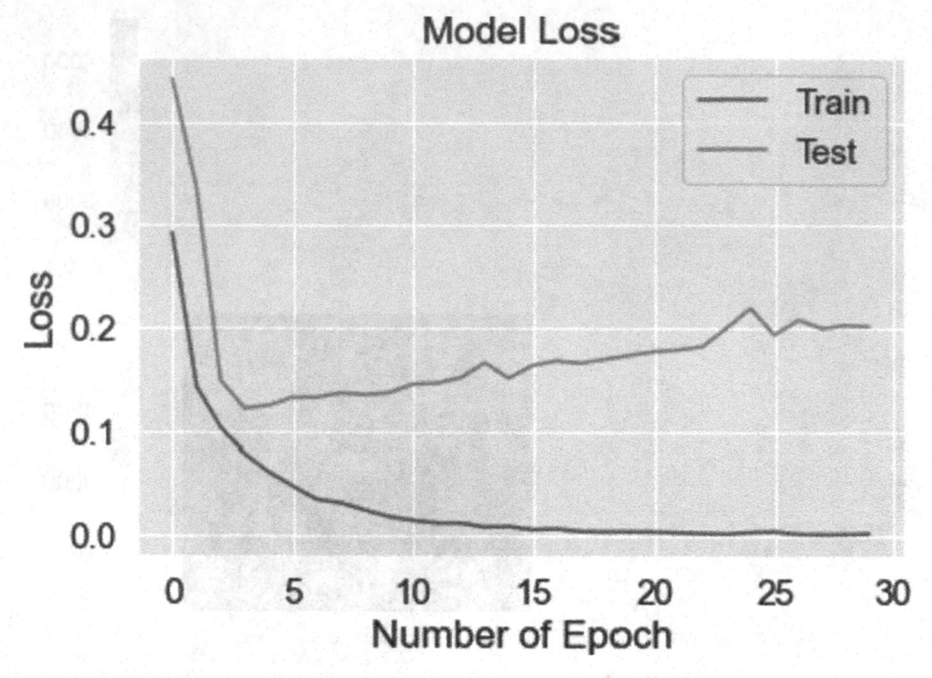

Fig. 5. Loss of proposed model

Table 1. Number of Hyper-parameters used in experiments

Model Parameters	Values
Activation function	ReLu
Loss Function	Categorical_crossentropy
Optimizer	Adam
Learning rate	0.0001
Batch size	64
Epochs	30
Images size	32 × 32

Table 2. Results comparison of proposed model with existing models

Method	Year	PS# (%)	RE# (%)	FS# (%)	ACC*(%)
Lightweight CNN [20]	2018	--	--	--	94.00
Fast Fuzy Pattern Tree [21]	2019	82.78	81.31	79.25	93.13
DexCRNN_BiGRU [22]	2020	95.40	96.20	95.80	95.80
Deep Neural Network [23]	2021	95.24	96.15	95.69	95.83
A-M-CNN-B-LSTM [14]	2022	96.00	95.00	95.00	95.00
Proposed *CNN-INCA-MD* Model		**96.68**	**96.67**	**96.66**	**96.98**

6 Conclusion

The increasing growth in *IoMT*-based services in smart healthcare is most targetable by malware attackers due to weak security, poor password, and unawareness of security by the medical staff, patients and doctors. Different number of *AI* (*ML* and *DL*) based malware detection framework have been proposed and achieved the better performance. However most of conventional methods based on signature pattern that are ineffective and incapable to detect new malware. In order to detect known and unknown malware. In this paper, we proposed a novel three-phase *CNN-INCA-MD* framework for *IoMT* environment. The *CNN* is used for deep feature extraction from input malware samples and *INCA* is used to select optimal feature or most discriminative features from feature extraction module to improve the accuracy of malware detection. The performance of the model can be affected by a smaller dataset which may degrades training and testing accuracy of the model. The future scope is to extend the work by using fog computing and different deep learning architecture for detecting new variants of malware.

References

1. Swarna Priya, R.M., et al.: An effective feature engineering for DNN using hybrid PCA-GWO for intrusion detection in IoMT architecture. Computer Communications **160**, 139–149 (2020)
2. Smmarwar, S.K., Gupta, G.P., Kumar, S.: Deep malware detection framework for IoT-based smart agriculture. Comput. Electr. Eng. **104**, 108410 (2022)
3. Kumar, P., Gupta, G.P., Tripathi, R.: An ensemble learning and fog-cloud architecture-driven cyber-attack detection framework for IoMT networks. Comput. Commun. **166**, 110–124 (2021)
4. Smmarwar, S.K., et al.: An optimized and efficient android malware detection framework for future sustainable computing. Sustainable Energy Technologies and Assessments **54**, 102852 (2022)
5. Smmarwar, S.K., Gupta, G.P., Kumar, S.: Design of a Fused Triple Convolutional Neural Network for Malware Detection: A Visual Classification Approach. International Conference on Advances in Computing and Data Sciences. Springer, Cham (2021)

6. Smmarwar, S.K., Gupta, G.P., Kumar, S.: A hybrid feature selection approach-based android malware detection framework using machine learning techniques. Cyber Security, Privacy and Networking, pp. 347–356. Springer, Singapore (2022)
7. Ravi, V., et al.: A multi-view attention-based deep learning framework for malware detection in smart healthcare systems. Computer Communications **195**, 73–81 (2022)
8. Wazid, M., et al.: IoMT malware detection approaches: analysis and research challenges. IEEE Access **7**, 182459–182476 (2019)
9. Alsubaei, F., et al.: IoMT-SAF: Internet of medical things security assessment framework. Internet of Things **8**, 100123 (2019)
10. Jeon, J., Park, J.H., Jeong, Y.-S.: Dynamic analysis for IoT malware detection with convolution neural network model. IEEE Access **8**, 96899–96911 (2020)
11. Liaqat, S., et al.: SDN orchestration to combat evolving cyber threats in Internet of Medical Things (IoMT). Computer Communications **160**, 697–705 (2020)
12. Saheed, Y.K., Arowolo, M.O.: Efficient cyber attack detection on the internet of medical things-smart environment based on deep recurrent neural network and machine learning algorithms. IEEE Access **9**, 161546–161554 (2021)
13. Khan, S., Akhunzada, A.: A hybrid DL-driven intelligent SDN-enabled malware detection framework for Internet of Medical Things (IoMT). Comput. Commun. **170**, 209–216 (2021)
14. Ravi, V., Pham, T.D., Alazab, M.: Attention-based multidimensional deep learning approach for cross-architecture IoMT malware detection and classification in healthcare cyber-physical systems. IEEE Transactions on Computational Social Systems (2022)
15. Naeem, H., Bin-Salem, A.A.: A CNN-LSTM network with multi-level feature extraction-based approach for automated detection of coronavirus from CT scan and X-ray images. Applied Soft Computing **113**, 107918 (2021)
16. Naeem, H., et al.: Malware detection in industrial internet of things based on hybrid image visualization and deep learning model. Ad Hoc Networks **105**, 102154 (2020)
17. Karadal, C.H., et al.: Automated classification of remote sensing images using multileveled MobileNetV2 and DWT techniques. Expert Systems with Applications **185**, 115659 (2021)
18. Aslan, N., et al.: Multi-classification deep CNN model for diagnosing COVID-19 using iterative neighborhood component analysis and iterative Relief F feature selection techniques with X-ray images. Chemometrics and Intelligent Laboratory Systems **224**, 104539 (2022)
19. Dib, M., et al.: A multi-dimensional deep learning framework for iot malware classification and family attribution. IEEE Trans. Netw. Ser. Manage. **18**(2), 1165–1177 (2021)
20. Su, J., et al.: Lightweight classification of IoT malware based on image recognition. In: 2018 IEEE 42Nd annual computer software and applications conference (COMPSAC), Vol. 2. IEEE (2018)
21. Dovom, E.M., et al.: Fuzzy pattern tree for edge malware detection and categorization in IoT. Journal of Systems Architecture **97**, 1–7 (2019)
22. Ren, Z., et al.: End-to-end malware detection for android IoT devices using deep learning. Ad Hoc Networks **101**, 102098 (2020)
23. Lu, N., et al.: An efficient combined deep neural network based malware detection framework in 5G environment. Computer Networks **189**, 107932 (2021)

An Artificial Intelligence-Driven Deep Learning Model for Chest X-ray Image Segmentation

Nillmani and Neeraj Sharma(✉)

School of Biomedical Engineering, Indian Institute of Technology (Banaras Hindu University), Varanasi 221005, India

neeraj.bme@iitbhu.ac.in

Abstract. Artificial Intelligence and CAD systems are becoming highly popular in medical diagnosis. The application of AI and deep learning in radiology is revolutionizing the medical industry with fast and accurate diagnosis. The Chest X-ray is one of the most significant radiological diagnostic methods being used for its easy availability, cost effectivity, and low radiation doses. The application of deep learning methods in chest X-rays has shown tremendous success in lesion detection. However, the chest-X ray contains a large non-region of interest in the form of the background that interrupts the AI system for accurate lesion detection. Towards the motive of solving the problem, this work proposes a robust and accurate deep learning-based UNet segmentation model to segment the region of interest, i.e., the lung region, and remove the background present in the X-ray images. Our model can successfully and accurately segment chest X-ray images. The model performed with an accuracy of 96.35% with dice coefficient and Jaccard index of 94.88% and 90.38%, respectively. Performing with high accuracy, dice, and Jaccard, our system proves its efficacy and robustness for efficiently segmenting the lung region in purpose for further diagnosis of numerous lung diseases, including COVID-19 and other pneumonia.

Keywords: Artificial intelligence · Deep learning · Chest X-ray · Segmentation · UNet

1 Introduction

In recent years AI-CAD (Artificial Intelligence based Computer-aided diagnosis) systems are widely being used in the medical field for the diagnosis of several diseases [1]. The AI-CAD system shows very accurate diagnoses with fast results and easily accessible methods [2]. The system is becoming popular and growing exponentially in developing countries where the lack of radiologists and rush in hospitals are major challenges [3]. Several radiological assessments such as MRI, Ultrasound, CT scan, mammograms, and X-rays are widely used for tumor detection [4, 5]. Cardiovascular disease diagnosis [6, 7], Plaque detection [8, 9], pneumonia detection [10, 11], and much more. Comparing the radiological techniques, MRI and CT are costly and have higher radiation doses, whereas X-rays are comparatively cost-effective, easily available, and

B. K. Singh et al. (Eds.): ICBEST 2023, CCIS 2003, pp. 107–116, 2024.
https://doi.org/10.1007/978-3-031-54547-4_9

have low radiation doses [12]. Also, the X-ray provides quick and instant results compared to CT and MRI. Therefore the advantages of X-rays over CT and MRI make them more popular for medical imaging, AI, and Computer-based diagnosis, especially for pneumonia detection [13]. Traditionally Chest X-rays (CXRs) are widely used for the diagnosis of pneumonia, tuberculosis, and other lung infections. The CXR also has been successfully used for AI-derived deep learning approaches based detection of several diseases such as COVID-19, tuberculosis, viral/bacterial pneumonia, lung tumor, and much more [14]. However, the CXR also contains a large area of non-region of interest that may interrupt the AI models for the accurate diagnosis of diseases [15]. Therefore it is highly recommended to separate the lung area from the whole CXR before the AI-based classification or detection approach implementation [16]. In this work, we have presented an AI-based deep learning model for the automatic segmentation of CXR images to segment the lung region for further detection of COVID-19, pneumonia, and other diseases infecting the lung.

2 Related Works

Previously several researchers have introduced and implemented different methods to segment chest X-ray images. Many of them have reported significant outcomes. Candemir et al. [17] presented a nonrigid registration-based segmentation model. The system uses an image retrieval-driven patient-specific adaptive model to detect the lung boundaries in chest X-ray images. They demonstrated an accuracy of 95.4% on the JSRT dataset and 94.1% on the Montgomery dataset. Ngo et al. [18] implemented the blend of a distance- regulated level set and deep belief system for the segmentation of CXR images taken from the JSRT dataset. Their model demonstrated an accuracy of 96.5%. Mittal et al. [19] applied an encoder decoder-based deep neural network for the segmentation of the CXR images taken from the JSRT and Montgomery datasets. Their model demonstrated an accuracy and Jaccard index of 98.73% and 95.10%, respectively. Hooda et al. [20] formulated a new CNN-based deep neural network to segment the CXRs from the JSRT dataset. They got an accuracy and Jaccard index of 98.92% and 95.88%, respectively. Saidy et al. [21] implemented an encoder- decoder approach to derive deep CNN and segmented the CXRs from the JSRT dataset. They got the test results with a Dice coefficient of 96%. Gaal et al. [22] formulated a new neural network and implemented it to the JSRT dataset. Their model performed with a Dice coefficient of 97.5%. Zhang et al. [23] implemented a modified UNet model with dual encoders for the segmentation of the Shenzhen and Montgomery datasets. Their system demonstrated an accuracy of 98.04%. The model also revealed a Dice of 96.67% and an AUC of 0.98. Liu et al. [24] formulated a modified UNet by using pre-trained EfficientNetB4 as the encoder and LeakyReLU activation function in the decoder part. They tested the network on JSRT and Montgomery datasets separately. They achieved the accuracy, dice, and Jaccard of 98.55%, 97.92%, and 95.73% on JSRT and 98.94%, 97.82%, and 95.55% on Montgomery datasets, respectively. Chandra et al. [25] proposed a multistage superpixel classification-based method for disease localization and severity detection in CXR images. The method was tested on the Montgomery dataset. Their system performed with the average Jaccard Index of 82% and Pearson's correlation coefficient of 0.95. Chandra et al. [26] presented a context-aware-adaptive scan algorithm to scan and correct the artifacts present along inner lung boundaries in CXRs. The algorithm was tested on the Montgomery dataset with a significant average segmentation accuracy improvement of 2.5%.

3 Methodology

Figure 1 represents the step-wise schematic diagram showing the overall methodology opted in the study. First, we utilized the "Chest X-ray masks and labeled" dataset [27] for the training of the segmentation model that was UNet. We utilized the images for our experiment in a manner of 70:20:10 ratio. 70% for training, 20% for validation, and 10% for testing the model. We implemented a five-fold cross-validation approach for model training. After training, we evaluated our model using testing accuracy, dice, Jaccard, AUC (Area-under-the- curve), and ROC (Receiver operating characteristic). Thereafter we applied the model for the segmentation of the "COVID-19 radiography database" [28], having a collection of >18000 CXR images to test the efficacy and feasibility of the model on larger databases.

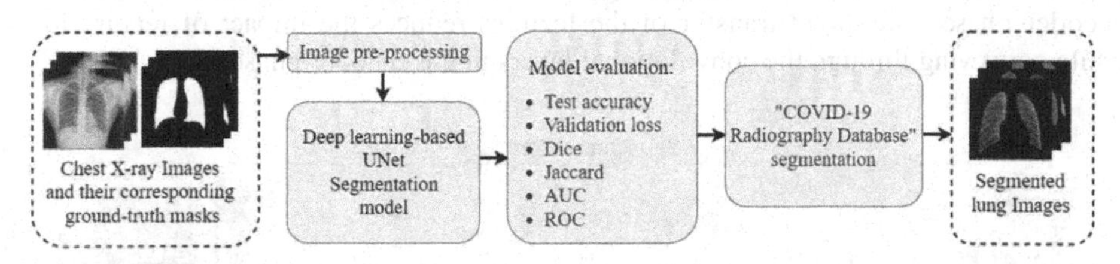

Fig. 1. The step-wise schematic diagram showing the overall methodology.

3.1 Dataset

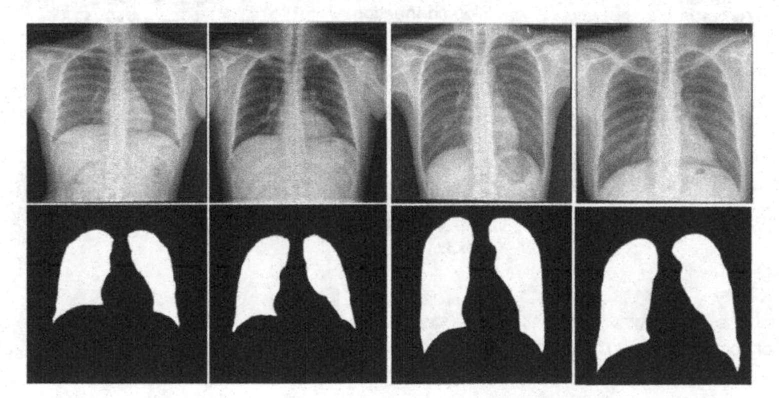

Fig. 2. Sample CXR images and their corresponding masks.

The dataset we utilized for the segmentation model training was “Chest X-ray masks and labeled” [27]. The dataset had a unique combination of CXRs and their corresponding masks. The total number of images within the dataset was 1408. Half, i.e., 704, are raw CXRs, and the left 704 are their corresponding masks. The dataset was taken from the open source, i.e., Kaggle. Fig. 2 represents the sample images of raw CXR and masks from the dataset. The second dataset utilized to test the performance of our model on larger datasets was the “COVID-19 radiography database”. The dataset contains >18000 CXRs in three classes: COVID-19, pneumonia, and normal.

3.2 The Proposed Model and Its Architecture

Figure 3 shows the complete architecture of the UNet model. The neural nodes in UNet are arranged in an encoder-decoder manner [29]. The encoder part of the network extracts the features from the images through several convolutional layers and transfers them to the next layer as input features for them. The decoder part of the network reconstructs the images into their original resolution with binary pixel value, i.e., 0 or 1, converting the features < 0.5 to 0 and < 0.5 to 1. In this way, the network generates the masks for each of the X-ray images. Both encoder decoder layers are connected through skip connections. The skip connection directly transfers the features to the corresponding decoder phase. The direct transfer of the features reduces the impact of feature loss while conveying through the convolutional layers in the encoder phase [30].

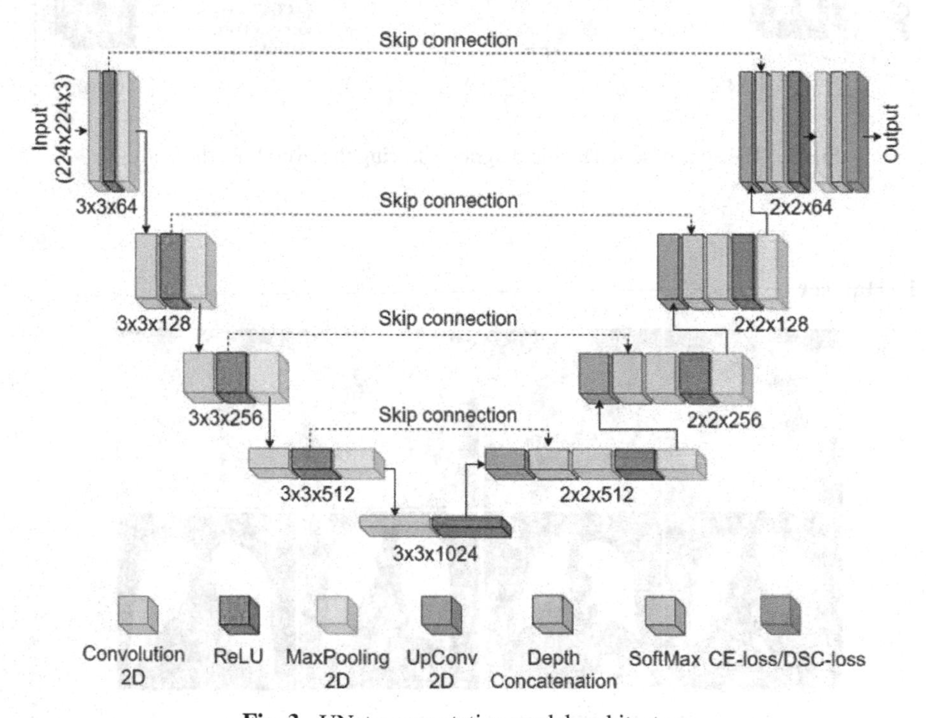

Fig. 3. UNet segmentation model architecture.

3.3 Training Parameters

For the segmentation model training, we applied and tested different combinations of parameters and loss functions and compared the performance. After training with different parameters, we selected the best combination providing higher accuracy and stable results. The best-performed learning rate for training the model was 0.001. The batch size for both training and validation was four images per batch. The loss function was the cross-entropy loss function and could be represented as the equation (1).

$$L_{ce} = [(y_i \times \log a_i) + (1 - y_i) \times \log(1 - a_i)] \tag{1}$$

Here, yi is the input GT label 1, (1 − yi) is GT label 0, and ai represents the Softmax classifier probability.

The whole experimentation was accompanied using Python 3.8. The workstation utilized for conducting the experiments has the following configurations: Processor = 8th generation Intel core i7, RAM = 16 GB, GPU = 8GB of NVIDIA Quadro P4000.

3.4 Performance Evaluation Metrics

The analysis of the performance of models is very important before establishing their reliability. The performance of the model was evaluated using several parameters to establish its reliability of the model. The ground truth (GT) masks and model-generated (AI) masks were compared to find out the test accuracy, Dice coefficient, Jaccard index, ROC, and AUC. The mathematical equation (2) - (4) represents each evaluation metric.

$$Accuracy = \frac{TP + TN}{(TP + FN) + (FP + TN)} \tag{2}$$

$$Jaccard\ index = \frac{TP}{(TP + FN + FP)} \tag{3}$$

$$Dice\ Coefficient = \frac{(2 * TP)}{(2 * TP + FN + FP)} \tag{4}$$

Here TP: True Positive, TN: True Negative, FP: False Positive, FN: False Negative.

4 Results and Discussion

We calculated the performance metrics of our segmentation model by comparing the ground truth (GT) masks and UNet-generated (AI) masks. The test accuracy achieved by the model was 96.35%. The validation loss during the model training was 0.15. The dice coefficient by the model was 94.88%. The Jaccard Index was 90.38%, and AUC was 0.99 with a p-value <0.001. The performance metrics show that our model performed outstandingly for the segmentation of CXRs.

Figure 4 represents the comparison of the GT and AI masks. As the figure reveals, row (a) shows the original CXR images from the test dataset. Row (b) shows the corresponding GT masks annotated by radiologists. Row (c) denotes the UNet-generated AI

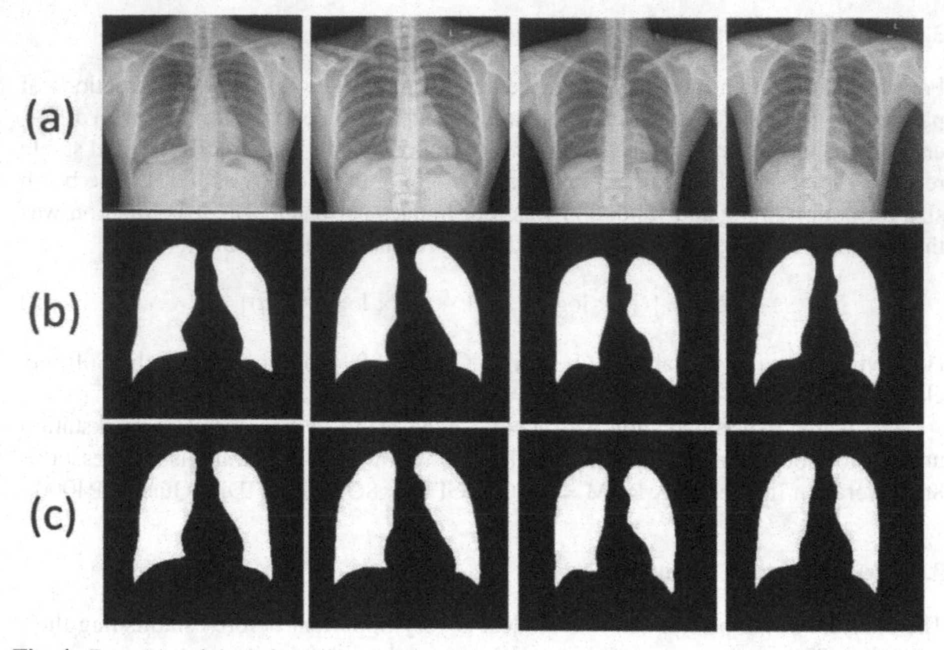

Fig. 4. Row (a): original chest X-ray images, (b): Corresponding ground truth masks, (c): Masks generated by the UNet model.

masks. The figure demonstrates the AI and GT masks are very similar, and difficult to find out differences between them.

Figure 5 represents the ROC and AUC for the UNet model. The ROC is the graphical representation of the true positive rate against the false positive rate. AUC is the numerical value for the area under the ROC curve. The value of AUC ranges between 0 and 1. The higher the AUC, means the model is performing better. The AUC by our segmentation model was 0.99, which establishes our model's excellence in performance and robustness.

Finally, we applied our segmentation model on the "COVID-19 radiography database" that has >18000 chest X-ray images to test our model's performance on a more extensive database. However, we did not have any ground truth masks for the dataset therefor; we were unable to do a mask comparison test to get the dice coefficient or Jaccard index. However, we analyzed the masks manually and compared them against the original CXR images. The results were excellent, with accurate and proper AI masks covering the lung region and eliminating the non-region of interest. Fig. 6 represents the sample of segmented lung region of CXR images from the "COVID-19 radiography database". First, we applied our trained UNet model for the mask generation of the raw CXR images from the database. Thereafter using the mask, we segmented the required lung region from all CXR images. The top row in the figure represents the raw CXR images. The middle and bottom rows represent the corresponding UNet-generated AI masks and segmented lung regions, respectively.

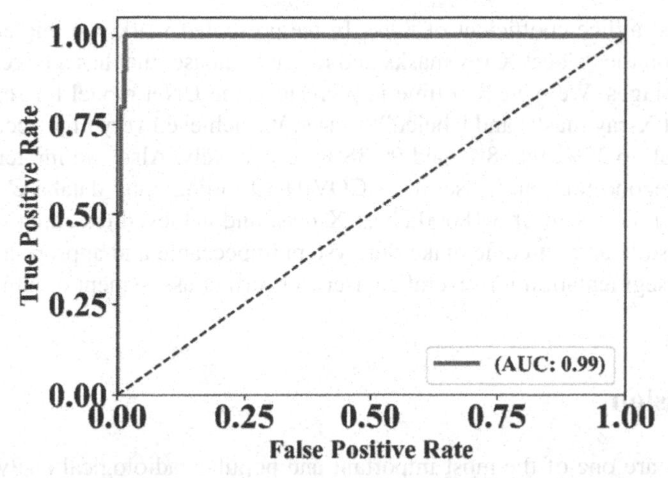

Fig. 5. The ROC and AUC curves for UNet model.

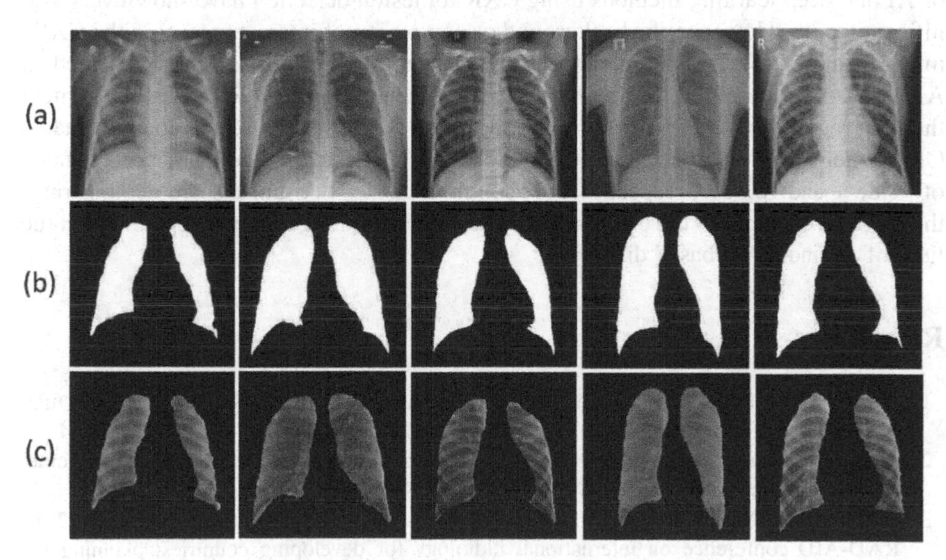

Fig. 6. The sample of segmented lung regions, row (a): original CXR image, (b): corresponding UNet-generated AI masks, (c): the segmented lung region.

The UNet network has always proven its efficacy for segmentation as having an encoder-decoder phase. Other researchers have also implemented the UNet for the segmentation of chest X-rays. However, they have usually used JSRT or Montgomery datasets that have a less number of images. Previously Souza et al. [31] applied AlexNet and ResNet-based networks to segment the chest X-rays from the Montgomery dataset, which contains a total of 138 CXR images. They achieved the Dice of 93.56% and Jaccard of 88.07%. Reamarron et al. [32] implemented the "Total variation-based active contour" method on JSRT and Montgomery datasets that have 247 and 138 CXRs, respectively.

They achieved a dice coefficient of 89%. In the presented work, we implemented the UNet model on the "Chest X-ray masks and labeled" dataset that has a large number of 704 X- ray images. We have first time implemented the UNet model for segmentation on the "Chest X-ray masks and labeled" dataset. We achieved very high accuracy, dice, and Jaccard of 96.35%, 94.88%, and 90.38%, respectively. Also, we implemented our model on a second unseen dataset, the "COVID-19 radiography database" which has an enormous image size of >18000 chest X-rays, and achieved excellent results. The significant results and outcome make our system impeccable and appropriate for chest X-ray image segmentation for several datasets for further assessment of numerous lung diseases.

5 Conclusion

Chest X-rays are one of the most important and popular radiological analyses for the detection of several diseases, specially COVID-19 or other pneumonia. The applications of AI and deep learning methods using CXR for lesion detection have shown very significant results. However, the background or non-region of interest present in the CXRs may misguide the AI system in detecting the lesion. In this work, we have presented an AI-based deep learning UNet model for the segmentation of CXR images to segment the lung region from the X-ray so that the accurate diagnosis of diseases could be made. Our model achieved an accuracy of 96.35% with a dice coefficient and Jaccard index of 94.88% and 90.38%, respectively. The high-accuracy dice and Jaccard demonstrates the model's efficacy for the efficient segmentation of the lung region and potential in the field of AI and CAD-based diagnosis.

References

1. Fujita, H.: AI-based computer-aided diagnosis (AI-CAD): the latest review to read first. Radiol. Phys. Technol. **13**(1), 6–19 (2020). https://doi.org/10.1007/s12194-019-00552-4
2. Yanase, J., Triantaphyllou, E.: A systematic survey of computer-aided diagnosis in medicine: Past and present developments. Expert Syst. Appl. **138**, 112821 (2019)
3. Mollura, D.J., Mazal, J., Everton, K.L., Group, R.-A.C.W.: White paper report of the 2012 RAD-AID conference on international radiology for developing countries: planning the implementation of global radiology. J. American Colle. Radiol. **10**(8), 618–624 (2013)
4. Tripathi, S., Verma, A., Sharma, N.: Automatic segmentation of brain tumour in MR images using an enhanced deep learning approach. Comp. Metho. Biomecha. Biomed. Eng. Imag. Visualiz. **9**(2), 121–130 (2021)
5. Tripathi, S., Sharma, N.: Computer-based segmentation of cancerous tissues in biomedical images using enhanced deep learning model. IETE Technical Review, 1–15 (2021)
6. Jain, P.K., Sharma, N., Kalra, M.K., Johri, A., Saba, L., Suri, J.S.: Far wall plaque segmentation and area measurement in common and internal carotid artery ultrasound using U-series architectures: an unseen artificial intelligence paradigm for stroke risk assessment. Comput. Biol. Med. **149**, 106017 (2022)
7. Jain, P.K., Sharma, N., Giannopoulos, A.A., Saba, L., Nicolaides, A., Suri, J.S.: Hybrid deep learning segmentation models for atherosclerotic plaque in internal carotid artery B-mode ultrasound. Comput. Biol. Med. **136**, 104721 (2021)

8. Jain, P.K., et al.: Attention-Based UNet deep learning model for plaque segmentation in carotid ultrasound for stroke risk stratification: an artificial intelligence paradigm. J. Cardiovascular Develop. Disea. **9**(10), 326 (2022)
9. Jain, P.K., et al.: Automated deep learning-based paradigm for high-risk plaque detection in B-mode common carotid ultrasound scans: An asymptomatic Japanese cohort study. Int. Angiol. **41**, 9–23 (2021)
10. Nillmani, et al.: Segmentation-based classification deep learning model embedded with explainable AI for COVID-19 detection in chest X-ray scans. Diagnostics **12**(9), 2132 (2022). [Online]. Available: https://www.mdpi.com/2075-4418/12/9/2132
11. Nillmani, et al.: Four types of multiclass frameworks for pneumonia classification and its validation in X-ray scans using seven types of deep learning artificial intelligence models. Diagnostics **12**(3), 652 (2022). [Online]. Available: https://www.mdpi.com/2075-4418/12/3/652
12. Yan, C., Hui, R., Lijuan, Z., Zhou, Y.: Lung ultrasound vs. chest X- ray in children with suspected pneumonia confirmed by chest computed tomography: A retrospective cohort study. Exp. Ther. Med. **19**(2), 1363–1369 (2020)
13. Wielpütz, M.O., Heußel, C.P., Herth, F.J., Kauczor, H.-U.: Radiological diagnosis in lung disease: factoring treatment options into the choice of diagnostic modality. Dtsch. Arztebl. Int. **111**(11), 181 (2014)
14. Ilyas, M., Rehman, H., Naït-Ali, A.: Detection of covid-19 from chest x-ray images using artificial intelligence: An early review (2020). arXiv preprint arXiv:2004.05436
15. Wang, C., Xu, Z., Liu, H., Wang, Y., Wang, J., Tai, R.: Background noise removal in x-ray ptychography. Appl. Opt. **56**(8), 2099–2111 (2017)
16. Teixeira, L.O., et al.: Impact of lung segmentation on the diagnosis and explanation of COVID-19 in chest X-ray images. Sensors **21**(21), 7116 (2021)
17. Candemir, S., et al.: Lung segmentation in chest radiographs using anatomical atlases with nonrigid registration. IEEE Trans. Med. Imaging **33**(2), 577–590 (2013)
18. Ngo, T.A., Carneiro, G.: Lung segmentation in chest radiographs using distance regularized level set and deep-structured learning and inference. In: 2015 IEEE International Conference on Image Processing (ICIP), pp. 2140–2143. IEEE (2015)
19. Mittal, A., Hooda, R., Sofat, S.: LF-SegNet: A fully convolutional encoder–decoder network for segmenting lung fields from chest radiographs. Wireless Pers. Commun. **101**, 511–529 (2018)
20. Hooda, R., Mittal, A., Sofat, S.: An efficient variant of fully-convolutional network for segmenting lung fields from chest radiographs. Wireless Pers. Commun. **101**, 1559–1579 (2018)
21. Saidy, L., Lee, C.-C.: Chest X-ray image segmentation using encoder-decoder convolutional network. In: 2018 IEEE International Conference on Consumer Electronics-Taiwan (ICCE-TW), pp. 1–2. IEEE (2018)
22. Gaál, G., Maga, B., Lukács, A.: Attention u-net based adversarial architectures for chest x-ray lung segmentation (2020). arXiv preprint arXiv:2003.10304
23. Zhang, L., Liu, A., Xiao, J., Taylor, P.: Dual encoder fusion u-net (defu-net) for cross-manufacturer chest x-ray segmentation. In: 2020 25th International Conference on Pattern Recognition (ICPR), pp. 9333–9339. IEEE (2021)
24. Liu, W., Luo, J., Yang, Y., Wang, W., Deng, J., Yu, L.: Automatic lung segmentation in chest X -ray images using improved U-Net. Sci. Rep. **12**(1), 8649 (2022)
25. Chandra, T.B., Singh, B.K., Jain, D.: Disease localization and severity assessment in chest x-ray images using multi-stage superpixels classification. Comput. Methods Programs Biomed. **222**, 106947 (2022)

26. Chandra, T.B., Verma, K., Jain, D., Netam, S.S.: Segmented lung boundary correction in chest radiograph using context-aware adaptive scan algorithm. In: Advances in Biomedical Engineering and Technology: Select Proceedings of ICBEST 2018, pp. 263–275. Springer (2021)
27. Pandey, N.: Chest X-ray Masks and Labels. https://www.kaggle.com/datasets/nikhilpandey360/chestxray-masks-and-labels, accessed 08 January 2022
28. Chowdhury, M.E.H.: COVID-19 Radiography Database. https://www.kaggle.com/datasets/tawsifurrahman/covid19-radiography-database, accessed 08 January 2022
29. Chen, H., Lin, H., Yao, M.: Improving the efficiency of encoder-decoder architecture for pixel-level crack detection. IEEE Access **7**, 186657–186670 (2019)
30. Li, X., Chen, H., Qi, X., Dou, Q., Fu, C.-W., Heng, P.-A.: H-DenseUNet: hybrid densely connected UNet for liver and tumor segmentation from CT volumes. IEEE Trans. Med. Imaging **37**(12), 2663–2674 (2018)
31. Souza, J.C., Diniz, J.O.B., Ferreira, J.L., da Silva, G.L.F., Silva, A.C., de Paiva, A.C.: An automatic method for lung segmentation and reconstruction in chest X-ray using deep neural networks. Comput. Methods Programs Biomed. **177**, 285–296 (2019)
32. Reamaroon, N., et al.: Robust segmentation of lung in chest x-ray: applications in analysis of acute respiratory distress syndrome. BMC Med. Imaging **20**(1), 1–13 (2020)

Deep Learning Approaches for Early Detection of Obstructive Sleep Apnea Using Single-Channel ECG: A Systematic Literature Review

Nivedita Singh(✉) and R. H. Talwekar

Government Engineering College, Sejbahar, Raipur, CG, India
nivedita.singhbit@gmail.com

Abstract. In this paper, we aimed to analyze various deep-learning approaches for obstructive sleep apnea (OSA) detection using single-channel ECG. The aim is to investigate an efficient and robust system for the early detection of OSA using single-channel ECG in different deep-learning approaches. The methodology we implemented was reviewing the literature in preferred reporting items for systematic reviews and meta-analyses (PRISMA) which includes research conducted during the last decade from 2012 to 2022. We explored various sources for collecting research articles relevant to OSA detection and then a total of 1110 papers are chosen. PRISMA framework facilitates the eligibility criteria to down-sample the articles which are most suitable for our review. Over a decade there is sharp growth in deep learning-based classification techniques for sleep apnea detection and particularly after the year 2017. In the year 2022 drastic increase in the using deep learning has been reported. During the years 2017 to 2022, deep learning approaches were used for classification where LSTM, CNN, RNN, pre-trained networks, and hybrid deep neural architectures were implemented.

It is explored thoroughly in the research reported during the year 2022 to detect SA using ECG in deep learning since it is simple and requires fewer computational tasks. This paper will lead future researchers to pursue their work by also procuring relevant background knowledge to pursue an efficient method and approach for OSA detection using ECG modality. The best deep learning models were the hybrid model which combines both RNN and CNN for the robustness of the proposed model.

Keywords: convolution neural network · deep learning · electrocardiogram · polysomnography · sleep apnea · systematic literature review

1 Introduction

Obstructive sleep apnea (OSA) is a respiratory abnormality that occurs during sleep and may cause comorbidity when it could not be diagnosed and treated beforehand. Sleep apnea may cause circulatory, neurological, or cardiovascular disease. From the

B. K. Singh et al. (Eds.): ICBEST 2023, CCIS 2003, pp. 117–130, 2024.
https://doi.org/10.1007/978-3-031-54547-4_10

latest study, it was investigated that more than 13% of Indians suffer from SA while the percentage of males who have OSA is more as compared to the percentage of females. PSG is the gold standard to detect OSA but it required skilled and well-trained experts to record the data.

According to the researchers (Kocak et al., 2012), sleep disorders are classified into two categories as parasomnia and dyssomnia. (Rizal et al., 2020) reviewed various physiological signals for detecting SA with different signal preprocessing and post-processing techniques and compared their performance. Authors (Novák et al., 2008) proposed sleep apnea classification using Heart Rate Variability (HRV) in Long Short-Term Memory (LSTM) with the advent of a deep neural network. They (Song et al., 2016) studied a discriminative hidden Markov model (DHMM) and experimented on the temporal dependence of ECG signals which was based on a new algorithm and performed better than other models. The authors (Cheng et al., 2017) proposed deep learning to Research Information Systems machine learning method. A study reported (Pathinarupothi et al., 2017) that RNN is a robust method too. The increasing popularity of deep learning in the field of biomedicine (Banluesombatkul et al., 2018) studied deep learning-based severity detection of SA and achieved high performance. The research (Urtnasan et al., 2018) presented the detection of SA using a convolutional neural network(CNN) with high precision and accuracy as compared to ML. Further work to characterize SA as normal, hypopnea, and apnea was investigated by them. (Cen et al., 2018) They (Warrick et al., 2019) discovered a hybrid scattering-LSTM model for non-apnea arousal and achieved the second-highest result overall. (Chaw et al., 2019) were studied detection of SA using the CNN model based on SPO2 and showed better performance. (Liang et al., 2019) were applied novel algorithm Nesterov accelerated gradient (NAG) algorithm in CNN and LSTM to detect SA.

(Erdenebayar et al., 2019) demonstrated various techniques of deep neural network (DNN) recurrent as well as CNN. They concluded that 1-D CNN and GRU had performed better than other classifiers. (T. Wang et al., 2019) studied According to the authors time window-based artificial neural network method has shown better performance than non-time window-based methods. (H. Y. Chang et al., 2020) worked on the data for both per record as well as per minute and compared the result while implementing deep learning and outperforming previous methods. (X. Wang et al., 2020) CNN is the most popular among all other algorithms of deep learning. (H. C. Chang et al., 2020) presented LSTM for SA detection as well as severity classification accurately. (Q. Chen et al., 2020) investigated transfer learning algorithm, multi-resolution ResNet which is more accurate and efficient than ResNet (Almutairi et al., 2021) studied SA detection using ECG signals in CNN, CNN with LSTM, and CNN with GRU. And concluded that the hybrid method is more robust than other methods. (Pombo et al., 2020) presented a study to detect SA using an S-Golay filter for feature extraction of HRV and ECG processed and analyzed in the minute-by-minute segments. (Urtnasan et al., 2020) studied achieved superior performance for GRU using ECG signal as compared to other conventional methods. (Steenkiste et al., 2020) investigated fusion methods with CNN as well as LSTM deep learning models with backward shortcut connections which performed better not only than another deep learning model but even other fusion methods. (Chen et al., 2020a, 2020b) studied CNN-LSTM algorithm for screening OSA and leave-one-subject-out (LOSO) cross-validation evaluation was conducted had improved the system performance remarkably. (Bai et al., 2021) proposed smart device for ECG data applied CNN to detect SA which was low cost and easy without any compromise in performance. (Huang & Ma, 2021) demonstrated combination technique of both DL and conventional methods for SA detection and achieved excellent performance. (Zhang et al., 2021) investigated the DL algorithm the CNN and LSTM for SA detection with Cohen's kappa coefficient of 0.92. (Loh et al., 2021) studied EEG-based 1D-CNN model for OSA detection and performed efficiently using cyclic alternating pattern (CAP) for robust and accurate classification. (X. Chen et al., 2021) had introduced a multi-scaled fusion network SE-MSCNN which is based on ECG signals acquired by wearable devices very easily in per-segment and per-recording accuracy is 90.64% and 100% respectively. (Mukherjee et al., 2021) worked on ensemble techniques with CNN and fusion of CNN and LSTM model to detect SA. (Faust et al., 2021) proved that Computer-Aided Diagnosis of OSA the model was designed with LSTM and validated with 10-fold cross-validation and hold-out validation was applied. (Rajawat et al., 2021) worked in new dimensions that is CNN-LSTM method for classification of contact based and non-contact based from subjects using torso and head for OSA detection. (Gupta et al., 2021) smoothed Gabor spectrogram (SGS) framework was applied to pre-trained networks viz Squeeze-Net, Res-Net50 called obstructive sleep apnea convolutional neural network (OSACN-Net) and achieved better accuracy. (Bahrami & Forouzanfar, 2021) had investigated CNN, LSTM, bi-LSTM, GRU, and hybrid models. The best performance is achieved by, a hybrid CNN and LSTM network. (Sharan et al., 2021) studied 1-D CNN with residual connections and for loss measurement weighted cross-entropy was used. Bayesian optimization for fine-tuning of hyperparameters. (Hedman et al., 2021) used bidirectional gated recurrent unit (GRU) model and a bidirectional long short-term memory (LSTM) and achieved better performance. (Bernardini et al., 2021) presented novel CNN obtained satisfactory and most of them outperforms current state-of-the-art solutions. (Qian et al., n.d.) studied different CNN, RNN, LSTM, residual neural network (ResNet). (Nasifoglu & Erogul, 2021) compared various transfer learning DL techniques AlexNet, GoogleNet and ResNet18 to predict SA. (Teng et al., 2022) developed DL model to detect SA using ECG sensor smartphone which is low cost and easy to use. (Salari et al., 2022) evaluated studies using ML and DL algorithms using ECG. Support-Vector Machine (SVM) and CNN. Unlike other studies it was observed that ML based techniques performed better than RNN. (Qin & Liu, 2022) worked to obtain RR interval of each 1-min using Christov algorithm and adaptive synthetic (ADASYN) sampling method to handle imbalance learning problem. 1DCNN-RLM and was used for feature extraction and finally BiGRU-TDM to detect OSA. (Bahrami & Forouzanfar, 2022a, b) studied occurrence of apnea from single-lead ECG in RNN and achieved high performance using wearable device.

1.1 Objectives of the Study

The main objective of this paper is to explore various possibilities of research in the field of SA detection in deep learning approaches using ECG modalities. The focus is also to replace the PSG which is the conventional gold standard of SA detection with any new methods which would be cheaper, easier and robust too. This paper will extend guidance to the researchers to work in this particular field. This will also include the comparison of the performance of various works conducted in the research area. It is aimed to categorize the research into various deep learning approaches and also to investigate which algorithm is better along with their limitations. The review will be focusing on SA detection in deep learning techniques using ECG signals.

1.2 Contribution of the Paper

The main focus of this SLR is to explore various methods of detecting and classifying OSA using ECG signal. Among various physiological signal the ECG signal is only chosen to detect OSA is for the reason that, ECG is reflecting the behavior of our heart and pumping and physiology of heart thus affect the supply of oxygen which ultimately determine the apnea during sleep.

Deep learning classifiers are investigated over machine learning to reduce the complexity of the classification and also to reduce the processing time.

Deep neural network is categorized into two types recurrent neural network and convolution neural network, for our problem the convolution neural network is more suitable as the classification is performed on image data.

Also, since the image data size is large in number so it is preferred to apply DNN rather than ML to get better performance.

Keeping on these above mention conditions the present SLR is performed to get the optimized classifier to detect and classify the OSA at early stage so that it could not affect any cardiac functioning direct or indirectly.

2 Material and Methods

2.1 PRISMA Framework

The first step of SLR is to search the relevant articles on the research topic chosen using the most suitable keywords. The main keywords for the review are sleep apnea, electrocardiogram, deep learning, convolution neural network, and polysomnography. Majorly the search was performed using an application from various sources such as google scholar, Crossref, PubMed, and google search engine. Initially, 1110 research papers were included as per the search strategy and keywords, as well as through various sources such as PubMed, MDPI, Springer, IOP, Web of Science, ScienceDirect, IEEE Xplore, ScienceDirect, and Elsevier, etc.

After assuring the search total of 1110 records were chosen which will later be scrutinized using PRISMA architecture. The process flow has already shown in Fig. 1 the very first stage in screening is to remove all irrelevant articles which do not cover the research area defined. Therefore, a total number of 603 records were included in the screening phase while 407 records were excluded since they do not relate to the research area. Now from 603 records that were considered for further screening due to multiple factors such as modality, state of art, or any other 400 more records were found not eligible for further study. The next phase of eligibility testing applied to 203 numbers of records and found they do not meet the search criterion so 90 more records were

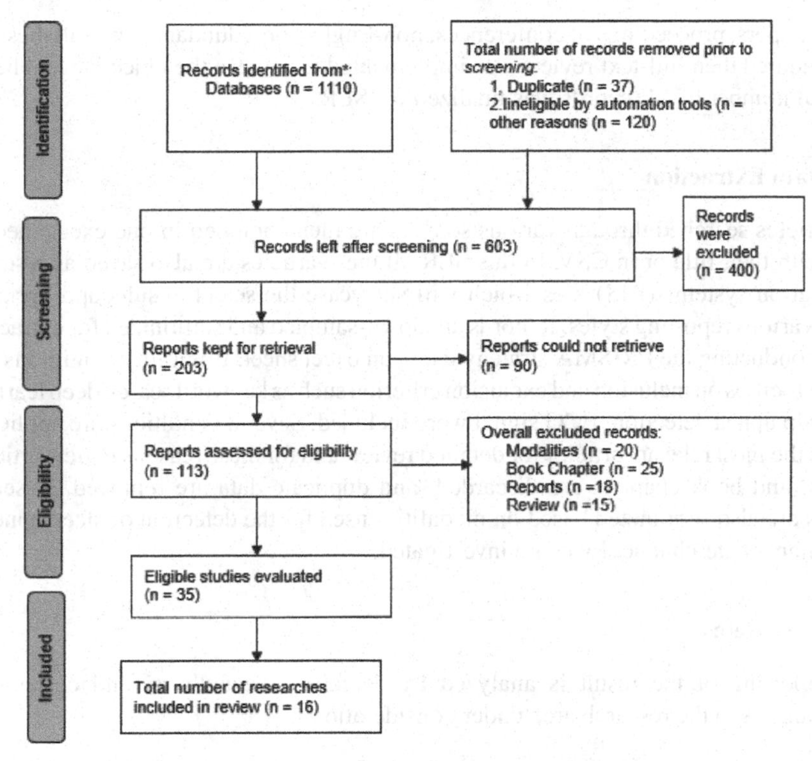

Fig. 1. The process flow of the PRISMA reporting style

excluded. The records which were found satisfactory for the next stage were 113. These records were investigated based on the title, abstract, and results achieved, and decided not to include 80 more records for the study.

A total of 33 works of literature are shortlisted after systematically reviewing the titles, abstracts, and detailed literature from all 113 articles and discarding irrelevant records. Records were found redundant, non-English, as well as review work so the last and final phase was applied to select an actual number of records that will be considered for a detailed review of research, which was 16 records. These numbers of records are satisfying quality assessment criteria to pursue SLR on them as full paper review.

2.2 Eligibility Criterion

The very first stage in this PRISMA methodology is to check whether a record selected fulfills the eligibility criterion sequentially. A total number of 1110 records were evaluated based on various aspects. The first concern in checking the eligibility is to decide whether the records are relevant to the topic or not. Once the topic is justified then the next step is to verify the narrow area of research matches the research area concerned. This is then followed by a systematic methodology to review in the best suitable manner to select the most relevant and specific records so that SLR will be conducted precisely which in return will lead the upcoming researcher in the field. The records excluded are review papers, proceedings of conferences, non-English or redundant. Once all these are implemented then full-text review was implemented to achieve the objective and hence the total number of 14 records was finalized for SLR.

2.3 Data Extraction

The articles searched through various sources are then compiled in one excel sheet as XLS with their title or in CSV. In this SLR all these articles are also saved as research information systems (RIS) files which will showcase the search results appropriately and in various reporting styles. Records are down-sampled and scrutinized for extraction while conducting the PRISMA strategy using an excel sheet. Eligibility conditions that mainly focuses on inclusion and exclusion criterion such as keyword search deep learning and sleep apnea detection, ECG signal were included. Several conditions are applied to choose the most relevant articles for detailed review and for these reasons review articles, reports, and book chapters are discarded, and duplicate data are removed. Research articles are also scrutinized based on modalities used for the detection of sleep apnea if other than single-channel ECG are investigated.

2.4 Data Items

The reporting of the result is analyzed by depicting the authors' and co-authors' contributions to the research area under consideration.

2.4.1 Search Based on Word Items

Figure 2 depicts the records based on word items popularly used in the records and hence filtered. The total number of records that are included in 106 to filter various word items and thus, the clusters formed in this way is 13. It is also very clear from the figure the basic word items are CNN, LSTM, ECG, SA, and deep learning.

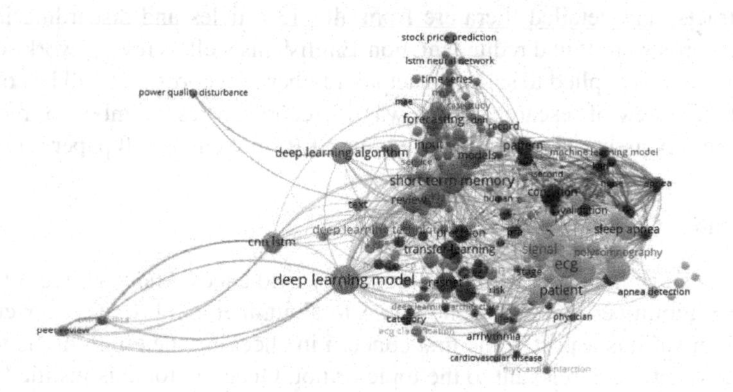

Fig. 2. Word item-based clusters

2.4.2 Clusters of Authors and Co-Authors

Figure 3 depicts various clusters formed according to authors and co-authors of selected records. This web of researchers is demonstrating the correlation and collaboration among them.

2.5 Quality Assessment or Risk of Bias

The quality assessment may be a risk of bias in this SLR article as the different sources of articles are limited and require a licensed version. These articles have been evaluated based on references such as physiological signals, methods, and accuracy. The reporting of results is also dependent on demographic and geographical area additionally, sample size, as well as age group. The risk of bias could be minimal while selecting inclusion and exclusion criteria critically. The main focus of this SLR is to perform a systematic review of SA detection using deep learning approaches.

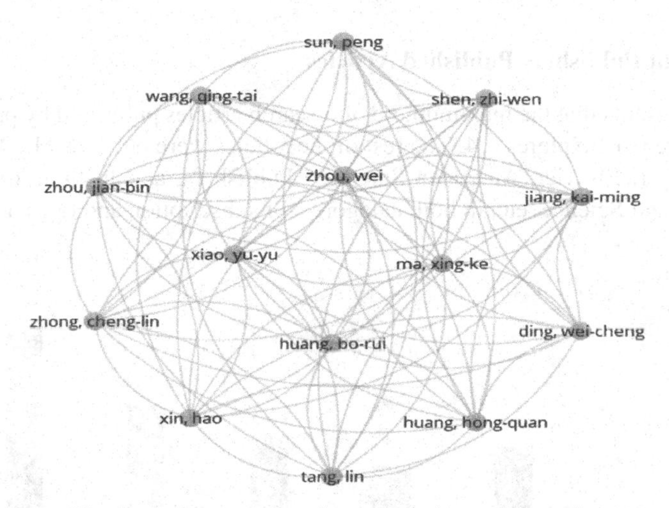

Fig. 3. Author and co-author correlation and clustering

3 Results

The advanced method of reviewing the research is very popular and very time-saving using the PRISMA methodology. In this paper, we implemented SLR based on PRISMA. The duration of the review under consideration is the last decade from 2012 to 2022. The popularity of research based on deep learning approaches in almost all domains of research has motivated researchers to explore the prospects in the area specified. This could be seen from the graphical representation from 2017 to 2021 and 2022. At present almost all domain researchers are working on deep learning.

The following are various criteria for reporting the results through descriptive analysis:

- Record distribution over the year.
- According to publishers.
- According to several citations.

3.1 Year-Wise Record Distribution

Figure 4 presents the no. of research reported for the present decade (2012–2022). From this graphical representation, it can be observed that the work done in the specific area has increased gradually from the year 2017 and increased remarkably from the year 2021 to 2022 till date. The tremendous popularity of deep learning has increased research work because of its ease, and robustness. OSA detection using ECG single lead signal in deep learning approaches requires extremely less computation and eventually involves less time.

3.2 Different Publishers Published Articles

Figure 5 represents that the maximum no of research articles published by publishers in the decade are IEEEXplorer is 4. The rest of the articles were published by MDPI, IOP, ACM, Springer, Elsevier, American Scientific Publishers, and Institute of Advanced Engineering and Science, etc. in both categories of subscription and open access.

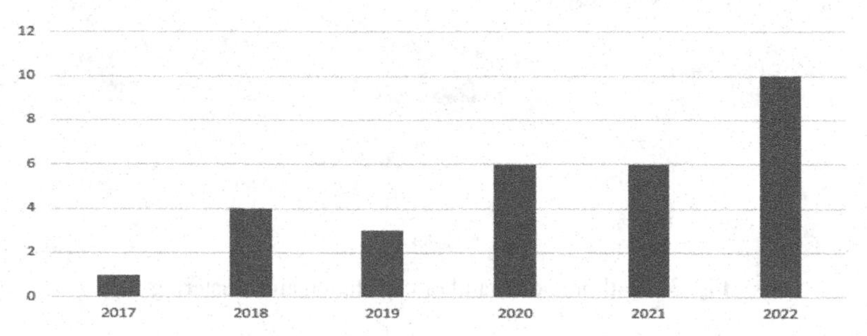

Fig. 4. Literature published during the decade

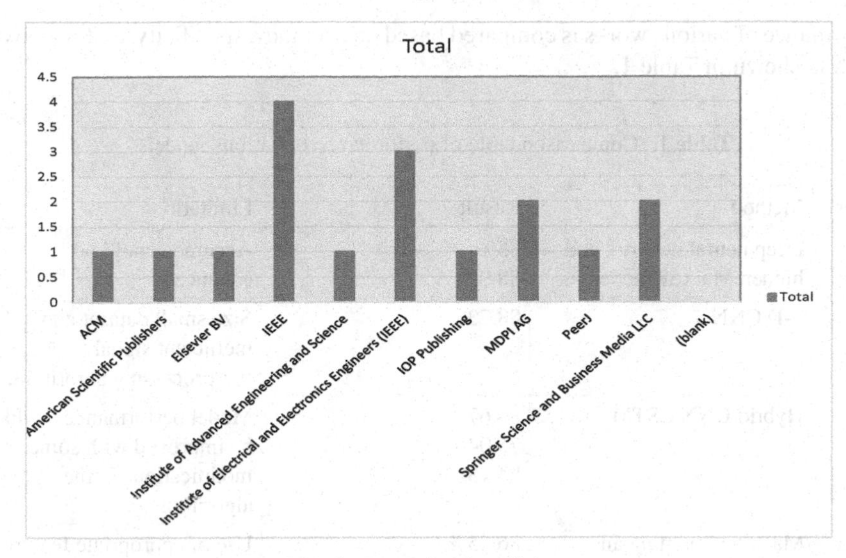

Fig. 5. Records of publishers

3.3 Number of Citations Per Record

Figure 6 depicts the total number of citations per record using deep learning approaches and in the detection of OSA using ECG modality. It may be depicted from Fig. 6 that one article has received a maximum of 73 citations similarly some other articles were cited 32, 22, 17, and 14 in number. The marginal number of citations 4, 3, 2, and 1 are shown for four to five articles. The last citation was shown for 9 articles amongst which most of the articles are published recently.

3.4 Result of Classifiers

In this section, we have reported key points of various deep learning approaches for SA detection with single-channel ECG modalities are discussed in this section. The

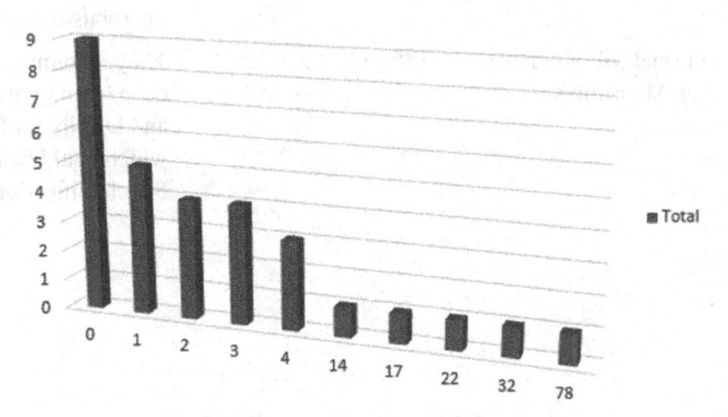

Fig. 6. Citation of records

performance of various works is compared based on accuracy, specificity, and sensitivity which is shown in Table 1.

Table 1. Comparison table of performance of various models

Paper	Method	Result	Limitations
[34]	Deep neural network and hidden Markov model	85% 88.9%	Accuracy could be enhanced
[44]	1-D CNN	88.23	Size small dataset and inefficient signal pre-processing algorithm,
[45]	Hybrid CNN LSTM	80.67 75.04 84.13	Model performance could be improved with some modifications in the algorithm
[46]	Machine Learning and Deep Learning Classifiers	86.25%	Use of appropriate features so the system performance can be improved
[13]	SE-MSCNN	90.64% per segment and 100% per record	Only OSA detection is done based on the investigation but not the severity
{10]	Filter Bank Decomposition and CNN	100% per record 87.5% per min	Small-size data limits the generalization of the performance of the model
[47]	Transfer learning AlexNet and GoogLeNet classifiers	99.87	Clinical dataset needs to be tested to verify robustness of model
[18]	OSACN-Net	94.81%	The system outperforms all other existing methodologies
[2]	EfficienNet-B0, B1, B2, Inception-v3, and Xception	93.33%	The 2-s segment is a very small size in limitation to the robustness of the system
[5]	Conventional ML and Deep Learning Algorithms	88.13% 84.26% 92.27%	They recommended hybrid deep learning using CNN and DRNN for SA detection which could be extended for SA classification

4 Discussion

This review paper has conducted a thematic review for OSA detection using ECG in various deep-learning approaches. A total 16 number of research articles, which were confined to the research area chosen, are reviewed systematically through PRISMA. We explored various sources for collecting research articles relevant to OSA detection and then 1110 total papers were chosen. PRISMA framework facilitates the eligibility criteria to down-sample the articles which are most suitable for our review. Over a decade there is sharp growth in deep learning-based classification techniques for sleep apnea detection and particularly after the year 2017. In the year 2022 drastic increase in the using deep learning has been reported. During the years 2017 to 2022, deep learning approaches were used for classification where LSTM, CNN, RNN, pre-trained networks, and hybrid deep neural architectures were implemented.

It is explored thoroughly in the research reported during the year 2022 to detect SA using ECG in deep learning since it is simple and requires fewer computational tasks. This paper will lead future researchers to pursue their work by also procuring relevant background knowledge to pursue an efficient method and approach for OSA detection using a single-channel ECG signal. The best deep learning models were the hybrid model which combines both RNN and CNN for the robustness of the proposed model.

5 Conclusion

The article represents a systematic literature review that provides a clear insight into choosing the most appropriate deep learning approach depending upon their research problem and application. This SLR also concludes observations from 16 selected articles and hence proposes that hybrid deep neural network models are the most accurate, efficient, and easy to implement. It concludes that feature engineering is the most time-consuming job in machine learning hence the state of art deep learning approaches perform better than the conventional machine learning approach. It was also found that the convolution neural network has shown better performance than long short-term memory or any other recurrent neural network in the detection of sleep apnea whereas the hybrid model is most suitable for our research topic. This SLR will give the insight to select a robust model to detect sleep apnea using a single-channel ECG signal as it requires less computation time and in ECG modality the noise issue is lesser as compared to any other physiological signal. We expect that this SLR will guide the researchers to choose a better model to pursue their research in the right direction which will eventually save them time. The research in sleep apnea detection using single-channel ECG in deep learning approaches is recent and most efficient. Studies were chosen as the specific work published in hybrid deep learning approaches needs less computation work. However, some of the literature may not fulfill the state of art deep learning approaches. This SLR is cited as a reference but does not claim to fulfill the direct observations.

References

Almutairi, H., Hassan, G.M., Datta, A.: Detection of obstructive sleep apnea by ECG signals using deep learning architectures. In: 2020 28th European Signal Processing Conference, pp. 1382–1386 (2021). https://ieeexplore.ieee.org/abstract/document/9287360/

Ayatollahi, A., Afrakhteh, S., Soltani, F., Saleh, E.: Sleep apnea detection from ECG signal using deep CNN-based structures. Evolving Syst. **14**(2), 191–206 (2022)

Bahrami, M., Forouzanfar, M.: Detection of sleep apnea from single-lead ECG: comparison of deep learning algorithms. In: 2021 IEEE International Symposium on Medical Measurements and Applications (2021). https://ieeexplore.ieee.org/abstract/document/9478745/

Bahrami, M., Forouzanfar, M.: Deep learning forecasts the occurrence of sleep apnea from single-lead ECG. Cardiovasc. Eng. Tech. **13**(6), 809–815 (2022). https://doi.org/10.1007/s13239-022-00615-5

Bahrami, M., Forouzanfar, M.: Sleep apnea detection from single-lead ECG: a comprehensive analysis of machine learning and deep learning algorithms. IEEE Trans. Instrum. Meas. **71**, 1–11 (2022)

Bai, Y., Zhang, L., Wan, D., Xie, Y., Deng, H.: Detection of sleep apnea syndrome by CNN based on ECG. J. Phys.: Conf. Ser. **1757**(1), 012043 (2021). https://doi.org/10.1088/1742-6596/1757/1/012043

Banluesombatkul, N., Rakthanmanon, T., et al.: Single channel ECG for obstructive sleep apnea severity detection using a deep learning approach. TENCON 2018–2018 (2018). https://ieeexplore.ieee.org/abstract/document/8650429/

Bernardini, A., Brunello, A., Gigli, G. L., Montanari, A., et al.: AIOSA: An approach to the automatic identification of obstructive sleep apnea events based on deep learning. In: Artificial Intelligence in Medicine (2021). https://www.sciencedirect.com/science/article/pii/S0933365721001263

Cen, L., Yu, Z.L., Kluge, T., Ser, W.: Automatic system for obstructive sleep apnea events detection using convolutional neural network. In: 2018 40th Annual International (2018). https://ieeexplore.ieee.org/abstract/document/8513363/

Chang, H.C., Wu, H.T., Huang, P.C., Ma, H.P., Lo, Y.L., et al.: Portable sleep apnea syndrome screening and event detection using long short-term memory recurrent neural network. Sensors **20**(21), 6067 (2020)

Chang, H.Y., Yeh, C.Y., Lee, C.T., Lin, C.C.: A sleep apnea detection system based on a one-dimensional deep convolution neural network model using single-lead electrocardiogram. Sensors **20**(15), 1–15 (2020). https://doi.org/10.3390/s20154157

Chaw, H.T., Kamolphiwong, S., Wongsritrang, K.: Sleep apnea detection using deep learning. Tehnički glasnik **13**(4), 261–266 (2019). https://doi.org/10.31803/tg-20191104191722

Chen, Q., Yue, H., Pang, X., Lei, W., Zhao, G., Liao, E., et al.: Mr-ResNeXt: a multi-resolution network architecture for detection of obstructive sleep Apnea. In: Zhang, H., Zhang, Z., Wu, Z., Hao, T. (eds.) Neural Computing for Advanced Applications: First International Conference, NCAA 2020, Shenzhen, China, July 3–5, 2020, Proceedings, pp. 420–432. Springer Singapore, Singapore (2020). https://doi.org/10.1007/978-981-15-7670-6_35

Chen, X., Chen, Y., Ma, W., Fan, X., et al.: SE-MSCNN: a lightweight multi-scaled fusion network for sleep apnea detection using single-lead ECG signals. In: 2021 IEEE International (2021). https://ieeexplore.ieee.org/abstract/document/9669358/

Chen, Z., Wu, M., Cui, W., Liu, C., Li, X.: An attention based CNN-LSTM approach for sleep-wake detection with heterogeneous sensors. IEEE J. Biomed. **25**(9), 3270–3277 (2020). https://ieeexplore.ieee.org/abstract/document/9130029/

Cheng, M., Sori, W.J., Jiang, F., Khan, A., Liu, S.: Recurrent neural network based classification of ECG signal features for obstruction of sleep apnea detection. In: IEEE International Conference on Computational Science and Engineering and IEEE/IFIP International Conference on Embedded and Ubiquitous Computing, vol. 2, pp. 199–202 (2017). https://doi.org/10.1109/CSE-EUC.2017.220

Erdenebayar, U., Kim, Y.J., Park, J.U., Joo, E.Y., Lee, K.J.: Deep learning approaches for automatic detection of sleep apnea events from an electrocardiogram. Comput. Methods Programs Biomed. **180**, 105001 (2019). https://doi.org/10.1016/j.cmpb.2019.105001

Faust, O., Barika, R., Shenfield, A., Ciaccio, E.J., et al.: Accurate detection of sleep apnea with long short-term memory network based on RR interval signals. Knowl.-Based **212**, 106591 (2021)

Gupta, K., Bajaj, V., Ansari, I.A.: OSACN-Net: automated classification of sleep apnea using deep learning model and smoothed Gabor spectrograms of ECG signal. IEEE Trans. Instrum. Meas. **71**, 1–9 (2022). https://doi.org/10.1109/TIM.2021.3132072

Hedman, M., Rojas, A., Arora, A., Ola, D.: Developing and comparing machine learning models to detect sleep apnoea using single-lead electrocardiogram (ECG) monitoring. MedRxiv (2021). https://doi.org/10.1101/2021.04.19.21255733.abstract

Huang, G., Ma, F.: ConCAD: contrastive learning-based cross attention for sleep Apnea detection. In: Dong, Y., Kourtellis, N., Hammer, B., Lozano, J.A. (eds.) ECML PKDD 2021. LNCS (LNAI), vol. 12979, pp. 68–84. Springer, Cham (2021). https://doi.org/10.1007/978-3-030-86517-7_5

Kocak, O., Bayrak, T., Erdamar, A., Ozparlak, L., Telatar, Z., Erogul, O.: Automated detection and classification of sleep apnea types using electrocardiogram (ECG) and electroencephalogram (EEG) features. In: Advances in Electrocardiograms – Clinical Applications. InTech. (2012). https://doi.org/10.5772/22782

Liang, X., Qiao, X., Li, Y.: Obstructive sleep apnea detection using a combination of CNN and LSTM techniques. In: 2019 IEEE 8th Joint International (2019). https://ieeexplore.ieee.org/abstract/document/8785833/

Loh, H.W., Ooi, C.P., Dhok, S.G., Sharma, M., Bhurane, A.A., et al.: Automated detection of cyclic alternating pattern and classification of sleep stages using deep neural network. Appl. Intell. (2021). https://doi.org/10.1007/s10489-021-02597-8

Mukherjee, D., Dhar, K., Schwenker, F., Sarkar, R.: Ensemble of deep learning models for sleep apnea detection: an experimental study. Sensors **21**(16), 5425 (2021). https://doi.org/10.3390/s21165425

Nasifoglu, H., Erogul, O.: Obstructive sleep apnea prediction from electrocardiogram scalograms and spectrograms using convolutional neural networks. Physiol. Meas. (2021). https://doi.org/10.1088/1361-6579/ac0a9c

Novák, D., Mucha, K., Al-Ani, T.: Long short-term memory for apnea detection based on heart rate variability. In: Proceedings of the 30th Annual International Conference of the IEEE Engineering in Medicine and Biology Society (2008). https://doi.org/10.1109/iembs.2008.4650394

Pathinarupothi, R.K., Vinaykumar, R., Rangan, E., Gopalakrishnan, E., Soman, K.P.: Instantaneous heart rate as a robust feature for sleep apnea severity detection using deep learning. In: 2017 IEEE EMBS International Conference on Biomedical and Health Informatics, pp. 293–296 (2017).https://doi.org/10.1109/BHI.2017.7897263

Pombo, N., Silva, B.M.C., Pinho, A.M., Garcia, N.: Classifier precision analysis for sleep apnea detection using ECG signals. IEEE Access **8**, 200477–200485 (2020)

Qian, X., Qiu, Y., He, Q., Lu, Y., Lin, H., Xu, F., Zhu, F., et al.: A review of methods for sleep arousal detection using polysomnographic signals. Brain Sci. **11**(10), 1274 (2021). https://doi.org/10.3390/brainsci11101274

Singh, N., Talwekar, R.H.: Automatic detection of sleep breathing disorder using Bayesian optimization algorithm from single-lead electrocardiogram. Int. J. Health Sci. (2022). https://doi.org/10.53730/ijhs.v6nS1.6965

Qin, H., Liu, G.: A dual-model deep learning method for sleep apnea detection based on representation learning and temporal dependence. Neurocomputing **473**, 24–36 (2022)

Rajawat, A. S., Rawat, R., Barhanpurkar, K., et al.: Sleep Apnea detection using contact-based and non-contact-based using deep learning methods. In: Computationally Intelligent Systems and their Applications (2021).https://doi.org/10.1007/978-981-16-0407-2_7

Rizal, A., Barus, D.T., Khatami, A.A., Sebayang, M.A.C.: Obstructive Sleep Apnea Detection using ECG Signal: A Survey. Researchgate.Net, 62 (2020). https://www.researchgate.net/profile/Achmad-Rizal

Salari, N., Hosseinian-Far, A., Mohammadi, M., et al.: Detection of sleep apnea using Machine learning algorithms based on ECG Signals: a comprehensive systematic review. Expert Syst. Appl. **187**, 115950 (2022). https://www.sciencedirect.com/science/article/pii/S0957417421013038

Sharan, R.V., Berkovsky, S., Xiong, H., Coiera, E.: End-to-end sleep apnea detection using single-lead ECG signal and 1-D residual neural networks. J. Med. Biol. Eng. (2021). https://doi.org/10.1007/s40846-021-00646-8

Song, C., Liu, K., Zhang, X., Chen, L., Xian, X.: An obstructive sleep apnea detection approach using a discriminative hidden Markov model from ECG signals. IEEE Trans. Biomed. Eng. **63**(7), 1532–1542 (2016). https://doi.org/10.1109/TBME.2015.2498199

Van Steenkiste, T., Deschrijver, D., et al.: Sensor fusion using backward shortcut connections for sleep apnea detection in multi-modal data. Machine Learning for Research (2020). http://proceedings.mlr.press/v116/van-steenkiste20a.html

Teng, F., Wang, D., Yuan, Y., Zhang, H., Singh, A., et al.: Multimedia monitoring system of obstructive sleep apnea via a deep active learning model. IEEE MultiMedia **29**(3), 48–56 (2022)

Urtnasan, E., Park, J.U., Joo, E.Y., Lee, K.J.: Automated detection of obstructive sleep apnea events from a single-lead electrocardiogram using a convolutional neural network. J. Med. Syst. **42**(6), 104 (2018). https://doi.org/10.1007/s10916-018-0963-0

Urtnasan, E., Park, J.U., Lee, K.J.: Automatic detection of sleep-disordered breathing events using recurrent neural networks from an electrocardiogram signal. Neural Comput. Appl. (2020). https://doi.org/10.1007/s00521-018-3833-2

Wang, T., Lu, C., Shen, G.: Detection of sleep apnea from single-lead ECG signal using a time window artificial neural network. Biomed. Res. Int. **2019**, 1–9 (2019). https://doi.org/10.1155/2019/9768072

Wang, X., et al.: Obstructive sleep apnea detection using ECG-sensor with convolutional neural networks. Multimedia Tools Appl. **79**(23–24), 15813–15827 (2020). https://doi.org/10.1007/s11042-018-6161-8

Warrick, P.A., Lostanlen, V., et al.: Hybrid scattering-LSTM networks for automated detection of sleep arousals. Physiological (2019). https://doi.org/10.1088/1361-6579/ab2664

Zhang, J., et al.: Automatic detection of obstructive sleep apnea events using a deep CNN-LSTM model. Comput. Intell. Neurosci. (2021). https://doi.org/10.1155/2021/5594733

Roneel et al.: ECG-Derived HRV Interpolation and 1-D Convolutional Neural Networks for Detecting Sleep Apnea (2020)

Bahrami, M.: Detection of Sleep Apnea from Single-Lead ECG: Comparison of DL Algorithms (2021)

Sheta, A., Turabieh, H., Thaher, T., Too, J., Mafarja, M., Hossain, M.S., Surani, S.R.: Diagnosis of obstructive sleep apnea from ECG Signals using machine learning and deep learning classifiers. Appl. Sci. **11**(14), 6622 (2021). https://doi.org/10.3390/app11146622

Performance Evaluation of Vanilla, Residual, and Dense 2D U-Net Architectures for Skull Stripping of Augmented 3D T1-Weighted MRI Head Scans

Anway S. Pimpalkar(✉), Rashmika K. Patole, Ketaki D. Kamble, and Mahesh H. Shindikar

COEP Technological University, Pune, MH 411005, India
{pimpalkaras19.extc,rkp.extc,kamblek.appsci, smh.appsci}@coep.ac.in

Abstract. Skull Stripping is a requisite preliminary step in most diagnostic neuroimaging applications. Manual Skull Stripping methods define the gold standard for the domain but are time-consuming and challenging to integrate into processing pipelines with a high number of data samples. Automated methods are an active area of research for head MRI segmentation, especially deep learning methods such as U-Net architecture implementations. This study compares Vanilla, Residual, and Dense 2D U-Net architectures for Skull Stripping. The Dense 2D U-Net architecture outperforms the Vanilla and Residual counterparts by achieving an accuracy of 99.75% on a test dataset. It is observed that dense interconnections in a U-Net encourage feature reuse across layers of the architecture and allow for shallower models with the strengths of a deeper network.

Keywords: Skull Stripping · MRI · Brain Segmentation · Semantic Segmentation · Deep Learning · U-Net

1 Introduction

Neuroimaging is a prevalent field for diagnostic assessments of neuroanatomy, neurophysiology, and internal functions such as cognition and control. Numerous techniques are employed in neuroimaging, such as Computed Tomography (CT), Magnetic Resonance Spectroscopy, Magnetization Transfer Imaging, Cerebral Perfusion Imaging, Single Photon Emission Computed Tomography (SPECT), Positron Emission Tomography (PET), Magnetic Resonance Imaging (MRI), Diffusion Tensor Imaging (DTI), and Ultrasound Imaging [1]. MRI scanning is used extensively in medical diagnostic settings due to its noninvasive and nondestructive nature. It employs static and variable external magnetic fields to perturb hydrogen atoms in the brain tissue. The alignment of the magnetic field of the hydrogen atoms changes to and from its original position once the external field is cut out or induced, causing the emission of a signal that receivers can pick up. The intensities of the received signals represent the different tissues present in the scan [2].

B. K. Singh et al. (Eds.): ICBEST 2023, CCIS 2003, pp. 131–142, 2024.
https://doi.org/10.1007/978-3-031-54547-4_11

Different image types are created by varying the sequence of the radio pulse frequency applied. Therefore, MRI scans can be imaged under different contrasts, namely (1) T1 weighted; (2) T2 weighted; and (3) Fluid Attenuated Inversion Recovery (FLAIR). T1-weighted MRIs rely upon the longitudinal relaxation of the magnetic vectors of a tissue after a pulse frequency is cut-off. The time taken for the vectors to return to the original vector direction is different for the different tissues present in the brain, leading to a variation in the signal intensities emitted [3]. T2-weighted MRIs measure the transverse relaxation of the magnetic vectors when a pulse frequency is induced, the different rates of which correspond to the different neural tissues [4]. FLAIR MRI scans are similar to T2-weighted techniques, with the additional removal of Cerebrospinal Fluid (CSF) from the resulting images. This sequence makes the differentiation between CSF and abnormalities much easier [5].

A structural MRI images the entirety of the tissue present in the head. Most neuroscience studies involving the brain require a segmentation step, which extracts the brain tissue from the scan, often referred to as Skull Stripping. It involves removing non-cerebral tissue from a head scan. It may subtract any non-cerebral tissue from the scan, including the skull, meninges, and scalp [6]. It is a critical step in most pre-processing workflows, as rectifications in the segmentation map cannot be performed in subsequent steps (Fig. 1).

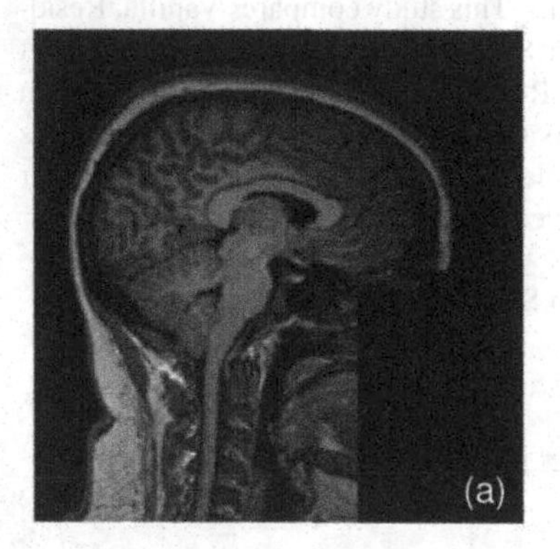

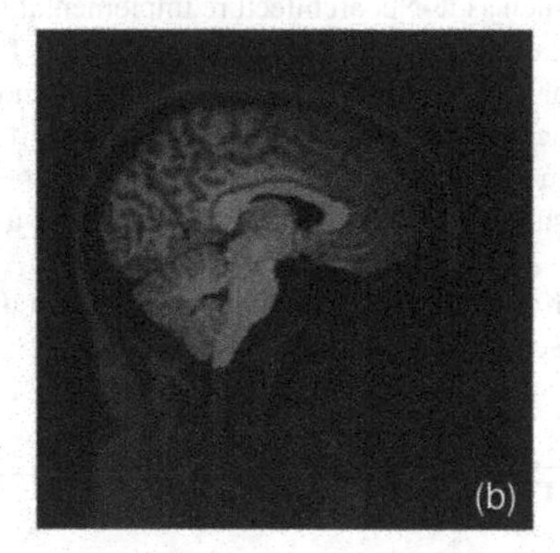

Fig. 1. A representation of (a) Sagittal 3D T1-w MRI slice from the NFBS repository [7]; and (b) corresponding Skull Stripped mask superimposed on the MRI scan.

Manual segmentation held prominence in the initial days of Skull Stripping. However, manual delineation methods are time-consuming and cannot be used for developing pipelines that involve multiple pre-processing steps. Thus, research into semi-automated and automated methods began to rise, and since, these methods have taken prevalence in the field. However, manual segmentation is still employed to define the gold standard or ground truth for comparison metrics to the advanced methods used today [8].

Conventional methodologies used for Skull Stripping are classified into five types: (1) morphology-based; (2) deformable surface-based; (3) intensity-based; (4) template and atlas-based; and (5) hybrid. Modern methods can be classified into (1) machine learning-based; and (2) deep learning-based.

Morphology-based Skull Stripping methods employ mathematical morphological operations and thresholding techniques for edge detection to extract image features and identify brain surfaces [9]. Deformable surface-based methods evolve and deform a dynamic curve driven by a function of energy towards the active contour. This contour shrinks to take the shape of the brain surface, stripping the skull from the scan. Brain Extraction Tool (BET) by Smith et al. [10] and BET2 by Jenkinson et al. [11] are the most common deformable surface-based methods due to their inclusion in the FMRIB Software Library (FSL) [12]. Intensity-based methods differentiate regions of the brain by focusing on pixel-wise intensities. The main disadvantage of such an approach is that it is sensitive to intensive bias present in the scan, which may be caused due to fluctuations in the magnetic field while scanning [13]. Template and atlas-based methods strip the skull using a template or atlas registration of the brain. It can distinguish between brain tissues even when no pre-defined relationship between the values has been defined [14]. Hybrid methods combine one or more of the abovementioned approaches to improve upon the accuracy of the results [15].

Machine and deep learning methods carry great potential for procedures such as Skull Stripping. Unlike conventional methods, these methods do not require many user-dependent parameters and can be optimized to work with datasets to achieve above-par results. Unsupervised methods such as the Fuzzy Active Surface Model by Kobashi et al. [16] have been experimented with and have yielded good outcomes. ROBEX, a hybrid generative-discriminative algorithm, explores the MRI image and identifies the most likely contour suggested by the discriminative model [17]. Deep learning methods are carried out using Convolutional Neural Networks (CNNs) and are a highly active research field. These methods are usually implemented through two main approaches: (1) voxel-wise network; and (2) segmenting the complete image as one single feed-forward step. The 3D-CNN method proposed by Kleesiek et al. in 2016 is often referred to as one of the bases of the deep learning methods for skull-stripping [18]. Ronneberger et al. [19] introduced U-Net convolutional networks for biomedical image segmentation, which are detailed in this study. Salehi et al. experimented with auto-context analysis algorithms for brain extraction [20]. Chen et al. [21] presented an auto-context VoxResNet approach, a residual network architecture [22] for brain segmentation.

In the extensive literature reviewed, no studies focused on data augmentation techniques to improve the performance of U-Net architectures for Skull Stripping. The motivation of this study is to employ data augmentation transforms to mimic the conditions posed by the different scanning parameters associated with each scan. This research evaluates the performance of Vanilla, Residual, and Dense 2D U-Net architectures for Skull Stripping. Dense 2D U-Net architectures have not previously been employed for this task and hence are emphasized and focused on in the scope of this study.

The research conducted is presented in three sections. The following section describes the methodology of the study, including the overview, network architectures, pre-processing, augmentation, and implementation. The subsequent section elaborates on the results and discusses them as a comparative study, followed by the conclusion.

2 Methodology

Skull Stripping involves separating brain tissue, including grey matter and white matter, from non-brain voxels such as the skull, scalp, and dura mater. Many assessments require Skull Stripping in the initial stages of the processing pipeline, such as volumetric or longitudinal analysis for diagnostics [23, 24], analysis of the progression of multiple sclerosis [25], cortical and sub- cortical analysis [26], assessment of mental disorders [27], and for the planning of surgical interventions in neurology [28]. The methodology employed in this study is described in the next section and can be summarized as shown in Fig. 2.

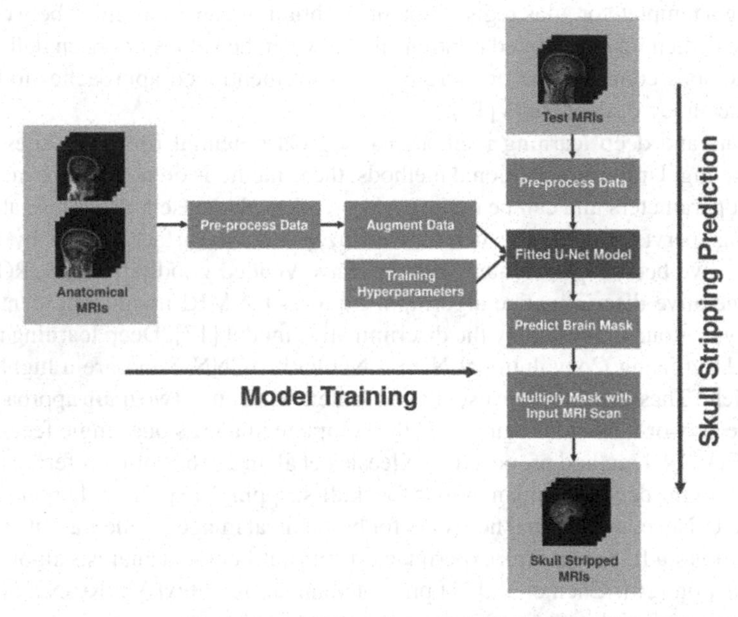

Fig. 2. Methodology for Skull Stripping using U-Net Architectures.

2.1 Overview

The general methodology employed for this study is divided into multiple steps as a part of two main pipelines. The main pipelines are (1) model training; and (2) Skull Stripping prediction on a test MRI. Both pipelines are initialized by a pre-processing step to z-normalize the data, detailed in Sect. 2.3. MRI scans are highly susceptible to bias variations due to inhomogeneity in the magnets [29]. Normalizing the data ensures that the mean of the data lies at 0 and that no particular regions or scans are given a higher priority than others. The training data is augmented to increase the samples fivefold. After defining the training hyperparameters for the chosen U-Net architecture, the models are fit to the augmented data, which are then used to predict Skull Stripped masks of unseen MRI head scans. The mask generated is then multiplied with the input

MRI to generate a Skull Stripped scan. The following section details the data used for the research study.

2.2 Dataset

The Neurofeedback Skull-stripped (NFBS) repository dataset [7] is used to train the three architectures of U-Net in this repository. The repository is available to researchers through the Pre-processed Connectomes Project. The database comprises 125 T1-weighted structural MRI scans and brain masks for each. The 125 MRIs are first segmented using the semi-automated BEaST method [30] and then manually correcting the improper results. In total, 85 of the resultant images are further manually rectified, while the rest are perfectly segmented without human assistance.

Each MRI in the dataset is formed by a sequence of 192 sagittal slices with 256 $\times$ 256 mm^2 field of view and an acquisition resolution of 1 $\times$ 1 $\times$ 1 mm^3 [31]. This anatomical data for each scan and mask is stored as NIfTI files, a standard imaging format for neuroradiology research. Of these scans, 110 are used for training, and the remaining 15 are used for testing.

2.3 Pre-processing and Data Augmentation

Z-normalization of intensity values can increase the mean accuracy of T1-weighted classifiers, as shown in the study by Carré et al. [32]. This method normalizes images by subtracting the mean intensity value of the brain pixels (μ_{brain}), from each pixel intensity ($I(x)$) and dividing the result by the standard deviation of the brain pixels (σ_{brain}), shown in Eq. (1):

$$I_{Z-norm}(x) = \frac{I(x) - \mu_{brain}}{\sigma_{brain}} \tag{1}$$

Data augmentation is a standard procedure employed in deep learning techniques to enhance the size and quality of training datasets such that better models can be built using them. It can help increase the size of the dataset manifold and tune the model to be robust to multi-scanner variations in the data. This study uses a combination of spatial and intensity transforms to expand the training MRI dataset size fivefold, from 110 initial scans to 550 augmented scans of 192 sagittal slices each. This increases the total slices available for training to 105,600 individual sagittal slices.

2.4 Neural Network Architectures

In this study, three neural network architectures are evaluated for Skull Stripping of augmented MRI head scans: (1) Vanilla 2D U-Net Architecture; (2) Residual 2D U-Net Architecture; and (3) Dense 2D U-Net Architecture.

Vanilla 2D U-Net Architecture. The contractive path of the Vanilla 2D U-Net architecture employs basic 2D convolutions. It consists of two ReLU-activated 3 $\times$ 3 convolutional layers with same padding, followed by a 2 $\times$ 2 max-pooling layer with stride (2, 2).

This block is repeated four times before beginning the expansive path. This path consists of 2 × 2 up-convolutions or transpose convolutions to generate an up-sampled feature map created by the contractive path and a shortcut concatenation with the corresponding feature map from the contracting path. This is followed by two 3 × 3 convolutional layers with Rectified Linear Unit (ReLU) activation. At the final layer, a unit convolution with sigmoid activation is used to classify each intensity vector as either brain or non-brain tissue. This architecture has a total of 7,759,521 trainable parameters.

Residual 2D U-Net Architecture. Residual connections improve the flow of information in the network. It encourages feature reuse through the layers in the form of learning residual functions to the layer inputs [22]. There are four residual block connections in the Residual 2D U-Net Architecture contractive path. In each of these blocks, the input to the two 3 × 3 convolutional layers is concatenated with their output. This concatenated result is then passed through a max-pooling layer with a 2 × 2 kernel and a stride of (2, 2). The same residual block pattern is also applied to the expansive path, with the difference that instead of max-pooling layers, up-convolutions or transpose convolutions are used to upscale the feature image. Similar to the Vanilla 2D U-Net, shortcut connections are made across the contractive and expansive paths at the same levels, and the expansive path ends with a unit convolution and a sigmoid activation function for multi-class segmentation. The total number of training parameters for this architecture is 9,895,073.

Dense 2D U-Net Architecture. According to the literature reviewed, the Dense 2D U-Net architecture for Skull Stripping proposed in this research study has not been explored previously. The groundwork for building Dense U-Net architectures is laid by Huang et al. in their work on Dense Convolutional Networks [33]. It allows for feature reuse throughout the network and strengthens shallower models with the benefits of deeper networks. They encourage identity mappings, deep supervision, and diversified depth for the models. The dense skip connections allow for improved aggregation of features across different semantic scales [34]. In the Residual 2D U-Net architecture, the input to the two 3 × 3 convolutional layers is concatenated with their output. In the Dense 2D U-Net architecture, along with these residual connections, additional dense connections are established that run within these blocks of convolutional layers. The concatenation of the output from the first convolutional layer and the input is fed through another convolutional layer. The total number of trainable parameters in this network is 15,479,681 (Fig. 3).

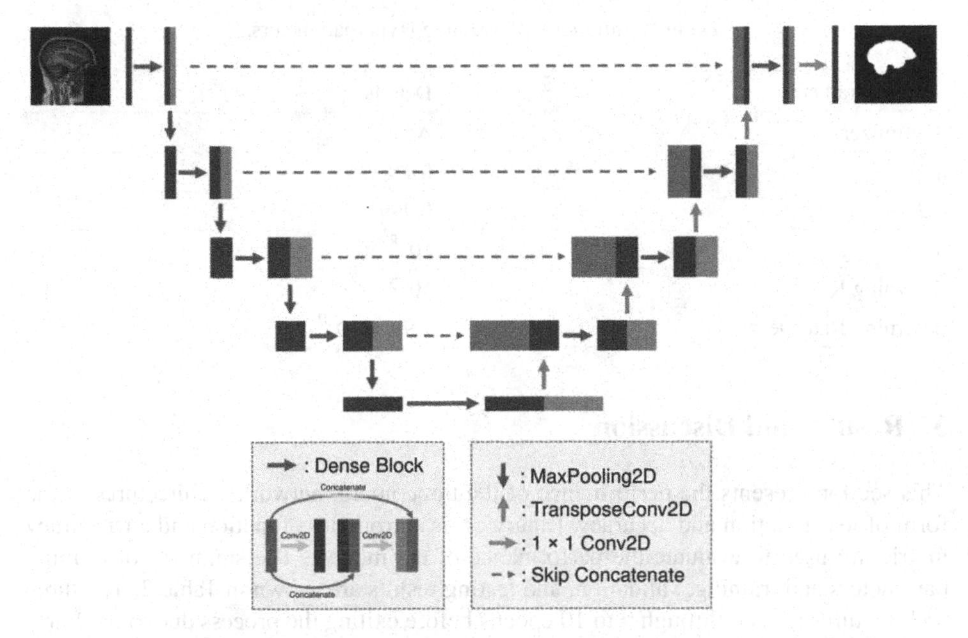

Fig. 3. Diagrammatic Representation of the Dense 2D U-Net Architecture for Skull Stripping.

2.5 Implementation

Each augmented MRI scan can be realized as an array of shape $192 \times 256 \times 256$. To optimize the training for our available resources, the data is converted from Float64 to Float16. The masks are converted from Float64 to Int8. The processed data is saved in the standard binary NumPy file format (NPY). This format stores the shape and data type information necessary to reconstruct the data on any machine, regardless of the system architecture. The dataset is loaded as a hashed memory map to preserve RAM for training at the expense of computational latency.

The project is implemented with a TensorFlow [35] backend, using the Keras APIs [36] included in the package. All models are trained on Google Colab[1], using an NVIDIA A100 Tensor Core GPU (40GB)[22] for hardware acceleration. The Adam optimizer [37] is employed for network training. The summary of hyperparameters used for training the models is shown in Table 1.

The train/validation split is set to 90/10, with a repeated holdout methodology. The EarlyStopping callback in TensorFlow is used to prevent overfitting of the models to the training dataset. The callback monitors the validation loss of the model and stops training when the validation loss starts increasing over epochs, with a patience value of 2. The batch size for each model is chosen according to memory limitations. In the next section, the training progress results are summarized and discussed.

[1] Google Corp., Mountain View, CA USA, 2018.
[2] NVIDIA Corp., Santa Clara, CA USA, 2021.

Table 1. Summary of Training Hyperparameters.

Hyperparameter	Details
Optimizer	Adam
β1	0.9
β2	0.999
ε	10^{-8}
Learning Rate	10^{-5}
Learning Rate Decay	1.99×10^{-7}

3 Results and Discussion

This section presents the performance of the three neural network architectures in the form of loss function and accuracy. Binary cross entropy loss function and an accuracy metric are used to evaluate the performance of the models. The summary of training parameters and training, validation, and testing results are shown in Table 2. The three architectures trained through 9 to 10 epochs before exiting the process due to the Early Stopping callback. The loss functions of the Residual and Dense 2D U-Net architectures converged in 9 epochs compared to the 10 epochs required by the Vanilla 2D U-Net architecture.

Table 2. Summary of Training Process Results.

Architecture	Epochs	Batch Size	Training Loss	Training Accuracy	Validation Loss	Validation Accuracy	Testing Loss	Testing Accuracy
Vanilla 2D U-Net	10	32	0.0066	99.72%	0.0093	99.63%	0.0065	99.73%
Residual 2D U-Net	9	32	0.0066	99.72%	0.0092	99.63%	0.0067	99.72%
Dense 2D U-Net	9	16	0.0053	99.77%	0.0085	99.67%	0.0062	99.75%

The epoch-wise loss and accuracies for the three neural network architectures are monitored during training and validation and represented graphically in Fig. 4 (a), (b), and (c). Upon training completion, the models are tested on the 15 pre-processed test MRIs kept aside initially.

3D U-Net architectures are also common along with their 2D counterparts. Theoretically, 3D convolutions leverage the spatiotemporal information between individual slices to add another correlational element to the model [38]. Research activities by Hwang et al. [39] and Kolařík et al. [40] have used 3D U-Net architectures for Skull Stripping and provided results better than conventional methods. Hsu et al. employed.

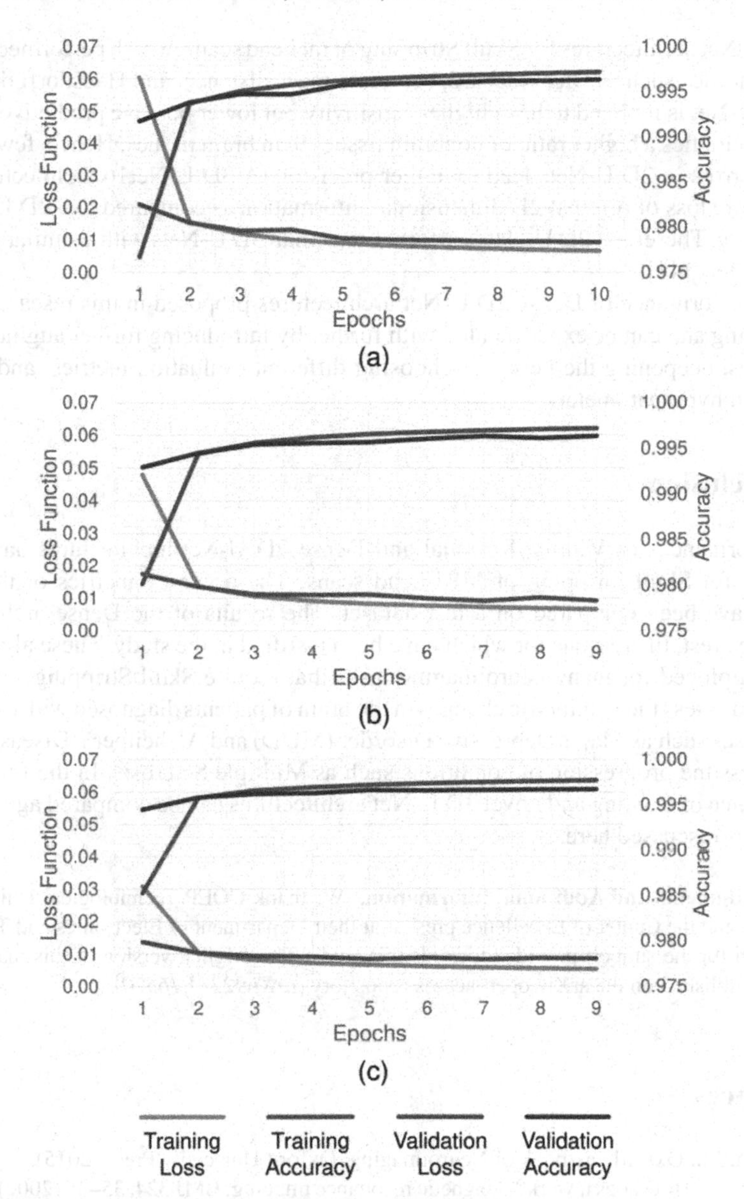

Fig. 4. Graphical representation of training and validation accuracy and losses for (a) Vanilla; (b) Residual; and (c) Dense 2D U-Net Architectures.

3D U-Net architectures for Skull Stripping of rat head scans, which performed well in various metrics such as Dice, Jaccard, center-of-mass distance, and Hausdorff distance. The 3D U-Net is inclined to have higher sensitivity but lower positive predictive values, as it misclassifies a higher ratio of nonbrain tissues than brain tissues. Hence, fewer false positive errors in 2D U-Nets lead to higher precision. A 3D U-Net16 architecture also causes more loss of original 2D dimensional information as compared to a 2D U-Net64 architecture. Therefore, 2D U-Nets perform better than 3D U-Nets with a similar number of parameters [41].

The performance of Dense 2D U-Net architectures proposed in this research study is promising and can be experimented with further by introducing further augmentation techniques, deepening the network, choosing different evaluation metrics, and tuning the chosen hyperparameters.

4 Conclusion

The performances of Vanilla, Residual and Dense 2D U-Net architectures have been evaluated for Skull Stripping of MRI head scans. The accuracy metrics of the three models have been compared on a test dataset. The results of the Dense architecture surpass the rest, the reasons for which have been justified in the study. These algorithms can be employed for many neuroimaging tasks that require Skull Stripping, such as a pipeline to assess the volumetric changes in the brain of patients diagnosed with disorders and diseases such as Major Depressive Disorder (MDD) and Alzheimer's Disease (AD), or to assess the progression of conditions such as Multiple Sclerosis. In the future, the performance of existing and novel 3D U-Net architectures can be compared against that of the ones discussed here.

Acknowledgments and Additional Information. We thank COEP Technological University's authorities and the Center of Excellence present at their Department of Electronics and Telecommunication for the support provided towards this study. The original version of this manuscript has been published on the arXiv open-access repository *(arXiv:2211.16570).*

References

1. Filippi, M.: Oxford Textbook of Neuroimaging. Oxford University Press (2015)
2. Berger, A.: How does it work?: magnetic resonance imaging. BMJ **324**, 35–35 (2002). https://doi.org/10.1136/bmj.324.7328.35
3. Baba, Y., Jones, J.: T1 Weighted Image. In: Radiopaedia.org. Radiopaedia.org (2005)
4. Haouimi, A., Jones, J.: T2 Weighted Image. In: Radiopaedia.org. Radiopaedia.org (2005). https://doi.org/10.53347/rID-6345
5. Preston, D.: MRI Basics. https://case.edu/med/neurology/NR/MRI%20Basics.htm (2016). Accessed 3 Nov 2022
6. Hahn, H.K., Peitgen, H.-O.: The skull stripping problem in MRI solved by a single 3D watershed transform. In: Delp, S.L., DiGoia, A.M., Jaramaz, B. (eds.) Medical Image Computing and Computer-Assisted Intervention – MICCAI 2000, pp. 134–143. Springer Berlin Heidelberg, Berlin, Heidelberg (2000). https://doi.org/10.1007/978-3-540-40899-4_14

7. Puccio, B., Pooley, J.P., Pellman, J.S., et al.: The pre-processed connectomes project repository of manually corrected skull-stripped T1-weighted anatomi- cal MRI data. Gigascience **5**, 45 (2016). https://doi.org/10.1186/s13742-016-0150-5
8. Lucena, O., Souza, R., Rittner, L., et al.: Convolutional neural networks for skull-stripping in brain MR imaging using silver standard masks. Artif Intell Med **98**, 48–58 (2019). https://doi.org/10.1016/j.artmed.2019.06.008
9. Brummer, M.E., Mersereau, R.M., Eisner, R.L., Lewine, R.R.J.: Automatic detection of brain contours in MRI data sets. IEEE Trans Med Imaging **12**, 153–166 (1993). https://doi.org/10.1109/42.232244
10. Smith, S.M.: Fast robust automated brain extraction. Hum Brain Mapp **17**, 143–155 (2002). https://doi.org/10.1002/hbm.10062
11. Jenkinson, M., Pechaud, M., Smith, S., et al.: BET2: MR-based estimation of brain, skull and scalp surfaces. In: Eleventh Annual Meeting of the Organization for Human Brain Mapping, p. 167 (2005)
12. Smith, S.M., Jenkinson, M., Woolrich, M.W., et al.: Advances in functional and structural MR image analysis and implementation as FSL. Neuroimage **23**, S208–S219 (2004). https://doi.org/10.1016/j.neuroimage.2004.07.051
13. Tao, X., Chang, M.-C.: A skull stripping method using deformable surface and tissue classification. In: Dawant, B.M., Haynor, D.R. (eds). p. 76233L (2010)
14. Ashburner, J., Friston, K.J.: Voxel-based morphometry—the methods. Neuroimage **11**, 805–821 (2000). https://doi.org/10.1006/nimg.2000.0582
15. Rex, D.E., Shattuck, D.W., Woods, R.P., et al.: A meta-algorithm for brain extraction in MRI. Neuroimage **23**, 625–637 (2004). https://doi.org/10.1016/J.NEUROIMAGE.2004.06.019
16. Kobashi, S., Fujimoto, Y., Ogawa, M., et al.: Fuzzy-ASM based automated skull stripping method from infantile brain MR images. In: 2007 IEEE International Conference on Granular Computing (GRC 2007), pp. 632–632. IEEE (2007)
17. Iglesias, J.E., Cheng-Yi Liu, P.M., Thompson, ZTu.: Robust brain extraction across datasets and comparison with publicly available methods. IEEE Trans. Med. Imaging **30**(9), 1617–1634 (2011). https://doi.org/10.1109/TMI.2011.2138152
18. Kleesiek, J., Urban, G., Hubert, A., et al.: Deep MRI brain extraction: a 3D convolutional neural network for skull stripping. Neuroimage **129**, 460–469 (2016). https://doi.org/10.1016/j.neuroimage.2016.01.024
19. Ronneberger, O., Fischer, P., Brox, T.: U-Net: convolutional networks for biomedical image segmentation. In: Navab, N., Hornegger, J., Wells, W.M., Frangi, A.F. (eds.) Medical Image Computing and Computer-Assisted Intervention – MICCAI 2015: 18th International Conference, Munich, Germany, October 5-9, 2015, Proceedings, Part III, pp. 234–241. Springer International Publishing, Cham (2015). https://doi.org/10.1007/978-3-319-24574-4_28
20. Mohseni Salehi, S.S., Erdogmus, D., Gholipour, A.: Auto-context convolutional neural network (Auto-Net) for brain extraction in magnetic resonance imaging. IEEE Trans. Med. Imaging **36**, 2319–2330 (2017). https://doi.org/10.1109/TMI.2017.2721362
21. Chen, H., Dou, Q., Yu, L., et al.: VoxResNet: deep voxelwise residual networks for brain segmentation from 3D MR images. Neuroimage **170**, 446–455 (2018). https://doi.org/10.1016/j.neuroimage.2017.04.041
22. He, K., Zhang, X., Ren, S., Sun, J.: Deep residual learning for image recognition. In: 2016 IEEE Conference on Computer Vision and Pattern Recognition (CVPR), pp. 770–778 . IEEE (2016)
23. Dubey, R.B.: Region growing for MRI brain tumor volume analysis. Indian J. Sci. Technol. **2**(9), 26–31 (2009). https://doi.org/10.17485/ijst/2009/v2i9.10
24. Li, G., Wang, L., Shi, F., et al.: Mapping longitudinal development of local cortical gyrification in infants from birth to 2 years of age. J. Neurosci. **34**, 4228–4238 (2014). https://doi.org/10.1523/JNEUROSCI.3976-13.2014

25. Ganiler, O., Oliver, A., Diez, Y., et al.: A subtraction pipeline for automatic detection of new appearing multiple sclerosis lesions in longitudinal studies. Neuroradiology **56**, 363–374 (2014). https://doi.org/10.1007/s00234-014-1343-1
26. Sowell, E.R., Trauner, D.A., Gamst, A., Jernigan, T.L.: Development of cortical and subcortical brain structures in childhood and adolescence: a structural MRI study. Dev. Med. Child Neurol. **44**, 4 (2002). https://doi.org/10.1017/S0012162201001591
27. Tanskanen, P., Veijola, J.M., Piippo, U.K., et al.: Hippocampus and amygdala volumes in schizophrenia and other psychoses in the Northern Finland 1966 birth cohort. Schizophr. Res. **75**, 283–294 (2005). https://doi.org/10.1016/j.schres.2004.09.022
28. Leote, J., Nunes, R.G., Cerqueira, L., et al.: Reconstruction of white matter fibre tracts using diffusion kurtosis tensor imaging at 1.5T: pre-surgical planning in patients with gliomas. Eur. J. Radiol. Open **5**, 20–23 (2018). https://doi.org/10.1016/j.ejro.2018.01.002
29. Yip, S.S.F., Aerts, H.J.W.L.: Applications and limitations of radiomics. Phys. Med. Biol. **61**, R150–R166 (2016). https://doi.org/10.1088/0031-9155/61/13/R150
30. Eskildsen, S.F., Coupé, P., Fonov, V., et al.: BEaST: Brain extraction based on nonlocal segmentation technique. Neuroimage **59**, 2362–2373 (2012). https://doi.org/10.1016/j.neuroimage.2011.09.012
31. Mugler, J.P., Brookeman, J.R.: Three-dimensional magnetization-prepared rapid gradient-echo imaging (3D MP RAGE). Magn. Reson. Med. **15**, 152–157 (1990). https://doi.org/10.1002/mrm.1910150117
32. Carré, A., Klausner, G., Edjlali, M., et al.: Standardization of brain MR images across machines and protocols: bridging the gap for MRI-based radiomics. Sci Rep **10**, 12340 (2020). https://doi.org/10.1038/s41598-020-69298-z
33. Huang G, Liu Z, van der Maaten L, Weinberger KQ (2016) Densely Connected Convolutional Networks
34. Zhou, Z., Rahman Siddiquee, M.M., Tajbakhsh, N., Liang, J.: UNet++: a nested U-Net architecture for medical image segmentation. In: Stoyanov, D., et al. (eds.) DLMIA/ML-CDS -2018. LNCS, vol. 11045, pp. 3–11. Springer, Cham (2018). https://doi.org/10.1007/978-3-030-00889-5_1
35. Abadi, A.A., Barham, P., et al.: TensorFlow: Large-Scale Machine Learning on Heterogeneous Systems (2015)
36. Chollet, F., et al.: Keras (2015)
37. Kingma, D.P., Ba, J.: Adam: A Method for Stochastic Optimization (2014)
38. Tran, D., Bourdev, L., Fergus, R., et al.: Learning Spatiotemporal Features with 3D Convolutional Networks (2014)
39. Hwang, H., Ur Rehman, H.Z., Lee, S.: 3D U-Net for skull stripping in brain MRI. Appl. Sci. **9**, 569 (2019). https://doi.org/10.3390/APP9030569
40. Kolařík, M., Burget, R., Uher, V., et al.: Optimized high resolution 3D dense-U-net network for brain and spine segmentation. Appl. Sci. **9**, 404 (2019). https://doi.org/10.3390/APP9030404
41. Hsu, L.-M., Wang, S., Walton, L., et al.: 3D U-Net improves automatic brain extraction for isotropic rat brain magnetic resonance imaging data. Front Neurosci (2021). https://doi.org/10.3389/fnins.2021.801008

EEG Based Classification of Learning Disability in Children Using Pretrained Network and Support Vector Machine

Sneha Agrawal[1], Guhan Seshadri N. P.[1], Bikesh Kumar Singh[1(✉)], Geethanjali B.[2], and Mahesh V.[2]

[1] National Institute of Technology Raipur, Raipur, Chhattisgarh 492001, India
bsingh.bme@nitrr.ac.in

[2] SSN College of Engineering, Chennai, Tamil Nadu 603110, India

Abstract. Learning disability (LD) is a commonly acknowledged neurological disorder that causes learners to struggle to decode, read, and write and in performing a mathematical task. If a child's learning disabilities go untreated, they may develop lifelong social and emotional issues, which may affect their future success in all spheres of their lives. With the help of early intervention, we can bridge the gap between generally normal developing and children with LD. Electroencephalogram (EEG) can be used to investigate the electrical activity of the brain in order to automatically detect and recognize the disability in young children. This paper proposes a model to classify from the rest EEG signals of normal and LD children. Before feeding the 19-channel EEG signal to convolution neural network (CNN), it was preprocessed, segmented and converted into spectrogram from the alpha, beta, delta and theta bands extracted using wavelet decomposition. The two different modalities proposed in this work were: (1) using pre-trained network for transfer learning approach (2) pretrained network to extract the image features and classification with the help of support vector machine (SVM). In this experiment, the networks like Alexnet, VGG16, and Resnet-18 were compared to compute the results of both the modalities. The highest classification accuracy of 98.3% was obtained using image features extracted from Alexnet and classifying it further using SVM. The use of pretrained network for extraction of image features approach resulted in increased accuracy. The results of the comparison showed that feature-based technique outperformed traditional CNN approach and it may be used for the development of intelligent automated diagnosis system.

Keywords: Learning disability · Electroencephalogram · Convolutional neural network · Wavelet decomposition · Transfer learning

1 Introduction

Cognitive function is the ability of the human brain to process the information from the external and internal world. This capability entails the ability to maintain contact with the external world, helps in selecting the information and memorizing the data. In this

B. K. Singh et al. (Eds.): ICBEST 2023, CCIS 2003, pp. 143–153, 2024.
https://doi.org/10.1007/978-3-031-54547-4_12

way, cognitive function allows humans to become aware of them and solve complex problems, which we refer to as intelligence. Any kind of impairment in cognitive ability reduces the capacity of learning in children which is commonly referred to as learning disability [1]. Delay in maturation, specific nervous system disorders, or injuries sustained prior to, during, or shortly after birth may all contribute to LD [2]. Children who had experienced medical issues soon after birth or who were born prematurely may inherit this disorder [2]. The most common type of disability found in the people is dyslexia (difficulty in reading). Other than dyslexia, dysgraphia (difficulty in writing) and dyscalculia (difficulty in performing arithmetic problems) are the common LD which is observed in children. Among these the neurological condition dyslexia has affected about 5–17% of the total population and is represented by slow word comprehension and phonological impairment [3]. The suspected child will be sent initially to child psychologist, who will make a primary diagnosis through standard test. The current method for LD diagnosis is through series of behavioral questionnaires and standard neuropsychological battery test. Standardized tests like the Comprehensive Test of Phonological Processing (CTOPP) [4], Woodcock-Johnson (WJ) [5], and Wechsler Individual Achievement Test (WIAT) [6] are used to assess the phonological processing abilities, writing, reading, and intelligence quotient of the people. When the results of these standard tests are combined with the family history and biographical information, people with any of the LD can be identified. The major limitation with the current method is that the diagnosis report is based on experts and requires children cooperation over a long time. In addition the diagnosis is possible only after children started showing symptoms. This urges researches to development of an automated system for an early detection of LD using the neurophysiological markers. Electroencephalography (EEG) markers are one kind plays an important role in the study of brain, diagnosis of different mental disorders, and in the treatment of medical fields [7]. EEG being the non-invasive mode of brain signal recording and the analysis of EEG signal has been identified as the most advanced approach to the problem of gaining knowledge of brain dynamics. EEG recordings can be used for detection of LD in children using brain computer interfaces (BCIs). Researchers have proposed various methods for detection and classification of children with LD. Pooja et al. [8] implemented learning vector quantization and rule based method to categorize the LD children based on the test scores and achieved classification accuracy about 91.8%. Al-Barhamtoshy et al. [9] examined the hemispherical activation in dyslexics using EEG signals. Mahmoodin et al. [10] studied the power spectral densities in the dyslexic brain using EEG signals during rest time and writing task. In summary, authors have implemented several methods to study the differences in the brain activity to understand the LD but the literatures are very small in the case of LD identification using EEG signals. Therefore our study aims to briefly discuss the approach towards the identification of LD at the earliest. We implemented two methods in the present study to the efficiency in LD classification using rest EEG signals. First method used pretrained networks such as Alexnet, VGG16 and Resnet-18 to classify the LD children and the second method is to use the features extracted from the pretrained network and classify using the traditional ML classifier.

2 Material and Methods

An overview of the methodology is presented in this section. The experiment was performed in two stages. Firstly, this experiment uses the spectrogram generated by the alpha beta delta and theta bands from the EEG signal to classify the normal developing children with LD through different pretrained CNN architectures. Secondly the features were extracted from the pretrained networks and classified using machine Learning classifier.

2.1 Data Acquisition

Fifteen LD children and fifteen normal developing children (NDC) were recruited from special and primary schools of the city respectively. The children were assessed for disability symptoms using a benchmark test scale to divide them into experimental and control group after being evaluated by school psychologist. In this experiment, LD children who scored at least 80 on the nonverbal Intelligence Quotient (IQ) were included. The children were free from any kind of abnormalities and had no record of seizures. Before starting the recording procedure, the consent form was duly signed by the participant's parents/guardians. The selected participants' EEG signals were recorded for the classification Table 1. In an equally lit and soundproof room, the children comfortably sat on the chair. The raw EEG signal was recorded at a sampling rate of 256 Hz/channel for two minutes with eyes closed condition with the help of RMS SuperSpec system. In accordance to International 10–20 system, the signal was recorded from 19 channels namely Fp1, F3, F7, Fz, Fp2, F4, F8 P3, Pz, P4, O1, O2, T3, T5, T4, T6 C3, Cz and C4. While acquiring the data, a notch filter was used to remove 50 Hz power line interference and an EMG filter for the removal of involuntary muscle artifacts. The EEG impedance was kept below 5 kΩ by applying a conductive gel. Necessary approval from Institutional Ethical Committee NIT Raipur (NITRR/IEC/2019/02) was taken to proceed with the experimental work over children.

Table 1. EEG dataset used in this study.

EEG Dataset	No. of patients	Length of the signal
NDC	15	10 s
Children with Learning Disability	15	10 s

2.2 Preprocessing and Data Segmentation

Preprocessing is the first step in analyzing an EEG signal. The raw EEG signal is depicted in Fig. 1. All the channels were applied with a smoothing filter having a triangular window (moving average) of order 5. Smoothened EEG signal is shown in Fig. 2. Eyeblink masks the essential information carried by the signal, therefore removal of these artifact is an immediate requirement. Various researchers, was developed after several trials with various types of noise removal techniques. In the present study we have applied wavelet denoising to remove the eyeblink in the recorded EEG signals [11]. Figure 3 shows the noise free EEG signals.

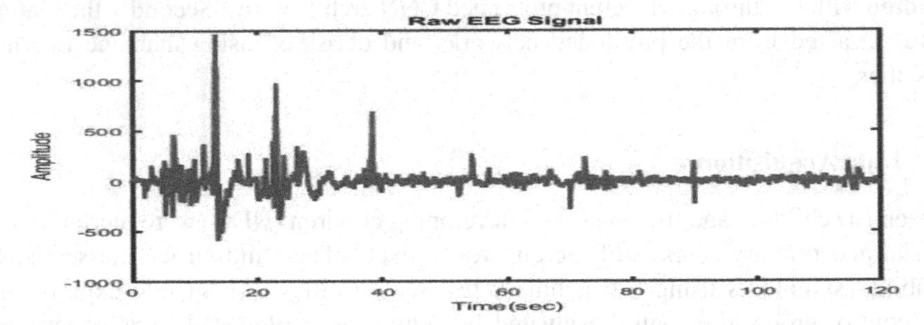

Fig. 1. Raw EEG during Rest condition

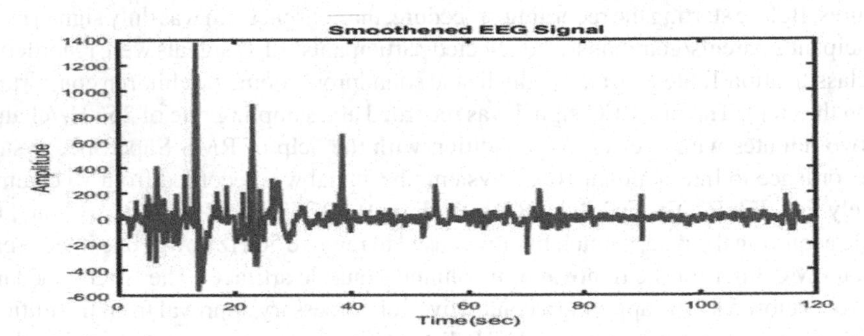

Fig. 2. Smoothened EEG Signal.

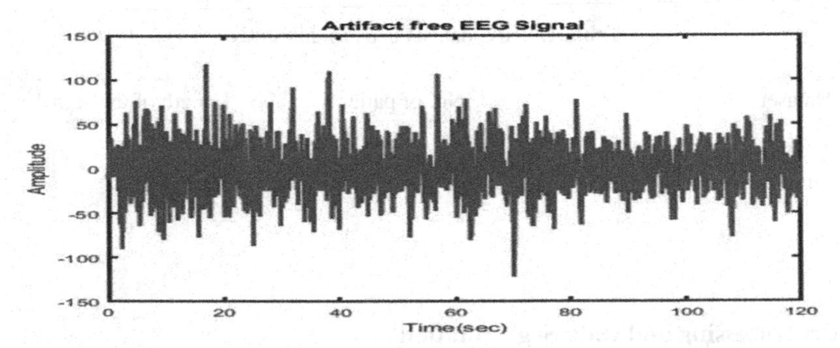

Fig. 3. Artifact free EEG Signal.

2.3 Wavelet Decomposition

When it comes to analyzing time domain signals, wavelet transform plays a significant role as it provides data on both frequency and time of the signal. The discrete wavelet transform (DWT) is simple and reduces the amount of time and resources required for computation [12, 13]. The signal to be analyzed is passed through a series of filters with varying cutoff frequencies and scales. The decomposition is calculated by filtering the discrete signal multiple times up to a predetermined level of N. A signal is split into two bands where the low pass filter extracts the signal's approximation coefficients and a high pass filter, providing detail coefficients. Table 2 shows the EEG bands obtained using wavelet decomposition method.

Table 2. EEG bands and decompositional levels.

Wavelet Coefficients	Frequency (in Hz)	Bands of EEG Signal
D4	16–32	Beta
D5	8–16	Alpha
D6	4–8	Theta
A6	0–4	Delta

2.4 Spectrogram Generation

The dataset consists of 19-channel EEG recordings of 15 NDC and 15 children with LD. The sample is broken down into 10-s segments and the spectrogram images were generated using the short-time Fourier transform (STFT) for all the bands of the segmented part. Each of the 10-s segments generates 4 bands in all the 19 channels of 30 subjects. Therefore, total of 27,360 spectrograms were generated including both the groups. In the

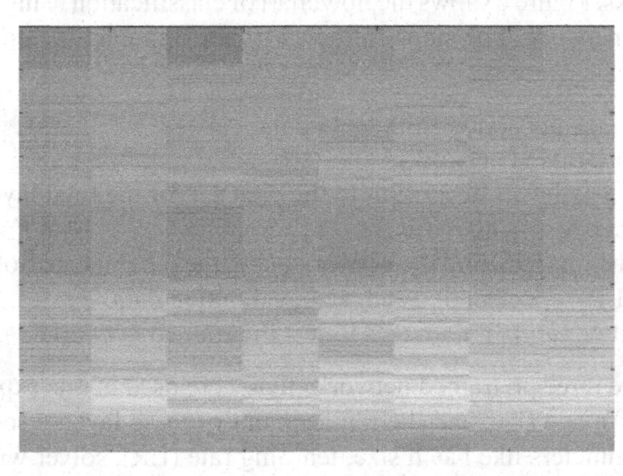

Fig. 4. Sample spectrogram image (alpha band) of LD children.

final step, the images saved were given as an input separately to the pretrained network for classification and feature extraction respectively. Figures 4 and 5 shows the sample of spectrogram images generated for the LD and NDC group.

Fig. 5. Sample spectrogram image (alpha band) of NDC.

2.5 Transfer Learning Using Pre-trained Network

Due to its self-feature learning capability and accurate classification results on multiclass classification problems, CNN is found to be an ideal for image-based classification [14]. A CNN is made up of a convolution layer, rectified linear unit (ReLU), activation function, a pooling layer, and batch normalization. Further, at the final layers it has fully connected layer, drop out, softmax, and classification output layers. The convolution layer consists of filters that help in detecting the patterns of images like edges, shapes, textures, and objects. In this experiment, three pretrained CNN models were used to find the best model for the proposed 2-class problem: Alexnet, VGG16 and Resnet18 [15]. The methodology proposed was tested using two methods: (1) Classification using transfer learning with a pretrained network (2) Extraction of image features using these pretrained networks. Figure 6 shows the flowchart of classification using transfer learning with a pretrained network. The steps followed to perform classification using transfer learning:

1. Spectrogram image generated from EEG data.
2. Selection of a pretrained network.
3. Substitution of new layers according to the data set for the final layers and specifies the number of classes (in our case, 2).
5. After resizing the images, train the network using the parameters: Solver name, initial learning rate (ILR), epochs no., batch size, validation data.
6. Classify the test or validation images to test the trained network.

Table 3 lists the three pretrained networks that were used in this experiment with the number of layers, image size, and the replacement of three layers used for pre-trained networks. The parameters like batch size, learning rate (LR), solver with 64, 0.001 and stochastic gradient descent with momentum (sgdm) were used.

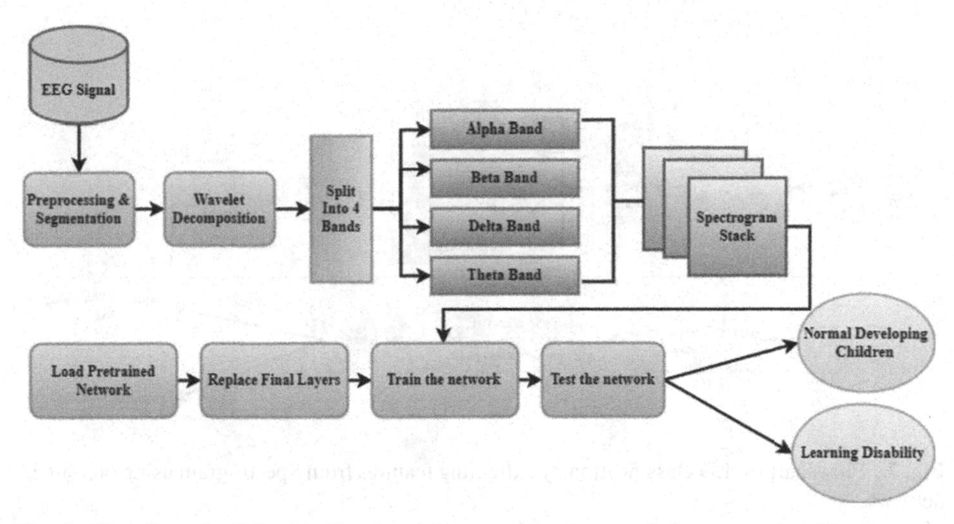

Fig. 6. Flowchart for LD classification using pretrained network by generating Spectrogram images from the bands.

Table 3. Pre-trained networks used in this study.

Pretrained Network	Depth	Number of Layers	Input Image Size	Final Layer Replaced
Alexnet	8	25	227 x 227	fc8, prob, output
VGG-16	16	41	224 x 224	fc8, prob, output
Resnet 18	18	72	224 x 224	fc1000, prob, ClassificationLayer_predictions

2.6 Feature Extraction Using Pre-trained Network

In this experiment, we used a pretrained network as a feature extractor by utilizing the activation feature of layers and further classifying with SVM classifier of different kernel functions. The extraction of features is the simple and an efficient method from deep network as it needs to pass the data only once. In the present study, fc8 layer for Alexnet, fc8 for VGG-16 and fc1000 for Resnet-18 were used. The flow for extraction of image features using these deep networks is shown in Fig. 7. The features were extracted from the deeper layers of these pretrained networks as it consists of high-level features that were built using the features from previous layers [16, 17].

2.7 Classification

The dataset was split into 70% for training and 30% for testing. Firstly, the classification of NDC and children with LD was computed using Alexnet, VGG16 and Resnet-18. Then secondly, features extracted from spectrogram images using pretrained networks were

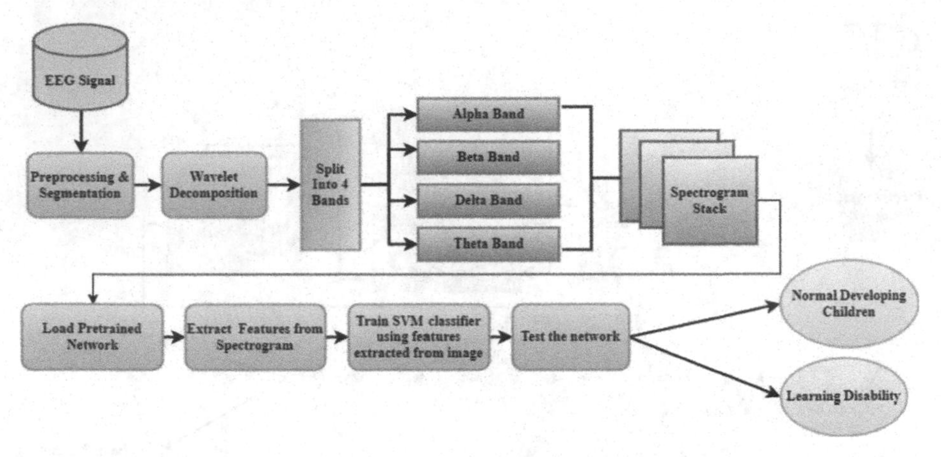

Fig. 7. Flowchart for LD classification by extracting features from Spectrogram using pretrained network.

fed to SVM classifier with various kernel functions to test the model. The classification accuracy was evaluated to assess the performance of model. The experiment was carried out multiple times, with the average results being reported.

3 Experimental Results and Discussion

This paper's classification task was completed using the two strategies described in Sect. 2. The first step is to feed the spectrogram images generated from the bands of 10 s segment length into AlexNet, VGG16, ResNet-18. The second approach was to extract the features using the same pretrained network and classifying it with the help of SVM. The entire experiment was performed using MATLAB 2021a. The results obtained for two different modalities were presented and discussed further in this section.

Table 4. Classification accuracy of different pre-trained network.

Spectrogram generated from bands	Accuracy		
	Alexnet	VGG-16	Resnet-18
Alpha Band	80.52%	77.3%	76.27%
Beta Band	63.27%	58.32%	60.74%
Delta Band	77.57%	72.67%	70.34%
Theta Band	75.33%	74.89%	75.20%
All bands combined	84.59%	85.98%	81.56%

The classification results obtained from the pre-trained network was given in Table 4. Using pre-trained model we were able to achieve maximum accuracy of 85.98% with all

Table 5. Classification accuracy obtained using SVM after extracting the features from Alexnet.

Classifier	Accuracy			
	Alpha	Beta	Delta	Theta
Linear SVM	86.8%	67.2%	69.2%	80.7%
Quadratic SVM	85.7%	77.1%	85.6%	83.4%
Cubic SVM	97.7%	84.9%	90.3%	84.6%
Fine Gaussian SVM	98.3%	85.6%	92.5%	86.2%
Medium Gaussian SVM	86.5%	64.1%	88.6%	79.7%

Table 6. Classification accuracy obtained using SVM after extracting the features from VGG-16.

Classifier	Accuracy			
	Alpha	Beta	Delta	Theta
Linear SVM	81.3%	65.2%	65.39%	73.56%
Quadratic SVM	80.7%	73.3%	82.67%	86.5%
Cubic SVM	95.67%	81.45%	87.38%	81.45%
Fine Gaussian SVM	93.26%	82.38%	89.45%	84.67%
Medium Gaussian SVM	84.35%	62.37%	88.6%	78.63%

Table 7. Classification accuracy obtained using SVM after extracting the features from Resnet-18.

Classifier	Accuracy			
	Alpha	Beta	Delta	Theta
Linear SVM	83.33%	63.35%	60.28%	78.25%
Quadratic SVM	92.27%	73.30%	76.56%	82.78%
Cubic SVM	91.6%	85.60%	89.45%	80.2%
Fine Gaussian SVM	82.74%	80.56%	87.23%	86.2%
Medium Gaussian SVM	84.35%	65.89%	83.45%	86.56%

bands spectrogram using VGG-16. When using spectrogram images of EEG bands separately we observed highest accuracy of 80.52% with alpha band using Alexnet, 77.3% with alpha band using VGG16 and 76.27% with alpha band using Resnet-18 classifiers. It is observed that the alpha band showed highest accuracy for classifying the data for all classifiers. When used the combined band images the accuracies were increased to 84.59%, 85.98% and 81.56% using Alexnet, VGG16 and Resnet-18 respectively. With the second method (features extraction from the pretrained network and SVM classifiers) we observed highest accuracy of 98.3% in fine Gaussian SVM classifier with alpha band features extracted using Alexnet (see Table 5). Next to Alexnet, we observed classification accuracies of 95.6% and 92.27% using cubic SVM and quadratic SVM with VGG16 and Resnet-18 features respectively (see Tables 6 and 7). It is observed that alpha band showed highest classification accuracies compared to other bands in all conditions. Our results can be compared with previous literatures on classification of LD or sub groups of LD such as dyslexia or dyscalculia. Perara et al. [18] achieved accuracy of 78.2% using cubic SVM when classifying the dyslexics and control groups using EEG signals during writing and typing. Jothi Prabha et al. [19] obtained classification accuracy of 90% with linear SVM when classifying dyslexics using EEG eye movements. In other study Rezvani et al. [20] used local graph network features to classify the dyslexic group using ML classifiers and obtained maximum classification accuracy of 95.4% with SVM classifier. Compare to other studies in the literature the present study method of image feature extraction from pretrained network with SVM classifier showed classification accuracy of 98.3% which highest among the existing literatures. The present study includes certain limitation which includes fewer subjects which is due to practical difficulty in recruiting children for the study. The present study suggests including more samples and advanced deep learning implementation for classification of LD in future work.

4 Conclusion

In conclusion, the present study aimed at classifying the LD children using rest-EEG signals. EEG signals were segmented into 10 s epoch and EEG bands were extracted. The 10 s EEG data then converted into spectrogram images and used for classification. Two approaches were implemented for the classification of LD children. Firstly, classification of LD children with EEG spectrogram images using pretrained networks such as Alexnet, VGG16 and Resnet-18. Secondly, the spectrogram image features were extracted from the pretrained networks and used for classification using SVM classifier. Our results showed highest accuracy 98.3% in fine Gaussian SVM classifier with alpha band features extracted using Alexnet. The present study concludes the image features extracted from the pretrained network with SVM classifier out-performed the transfer learning approach using Alexnet, VGG16 and Resnet-18.

Acknowledgement. This work is supported by the Science and Engineering Research Board, Government of India (Grant no: SERB/F/3051/2022–2023, Sanction no: CRG/2020/004974) and the authors are grateful to them. Also, the authors are thankful to all the parents, Principals and psychologists of special schools who had participated in this study.

References

1. Fajariyanti, F.M., Agata, D., Harsono, T.: Expert system for learning disability classification in school-age children. In: International Conference on Applied Science and Technology on Social Science 2021 (iCAST-SS 2021), pp. 231–237. Atlantis Press (2022)
2. Hulme, C., Mackenzie, S.: Working Memory and Severe Learning Difficulties (PLE: Memory). Psychology Press (2014)

3. Tamboer, P., Vorst, H.C.M., Ghebreab, S., Scholte, H.S.: Machine learning and dyslexia: classification of individual structural neuro-imaging scans of students with and without dyslexia. NeuroImage Clin. **11**, 508–514 (2016)
4. Bruno, R.M., Walker, S.C.: Comprehensive test of phonological processing (CTOPP). Diagnostique **24**, 69–82 (1999)
5. Woodcock, R.W.: The Woodcock-Johnson tests of cognitive ability—Revised (1997)
6. Treloar, J.M.: Wechsler individual achievement test (WIAT). Interv. Sch. Clin. **29**, 242–246 (1994)
7. Popa, L.L., Dragos, H., Pantelemon, C., Rosu, O.V., Strilciuc, S.: The role of quantitative EEG in the diagnosis of neuropsychiatric disorders. J. Med. Life **13**, 8–15 (2020)
8. Manghirmalani, P., Panthaky, Z., Jain, K.: Learning disability diagnosis and classification-A soft computing approach. In: 2011 World Congress on Information and Communication Technologies. pp. 479–484. IEEE (2011)
9. Al-Barhamtoshy, H.M., Motaweh, D.M.: Diagnosis of Dyslexia using computation analysis. In: 2017 International Conference on Informatics, Health & Technology (ICIHT), pp. 1–7. IEEE (2017)
10. Mahmoodin, Z., Mansor, W., Lee, K.Y., Mohamad, N.B.: An analysis of EEG signal power spectrum density generated during writing in children with dyslexia. In: 2015 IEEE 11th International Colloquium on Signal Processing & Its Applications (CSPA), pp. 156–160. IEEE (2015)
11. Balasubramanian, G., Kanagasabai, A., Mohan, J., Seshadri, N.P.G.: Music induced emotion using wavelet packet decomposition—An EEG study. Biomed. Signal Process. Control **42**, 115–128 (2018)
12. Oltu, B., Akşahin, M.F., Kibaroğlu, S.: A novel electroencephalography based approach for Alzheimer's disease and mild cognitive impairment detection. Biomed. Signal Process. Control **63**, 102223 (2021). https://doi.org/10.1016/j.bspc.2020.102223
13. Sharmila, A., Geethanjali, P.: DWT based detection of epileptic seizure from EEG signals using naive bayes and k-NN classifiers. IEEE Access **4**, 7716–7727 (2016). https://doi.org/10.1109/ACCESS.2016.2585661
14. Raghu, S., Sriraam, N., Temel, Y., Rao, S.V., Kubben, P.L.: EEG based multi-class seizure type classification using convolutional neural network and transfer learning. Neural Netw. **124**, 202–212 (2020)
15. Deep Learning-Pretrained Deep Neural Networks. https://in.mathworks.com/help/deeplearning/ug/pretrained-convolutional-neural-networks.html. Last accessed 10 Oct 2022
16. Deep Learning-Extract Image Features Using Pretrained Network. https://in.mathworks.com/help/deeplearning/ug/extract-image-features-using-pretrained-network.html;jsessionid=e57284ef4483914f39cceb2c12f5. Last accessed 10 Oct 2022
17. Zeiler, M.D., Fergus, R.: Visualizing and understanding convolutional networks. In: Fleet, D., Pajdla, T., Schiele, B., Tuytelaars, T. (eds.) ECCV 2014. LNCS, vol. 8689, pp. 818–833. Springer, Cham (2014). https://doi.org/10.1007/978-3-319-10590-1_53
18. Perera, H., Shiratuddin, M.F., Wong, K.W., Fullarton, K.: EEG signal analysis of writing and typing between adults with dyslexia and normal controls. Int. J. Interact. Multimed. Artif. Intell. **5**, 62–67 (2018)
19. Jothi Prabha, A., Bhargavi, R.: Prediction of dyslexia from eye movements using machine learning. IETE J. Res. **68**, 814–823 (2022)
20. Rezvani, Z., Zare, M., Žarić, G., Bonte, M., Tijms, J., Van der Molen, M.W., González, G.F.: Machine learning classification of dyslexic children based on EEG local network features. BioRxiv. 569996 (2019)

A Two-Level Classifier for Prediction of Healthy and Unhealthy Lung Sounds Using Machine Learning and Convolutional Neural Network

Vaibhav Koshta and Bikesh Kumar Singh(✉)

Department of Biomedical Engineering, National Institute of Technology, Raipur, India
{vkoshta.srf2022.bme,bsingh.bme}@nitrr.ac.in

Abstract. The ability to detect lung disorders with the aid of lung sound auscultation has become very common these days. The early diagnose of these diseases can cause a reduced risk to loss of life. Auscultation has a disadvantage that it requires a skilled practitioner to listen and remember the lung sound from different parts of the posterior chest and predict the kind of lung disease. In this paper, a comparative study of one-dimension convolutional neural network (1D CNN), and machine learning methods is demonstrated on the lung sound database obtained from Kaggle. SMOTE (Synthetic Minority Over-Sampling Technique) is applied to treat the imbalance in the unhealthy recordings. The sound features like Mel spectrogram, MFCC (Mel Frequency Cepstral Coefficient), MFCC Delta, MFCC Delta-Delta, pitch, spectral centroid, spectral entropy, spectral flux, spectral kurtosis, spectral skewness and spectral spread was obtained and tested on different machine learning models using 10-fold, 5-fold cross validation and 33% hold out. The results obtained under normal and abnormal classification for 10-fold, 5-fold and 33% hold out showed an accuracy of 78.6%, 77.5% and 76.4% for MFCC Delta-Delta features respectively for fine KNN machine learning model. Under Asthma and COPD binary classification, proposed study showed an accuracy of 91.6%, 90.6%, 89.1% respectively for 10-fold, 5-fold and 33% hold out for subspace KNN and fine KNN models. The 1D CNN neural network implements the Normal-Abnormal and Asthma-COPD two level binary classification showed an accuracy of 94.67% and 75.92%.

Keywords: Lung sound · auscultation · MFCC · MFCC Delta-Delta · MFCC Delta · Mel Spectrogram · 1D-CNN · CNN · asthma · COPD · binary classification · spectral skewness · pitch · spectral spread · spectral entropy · spectral kurtosis · neural network etc.

1 Introduction

Some of the most common diseases with lung include Asthma, Chronic Obstructive Pulmonary Disease (COPD), upper and lower respiratory tract infection, pneumonia etc. Since the invention of stethoscope in 1816 auscultation is the most used and only method for listening the respiratory sounds. The primary method of detecting lung disease is the

B. K. Singh et al. (Eds.): ICBEST 2023, CCIS 2003, pp. 154–168, 2024.
https://doi.org/10.1007/978-3-031-54547-4_13

auscultation technique, where a doctor or medical practitioner listens to the lung sound from the posterior chest and detects the disorder or associated lung disease [23]. It is a noninvasive method, economical and does not poses any harm to the patients being treated. Auscultation requires hearing and recording the lung sounds for the purpose of analysis to classify the patient as having Asthma and COPD. It requires the doctors to remember the lung sound thus creating the demand for computer-aided technologies enabling the doctors to record and store the sounds and run the deep learning model to predict asthma and COPD. Lung sounds can be classified as wheeze, rhonchi, crackles etc. Wheezing occurs when the bronchial tubes are narrowed, is characterized by the high-pitched whistling sound made during the inhale or exhale phase of the respiratory cycle and is associated with Asthma and COPD. Rhonchi is characterized by the continuous low-pitched sound caused during fluid movement in the large airways. On the other hand, crackles are the discontinuous short sounds occurring due to the air sacs filling the fluid in the lung. It is most seen in persons with pneumonia or heart failure when there's any air sac movement while you are breathing.

An adventitious respiratory sound classification network with channel spatial was proposed. The proposed method extract Mel Frequency Cepstral Coefficient (MFCC) and Mel spectrogram from the lung sounds available in ICBHI 2017 database showed an accuracy of 80% and 92.4% for abnormal-normal and crackles-wheeze classification [1]. A light weight CNN was designed and tested on ICBHI 2017 database using the scalogram images of the lung sounds. The method produces an accuracy of 98.92% and 98.70% for three class classification and six disease classes for pathological classification [2]. A Fourier Bessel Series Expansion based flexible time frequency wavelet transform was proposed for focal epilepsy classification using mixture entropy and exponential energy. The proposed method obtained an accuracy of 95.85% for 10 fold classification with least square support vector machine (LS-SVM) classifier [3]. A deep CNN-RNN model was proposed with Mel spectrogram feature extraction classifying the respiratory sounds in ICBHI 2017. The proposed method achieves an accuracy of 66.31% for four class classification problem. A patient independent specific data was introduced and it was shown that the patient specific re-training of the model performance was better than original train test validation method an achieves an score of 71.81% for leave one out validation [4]. An ICBHI 2017 database was utilized to test the pre-trained deep CNN model with support vector machine (SVM) classifier. A pre-trained deep CNN with transfer learning utilizing spectrogram lung sound for classification was proposed and the accuracy obtained under CNN and SVM classifiers are 65.5% and 63.09% respectively [5].

The lung sounds were recorded from four-channel data acquisition system over the posterior chest. The power spectral density was calculated from the Welch method and decomposed into sub bands to work in artificial neural network to solve classification using power spectral density and SVM. The results obtained showed an accuracy of 89.2% and 93.3% for 2-channel and 3-channel combination [6]. A novel approach for EEG signal classification using Fourier Bessel coefficient and parametric representation was proposed [7]. An Empirical Mode Decomposition (EMD) and Ensemble Empirical Mode Decomposition (EEMD) methods were applied to classify the epilepsy patients.

A hybrid IMF (intrinsic mode selection) strategy was discussed utilizing energy, correlation, power spectral density and statistical significant measures to select the most informative IMF.

The proposed hybrid IMF selection method was implemented under EMD and EEMD. The accuracies obtained for different classifiers were listed [8].

A Fourier Bessel Series Expansion in addition with Empirical Wavelet Transform (FBSE-EWT) for the detection of valvular heart disease using PCG (Phonocardiogram) signal was implemented. A two dimension phase space reconstruction was utilized to extract the features from the IMF's obtained from EWT. Three metaheuristic optimization techniques viz. SSOA, EPOA and TGOA optimizes the feature set dimension reduction. The results showed that the proposed method obtained the highest accuracy of 98.53%, 98.84%, 99.07%, and 99.70% for five, four, three and two class classification problem respectively [9]. Decision trees and discriminant classifier were utilized to classify lung sound recordings. Shannon entropy, logarithmic energy entropy and spectral entropy was calculated [10]. The recognition of wheeze sound in asthmatic and pneumonia affected subjects was done over 300 subjects including asthma, pneumonia and normal subjects. The classification problem was solved using support vector machine and achieved an accuracy of 96% [11]. A deep learning convolutional neural network (CNN) classifying lung sounds into crackle, wheeze and ronchi was demonstrated, and it shows an accuracy of 85.7% [12]. A neonatal heart and lung sound quality assessment was done in order to store the high quality heart and lung sounds which can be transmitted for remote monitoring and diagnosis. The proposed study was capable of distinguishing between high and low quality signal with an accuracy of 93% and 82% for heart and lung sound respectively [13]. An adaptive technique based on discrete wavelet transform and artificial neural network to remove the noise from the lung sound. The results showed that the model performs the denoising process well in the range of −2 to 20 dB [14].

A CNN model was trained and tested on the ICBHI 2017 lung sound database utilizing the Mel spectrogram images and achieved an accuracy of 43% [15]. Empirical mode decomposition method was demonstrated on asthma, bronchitis and normal patients at preprocessing stage. Standard deviation, Shannon energy, peak to peak and root mean square features were extracted and tested on different KNN classifier has an accuracy of 99.3% [16]. Lung sound identification algorithm, VGGish-BiGRU designed utilizing the concept of transfer learning was demonstrated. The features obtained from VGGish pre-training were transferred to the target layer of the VGGish network layer and the parameters of BiGRU were fine tuned which showed an improvement in the accuracy of recognition [17]. MFCC with SVM and spectrogram images in the CNN were utilized and tested on four different sets viz. Healthy-pathological classification, rale-rhonchus-normal classification, singular respiratory sound and audio classification. The accuracy obtained under CNN and SVM for four categories were listed [18].

A quad binary pattern with wavelet packet decomposition to extract the histogram and non linear features from EEG signals to classify it into focal and non focal signals using ANN was shown and classification accuracy of 95.74% was obtained [19]. A convolutional recurrent neural network to classify the lung sounds recorded from 16 channel lung sound recording developed under study was demonstrated. Different deep neural network were compared and it was shown that the proposed method outperforms all the networks with an F1 score of 92% [20]. Continuous wavelet transform, short time Fourier transform and pseudo Wigner-Ville distribution (SPWVD) to convert EEG signals into scalogram, spectrogram and SPWVD time frequency plots was utilized. These plots were fed to AlexNet, VGG16, ResNet50 and CNN and highest accuracy recorded is 93.36% under SPWVD time frequency and CNN model [21]. Lung sound classification using entropy was done where 7 different types of entropies viz. Shannon entropy, spectral entropy, renyi entropy, tsallis entropy, wavelet entropy, approximate entropy and sample entropy was calculated and tested individually and combination of entropies on multilayer perceptron, the results obtained were validated using three fold cross validation. Accuracy obtained was 94.95% for five classes of data [22].

The effectiveness of EMD in the classification of lung sounds into chronic and non-chronic categories was demonstrated. The 2D-PSR (Phase space representation) and higher order PSR were constructed. It was shown that the Ensemble bagged trees model provides an accuracy of 97.14% computed over feature space constituted by 10 dimensional-PSR [23]. Mel spectrogram, chroma, chroma (constant Q), chroma CENS features were utilized to test and validate the convolutional neural network (CNN). The ICBHI database was used to classify the respiratory audio data into Chronic Obstructive Pulmonary Disease (COPD) [24].The contribution of the present study are as follows:

- A one dimension convolutional neural network is implemented for automatic feature extraction and classification of the lung sound signals into two classes viz. Normal-abnormal and asthma-COPD.
- SMOTE technique was applied for the purpose of data balancing on the online available dataset in Kaggle.
- Comparison of various machine learning models and 1D CNN was done and it was shown that the 1D CNN performance is better than the machine learning models.

2 Methodology

2.1 Dataset Description and Preprocessing

The dataset used under study is a respiratory dataset obtained from Kaggle. It consists of 228 r ecordings, including 105 healthy and 123 unhealthy observations from 141 males and 87 females belonging to the age group of 21 to 80 years. The dataset consists of normal, asthma and COPD patient's lung sound recording after applying three different types of filtration methods viz. Bell mode filtration which amplifies the signal in the frequency range of 20 Hz to 1 kHz, diaphragm mode filtration is applied to the signals ranging from 20 Hz to 2 kHz and extended mode filtration amplifies frequencies in the range 20 Hz to 1 kHz while focusing on the frequencies from 50 Hz to 500 Hz. The unhealthy observations have 96 and 27 lung sound recordings of asthma and COPD patient's respectively, it indicates the imbalance in the dataset. To treat the unbalancing, synthetic minority over sampling technique (SMOTE) was utilized resulting in 12582 observations under each unhealthy patient category. In SMOTE, it creates the additional data points which are not the duplicates of the original data rather then it shifts the original data slightly towards the nearest neighbor such that it resembles the original data, hence it does not add perturbations in to it. Segmentation technique was employed during the designing of one dimensional convolutional neural network as it requires large dataset for training of its different layers. To segment the audio files we have used MIR tool box in MATLAB utilizing mirsegment() function to divide the audio recordings into smaller segments of 5 s each (Fig. 1).

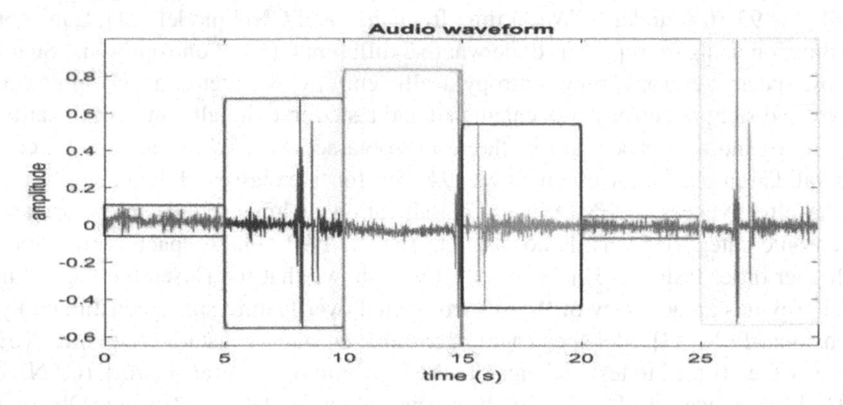

Fig. 1. Segmentation

2.2 Feature Extraction

It is process of extracting the important information from a signal represented in time domain and frequency domain. These extracted features are helpful in classifying the signal as normal and abnormal. Further abnormal signals are classified two classes as asthma and COPD using machine learning techniques. Following features are extracted from the signals under study using a hamming window of 1024 sample size:

a. **Mel Spectrogram**
   Mel spectrogram is generally used to map the frequencies in the audio signal to a Mel scale which resembles the human hearing to be equal distance from each other. It is a nonlinear visualization of the frequency components presents in the signals plotted with respect to time.
b. **MFCC**
   MFCC calculation is a compact representation of an audio signal when expressed as a sum of infinite sinusoids. The audio signals by nature are nonlinear signal i.e., there fr equencies vary with time hence taking Fast Fourier Transform (FFT) of the entire signal is not a feasible solution as it would lose some frequencies over time. To calculate the MFCC feature, a Short Time Fourier Transform (STFT) is utilized which uses a short time frame of typically 25ms and integrating a hamming window with 1024 samples. A 512-point FFT is calculated over each time frame with application to Mel spaced filter bank with 40 set of triangular filters. These obtained filter bank coefficients have a disadvantage of high correlation among them and to avoid this discrete cosine transform is calculated.

c. **MFCC Delta**
It is obtained from the first derivative of MFCC feature and used to represent the changes in cepstral features over time. The MFCC delta features have an advantage over MFCC that it is.
used to represent the temporal information present among various frames to differentiate the overlapping trajectories.

d. **MFCC Delta-Delta**
MFCC Delta-Delta features are obtained by differentiating the MFCC Delta features and are used to represent the changes between the frames in the MFCC Delta features. They carry information about the peaks and valleys in the given trajectories.

e. **Spectral Centroid**
It is one of the most common features which indicate the central mass of magnitude spectrum of an audio signal. It is the weighted mean of the frequencies present in the signal.

e. **Spectral Entropy**
It represents the amount of disorder in the spectrum. Higher disorder corresponds to higher entropy and vice versa. The spectral entropy is calculated as (1), s(f) is the fourier transform of signal s(t):

$$SE = \sum_{f}^{\infty} s(f) * \ln(\frac{1}{s(f)}) \quad (1)$$

f. **Spectral Flux**
It is used to show the rate of change of power spectrum in one frame with respect to the previous frame. It calculates the Euclidean distance between the two normalized spectrum and by comparing the power spectrum of two frames, it determines the rate of change of power.

g. **Spectral Kurtosis**
Kurtosis is a feature of audio signal which is used to measure the outliers present in the data. It is a most common feature used to indicate the non-stationarity and non-gaussian characteristics in frequency domain representation of a signal.

h. **Spectral Skewness**
Skewness shows the difference in the shape of the spectrum above and below the mean fr equency. It provides information about right and left skewness.

i. **Spectral Spread**
Spectral spread is the property of an audio signal which provides the information about the spread in the power spectrum around the spectral centroid.

j. **Pitch**
It gives the quality of sound produced by the vibration generating it. It is used to distinguish between an acute and flat note of a sound.

2.3 One Dimensional Convolutional Neural Network (1D CNN)

Convolutional neural networks are the subset of deep learning which is a branch of machine learning and used in multiple areas for solving classification problem. Major areas of application include object detection, image processing, text analysis, speech recognition etc. Most commonly, CNN consist of convolutional layer and multi-layer feed forward layer. A convolutional layer is generally located after the input layer, it does convolution by sliding the kernel through a defined steps known as stride. Batch size is also one of the important deciding factors for training of CNN model, it represents the total number of samples which can be processed for training. Pooling layer is used to reduce the size of feature vector matrix while highlighting the prominent features.

A multi-layer feed forward or fully connected layer is used to connect the neurons in the layer to the previous layer neurons. The multi-layer feed forward layer (MLFF) operates by predicting either a class or value from the features extracted through convolution operation. On the other hand, back propagation method can also be used in the designing of CNN network. It calculates the loss function which provides information about how much th e predicted output differs from the original output.

1D CNN network designed under study consists of convolutional 2D layer in which the first layer is kept as '1', batch normalization layer, max pooling layer and relu activation function is applied to all the layers except the output layer which uses the softmax function for binary classification. Basic block diagram and model of the convolutional neural network is shown (Figs. 2 and 3).

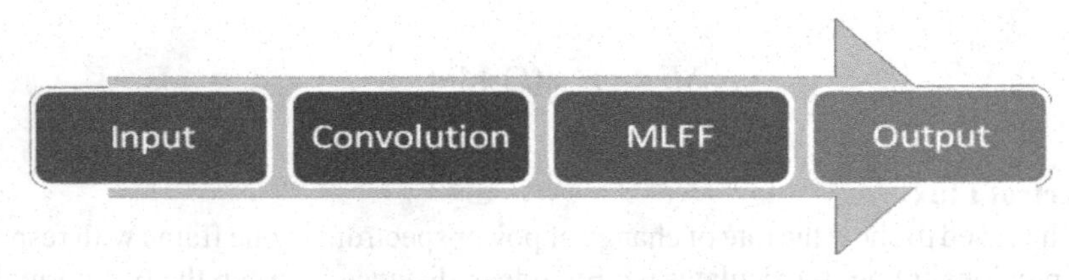

Fig. 2. Block diagram of CNN

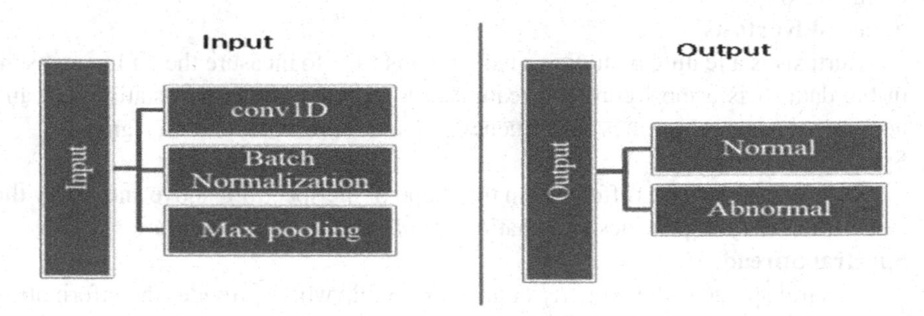

Fig. 3. CNN model

3 Result and Discussion

The dataset used under study is a respiratory dataset obtained from Kaggle. It consists of 228 r ecordings, including 105 healthy and 123 unhealthy observations from 141 males and 87 females belonging to the age group of 21 to 80 years. Under study, we have used MATLAB R2021a software and classification learner app to solve two stage classification problem as Normal-Abnormal and Asthma-COPD respectively. The inbuilt machine learning models in the classification learner app is used with default parameters values along with the dataset of spectral features as mentioned in the Sect. 2.2. The results obtained from the first stage of classification i.e., Normal and Abnormal using 10-fold, 5-fold and 33% hold out is tabulated (Table 1). It is evident from the observations that fine KNN model gives highest accuracy as 78.6%, 77.5% and 76.4% corresponding to MFCC Delta-Delta feature. The region operating characteristics (ROC) were plotted and area under the curve (AUC) is calculated for each of the 30 machine learning models in the classification learner app. The ROC curve obtained for the highest accurate model shows the area under curve of 79%, 78% and 77% for 10-fold, 5-fold and 33% hold out fine KNN model for MFCC Delta-Delta feature is shown in Figs. 4, 5 and 6 respectively.

Table 1. Normal-Abnormal classification cross validation and hold out result

Method	Model	Feature	Accuracy	Sensitivity	Specificity	Precision
10-fold cv	Fine KNN	MFCC Delta-Delta	78.6%	76%	81%	79.6%
5-fold cv			77.5%	74.9%	80%	78.4%
33% hold			76.4%	73.5%	79.3%	77.9%

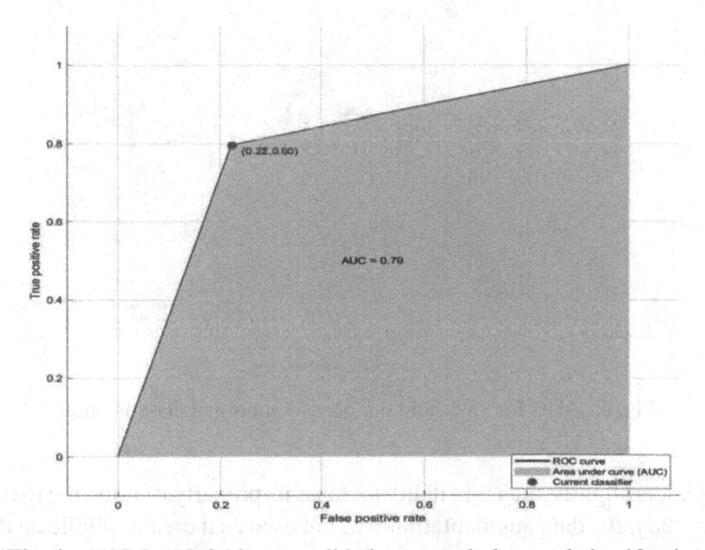

Fig. 4. AUC for 10-fold cross validation normal-abnormal classification

The second level of binary classification has been done for classifying the patients into asthma and COPD affected. Here, it has been observed that class imbalance was present in the dataset with 96 and 27 lung sound recordings from asthma and COPD patients respectively. Hence SMOTE technique was utilized to treat the unbalancing, in SMOTE we generate the synthetic data points by shifting the original data points to its

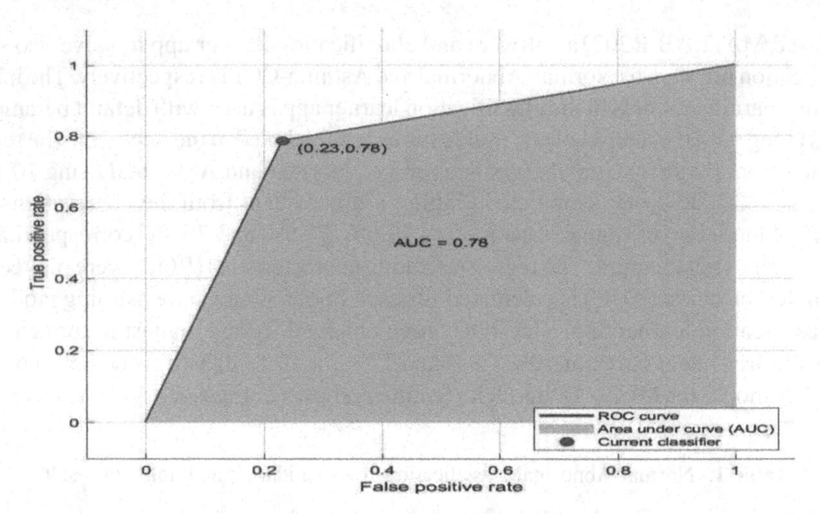

Fig. 5. AUC for 5-fold cross validation normal-abnormal classification

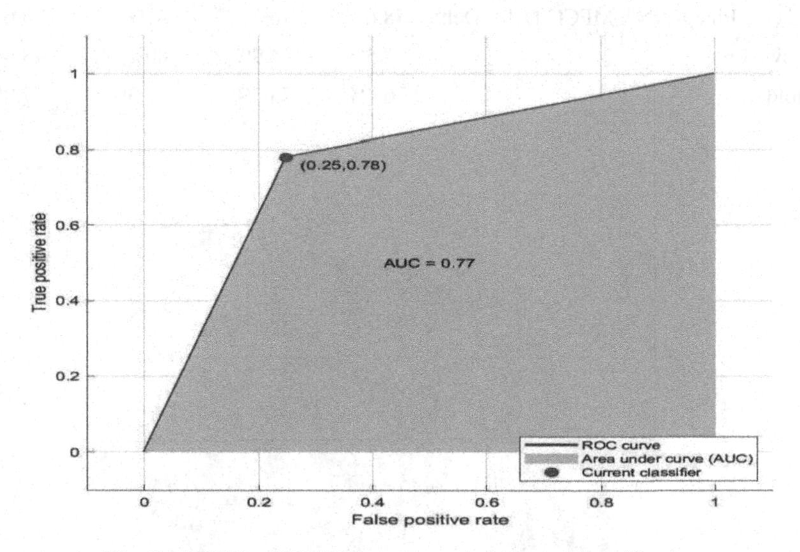

Fig. 6. AUC for 33% hold out normal-abnormal classification

nearest neighbors slightly such that it did not loses its property of referring to the original data. Under study, the data augmentation was not used as it creates duplicate data points thereby adding perturbation in the original dataset, however this is not a problem with the SMOTE (synthetic minority over sampling technique). For classification, different machine learning models in MATLAB classification learner app were trained and tested with default parameters values on the balanced dataset. The results obtained from all the models showed the best accuracy of 91.6% and 90.6% by subspace KNN for 10-fold and 5-fold cross validation respectively. However, 33% hold out method showed the best accuracy of 89.1% by fine KNN machine learning model (Table 2). The region operating characteristic (ROC) under each model was obtained and area under curve is calculated. The ROC for the highest accurate model corresponding to 10-fold, 5-fold and 33% hold out is shown (Fig. 7).

Table 2. Asthma-COPD classification cross validation and hold out result

Method	Model	Feature	Accuracy	Sensitivity	Specificity	Precision
10-fold cv	Subspace KNN	MFCC DeltaDelta	91.6%	97.8%	85.4%	87.04%
5-fold cv			90.6%	97.2%	83.9%	85.8%
33% hold	Fine KNN		89.1%	96.29%	82%	84.25%

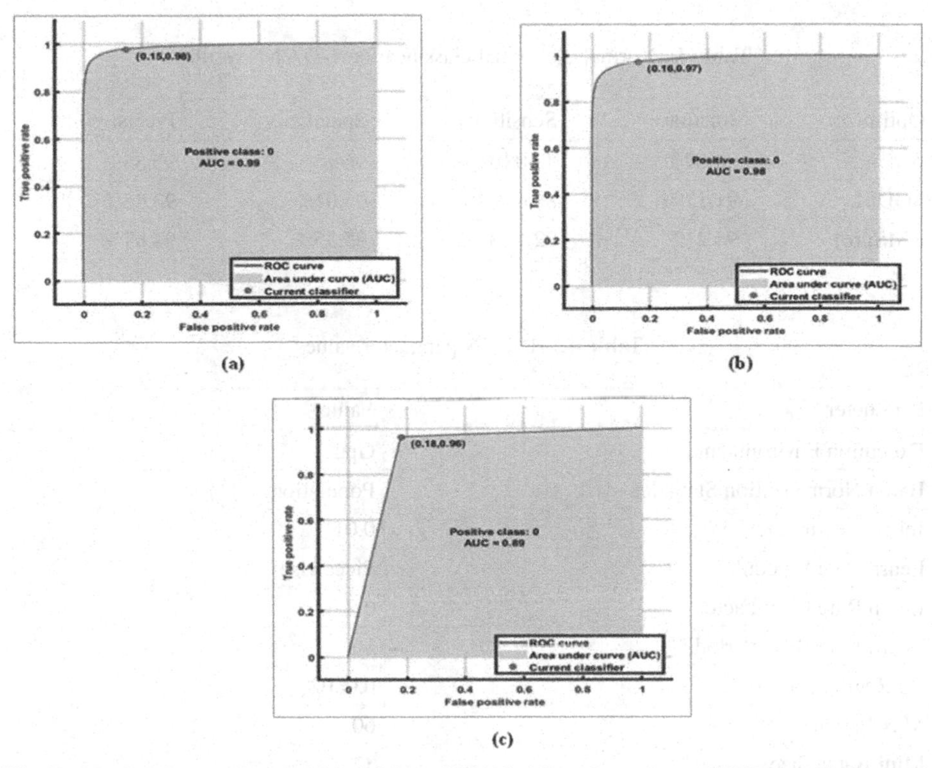

Fig. 7. The ROC curve for 10-fold, 5-fold and 33% hold out for asthma-COPD classification. (a) 10-fold cross validation, (b) 5-fold cross validation, (c) 33% hold out

In addition to deploying machine learning models for predicting the patients into normal-abnormal and asthma-COPD binary classes, under study a one-dimensional convolutional neural network (1D-CNN) is also utilized. Since the total dataset available has 228 recordings including healthy and unhealthy patients, hence it is difficult to train a deep neural network. To increase the dataset, we have integrated segmentation method to break each audio file into a smaller segment of 5 s (Fig. 1). A 1-D CNN is designed using convolutional 2d layer where the first dimension is kept as '1', batch normalization layer, max pooling layer with relu activation function. Three such convolution layers were introduced in the network including 16 filters and stride of 20, 8 filters with stride of 10 and 8 filter with stride of 6 respectively for three convolutional layers. The output layer consists of a single fully connected layer and softmax function with two classes to classify the signals into normal and abnormal.

A CNN model with above layers is designed and tested against the parameter values as listed (Table 4) and the results obtained for different optimizers is listed (Table 3, Table 5). The best performance under normal-abnormal classification was obtained with adam optimization and the highest accuracy recorded is 94.67%. However, for asthma-COPD classification, CNN model recoded a highest accuracy of 75.92% with sgdm optimization.

Table 3. Normal-Abnormal classification 1-D CNN result

Optimizer	Accuracy	Sensitivity	Specificity	Precision
Adam	94.67%	93.08%	96%	95.35%
SGDM	91.17%	86.69%	95.01%	93.85%
RMSprop	94.21%	92.67%	95.55%	94.87%

Table 4. 1D CNN parameter values

Parameter	Value
Execution Environment	Gpu
Batch Normalization Statistics	Population
Initial Learn Rate	0.01
Learn Rate Schedule	Piecewise
Learn Rate Drop Factor	0.2
Learn Rate Drop Period	16
L2 Regularization	0.0002
Max Epoch	60
Mini Batch Size	32
Shuffle	Every epoch
Verbose	false

Table 5. Asthma-COPD classification 1-D CNN result

Optimizer	Accuracy	Sensitivity	Specificity	Precision
Adam	74.65%	36.81%	81.39%	27.1%
SGDM	75.92%	43.67%	81%	22.5%
RMSprop	74.36%	50.83%	82.07%	30.42%

4 Comparison of Results Obtained from Proposed Work

The results obtained from different machine learning (ML) models under normal-abnormal and asthma-copd classification corresponding to 10-fold, 5-fold and 33% hold out is listed in Table 6. The 1D CNN model designed under study is also deployed for solving the binary classification problem for each category of combination viz. Normal – abnormal and asthma – COPD. The results obtained are listed in Table 6 for the purpose of comparison and hence the performance of ML model and 1D CNN is concluded.

The results showed that for classifying the subject into normal and abnormal classes, 1D convolutional neural network (CNN) outperforms all the machine learning models in classification learner app achieving the highest accuracy of 94.67% and sensitivity of 93.08%. However for classifying patients into asthma and COPD binary classes the machine learning models outperforms the 1D CNN network scoring a highest accuracy of 91.6% and sensitivity of 97.8%.

It is also observed from the Table 6 that the sensitivity of the proposed 1D CNN model is bit low while categorizing asthma and COPD subjects. One of the reasons behind this may be the similar respiratory symptoms present in airways of asthmatic and COPD patients, which confuses the decision network. However, more work is needed in the future to minimize the said limitation.

Table 6. Comparison of different models

Class	Method	Model/optimizer	Accuracy	Sensitivity	Specificity	Precision
Normal vs Abnormal	10-fold	Fine kNN	78.6%	76%	81%	79.6%
	5-fold	Fine kNN	77.5%	74.9%	80%	78.4%
	33%	Fine kNN	76.4%	73.5%	79.3%	77.9%
	1D CNN	Adam	94.67%	93.08%	96%	95.35%
Asthma vs COPD	10-fold	Subspace kNN	91.6%	97.8%	85.4%	87.04%
	5-fold	Subspace kNN	90.6%	97.2%	83.9%	85.8%
	33%	Fine kNN	89.1%	96.29%	82%	84.25%
	1D CNN	SGDM	75.92%	43.67%	81%	22.5%

5 Conclusion

The proposed study focuses on the analysis of the lung sounds classification under two levels as normal-abnormal and asthma-COPD. The dataset consists of 228 recordings, including 105 healthy and 123 unhealthy observations from 141 males and 87 females belonging to the age group of 21 to 80 years. Here, the comparative study of different machine learning models and one dimensional convolutional neural network was done, and results obtained were tabulated (Tables 1, 2, 3 and 5). Various machine models (viz. SVM, fine KNN, subspace KNN, linear discriminant, quadratic discriminant etc.) were trained and accuracy was calculated. It is observed that the fine KNN model performs the best under MFCC Delta-Delta feature for 10-fold, 5-fold and 33% hold out, classifying the lung sound as normal and abnormal with accuracies as 78.6%, 77.5% and 76.4% respectively (Table 1). However, for classifying patients into asthma and COPD affected, the best performance was shown by the subspace KNN and fine KNN machine learning models with MFCC Delta-Delta feature selected. The accuracy obtained was 91.6%, 90.6% and 89.1% for 10-fold, 5-fold and 33% hold out respectively (Table 2).

A 1D CNN model was designed and trained after doing the segmentation of the dataset under study. The segmentation has been done for each audio signal to break it into smaller segments of 5 s, it increases the total number of samples which help to train the network more efficiently. The parameter values used under study network is listed (Table 4) and the results obtained were tabulated (Tables 3 and 5). It is observed that the 1D CNN model gives highest accuracy of 94.67% for adam optimizer under normal-abnormal classification and 75.92% for sgdm optimizer under asthma-COPD classification. The comparative study shows that the 1D CNN model performances is better than the machine learning models.

Acknowledgement. The authors would like to express their gratitude to the funding agency ICMR for the grants provided to carry out this research ((R.11013/15/2021-GIA/HR)).

References

1. Xu, L., Cheng, J., Liu, J., Kuang, H., Wu, F., Wang, J.: ARSC-Net: adventitious respiratory sound classification network using parallel paths with channel-spatial attention. In: 2021 IEEE International Conference on Bioinformatics and Biomedicine (BIBM), pp. 1125–1130. IEEE (2021)
2. Shuvo, S.B., Ali, S.N., Swapnil, S.I., Hasan, T., Bhuiyan, M.I.H.: A lightweight CNN model for detecting respiratory diseases from lung auscultation sounds using EMD-CWT-based hybrid scalogram. IEEE J. Biomed. Heal. Informatics. **25**, 2595–2603 (2020)
3. Gupta, V., Pachori, R.B.: Classification of focal EEG signals using FBSE based flexible time – frequency coverage wavelet transform. Biomed. Signal Process. Control **62**, 102124 (2020)
4. Acharya, J., Basu, A.: Deep neural network for respiratory sound classification in wearable devices enabled by patient specific model tuning. IEEE Trans. Biomed. Circuits Syst. **14**, 535–544 (2020)
5. Demir, F., Sengur, A., Bajaj, V.: Convolutional neural networks based efficient approach for classification of lung diseases. Heal. Inf. Sci. Syst. **8**, 1–8 (2020)

6. Islam, M.A., Bandyopadhyaya, I., Bhattacharyya, P., Saha, G.: Multichannel lung sound analysis for asthma detection. Comput. Methods Programs Biomed. **159**, 111–123 (2018)
7. Pachori, R.B., Sircar, P.: EEG signal analysis using FB expansion and second-order linear TVAR process. Signal Process. **88**, 415–420 (2008)
8. Karabiber Cura, O., Kocaaslan Atli, S., Türe, H.S., Akan, A.: Epileptic seizure classifications using empirical mode decomposition and its derivative. Biomed. Eng. Online **19**, 1–22 (2020)
9. Khan, S.I., Qaisar, S.M., Pachori, R.B.: Automated classification of valvular heart diseases using FBSE-EWT and PSR based geometrical features. Biomed. Signal Process. Control **73**, 103445 (2022)
10. Fraiwan, L., Hassanin, O., Fraiwan, M., Khassawneh, B., Ibnian, A.M., Alkhodari, M.: Automatic identification of respiratory diseases from stethoscopic lung sound signals using ensemble cl assifiers. Biocybern. Biomed. Eng. **41**, 1–14 (2021)
11. Naqvi, S.Z.H., Arooj, M., Aziz, S., Khan, M.U., Choudhary, M.A.: Spectral analysis of lungs sounds for classification of asthma and pneumonia wheezing. In: 2020 International Conference on Electrical, Communication, and Computer Engineering (ICECCE), pp. 1–6. IEEE (2020)
12. Kim, Y., et al.: Respiratory sound classification for crackles, wheezes, and rhonchi in the clinical field using deep learning. Sci. Rep. **11**, 1–11 (2021)
13. Grooby, E., He, J., Kiewsky, J., Fattahi, D., Zhou, L., King, A., Ramanathan, A., Malhotra, A., Dumont, G.A., Marzbanrad, F.: Neonatal heart and lung sound quality assessment for robust heart and breathing rate estimation for telehealth applications. IEEE J. Biomed. Heal. Informatics **25**, 4255–4266 (2020)
14. Pouyani, M.F., Vali, M., Ghasemi, M.A.: Lung sound signal denoising using discrete wavelet transform and artificial neural network. Biomed. Signal Process. Control **72**, 103329 (2022)
15. Faustino, P., Oliveira, J., Coimbra, M.: Crackle and wheeze detection in lung sound signals using convolutional neural networks. In: 2021 43rd Annual International Conference of the IEEE Engineering in Medicine & Biology Society (EMBC), pp. 345–348. IEEE (2021)
16. Naqvi, S.Z.H., Choudhry, M.A., Khan, A.Z., Shakeel, M.: Intelligent system for classification of pulmonary diseases from lung sound. In: 2019 13th International Conference on Mathematics, Actuarial Science, Computer Science and Statistics (MACS), pp. 1–6. IEEE (2019)
17. Shi, L., Du, K., Zhang, C., Ma, H., Yan, W.: Lung sound recognition algorithm based on vggishbigru. IEEE Access **7**, 139438–139449 (2019)
18. Aykanat, M., Kılıç, Ö., Kurt, B., Saryal, S.: Classification of lung sounds using convolutional neural networks. EURASIP J. Image Video Process. **2017**, 1–9 (2017)
19. Sairamya, N.J., Subathra, M.S.P., Suviseshamuthu, E.S., George, S.T.: A new approach for automatic detection of focal EEG signals using wavelet packet decomposition and quad binary pattern method. Biomed. Signal Process. Control **63**, 102096 (2021)
20. Messner, E., et al.: Multi-channel lung sound classification with convolutional recurrent neural networks. Comput. Biol. Med. **122**, 103831 (2020)
21. Khare, S.K., Bajaj, V., Acharya, U.R.: SPWVD-CNN for automated detection of schizophrenia patients using EEG signals. IEEE Trans. Instrum. Meas. **70**, 1–9 (2021)
22. Rizal, A., Hidayat, R., Nugroho, H.A.: Entropy measurement as features extraction in automatic lung sound classification. In: 2017 International Conference on Control, Electronics, Renewable Energy and Communications (ICCREC), pp. 93–97. IEEE (2017)

23. Khan, S.I., Pachori, R.B.: Automated classification of lung sound signals based on empirical mode decomposition. Expert Syst. Appl. **184**, 115456 (2021)
24. Srivastava, A., Jain, S., Miranda, R., Patil, S., Pandya, S., Kotecha, K.: Deep learning based respiratory sound analysis for detection of chronic obstructive pulmonary disease. PeerJ Comput. Sci. **7**, e369 (2021)

Classification of Meditation Expertise from EEG Signals Using Shallow Neural Networks

Katinder Kaur(✉), Padmavati Khandnor, and Ashima Khosla

Punjab Engineering College, Chandigarh 160012, India
{katinderkaur.mtcse,padmavati}@pec.edu.in

Abstract. Recently machine-learning and deep-learning approaches have been widely adopted for automatic inter-group and inter-state classification of the brain states. Traditional machine learning-based classification requires a complicated pipeline of signal processing, artifact removal, feature extraction, and selection, which is time-consuming and requires human intervention. Whereas deep learning-based classifiers attempt to overcome this manual overhead by allowing end-to-end processing however at the cost of training time and model complexity. In this work, we have explored the use of different Shallow Convolutional Neural Networks (SCNN) based classifiers for inter-group (expert and non-expert) classification of EEG signals during Himalayan Yoga meditation. Several experiments were carried out to record the effects on the classification performance under varied conditions. We experimented with input representation of the signal, window size for signal segmentation, and various model parameters. It was observed that a very simple time domain representation of the raw signal, segmented by a window of 5s, when combined with a shallow 1D-CNN showed the best performance, with an accuracy score of 99.48%. The performance of all the shallow networks was found to be at par with the pre-trained state-of-the-art deep-CNN model, VGG-16, fine-tuned on the same data. Such lightweight models can be particularly useful for the on-the-go wearable and personal EEG devices which are latency sensitive, and highly susceptible to artifactual noise.

Keywords: EEG Classification · Shallow Neural Networks · Expert vs Non-expert meditators · Biomedical Signals

1 Introduction

Recently there has been a growing body of clinical evidence demonstrating the benefits of meditation in the reduction of stress [15], anxiety [18], depression [13], and chronic physical pain [36], thereby leading to a renewed interest in the field of meditation research.

Several previous studies [2, 5, 6, 31] have concluded that the brain activity of an expert meditator is different from that of a non-expert meditator, especially during core meditation. It has also been found that expert meditators tend to have increased focus [8], decreased mind wandering [5], and enhanced emotional stability [19]. Meditation

B. K. Singh et al. (Eds.): ICBEST 2023, CCIS 2003, pp. 169–185, 2024.
https://doi.org/10.1007/978-3-031-54547-4_14

helps practitioners attain more control over thoughts and emotions, thus reducing the chances of having prolonged negative emotions. It also helps in developing invaluable skills such as "trained attention" [6]. Moreover, along with the functional changes, several studies have also reported constructive structural changes to the brains of long-term meditators. These include increased cortical thickness in regions associated with attention, interoception and sensory processing [21], increased grey matter volume [17], as well as strengthening of brain connections [11]. Regular meditation has been associated with an increase in overall professional performance [7] too.

Therefore, an amateur practicing meditation may desire to measure their progress and direct their future training in a particular direction in order to achieve the same mental state as that of experts.

Now, for studying brain activity one of the most effective tools is Electroencephalography (EEG). By capturing the electric signals emitted by the brain, it provides a non-invasive and inexpensive method to get an insight into the brain's internal functioning [20].

Most previous EEG based meditation classification studies focus on manually extracting features and then combining them with traditional machine learning algorithms for classification. But it is to be noted that manual feature extraction is an unsupervised process that is independent of the output classes. The extracted features may not always be able to reflect the components that can effectively discriminate the inputs into respective classes.

The concept of automatic feature extraction and classification with deep learning is still relatively new to the field of meditation classification. Therefore, there is a wide scope for research and experimentation in this area. Some recent studies [26] have suggested a deep learning-based end-to-end pipeline that somewhat eliminates the problems with feature extraction and subjectivity due to human intervention. Another need of the hour is lightweight models that can be trained and tested on the go for real-time applications like wearable neurofeedback devices.

In this study we investigate the application of light-weight Shallow Convolutional Neural Networks (SCNNs) to sufficiently discriminate between the mental states of expert and non-expert meditators from their EEG signatures. We experiment with both, time domain representations (segment frames) and spatial domain representations (topographic plots) of the EEG signals for this purpose.

We also investigate the effects of window size for segmenting the EEG signals as well as various model parameters of the proposed classifier. We compare the performance of our model with the existing State-of-the-Art (SOTA) deep-CNN, VGG-16 [30].

1.1 Related Work and Motivation

A large share of the existing EEG-based studies [2, 4, 5, 31] in meditation research focus only on a statistical analysis of EEG correlates of meditators, in an attempt to find significant state and trait effects of meditation. They commonly compare frequency sub-band powers for analyzing the inter-group or inter-state differences with the help of statistical models.

Our focus is on automated classification. Automated classification of groups and states has only been introduced recently in this domain. One of the very first studies [29] to attempt automated discrimination between meditative and non-meditative brain features used statistical features of EEG signals like variance, kurtosis, band energy, and entropy for extracting information from the signals. For analysing the underlying patterns, the Principal Component Analysis (PCA) algorithm was used. The researchers observed a clear distinction between both classes. This study was further extended into another work [28] which used Support Vector Machine (SVM) to automatically classify the features and obtained an average accuracy of 90.83%. Both these studies focused on Kriya Yoga meditation. Another study [32] used discrete wavelet transform to extract features for detecting the transition between meditative and non-meditative states. The authors used binomial logistic regression for classifying EEG signal segments into meditation and transition states. A recent study [24] also used common spatial pattern with Tikhonov regularisation (TR-CSP) for extracting spatial features for intra as well as inter-subject binary classification (meditation vs rest state) in Rajayoga practitioners.

One exploratory study [27] has used five different wavelet families to extract features from the EEG data produced by a previous study [5]. It further utilizes twelve different machine learning classifiers to classify the EEG signals as belonging to an expert or a non-expert meditator. The authors exhaustively experimented with feature combinations and obtained the best performance by using Quadratic Discriminant Analysis.

It can be noted that all the above studies, using traditional machine learning for the classification, require extensive data cleaning followed by feature extraction and selection. Manual feature extraction requires a great degree of domain knowledge, otherwise, the chosen features may not be able to discriminate well between the required classes [16]. Different experts may suggest extracting different feature sets from the same data as per their expertise. And feature selection can be computationally expensive when there are a very large number of features. Many commonly used feature selection algorithms fail to report all solutions when there are multiple optimal solution sets [3].

Deep learning models are capable of extracting relevant features automatically from raw data [22], and that too in a supervised manner keeping the desired output classes in consideration. However, in reality, we observed that most of the studies [24, 25] still use at least some manual feature extraction techniques. In their review paper [10], Craik et al. also question the need to manually filter data when using deep neural networks. Therefore, in our study we use an end-to-end model with no explicit feature extraction.

In a recent deep learning-based study [26] conducted on the same dataset as our study (along with control data from [4]), the authors have proposed a pipeline to convert EEG signals into band-specific topographic plots for a 2-dimensional spatial representation of EEG signals, and a 4-block 2D-CNN model for three class classification (expert meditators, non-expert meditators, and controls). The work compares the performance of their proposed lightweight model with 4 SOTA 2D-CNN models (VGG-16, Resnet50, MobileNetv1 and MobileNetv2), and finds the performance of their proposed model comparable to the SOTA architectures. This work has majorly studied the effects of theta and alpha frequency bands.

Since the above study has not mentioned any particular event of interest for segment extraction, it is concluded that the entire signal was used to generate segments. The problem with the above method is that the data for control subjects have been taken from a separate study [4]. Although the preprocessing techniques are the same for all three classes, the data collection paradigm and the instructions given to the subjects were different for both the experiments.

The first study [5] includes interrupting the participants repeatedly during their meditation practice for probing, whereas for the second study [4] the participants were left to meditate without interruptions. This inconsistency could lead to a bias in the feature learning process of the classification model. For our study, we have used only the portion of the signal where the participant is in a relatively undisturbed meditation state, rather than using the entire signal.

For our end-to-end classification system, we chose to experiment with Shallow Convolutional Neural Networks (SCNNs). The choice of this neural network is attributed to the following reasons:

1. Many recent studies [20, 26] highlight the importance of lightweight and portable models for wearable EEG devices. For any kind of on-device analysis such devices require the algorithms to be robust (to handle noisy input data) and have low latency (no significant delays).
2. Irrespective of the EEG task, a large number of studies utilising deep learning for EEG signal classification suggest the use of Convolutional Neural Networks (CNNs) [10]. CNNs are known to be capable of modelling the spatio-temporal characteristics of EEG signals as well as handling the high dimensional EEG data better as compared to other deep learning algorithms [10].
3. EEG data is not as abundant and openly available as data in other application domains of deep learning therefore there is a need for SCNN-like architectures which are less data hungry and faster to train due to their small size.

Under these conditions, with their light-weight nature, a performance practically comparable to the deeper networks, and increased interpretability, SCNNs offer a favourable prospect.

2 Materials and Methods

Our research study compares the performance of classification pipelines obtained by combining different input signal representations, segment lengths, and SCNN-based classifiers. An overview of the proposed methodology and experimentation parameters has been shown in Fig. 1 below.

2.1 Data Collection

For our work we have used the publicly available data from a previous study, [5]. The available data consists of EEG recordings of 24 subjects (12 long-term meditators and 12 novice meditators). According to the details provided in [5], the data was recorded via a BioSemi 10–20 head cap with 64 EEG channels and 14 other channels, at a sampling rate of 2048 Hz (later down sampled to 256 Hz).

The data collection paradigm consisted of regular probing of subjects for noting down their first-hand meditation experience. The length of the session for each subject

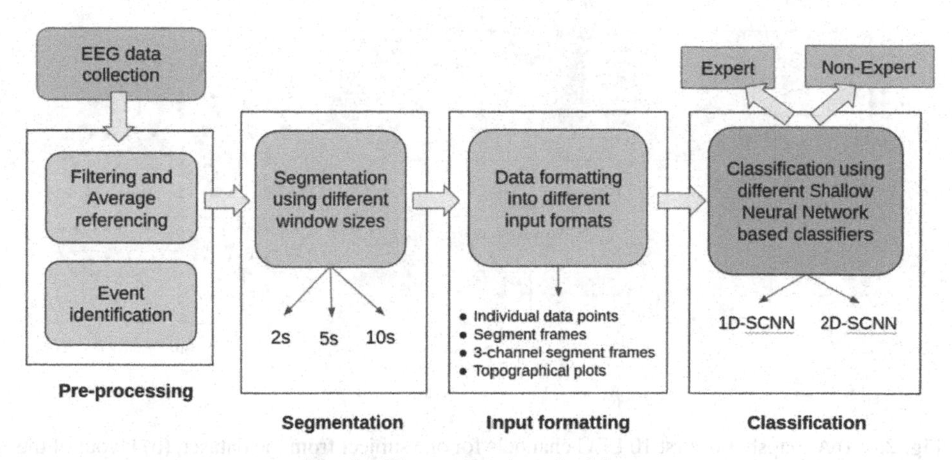

Fig. 1. The proposed methodology

lasted from 45 min to 1 h 30 min, and the minimum number of probes each participant received was 30 [5].

The same study also assigned a level of expertise to each subject according to the number of hours and regularity of meditation practice. Individuals who engaged in a daily practice of minimum 2 h were grouped into the Expert group whereas those who reported irregular practice but were familiar with the techniques, were grouped under the Non-Experts group. More data-related information can be found in [5]. A 5 s snapshot of a processed signal for one of the subjects from the dataset as well as the surface map of electrode locations are given in Fig. 2.

2.2 Pre-processing of EEG Signals

For the processing of data, the MNE library of Python [14] was used. The preprocessing steps involved removing non-scalp EEG channels, applying a high pass filter of 1 Hz and then average referencing the signals. Non-scalp and unlabeled channels were dropped, retaining only 64 channels for further analysis. The high-pass filter removes the slow drifts. It also removes some low-frequency content but since lower frequency delta band does not have a well-described role in meditation [23] it is an acceptable trade-off. A Finite Impulse Response (FIR) filter was employed due to its steady nature and stable performance. No low-pass filtering was done in order to retain the higher frequencies of the gamma band, as it is an important and characteristic indicator of meditation [4, 34]. After performing these steps, most of the visible signal irregularities were removed.

The pre-processing was kept to a minimum in order to study the effectiveness of SCNN in capturing distinguishing features from near-raw signals as done in [12, 35].

2.3 Segmentation of Signals Using Different Window Sizes

Segmentation of EEG signal is necessary because, by nature, it is a non-stationary signal. For conducting research analysis we divide the signal into quasi-stationary segments. As

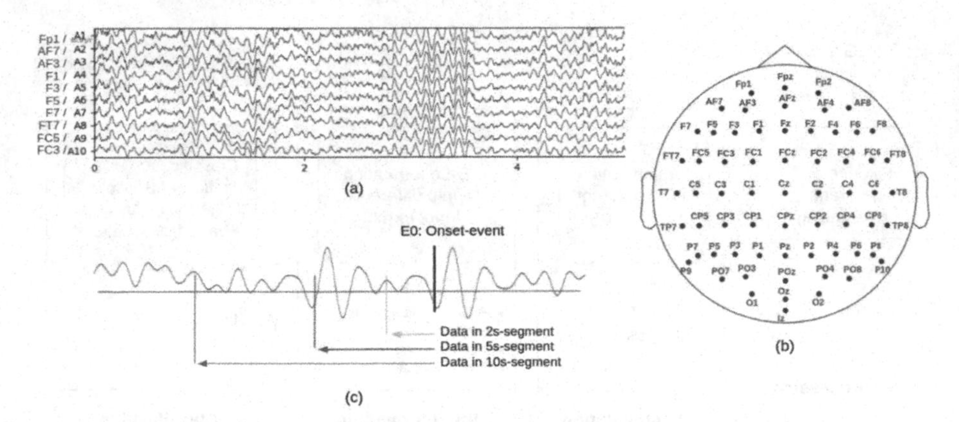

Fig. 2. (a) A snapshot of first 10 EEG channels for one subject from the dataset, (b) layout of the electrodes, (c) visual representation of onset-event and data segments selected for our work

the data collection method used in the original study [5] involves repeated experiential probing, this leads to an interruption in meditation. Therefore, to obtain the signals for undisturbed meditation, only the signal segments before the onset of the probing routine are used. Overall, 960 such probing events were identified from the multi-session data of the 24 subjects.

The signals were then divided into non-overlapping segments with varying window sizes (2 s, 5 s, and 10 s) before the onset of the event (See Fig. 2(c)). It has been done keeping in mind that the information included in a window is directly proportional to the size of the window.

2.4 Data Representation with Different Input Formats

In this study, we experimented with four different input representations for our EEG signals. The input formulations used in our study have been discussed below:

1. Individual data points: Under this input type, each 1×64 sample data point of the segment was treated as an individual training example. The number of such data instances varies according to the size of the window for segmentation. The number of training examples can be computed using the Eq. (1):

$$\mathrm{Pt} = \mathrm{l} * \mathrm{fs} * \mathrm{ne} \tag{1}$$

   where Pt represents the number of training pairs, l represents the length of epoch in seconds, fs represents the sampling frequency of the signal and ne represents the number of instances of the event of interest.
2. Segment frames: Multi-dimensional segment frames have been used directly in several studies [1, 33] for obtaining a 2D view of the EEG signals. Each window size generated segment frames with varying heights, whereas the width of the frame remained fixed (equal to the number of EEG channels). Each point in this 2D matrix represents the amplitude of the signal at a given channel and offset in a segment. Graphical visualization of one 2s segment is depicted in Fig. 3(a).

3. 3-channels segment frames: The transfer learning model was originally pretrained on 3-channel images. Therefore, we converted a 2D segment frame matrix into a 3D matrix, similar to an image with 3 channels. For obtaining a 3-channel format, each segment matrix was reshaped to (x,y,3), where x and y represent the resultant height and width. Minimal zero-padding was done to facilitate this conversion. Such a conversion affects the temporal and spatial information represented in the original data matrix. Graphical visualization of one such segment frame has been depicted in Fig. 3(b).
4. Topographic maps: These maps provide a spatial domain representation of the EEG signals. For generating the topographic plots in this study, all 64 channels and the entire frequency spectrum were included. Examples from the training set of topographic plots of expert and non-expert meditators have been depicted in Fig. 3(c) and (d) respectively.

Table 1 depicts the size of one input instance for every input formulation along with the total number of such instances generated.

2.5 Classification Using Different Neural Network Based Classifiers

The architectures of the different Shallow Convolutional Neural Networks and the transfer-learning model (used for comparison) are discussed below.

Classification Using Shallow Neural Networks. We have investigated the use of 1D-SCNNs, and 2D-SCNNs and also analysed the impact of model hyperparameters (namely filter-size and number of layers) on classification performance. To keep the models simple and as interpretable as possible we have used only the standard convolutional layers. All model architectures have been briefly discussed below.

1. 1D-SCNN architectures: Four 1D-SCNN architectures that were considered for experimentation are explained in Table 2 with their detailed configurations. The input to all these networks is IF1 along with the segment offset.

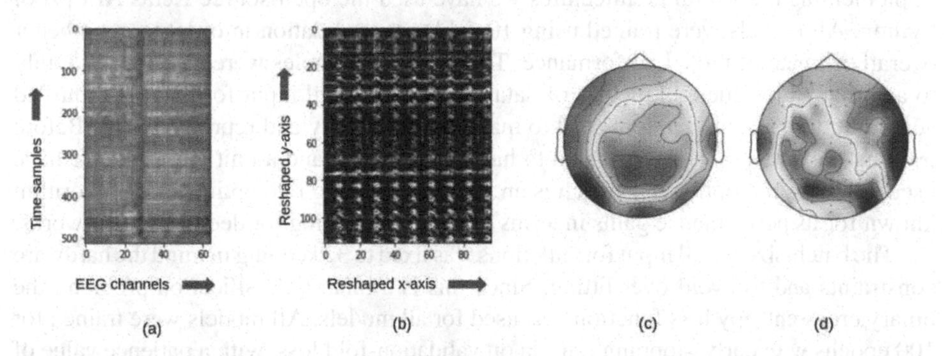

Fig. 3. Visual representation of (a) 2-d segment frame for 2s window (b) 3-channel segment frame for 2s window, (c) topographic plot for expert meditator while in meditation state and (d) topographic plot for non-expert meditator while in meditation state

Table 1. Size of one training instance and the total number of such instances in the generated data, corresponding to each window size and input formulation

Code	Input Type	2s Window	5s Window	10s Window
IF1	Individual data points	(1,65), 492480	(1,65), 1229760	(1,65), 2458560
IF2	Segment frames	(513,64), 960	(2181,64), 960	(2560,64), 960
IF3	3-channel frames	(114,96,3), 960	(122,224,3), 960	(244,224,3), 960
IF4	Topographic maps	(100,100,3), 960	-	-

2. 2D-SCNN architectures: To compare the performance of 1D-CNN with 2D-CNN architectures, four different models with two-dimensional convolutional layer were included in the analysis. IF2, IF3 and IF4 (see Table 1 for codes) were used as inputs for these models. The model architectures are described in Table 3.

Classification Using VGG-16. As seen previously in [26], for classification of meditation expertise via topographic maps, VGG-16 showed the best results among all the experimented SOTA networks. Therefore, for this study, we have used the VGG-16 network for gauging the performance of our shallow networks.

We have employed a VGG-16 net pre-trained on the ImageNet dataset [30]. VGG-16 contains 5 blocks with 13 convolutional layers and 3 fully connected layers. In this architecture, the size of kernel is fixed to 3×3 to minimize the number of hyper-parameters. For this study, the final classification layer of the model was dropped and a dense layer followed by a single-unit output layer was added.

3 Performance Evaluation

3.1 Implementation Details

All experiments were carried out on GPU NVIDIA Quadro P4000 (8GB RAM). For implementing the neural architectures we have used the open-source Keras API [9] of Python. All models were trained using 10-fold cross-validation in order to get a better overall estimate of model performance. The training examples were shuffled randomly to avoid any bias due to structure of data. Also data for all input formats was shuffled according to the same random seed to maintain uniformity and reproducibility. Before training the data points were scaled to have a mean of 0 and a unit variance. We have used Adam as the optimizer, which is an adaptive learning rate optimization algorithm known for its performance gains in terms of speed of training for deep neural networks.

The batch-size for all input formulations was fixed to 32 keeping in mind the hardware constraints and to avoid over-fitting. Since this is a binary classification problem, the binary cross-entropy loss function was used for all models. All models were trained for 100 epochs with early-stopping criteria on validation-fold loss, with a patience value of 10 epochs. Model checkpoints were created for each epoch and at the end of every fold the best performing parameters were restored from the saved chec k-points and used for prediction over the validation split. The reported performance metrics are evaluated over the validation splits.

Table 2. 1D CNN architectures used in our research study.

Model 1	Model 2	Model 3	Model 4
Conv1D	**Conv1D**	**Conv1D**	**Conv1D**
(filters = 16, size = 3)	(16, 3)	(16, 3)	(16, 3)
MaxPool1D	**MaxPool1D**	**MaxPool1D**	**MaxPool1D**
(poolsize = 2, stride = 2)	(2, 2)	(2, 2)	(2, 2)
Dense	**Conv1D**	**Conv1D**	**Conv1D**
(units = 64, act = ReLU)	(16, 3)	(16, 3)	(16, 3)
Dense	**MaxPool1D**	**MaxPool1D**	**MaxPool1D**
(units = 1, act = sigmoid)	(2, 2)	(2, 2)	(2, 2)
	Dense	**Conv1D**	**Conv1D**
	(64, ReLU)	(16,3)	(16,3)
	Dense	**MaxPool1D**	**MaxPool1D**
	(1, sigmoid)	(2, 2)	(2, 2)
		Dense	**Conv1D**
		(64, ReLU)	(16, 3)
		Dense	**MaxPool1D**
		(1, sigmoid)	(2,2)
			Dense
			(64, ReLU)
			Dense
			(1,sigmoid)

Table 3. 2D CNN architectures used in our research study.

Model 5	Model 6	Model 7	Model 8
Conv2D	**Conv2D**	**Conv2D**	**Conv2D**
(filters = 16, size = 3 × 3)	(16, 3 × 3)	(16, 5 × 5)	(32, 5 × 5)
MaxPool2D	**MaxPool2D**	**MaxPool2D**	**MaxPool2D**
(poolsize = 2 × 2, stride = 2)	(2 × 2, 2)	(2 × 2, 2)	(2 × 2, 2)
Conv2D	**Conv2D**	**Conv2D**	**Conv2D**

(*continued*)

Table 3. *(continued)*

Model 5	Model 6	Model 7	Model 8
(filters = 16, size = 3 × 3)	(32, 3 × 3)	(16, 5 × 5)	(64, 5 × 5)
MaxPool2D	**MaxPool2D**	**MaxPool2D**	**MaxPool2D**
(poolsize = 2 × 2, stride = 2)	(2 × 2, 2)	(2 × 2, 2)	(2 × 2, 2)
Dense	**Dense**	**Dense**	**Dense**
(units = 64, act = ReLU)	(64, ReLU)	(64, ReLU)	(64, ReLU)
Dense	**Dense**	**Dense**	**Dense**
(units = 1, act = sigmoid)	(1, sigmoid)	(1, sigmoid)	(1, sigmoid)

3.2 Performance Metrics

The performance metrics used to evaluate the proposed models are briefly introduced as follows:

$$\text{Accuracy} = (\text{TP} + \text{TN}) / (\text{TP} + \text{TN} + \text{FP} + \text{FN}) \tag{2}$$

$$\text{Precision} = (\text{TP}) / (\text{TP} + \text{FP}) \tag{3}$$

$$\text{Recall(or Sensitivity)} = (\text{TP}) / (\text{TP} + \text{FN}) \tag{4}$$

$$\text{Specificity} = (\text{TN}) / (\text{TN} + \text{FP}) \tag{5}$$

$$\text{F1score} = (2 * \text{Precision} * \text{Recall}) / (\text{Precision} + \text{Recall}) \tag{6}$$

where TP, TN, FP, FN represent the true positives, true negatives, false positives and false negatives respectively. The dataset in this study is balanced, therefore all the above metrics can be seen as relevant indicators of performance. The performance of all the models corresponding to different window sizes is summarised in Tables 4, 5, 6 and 7 (where the best value for each metric is highlighted in bold). And the confusion matrices for the best performing models for each of the window sizes are given in Fig. 4.

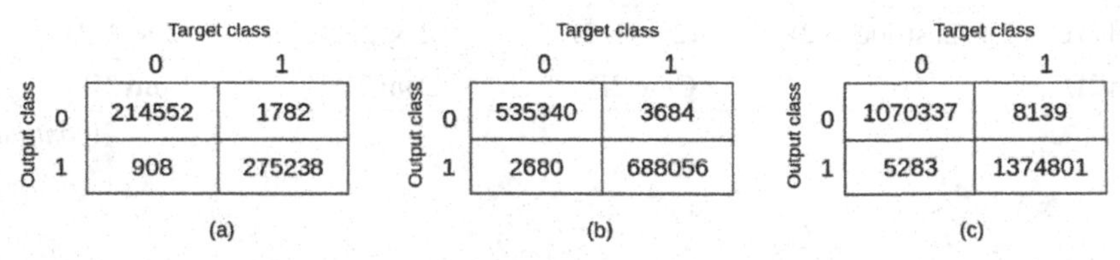

Fig. 4. Confusion matrices for best performing models for (a) 2s window, (b) 5s window, (c) 10 s window.

3.3 Results and Analysis

1D-SCNN models trained on individual data points (IF1) achieved the best performance scores across all metrics. This can be attributed to the amount of data available for training as compared to other input formulations which combine several data points into a single training example, along with the inherent nature of 1D-CNN to work well with time series data.

With a slight increase in model complexity, it can be seen that model performance on the validation set starts decreasing. This trend can be seen for all input formats and all segment lengths (see Fig. 5) and can be attributed to the phenomenon of overfitting. This belief is supported by the trend seen in the training accuracy which gets very close to 1 while the training algorithm keeps trying to reduce the loss.

Table 4. Performance scores of the different models for 2 s segments

Model	Input	Acc	Prec	Recall	F1-score	Spec	Loss
M1	IF1	**99.45%**	**99.18%**	**99.58%**	**99.38%**	**99.36%**	**0.0183**
M2	IF1	98.80%	98.46%	98.80%	98.63%	98.80%	0.0397
M3	IF1	95.28%	94.46%	94.77%	94.62%	95.68%	0.1186
M4	IF1	91.88%	90.64%	90.83%	90.73%	92.70%	0.1911
M5	IF2	94.58%	92.59%	95.24%	93.90%	94.07%	0.3892
M6	IF2	**95.42%**	**93.52%**	**96.19%**	**94.81%**	**94.84%**	**0.2046**
M7	IF2	95.31%	93.30%	**96.19%**	94.72%	94.63%	0.2272
M8	IF2	92.71%	89.77%	94.05%	91.86%	91.67%	0.4790
M5	IF3	94.58%	92.59%	95.24%	93.90%	94.07%	0.2214
M6	IF3	**95.31%**	**93.30%**	**96.19%**	**94.72%**	**94.68%**	**0.2056**
M7	IF3	94.06%	90.79%	**96.19%**	93.41%	92.41%	0.2698
M8	IF3	80.52%	88.20%	64.05%	74.21%	93.33%	0.4885
VGG	IF3	92.29%	88.79%	94.29%	91.45%	90.74%	0.3439

For IF1, the model with a single 1D-convolutional layer performs better. For segment frames (IF2), the performance of model 8 was found to be consistently worse than its less complex counterparts. A similar trend was seen for reshaped-segment frames (IF3) with its performance being close to that of chance. Also, for this input type, the SCNN models outperform the deep VGG16 net. The VGG net has the best accuracy score of 92.71% but the precision of the model is quite low, almost around chance values for 10 s segments.

Similarly, for topographic maps (IF4) simpler models show lower loss values, though this input format does not perform as well as the other tested input formats. The topographic maps which are very widely used and easy for humans to understand do not prove to be particularly intuitive for the SCNN networks.

Table 5. Performance scores of the different models for 5 s segments

Model	Input	Acc	Prec	Recall	F1-score	Spec	Loss
M1	IF1	**99.48%**	**99.31%**	**99.50%**	**99.41%**	**99.47%**	**0.0176**
M2	IF1	98.64%	98.35%	98.54%	98.44%	98.71%	0.0383
M3	IF1	95.55%	94.68%	95.19%	94.93%	95.84%	0.1120
M4	IF1	92.32%	91.16%	91.31%	91.23%	93.12%	0.1827
M5	IF2	94.69%	93.91%	94.52%	93.96%	94.81%	0.3120
M6	IF2	**95.73%**	**94.59%**	**95.71%**	**95.15%**	**95.74%**	0.3090
M7	IF2	94.48%	92.38%	95.24%	93.79%	93.89%	**0.2606**
M8	IF2	93.54%	92.02%	93.33%	92.67%	93.70%	0.4176
M5	IF3	94.58%	93.19%	94.52%	93.85%	94.63%	0.2901
M6	IF3	95.21%	94.74%	94.29%	94.51%	95.92%	**0.2142**
M7	IF3	**95.62%**	94.58%	**95.48%**	**95.02%**	95.74%	0.2163
M8	IF3	72.08%	**95.78%**	37.88%	54.27%	**98.70%**	0.5452
VGG	IF3	92.08%	89.27%	93.09%	91.14%	91.30%	0.2951

Table 6. Performance scores of the different models for 10 s segments

Model	Input	Acc	Prec	Recall	F1-score	Spec	Loss
M1	IF1	**99.45%**	**99.24%**	**99.51%**	**99.38%**	**99.41%**	**0.0176**
M2	IF1	98.46%	98.09%	98.39%	98.24%	98.51%	0.0420
M3	IF1	95.59%	94.81%	95.12%	95.00%	95.95%	0.1117
M4	IF1	95.63%	94.98%	95.04%	95.01%	96.09%	0.1107
M5	IF2	95.52%	**94.56%**	95.23%	**95.74%**	94.90%	**0.1770**
M6	IF2	91.77%	92.95%	87.86%	90.33%	94.81%	0.2863
M7	IF2	**96.14%**	94.23%	**97.14%**	95.37%	**95.66%**	0.2217
M8	IF2	78.96%	87.07%	60.95%	92.96%	71.71%	0.5251
M5	IF3	95.10%	94.51%	94.28%	94.40%	95.74%	0.2400
M6	IF3	**95.42%**	**95.19%**	**94.32%**	**94.74%**	96.30%	**0.2014**
M7	IF3	93.96%	92.69%	93.57%	93.13%	94.26%	0.2941
M8	IF3	59.69%	83.67%	09.76%	17.48%	**98.52%**	0.6620
VGG	IF3	92.71%	60.51%	93.10%	91.78%	92.41%	0.2662

Though for our task we also found it hard to distinguish between the maps generated by the two types of meditators via visual inspection. It can also be argued that the training set was rather small for the models to be able to learn the distinguishing features well.

Table 7. Performance scores of the different models on IF4

Model	Input	Acc	Prec	Recall	F1-score	Spec	Loss
M5	IF4	**88.07%**	84.13%	97.61%	75.25%	**90.37%**	**0.2811**
M6	IF4	85.43%	**85.22%**	90.24%	**78.96%**	87.66%	0.3521
M7	IF4	71.60%	67.25%	**98.34%**	35.64%	79.88%	0.4982
M8	IF4	85.64%	83.64%	93.19%	75.50%	88.15%	0.4221
VGG	IF4	80.89%	82.09%	85.27%	75.00%	83.65%	0.6227

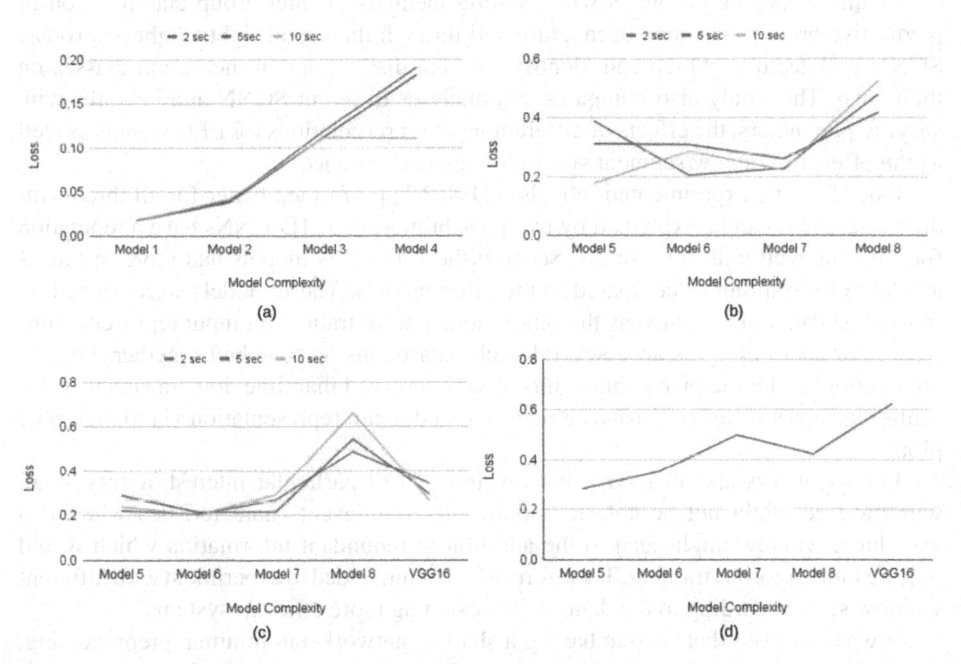

Fig. 5. Model performance (loss) vs Model complexity curves for each segment length for (a) IF1, (b) IF2, (c) IF3, and (d) IF4

It is noteworthy that the two-layered simple architectures performed at par with the SOTA VGG-16 net. The VGG-16 network achieved the highest accuracy of 92.71%, but it has 14 million parameters whereas the largest SCNN model in our study, model 8, has 5 million trainable parameters, which is only about 36% of the VGG-16 network. The model 1, which gives an accuracy of 99.48% on 5 s segments has only 15k parameters, which is 0.1% of that of the VGG-16 net.

Out of all the tested segment lengths, the models trained on 10 s signal segments showed overall lower loss values, whereas the best models on 5 s segment showed better accuracy values in 2 out of the first 3 input formulations. Intuitively, longer segments can capture lower frequency content better as compared to smaller segment lengths as low frequency signals have longer wavelengths.

Therefore, longer segment lengths not only have more time-points, but they also tend to have more detailed frequency information. But more information in a single training example also means more training time.

4 Conclusion

Mental health-related concerns have become an area of increased focus in recent years. Therefore, stress-relieving practices like meditation are being readily adopted. Long-term meditators show a distinct brain activity which is the goal state for many amateur meditators. Therefore, our work contributes to this research area.

In this work, the problems with existing methods of inter-group classification of meditative brain states have been addressed through the use of lightweight and robust SCNN architectures which can identify relevant distinctions in the output classes on their own. This study also compares and analyses different SCNN architectures with varying parameters, the effects of different input representations for EEG signals as well as the effect of window/segment size on model performance.

Out of all the experimented models, 1D-SCNN performed better for all three window sizes. This can be explained by two possibilities, first: 1D-CNNs have a reputation for working well with time series. Second: the 1D-SCNN models had more instances available for training as compared to the other models. These models were trained on individual data points, whereas the other models were trained on input representations where for each input instance several such data points are clubbed together. For our application, under the given constraints, it was observed that time domain signal representation shows better performance than spatial domain representation via topographic plots.

Finally, it was also observed that for an event of particular interest, a very small window size might not be able to capture all the relevant characteristics whereas a very large window might lead to the addition of redundant information which would require more rigorous training. Therefore, it is recommended to experiment with different window sizes according to the dataset for designing more efficient systems.

It was observed that despite having a shallow network and minimal preprocessing, the performance of the best model is comparable to the performance of previous studies [26, 27], conducted on the same dataset, which utilise extensive pre-processing, feature extraction, and feature selection. This study proves the effectiveness of convolutional neural networks to extract features of relevance on their own from the given raw data.

There is a large scope for research in analysis of EEG correlates of meditation, but we have felt that there is a lack of publicly available datasets of EEG recordings for meditation tasks as well as benchmarks for meditation classification. Therefore, to encourage research in this field more datasets need to be made available publicly and a standardised procedure for meditation data collection needs to be adopted.

References

1. Ansari, A.H., Cherian, P.J., Caicedo, A., Naulaers, G., De Vos, M., Van Huffel, S.: Neonatal seizure detection using deep convolutional neural networks. Int. J. Neural Syst. **29**(04), 1850011 (2019). https://doi.org/10.1142/S0129065718500119

2. Banquet, J.P.: Spectral analysis of the EEG in meditation. Electroencephalogr. Clin. Neurophysiol. **35**(2), 143–151 (1973). https://doi.org/10.1016/00134694(73)90170-3
3. Borboudakis, G., Tsamardinos, I.: Extending greedy feature selection algorithms to multiple solutions. Data Min. Knowl. Disc. **35**(4), 1393–1434 (2021)
4. Braboszcz, C., Cahn, B.R., Levy, J., Fernandez, M., Delorme, A.: Increased gamma brainwave amplitude compared to control in three different meditation traditions. PLoS ONE **12**(1), e0170647 (2017). https://doi.org/10.1371/journal.pone.0170647
5. Brandmeyer, T., Delorme, A.: Reduced mind wandering in experienced meditators and associated EEG correlates. Exp. Brain Res. **236**(9), 2519–2528 (2018). https://doi.org/10.1007/s00221-016-4811-5
6. Cahn, B.R., Polich, J.: Meditation states and traits: EEG, ERP, and neuroimaging studies. Psychol. Bull. **132**(2), 180 (2006)
7. Carter, K.S., Carter, R., III.: Breath-based meditation: a mechanism to restore the physiological and cognitive reserves for optimal human performance. World Journal of Clinical Cases **4**(4), 99 (2016)
8. Chan, D., Woollacott, M.: Effects of level of meditation experience on attentional focus: is the efficiency of executive or orientation networks improved? The J. Altern. Complement. Med. **13**(6), 651–658 (2007)
9. Chollet, F., et al.: Keras https://github.com/fchollet/keras (2015)
10. Craik, A., He, Y., Contreras-Vidal, J.L.: Deep learning for electroencephalogram (EEG) classification tasks: a review. J. Neural Eng. **16**(3), 031001 (2019). https://doi.org/10.1088/1741-2552/ab0ab5
11. De Filippi, E., Escrichs, A., Camara, E., Garrido, C., Marins, T., Sanchez-Fibla, M., Gilson, M., Deco, G.: Meditation-induced effects on whole-brain structural and effective connectivity. Brain Struct. Funct. **12**, 92 (2022)
12. Dose, H., Møller, J.S., Iversen, H.K., Puthusserypady, S.: An end-to-end deep learning approach to MI-EEG signal classification for BCIS. Expert Syst. Appl. **114**, 532–542 (2018)
13. González-Valero, G., Zurita-Ortega, F., Ubago-Jiménez, J.L., Puertas-Molero, P.: Use of meditation and cognitive behavioral therapies for the treatment of stress, depression and anxiety in students. a systematic review and meta-analysis. Int. J. Env. Res. Public Health **16**(22), 4394 (2019). https://doi.org/10.3390/ijerph16224394
14. Gramfort, A., et al.: MEG and EEG data analysis with MNE-Python. Front. Neurosci. **7**(267), 1–13 (2013)
15. Grossman, P., Niemann, L., Schmidt, S., Walach, H.: Mindfulness-based stress reduction and health benefits: a meta-analysis. J. Psychosom. Res. **57**(1), 35–43 (2004). https://doi.org/10.1016/S0022-3999(03)00573-7
16. Hemanth, D.J.: Automated feature extraction in deep learning models: a boon or a bane? In: 2021 8th International Conference on Electrical Engineering, Computer Science and Informatics (EECSI), p. 3 (2021). https://doi.org/10.23919/EECSI53397.2021.9624287
17. Hernandez, S.E., Suero, J., Barros, A., Gonzalez-Mora, J.L., Rubia, K.: Increased grey matter associated with long-term Sahaja yoga meditation: a voxel-based morphometry study. PLoS ONE **11**(3), e0150757 (2016)
18. Hofmann, S.G., Grossman, P., Hinton, D.E.: Loving-kindness and compassion meditation: potential for psychological interventions. Clin. Psychol. Rev. **31**(7), 1126–1132 (2011). https://doi.org/10.1016/j.cpr.2011.07.003
19. Kabat-Zinn, J., et al.: Influence of a mindfulness meditation-based stress reduction intervention on rates of skin clearing in patients with moderate to severe psoriasis undergoing phototherapy (UVB) and photochemotherapy (PUVA). Psychosom. Med. **60**(5), 625–632 (1998)

20. Khosla, A., Khandnor, P., Chand, T.: A comparative analysis of signal processing and classification methods for different applications based on EEG signals. Biocybernet. Biomed. Eng. **40**(2), 649–690 (2020)
21. Lazar, S.W., et al.: Meditation experience is associated with increased cortical thickness. NeuroReport **16**(17), 1893–1897 (2005)
22. LeCun, Y., Bengio, Y., Hinton, G.: Deep learning. Nature **521**(7553), 436–444 (2015)
23. Lee, D.J., Kulubya, E., Goldin, P., Goodarzi, A., Girgis, F.: Review of the neural oscillations underlying meditation. Front. Neurosci. **12**, 178 (2018)
24. Panachakel, J.T., Govindaiah, P.K., Sharma, K., Ganesan, R.A.: Binary classification of meditative state from the resting state using EEG. In: 2021 IEEE 18th India Council International Conference (INDICON), pp. 1–6. IEEE (2021)
25. Panachakel, J.T., Kumar, P., Ramakrishnan, A., Sharma, K.: Automated classification of EEG into meditation and non-meditation epochs using common spatial pattern, linear discriminant analysis, and LSTM. In: TENCON 2021–2021 IEEE Region 10 Conference (TENCON), pp. 215–218. IEEE (2021)
26. Pandey, P., Miyapuram, K.P.: Brain2depth: Lightweight CNN model for classification of cognitive states from EEG recordings. In: Papież, B.W., Yaqub, M., Jiao, J., Namburete, A.I.L., Alison Noble, J. (eds.) Medical Image Understanding and Analysis: 25th Annual Conference, MIUA 2021, Oxford, United Kingdom, July 12–14, 2021, Proceedings, pp. 394–407. Springer International Publishing, Cham (2021). https://doi.org/10.1007/978-3-030-80432-9_30
27. Pandey, P., Prasad Miyapuram, K.: Classifying oscillatory signatures of expert vs nonexpert meditators. In: 2020 International Joint Conference on Neural Networks (IJCNN), pp. 1–7 (2020)
28. Shaw, L., Routray, A.: A critical comparison between SVM and K-SVM in the classification of KRIYA yoga meditation state-allied EEG. In: 2016 IEEE International WIE Conference on Electrical and Computer Engineering (WIECON-ECE), pp. 134–138. IEEE (2016)
29. Shaw, L., Routray, A.: Statistical features extraction for multivariate pattern analysis in meditation EEG using pca. In: 2016 IEEE EMBS International Student Conference (ISC), pp. 1–4. IEEE (2016)
30. Simonyan, K., Zisserman, A.: Very deep convolutional networks for large-scale image recognition. arXiv preprint arXiv:1409.1556 (2014)
31. Stapleton, P., Dispenza, J., McGill, S., Sabot, D., Peach, M., Raynor, D.: Large effects of brief meditation intervention on EEG spectra in meditation novices. IBRO Reports **9**, 290–301 (2020). https://doi.org/10.1016/j.ibror.2020.10.006
32. Tee, J.L., Phang, S.K., Chew, W.J., Phang, S.W., Mun, H.K.: Classification of meditation states through EEG: a method using discrete wavelet transform. In: AIP Conference Proceedings. vol. 2233, p. 030010. AIP Publishing LLC (2020)
33. Van Putten, M.J., Olbrich, S., Arns, M.: Predicting sex from brain rhythms with deep learning. Sci. Rep. **8**(1), 1–7 (2018)
34. Vazquez, M.A., Jin, J., Dauwels, J., Vialatte, F.B.: Automated detection of paroxysmal gamma waves in meditation EEG. In: 2013 IEEE International Conference on Acoustics, Speech and Signal Processing, pp. 1192–1196. IEEE (2013)

35. Waytowich, N., et al.: Compact convolutional neural networks for classification of asynchronous steady-state visual evoked potentials. J. Neural Eng. **15**(6), 066031 (2018)
36. Zeidan, F., Martucci, K.T., Kraft, R.A., Gordon, N.S., McHaffie, J.G., Coghill, R.C.: Brain mechanisms supporting the modulation of pain by mindfulness meditation. J. Neurosci. **31**(14), 5540–5548 (2011). https://doi.org/10.1523/JNEUROSCI.5791-10.2011

Efficient Model for Prediction of Parkinson's Disease Using Machine Learning Algorithms with Hybrid Feature Selection Methods

Nutan Singh(✉) and Priyanka Tripathi

Department of Computer Applications, Raipur, Chhattisgarh, India
nsingh.phd2021.mca@nitrr.ac.in

Abstract. In recent years, The development of telemonitoring and tel-diagnosis tools for assessing and monitoring Parkinson's disease (PD) has assumed growing significance. In this study, a Parkinson's disease prediction model is put forth. The monotony and distortion of speech are two of the most apparent signs of PD. Approaches based on artificial intelligence can assist specialists and medical professionals in automatically identifying these disorders. This study proposed an effective method for detecting PD from features incorporating voice recordings of subjects who have already been diagnosed. The main objective of this paper is to analyze the effect of an imbalanced dataset and feature selection. We have used SMOTE technique to balance the imbalanced data on the efficiency of the voice-based PD detection system and use a learning-based classifier for this work. Feature selection using LassoCV, Recursive Feature Elimination (RFE) and Hybrid selection techniques. Various machine learning classifiers, including Random-Forest (RF), Logistic Regression (LR), Decision-Tree (DT), Gaussian NB (GB), SGD Classifier (SGDC), Nearest-Neighbors (NN), Support Vector Machine (SVM), Adboost (Adb), MLPClassifier(MLP) and BaggingClassifier(BC), have been Classified and explored for PD detection purposes. All other models were outperformed by the Bagging classifier results with Hybrid feature selection, which had high Accuracy (A) of 91.6%, Precision (P) of 91.8%, Recall (R) of 91.6%, and F1- score (F) of 91.7% and CVS of 97.4%. The suggested method results show that the Hybrid features selection approach with Bagging Classifier (BC) successfully enhances the overall performance of the Parkinson's disease detection model.

Keywords: Voice Disorder · Parkinson's Disease(PD) · Machine Learning(ML) · SMOTE · Feature Selection · Lasso CV · RFE · Hybrid Method

1 Introduction

Parkinson's disease (PD) [1] is a chronic central nervous system degenerative condition that manifests both motor and non-motor [2] symptoms. Parkinson's disease (PD), a neurological ailment that advances slowly, has a detrimental impact on people's daily life. Early diagnosis is crucial for PD to have as little of an impact as possible. The monotony

B. K. Singh et al. (Eds.): ICBEST 2023, CCIS 2003, pp. 186–203, 2024.
https://doi.org/10.1007/978-3-031-54547-4_15

and distortion of speech are two of the most crucial signs for the early diagnosis of PD disease. Parkinson's disease (PD) is the second most prevalent neurological disorder among people in their 60s. Bradykinesia, stiffness, and resting tremors are all too frequent symptoms. Parkinson's disease is caused by the death of dopaminergic neurons in the substantia nigra of the brain. Parkinson's disease (PD) is the second most common neurological chronic condition, affecting 10 million people globally. It has a prevalence of 1% in adults over the age of 60 [3]. They are developing a system that an IoT-based platform to automate PD diagnosis and predict clinical scores. These techniques can potentially aid in disease diagnosis and progression prediction in Parkinson's disease [4]. The two primary diagnostic approaches for Parkinson's disease at the moment are conventional empirical diagnosis by physicians and the use of artificial intelligence to identify the illness. A weekly outpatient clinic can only diagnose many patients since diagnosing patients is time-consuming and labour-intensive. Patients frequently require many tests. With the advancement of machine learning technology, it is frequently possible to obtain better test results and significantly increase disease diagnosis efficiency by modelling medical data and training the models. When classifying things, machine learning approaches seem to be quite effective. Most medical practices currently rely on doctors' judgments to determine the proper diagnosis.

A lack of resources makes it difficult for those who cannot afford to visit a professional. In many instances, the disease goes undiagnosed for a very long time, which results in an extremely low patient survival rate. Numerous technical developments have been made to decrease diagnostic errors and hasten the arrival of better results. The purpose of this essay is to highlight their advancements. All of this is done to gain a realistic understanding of the current situation and, as a result, develop a better plan for how we should each strive to get better results. Additionally, a suggestion is presented that can be compared to the current algorithms to advance the field of disease diagnosis. Several Supervised classifiers, including Random-Forest, Logistic Regression, Decision-Tree, Gaussian NB, SGD Classifier, Nearest Neighbors, Support Vector Machine (SVM), Adboost, MLP Classifier, and Bagging Classifier, were used in this study Bagging Classifier. Results are contrasted with those of prior research. The paper follows the following structure: Section 2 covers the purpose and reach of this research, while Section 1 describes the introduction. A review of related work is given in Section 3. Proposed Methodology in section 4, along with several performance measures. Section 5 explains the data preparation, and preprocessing, with machine learning approaches and a Discussion of the result's statistical significance. Section 6 includes some conclusions and suggestions for additional research.

2 Research Aim and Scope of the Paper

This research aims to create an efficient method for correctly predicting Parkinson's disease, precisely the voice problem. The necessary actions can be summed up as follows:

i. Working with Parkinson's dataset from the UCI repository or unbalanced data. To balance the dataset, we employed the SMOTE(Synthetic Minority Oversampling Technique) approach.

ii. Based on rank values (25, 50, 75, and 100, respectively) from the original dataset, the most pertinent features are chosen using two selection algorithms, RFE and LASSO. Additionally, this aids in addressing the machine learning issues of overfitting and underfitting.
iii. After that used Hybrid Features selection procedures, Based on rank values (25, 50, 75, and 100 features, respectively)
iv. The effectiveness of the various models such as bagging are also used to increase the testing rate and decrease execution time assessed overall findings using the All, RFE, LASSO and Hybrid features selected characteristics.

3 Literature

Numerous studies have been conducted to forecast Parkinson's disease in a patient. These pieces have made use of a variety of machine-learning methods. The second most prevalent form of neurodegenerative disorder to affect speaking and accurate voice is called Parkinson's disease (PD). There is no indication of a hereditary component to illness onset after age 50. To enhance functional communication in individuals with idiopathic PD, studies and investigations into the effectiveness of intensive voice therapy are now being conducted. The onset of Parkinson's disease ranges from 50 to 95 years of age [5]. In a survey by Hawi S et al. [5] Random Forest and decision trees were discovered to be the better classifiers in comparison to other methods for Based on the combination of long-term acoustic characteristics and Mel frequency cepstral coefficients, Parkinson's disease detection (MFCC). These methods received the highest accuracy score, i.e., 88.84%. In most of the reported research, Sakar et al. [6] Utilised the attributes collected from voice signals mRMR-50 with the Tunable Q-factor Wavelet Transform (TQWT). For predicting the severity of PD using the machine learning classifier Support Vector Machine (SVM)(RBF kernel), their accuracy score is 84%. Liu Y et al. [7] Parkinson's disease diagnosis based on the selection of SHAP value features When 50 characteristics are chosen, the SHAP-LightGBM model and LightGBM achieve classification accuracy and F1-score of 91.62% and 0.945, respectively. Jain D and others [9] Parkinson's disease was predicted automatically using machine learning and speech features, with an average prediction accuracy of 86.5%. To study deep neural network-based techniques, the training data size is additionally increased using the oversampling strategy. After oversampling, significant increases were seen, with average prediction accuracy increasing to 91.5.

4 The Proposed Method

The proposed techniques' flowchart is displayed in Fig. 1. In this essay, the speech signals were recorded using the dataset Sakar et al. produced in 2018 at t he Department of Neurology at the CerrahpaÅŸa Faculty of Medicine, Istanbul University [8]. The class distribution in the data set is unbalanced. Utilizing the SMOTE (Synthetic Minority Over-Sampling Technique) approach, an unbalanced dataset is converted into a balanced dataset. The Parkinson's disease dataset was then classified using a separate classification technique after being converted to a balanced class distribution. K Fold

Cross Validation is used with the Value of K set to ten. The following data preprocessing steps were performed in each cross-validation repetition. We have used LassoCV and Recursive feature elimination (RFE) [9] as a feature selection algorithm on the original feature sets. The most popular methods for feature selection and data reduction are LassoCV, Recursive Feature Removal (RFE) and Hybrid Feature selection. In LassoCV, the Use Select From Model meta-transformer and Lasso select the best features from the Parkinson's voice dataset [10]. The unnecessary variables are eliminated throughout the feature selection procedure. After applying the algorithms to the initial feature sets for both LassoCV and the Recursive Feature Elimination (RFE) Feature selection technique, we have chosen the top 25, 50, 75, and 100 features. We have used different classification approaches such as RF, LR, DT, NB, SGDC, K-NN, Support Vector Machine (SVM), Adaboost, MLP and Bagging. Bagging classifiers are giving good Accuracy. SVM is the most popular linear classification method. While performing categorization, it makes use of many kernel functions. Because many renowned data scientists recommend it,

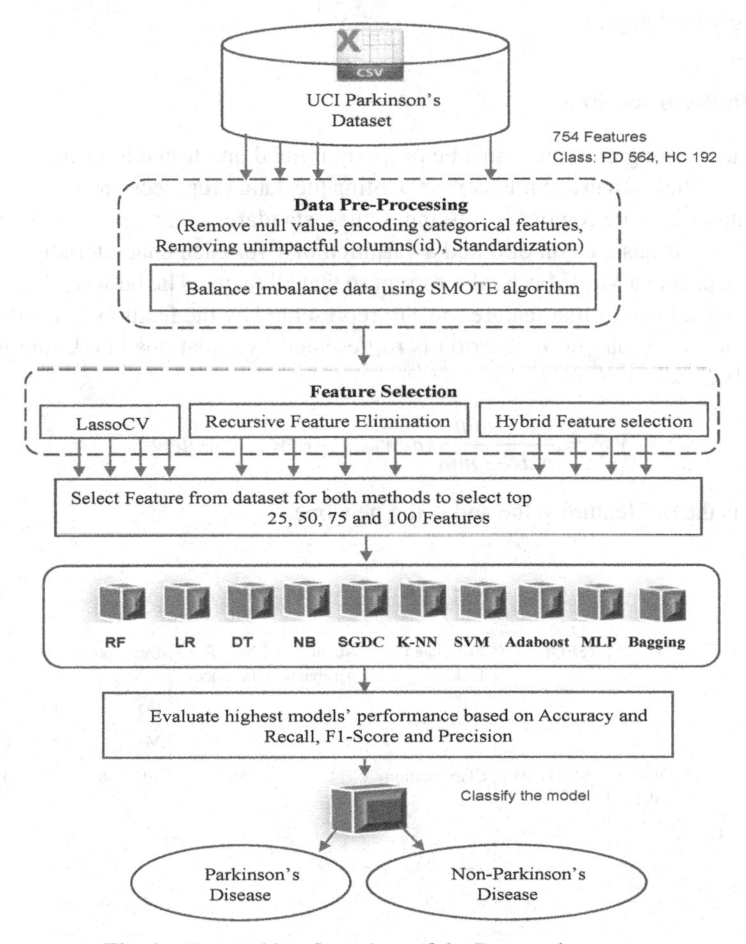

Fig. 1. The working flow chart of the Proposed system.

the radial basis function kernel function is used in this research. After applying various classification methods, we compared performance indicators like Accuracy, sensitivity, and specificity.

4.1 Dataset Details

The data for this study came from 188 PD patients (107 men and 81 women) from the Department of Neurology at the Cerrahpaşa Faculty of Medicine, Istanbul University, whose ages ranged from 33 to 87 (65.110.9) [8]. The control group consists of 64 healthy individuals (23 men and 41 women), ranging in age from 41 to 82 (61.1 8.9). Following the doctor's examination, each participant's sustained phonation of the vowel /a/ was recorded three times, with the microphone tuned to 44.1 kHz for the purpose of gathering data. Figure 2 displays the class distribution for the Parkinson's disease dataset. Two classes may be seen in this diagram: the green class, which represents the majority, and the yellow class (minority class). Making a distinction between these classes is exceedingly challenging.

4.2 Data Preprocessing

A machine-learning classifier must be properly trained and tested after pre-processing, which is a crucial step in accurately representing the data. Preprocessing techniques used on the dataset include removing missing values, standard scalar, and Min-Max Scalar. A typical Scalar has a mean of 0 and a variation of 1 for each characteristic. Similarly, arrange the data in a Min-Max Scalar format so that all features lie between 0 and 1, [13]. The missing values for that feature row are represented by the features in the dataset. In mathematics, min-max normalisation is represented by equations (1). Using min-max to normalise:

$$V^{-} = \frac{v - min}{max - min}(new_{max} - new_{min}) + new_{min} \tag{1}$$

where V is the old feature value and is the new one.

Table 1. Details of Dataset

Dataset Name	Author(s)	Year(s)	Associated Tasks	No. of Attributes	No. of Instances	Description				Ref.
						PD		HP.		
						M	F	M	F	
Parkinson's	C. Okan Sakar a et al.	2018-11-05	Classification	754	756	107	81	23	41	[8]
Disease										
Classification										
Data Set										

4.3 Feature Correlation Analysis

4.3.1 Smote

The class distribution in the data set is unbalanced. The SMOTE (Synthetic Minority Over- Sampling Technique) approach is employed to balance this unbalanced dataset [14]. The Parkinson's disease dataset was then classified using the Random Forests classification method after being converted to a balanced class distribution. The over-sampling phase of Fig. 2 also includes the Synthetic Minority Oversampling Technique (SMOTE). Consider a dataset that is being used for any machine learning activity and has lots of examples of one class and a lot fewer examples of the other. A dataset in this situation is referred to as a class-imbalanced dataset. By generating synthetic samples from the minority class, SMOTE oversampling addresses the issue of class imbalance. Each minority instance is taken into account by SMOTE, which then selects one of its k closest minority neighbours before creating a new minority instance between the neighbour and itself. This process gets rid of the class imbalance problem [15]. The approach then goes through identical stages as the first method to calculate the average performance of the trained models on the testing data that was previously set aside after the oversampling.

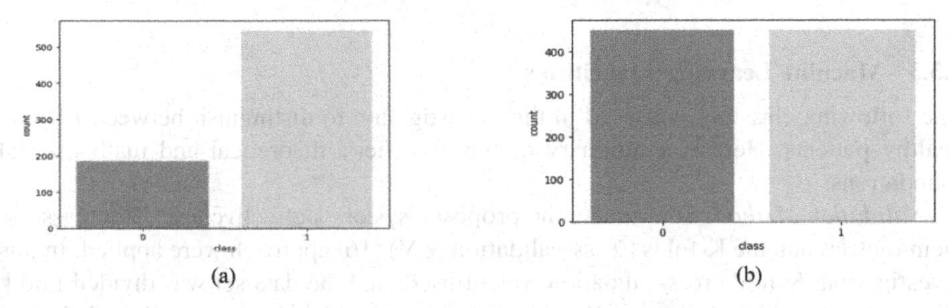

Fig. 2. Transform this (a) imbalanced dataset to (b) balanced dataset

4.3.2 Feature Selection

To exclude a feature that is not important to the output class or labels, a feature selection criterion must be used. From the perspective of machine learning, a system will utilise this information for fresh data if it employs irrelevant variables, which will result in poor generalisation. It is not appropriate to compare removing unnecessary variables to other approaches like LassoCV [16] and RFE [9, 10] as desirable traits may be independent of other data. Since it makes use of the input features themselves to decrease their quantity, feature removal does not produce any new features. After choosing a feature selection criterion, a method must be created to identify the subset of valuable characteristics [17, 18]. To eliminate duplicate data with manageable calculations, a suboptimal approach must be applied. In machine learning or pattern recognition applications, feature selection algorithms give us the means to shorten computation times, enhance prediction accuracy, and get a deeper knowledge of the data. To highlight the applicability of feature selection strategies, we also apply certain feature selection techniques to common datasets [18].

LassoCV: The ability to change the absolute value of the coefficient of functions is necessary for the minimal selection and shrinkage features of this operator. The subset of features can also exclude features with negative coefficients and features whose coefficient values are zero. When it comes to feature values with tiny coefficients, the LASSO [16] performs quite well. The selected feature subsets will contain features with high coefficient values. With LASSO, unnecessary features might be detected. Additionally, by repeatedly doing the previous technique, and finally considering the qualities that are discovered the most frequently to be the most significant, the dependability of this feature may be improved.

RFE: Following the computation of the missing values, it is necessary to select the critical aspects that strongly and favourably correlate with those critical to the diagnosis of a condition. A robust diagnostic model cannot be built since the vector characteristics must be extracted to exclude features that are irrelevant and unhelpful for prediction. To extract a prediction's most crucial properties, we employed the RFE approach. Due to its simplicity of use and setups, as well as its efficiency in picking features in training datasets important to predicting target variables and removing weak features, the recursive Feature Elimination (RFE) technique is particularly well-liked. By identifying a strong association between particular characteristics and the aim, the RFE approach is used to choose the most important features (labels) [19].

4.3.3 Machine Learning Algorithms

The following classifier was used in this investigation to distinguish between PD and healthy patients. Here is a summary of the classifier's theoretical and mathematical foundations.

Validation Method: To evaluate the proposed system's effectiveness, Three assessment metrics and the K-folds Cross-validation (CV) [16] approach were applied. In this investigation, K-fold cross-validation was utilised, and the data set was divided into k identical components following K-fold. In each stage, the k-1 groups and the leftovers for training were used for testing. The number of times the procedure iterates. The classifier's performance is then determined by computing the average of k outcomes. We chose a different cross-validation value of k, and in our studies, we utilised $k = 10$. 30% of the data was utilised for testing throughout the 10-fold CV procedure, while 70% was used for training [24]. The validation procedure was done ten times. Before choosing fresh training and test sets for the subsequent cycle, all samples in the training and test groups are randomly dispersed over the whole dataset during each fold. Finally, an average of all performance measures is calculated after 10-fold processes. As seen in Fig. 2, the estimated average performance E of the model was determined by adding the estimated performances Ei for each fold. Equation (2) represents the mathematical formula for estimated average performance:

$$E = \frac{1}{10} \sum_{i=1}^{10} E_i \tag{2}$$

4.3.4 Performance Evaluation

Accuracy, specificity, and sensitivity are just a few of the assessment measures that have been used to gauge how well the suggested automated system performs. Accuracy for LassoCV [16] provides details on patients that were accurately identified in the dataset. While specificity provides information on accurately diagnosed healthy participants, sensitivity represents patients who have been successfully classified. K Fold Cross Validation was carried out with a K value of 10 to assess the ML models. The performance of the algorithms above was evaluated using the four assessment metrics listed below to determine which algorithm best predicts Parkinson's disease [20]. We have shortened a few terms in the equations below, and they are as follows:

TN – True Negative, TP – True Positive, FN – False Negative, FP – False Positive

5 Experimental Analysis

On a system with an Intel(R) Core(TM) i5-3000 CPU running at 2.48 GHz and 16 GB of RAM, all experiments were carried out using the Python3 library. Scikit Learn, NumPy, Pandas, Matplotlib, and Seaborn are some of the Python 3.8.8 libraries that were used to construct the suggested method in Jupyter Notebook 6.4.5.

5.1 Outcomes of Feature Selection Processes Using LassoCV

The dataset was balanced using SMOTE as the initial stage in this study, and the extracted features were then fed into the created feature selection technique (LassoCV and RFE) to lower the dimensionality of feature subsets and identify the best feature set [20] (Fig. 3).

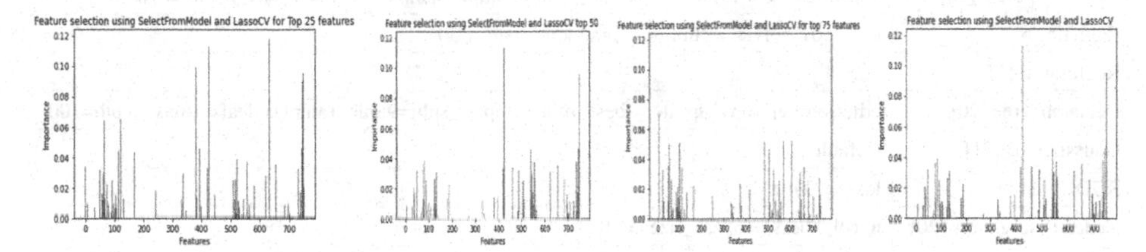

Fig. 3. illustrate the selected feature from the dataset with the top (n = 25, 50, 75, and 100).

Result and Discussion

5.1.1 Comparison of Various Classifiers Using LassoCV Features Selection

In this section. Table 5 and Table 6 give LassoCV validation results of the ten machine learning classifiers with the selected n features (n=25, 50, 75 and 100, respectively) by LassoCV. Table 5 demonstrates that the Bagging classifier outperforms the other classifier. In Table 5, the best outcomes for each statistic are underlined. Top n=25 features yielded results for the Bagging classifier of 83.7% Accuracy (A), 83.7% F1-score (F), 83.7% Recall (R), 83.8% Precision (P), and 95% Cross validation Score (CVS), while top n=50 features yielded values of 89.4%, 89.2%, 89.2%, and 97.0%, respectively. In Table 6. The Bagging classifier got the highest accuracy to other classifiers when we used the Top n= 75 features s had results of 91.6% Accuracy(A), 91.7% F1-score(F), 91.6% Recall(R), 91.7% Precision (P) and 96.1% Cross validation Score (CVS) and top n= 100 features scores 88.5%, 88.5%, 88.5%, 89.4% and 97.0% respectively. When we choose the top n=75 features, feature selection enhances the performance of classifiers. All assessment metrics represented in Fig. 4 have experienced considerable improvements.

Table 2. Number of Features in the dataset.

Feature	Number of Features
Measure Baseline features[11]	21
Time-frequency features	11
(MFCCs) [12]	84
Wavelet transform-based features	182
Vocal fold features	22
(Tqwt) [13]	432
TOTAL Features	752

Table 3. Hyperparameter settings of each ML classifier

ML Classifier	Tuning Parameters
Random-Forest [20]	n_estimators = 190,min_samples_split = 3,min_samples_leaf = 1,max_depth = 160
Logistic	C = 0.01, solver = 'liblinear', random_state = 30
Regression [17]	
Decision-Tree [20]	criterion='entropy',splitter='best',min_samples_split=9,min_samples_leaf=2,max_depth=100
GaussianNB [21]	Default
SGDClassifier	loss = "hinge"
Nearest-Neighbors [20]	n_neighbors = 5,leaf_size = 30
SVM [22]	C=100,gamma=0.1,kernel='rbf'
Adboost [23]	n_estimators= 50
MLPClassifier	hidden_layer_sizes=(100,100,100), max_iter=500, alpha=0.0001, solver='sgd', verbose=10, random_state=21,tol=0.000000001
Bagging Classifier	base_estimator=tree, n_estimators=1500, random_state=42

Table 4. Performance evaluation parameters and formulas

Measure	Formula	Description
Accuracy	$\frac{TP+TN}{TP+FP+FN+TN}$	Accurate measurements are made of the system's predictive capability
Precision	$\frac{TP}{TP+FP}$	The model accuracy displays the proportion of a set of accurate outputs to all of the model's results.
Recall	$\frac{TP}{TP+FN}$	It proves the patient has Parkinson's disease and that a diagnostic test returned a positive result.
F1 Score	$\frac{2*(\text{precision}*\text{recall})}{(\text{precision}+\text{recall})}$	This measure is produced by taking into account Precision and Recall and calculating their harmonic means. In other words, it combines Accuracy and recall and provides insight into the behaviour of both measurements.

Table 5. Performance Comparison of classifiers for selected features(n=25,50) by LassoCV

	LassoCV for top 25 features					LassoCV for top 50 features				
Models	A(%)	F(%)	P(%)	R(%)	CVS	A(%)	F(%)	P(%)	R(%)	CVS
Random-Forest	81.9	81.2	81.1	81.9	96.6	83.3	81.9	82.8	83.3	96.7
logistic Regression	80.2	81.1	83.9	80.2	80.1	86.8	87.2	88.3	86.8	83.4
Decision-Tree	75.3	75.6	75.9	75.3	90.3	81.1	81.5	82.4	81.1	90.8
GaussianNB	71.4	73.0	79.6	71.4	75.8	77.1	78.3	82.5	77.1	78.3
SGDClassifier	79.3	80.0	81.7	81.7	77.1	85.9	86.0	86.1	86.1	82.6
Nearest-Neighbors	77.1	78.3	82.5	77.1	83.8	80.2	81.1	84.3	80.2	87.4
SVM	81.9	80.4	81.2	81.9	96.1	76.7	70.1	77.8	76.7	97.4
Adboost	82.4	82.7	83.3	82.4	92.1	85.0	85.4	86.1	85.0	93.6
MLPClassifier	83.7	83.4	83.2	83.7	93.2	85.0	84.3	84.5	85.0	94.8
BaggingClassifier	**83.7**	**83.7**	**83.8**	**83.7**	**95.0**	**89.4**	**89.2**	**89.2**	**89.4**	**97.0**

5.2 Comparison of Various Classifier Algorithms on the Recursive Feature Elimination (RFE) Selected Features

In this section. Table 7 and Table 8 give Recursive Feature Elimination (RFE) results of the ten machine learning classifiers with the selected n features (n=25, 50, 75 and 100, respectively) by RFE. According to Table 7, Random Forest classifier overcomes the other classification algorithms. The best results for each metric are highlighted in Table 7. The Random Forest classifier using the Top n=25 features s had scores of 89.9% Accuracy(A), 89.7% F1-score(F), 89.9% Recall(R), 89.7% Precision (P) and 97.2% Cross validation Score (CVS) another hand got highest Accuracy to other classifiers when we using the top n=50 features scores 90.1% Accuracy(A), 90% F1-score(F), 90.1% Recall(R), 89.9% Precision (P) and 89.9% CSV respectively. In Table 8 the Random

Table 6. Performance Comparison of classifiers for selected features(n = 75,100) by LassoCV

	LassoCV for top 75 features					LassoCV for top 100 features				
Models	A(%)	F(%)	P(%)	R(%)	CVS	A(%)	F(%)	P(%)	R(%)	CVS
Random-Forest	88.1	87.1	88.9	88.1	97.0	85.9	85.0	85.7	85.9	97.2
logistic Regression	85.9	86.2	86.7	85.9	84.8	83.7	84.2	85.5	83.7	86.8
Decision-Tree	84.1	83.9	83.7	84.1	93.3	78.4	78.8	79.3	78.4	92.4
GaussianNB	77.1	78.2	81.2	77.1	76.2	71.8	73.4	79.3	71.8	85.5
SGDClassifier	83.3	83.8	85.3	85.3	84.8	81.1	81.5	82.1	82.1	85.5
Nearest-Neighbors	78.0	79.0	82.5	78.0	88.5	78.0	79.0	82.5	78.0	85.5
SVM	73.6	62.4	54.1	73.6	96.7	73.6	62.4	54.1	73.6	96.7
Adboost	87.2	87.3	87.5	87.2	93.1	85.9	86.0	86.2	85.9	95.0
MLPClassifier	85.5	85.0	85.0	85.5	96.2	85.0	84.6	84.5	85.0	95.2
BaggingClassifier	**91.6**	**91.7**	**91.7**	**91.6**	**96.1**	**88.5**	**88.5**	**88.5**	**88.5**	**96.5**

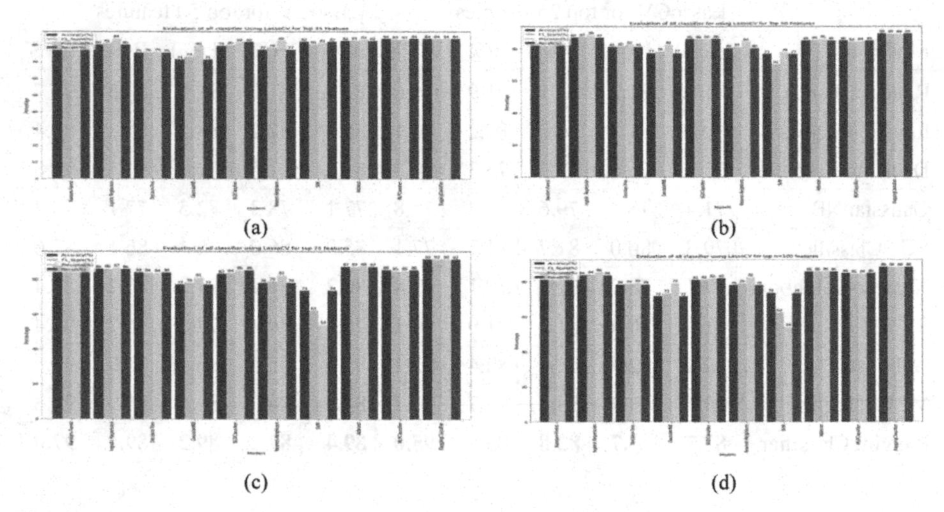

Fig. 4. Performance of different classifier algorithms using the LassoCV features selection Algorithm when features are (n = 25, 50, 75 and 100): (a) feature selected top 25 (b) feature selected top 50 (c) feature selected top 75 (d) feature selected top 100.

Forest classifier when we using the Top n= 75 features had results of 88.8% Accuracy(A), 88.5% F1-score(F), 88.8% Recall(R), 88.5% Precision (P) and 97.3% Cross validation Score (CVS) and top n= 100 features scores 88.2%, 87.9%, 88.2%, 87.9% and 97.2% respectively. Feature selection improves the performance of classifiers when we select the top n=50 features. There are significant improvements in all evaluation metrics depicted in Fig. 5.

Table 7. Performance Comparison of classifiers for selected features(n = 25,50) by RFE.

	RFE for top 25 features selected					RFE for top 50 features selected				
Models	A(%)	F(%)	P(%)	R(%)	CVS	A(%)	F(%)	P(%)	R(%)	CVS
Random-Forest	**89.9**	**89.7**	**89.7**	**89.9**	**97.2**	**90.1**	**90.0**	**89.9**	**90.1**	**89.9**
logistic Regression	74.9	76.2	79.6	74.9	81.2	77.6	78.5	80.4	77.6	74.9
Decision-Tree	84.1	84.0	84.0	84.1	92.5	81.6	81.7	81.9	81.6	84.1
GaussianNB	78.4	79.0	80.0	78.4	80.2	79.6	80.0	80.6	79.6	78.4
SGDClassifier	78.4	79.1	80.3	80.3	77.5	79.6	80.5	82.5	82.5	78.4
Nearest-Neighbors	81.5	82.2	84.0	81.5	92.1	86.2	86.8	89.2	86.2	81.5
SVM	89.4	89.2	89.2	89.4	98.6	90.8	90.2	91.3	90.8	89.4
Adboost	86.3	86.6	87.0	86.3	95.7	83.6	83.6	83.7	83.6	86.3
MLPClassifier	88.1	88.1	88.0	88.1	93.8	88.8	88.8	88.7	88.8	88.1
BaggingClassifier	87.2	87.5	87.9	87.2	96.3	88.2	88.2	88.2	88.2	87.2

Table 8. Performance Comparison of classifiers for selected features(n = 75 and 100) by RFE.

	RFE for top 75 features selected					RFE for top 100 features selected				
Models	A(%)	F(%)	P(%)	R(%)	CVS	A(%)	F(%)	P(%)	R(%)	CVS
Random-Forest	**88.8**	**88.5**	**88.5**	**88.8**	**97.3**	**88.2**	**87.9**	**87.9**	**88.2**	**97.2**
logistic Regression	77.0	78.0	80.1	77.0	86.3	77.0	77.8	79.5	77.0	86.8
Decision-Tree	85.5	85.6	85.8	85.5	93.0	85.5	85.6	85.8	85.5	92.4
GaussianNB	81.6	82.0	82.7	81.6	80.6	78.3	78.6	78.9	78.3	85.5
SGDClassifier	78.9	79.6	80.6	80.6	85.1	78.9	79.8	81.6	81.6	86.0
Nearest-Neighbors	86.8	87.4	89.6	86.8	91.9	88.2	88.7	90.8	88.2	91.7
SVM	82.9	80.2	83.6	82.9	97.1	81.6	77.2	85.2	81.6	97.4
Adboost	88.2	88.2	88.2	88.2	94.1	86.2	86.5	86.9	86.2	95.3
MLPClassifier	87.5	87.4	87.4	87.5	96.4	86.2	86.1	86.1	86.2	96.6
BaggingClassifier	87.5	87.6	87.6	87.5	96.9	88.2	88.2	88.2	88.2	96.9

5.3 Hybrid Features Selection

For feature selection in this part, we employ the Hybrid feature selection technique [20]. For hybrid selection, we use a union operation of both selected features (LassoCV RFE). First, we choose 25 characteristics from RFE and LassoCV. 47 out of 50 features were assigned to their shared characteristics; similarly, 87 out of 100, 129 out of 150, and 176 out of 200 features were all assigned to them. Bagging classifiers outperform the other classification techniques, as seen in Tables 9 and 10. The hybrid feature selection n=46 scores of 91.6% Accuracy(A), 91.7% F1-score(F), 91.6% Recall(R), 91.8% Precision(P),

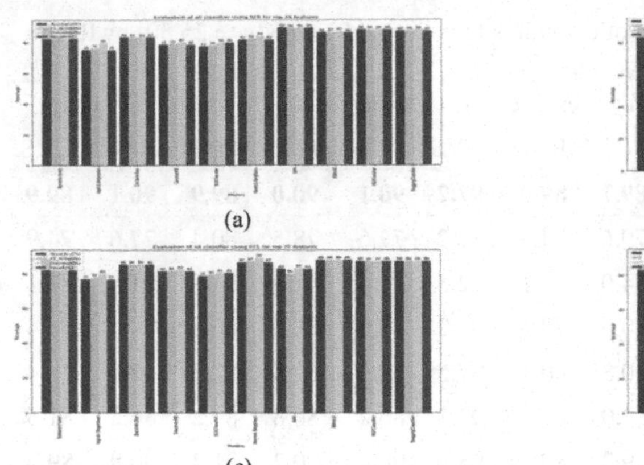

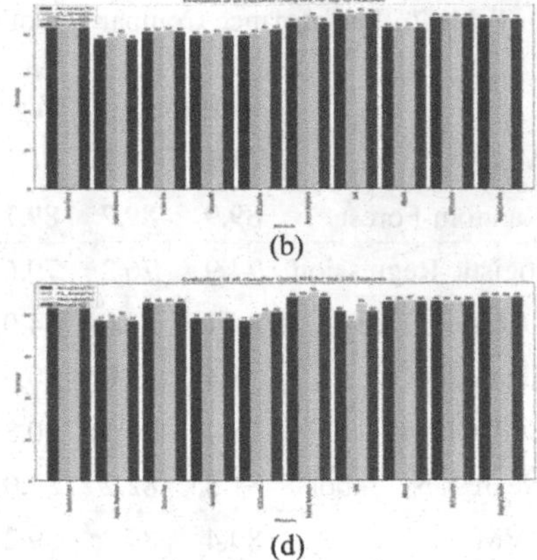

(a) (b)

(c) (d)

Fig. 5. Performance of different classifier algorithms using the Recursive Feature Elimination Algorithm when features are (n = 25, 50, 75 and 100): (a) feature selected top 25, (b) feature selected top 50, (c) feature selected top 75 (d) feature selected top 100.

and 97.4% Cross validation Score provide the greatest results for each measure (CVS) [20].

Table 9. Comparison of the performance of the method when (n=46 and 87 features) using Hybrid selection

	Hybrid selection using LassoCV REF for N = 46 features					Hybrid selection using LassoCV REF for N = 87 features				
Models	A(%)	F(%)	P(%)	R(%)	CVS	A(%)	F(%)	P(%)	R(%)	CVS
Random-Forest	89.4	88.9	89.4	89.4	97.6	88.1	87.4	88.1	88.1	97.1
logistic Regression	84.1	84.7	86.4	84.1	82.7	85.5	85.9	87.2	85.5	85.1
Decision-Tree	78.4	79.0	79.9	78.4	93.7	83.3	83.3	83.3	83.3	93.6
GaussianNB	76.2	77.5	82.1	76.2	79.8	75.8	77.0	80.6	75.8	79.8
SGDClassifier	83.3	83.9	85.6	85.6	81.2	81.1	81.5	82.1	82.1	87.1
Nearest-Neighbors	85.5	86.1	88.5	85.5	87.9	84.6	85.2	87.3	84.6	89.9
SVM	80.6	76.8	82.4	80.6	97.6	74.4	64.4	81.0	74.4	96.9
Adboost	83.7	83.6	83.5	83.7	95.0	87.7	87.6	87.5	87.7	95.6
MLPClassifier	86.3	85.8	85.9	86.3	95.1	88.1	87.5	88.0	88.1	96.2
BaggingClassifier	**91.6**	**91.7**	**91.8**	**91.6**	**97.4**	**89.4**	**89.2**	**89.2**	**89.4**	**97.4**

Table 10. Comparison of the performance of the method when (n = 129 and 176 features) using Hybrid selection

	Hybrid selection using LassoCV REF for n=129 features					Hybrid selection using LassoCV REF for n=176 features				
Models	A(%)	F(%)	P(%)	R(%)	CVS	A(%)	F(%)	P(%)	R(%)	CVS
Random-Forest	86.8	86.2	86.5	86.8	97.0	87.7	87.1	87.4	87.7	97.4
logistic Regression	83.3	83.7	84.7	83.3	85.3	85.0	85.4	86.3	85.0	86.6
Decision-Tree	81.1	81.0	81.0	81.1	92.6	83.3	83.2	83.1	83.3	91.9
GaussianNB	78.4	79.2	81.2	78.4	80.0	76.2	77.2	79.6	76.2	80.1
SGDClassifier	83.7	83.8	84.0	84.0	91.1	84.6	84.8	85.3	85.3	92.2
Nearest-Neighbors	84.6	85.2	87.0	84.6	89.3	80.6	81.6	85.4	80.6	89.9
SVM	73.6	62.4	54.1	73.6	96.7	73.6	62.4	54.1	73.6	96.7
Adboost	81.9	82.0	82.0	81.9	94.7	82.8	82.5	82.3	82.8	95.0
MLPClassifier	86.8	86.2	86.5	86.8	95.2	85.9	85.5	85.5	85.9	95.3
BaggingClassifier	**88.5**	**88.4**	**88.4**	**88.5**	**96.7**	**90.7**	**90.6**	**90.6**	**90.7**	**96.9**

To evaluate distinct subsets acquired for a specific number of runs, various measures are devised. For many datasets, a more reliable subset can be discovered using these metrics. To obtain a resilient subset based on merging many classifiers to increase Accuracy, the LassoCV technique was developed, which combines multiple feature selection algorithms to rank/score the features. The author also advises segmenting the input characteristics into separate classifiers based on their methods for feature extraction and combining the predictions to arrive at a final decision (Fig. 6).

Table 11. Comparing the results obtained with research published recently in the literature

Study	Number of subjects	Used Method	Used Classifier	System Accuracy (%)
Yücelbaş Ş et al., 2020 [17]	188 Pd 64 HC [a][8]	Using a straightforward greedy stepwise method, the logistic hybrid system (SLGS)	SLGS	For men with 11 features, 88.71, and 87.15 (for females with 9 features)
Sakar et al., 2019 [6]	188 Pd 64 HC [a][8]	mRMR-50 and the Tunable Q-factor Wavelet	SVM (RBF kernels)	86 (with 50 features)
Transform (TQWT)				
Al-Husban A et al. 2022 [25]	188 Pd 64 HC [a][8]	DT, SVM, and kNN	SVM	86%
Tallapureddy G et al., 2022 [26]	188 Pd 64 HC [a][8]	Feature Principal	Reduction Component	Voting 85 Stacking 81
Analysis (PCA)				
Yücelbaş Ş et 188 PD al., 2020 [17] 64 HC [a][8]		Simple Logistic hybrid system based on a greedy stepwise algorithm (SLGS)	SLGS	88.71 (for males with 11 features) 87.15 (for females with 9 features)
Our Approach	188 PD 64 HC [a][8]	-Feature selection: LassoCV and REF -classifiers: Random-Forest, Logistic Regression, Decision-Tree, GaussianNB, SGDClassifier, Nearest-Neighbors, SVM, Adboost, MLPClassifier, and Bagging Classifier	Hybrid Feature Selection with Bagging Classifier	**91.6% (46 features)**

[a]In these studies, same dataset was used

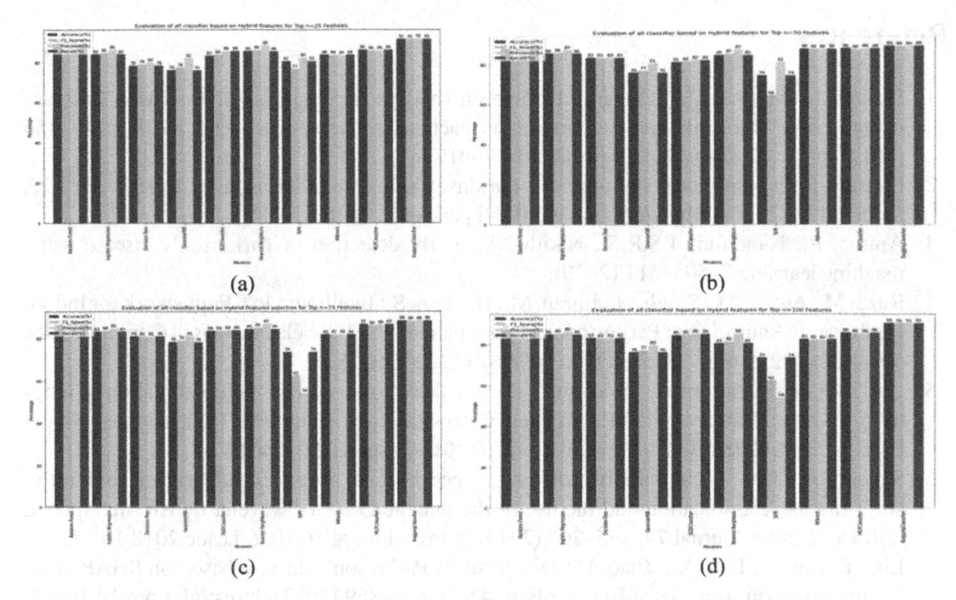

Fig. 6. Performance of various classifier methods employing hybrid features selection.

6 Conclusion

In this study, we combine the LassoCV and RFE feature selection methods to create a hybrid feature selection algorithm that can be used in many configurations. We evaluated the Parkinson's disease diagnosis study using ten ML classifiers: Random-Forest, Logistic Regression, Decision-Tree, GaussianNB, SGDClassifier, Nearest-Neighbors, SVM, Adboost, MLPClassifier, and Bagging Classifier, and we came to the following results. On the Parkinson's disease dataset, the Bagging model based on Hybrid feature selection outperforms previous feature selection methods with classification results of 91.6% Accuracy (A), 91.7% F1-score (F), 91.6% Recall (R), 91.7% Precision (P), and 96.1% Cross-Validation Score (CVS). On the Parkinson's disease dataset, the Random Forest model based on RFE feature selection has the high Accuracy compared to other classifiers when we use the top n=50 features, scoring 90.1% Accuracy (A), 90% F1-score (F), 90.1% Recall (R), 91.8% Precision (P), and 97.4% CSV, respectively. The Bagging model offers the best classification performance compared to other classification techniques, but its computing efficiency is inferior when calculating Accuracy. In light of the combined consideration of classification performance and computing efficiency, the Bagging model is more appropriate for real-world applications in diagnosing Parkinson's disease. In future work, on significant speech and voice datasets to test the suggested Methodology and enhance the performances to classify PD and enhance the performance of our recommended strategy.

References

1. Shahbakhi, M., Far, D.T., Tahami, E.: Speech analysis for diagnosis of parkinson's disease using genetic algorithm and support vector machine. J. Biomed Sci. Eng. **07**(04), 147–156 (2014). https://doi.org/10.4236/jbise.2014.74019
2. Pfeiffer, R.F.: Non-motor symptoms in Parkinson's disease. Parkinsonism Relat Disord **22**, S119–S122 (2016). https://doi.org/10.1016/j.parkreldis.2015.09.004
3. Anitha, R., Nandhini, T.S.R.S., Nikitha, V.: Early detection of parkinson's disease using machine learning **2**, 505–511 (2020)
4. Raza, M., Awais, M., Singh, N., Imran, M., Hussain, S.: Intelligent IoT Framework for Indoor Healthcare Monitoring of Parkinson's Disease Patient. IEEE J. Select. Areas in Commu. **39**(2), 593–602 (2021). https://doi.org/10.1109/JSAC.2020.3021571
5. Hawi, S., et al.: Automatic Parkinson's disease detection based on the combination of long-term acoustic features and Mel frequency cepstral coefficients (MFCC). Biomed Signal Process Control **78** (2022). https://doi.org/10.1016/j.bspc.2022.104013
6. Sakar, C.O., et al.: A comparative analysis of speech signal processing algorithms for Parkinson's disease classification and the use of the tunable Q-factor wavelet transform. Applied Soft Computing Journal **74**, 255–263 (2019). https://doi.org/10.1016/j.asoc.2018.10.022
7. Liu, Y., Liu, Z., Luo, X., Zhao, H.: Diagnosis of Parkinson's disease based on SHAP value feature selection. Biocybern Biomed Eng. **42**(3), 856–869 (2022). https://doi.org/10.1016/j.bbe.2022.06.007
8. UCI Machine Learning Repository: Parkinson's Disease Classification Data Set. https://archive.ics.uci.edu/ml/datasets/Parkinson%2527s%2BDisease%2BClassification, accessed 22 Aug. 2022
9. sklearn.feature_selection.RFE — scikit-learn 1.1.2 documentation. https://scikit-learn.org/stable/modules/generated/sklearn.feature_selection.RFE.html, accessed 15 Sep. 2022
10. Feature selection using SelectFromModel and LassoCV — scikit-learn 0.19.2 documentation. https://scikit-learn.org/0.19/auto_examples/feature_selection/plot_select_from_model_boston.html, accessed 15 Sep. 2022
11. Little, M.A., McSharry, P.E., Hunter, E.J., Spielman, J., Ramig, L.O.: Suitable of Dysphonia measurements for telemonitoring of Parkinson's Disease. IEEE Trans. Biomed. Eng. **56**(4), 1–20 (2009). https://doi.org/10.1109/TBME.2008.2005954
12. Benba, A., Jilbab, A., Hammouch, A., Sandabad, S.: Voiceprints analysis using MFCC and SVM for detecting patients with Parkinson's disease. In: Proceedings of 2015 International Conference on Electrical and Information Technologies, ICEIT 2015, pp. 300–304 (2015). https://doi.org/10.1109/EITech.2015.7163000
13. Selesnick, I.W.: Wavelet Transform with Tunable Q-Factor (2011)
14. Polat, K.: A hybrid approach to Parkinson disease classification using speech signal: The combination of SMOTE and random forests. In: 2019 Scientific Meeting on Electrical-Electronics and Biomedical Engineering and Computer Science, EBBT 2019 (2019). https://doi.org/10.1109/EBBT.2019.8741725
15. Jain, D., Mishra, A.K., Das, S.K.: Machine learning based automatic prediction of parkinson's disease using speech features. Advances in Intelligent Systems and Computing **1164**, 351–362 (2021). https://doi.org/10.1007/978-981-15-4992-2_33
16. Cantürk, İ., Karabiber, F.: A machine learning system for the diagnosis of parkinson's disease from speech signals and its application to multiple speech signal types. Arab. J. Sci. Eng. **41**(12), 5049–5059 (2016). https://doi.org/10.1007/s13369-016-2206-3
17. Yücelbaş, Ş.: Simple logistic hybrid system based on greedy stepwise algorithm for feature analysis to diagnose parkinson's disease according to gender. Arab. J. Sci. Eng. **45**(3), 2001–2016 (2020). https://doi.org/10.1007/s13369-020-04357-1

18. Chandrashekar, G., Sahin, F.: A survey on feature selection methods. Computers and Electrical Engineering **40**(1), 16–28 (2014). https://doi.org/10.1016/j.compeleceng.2013.11.024
19. Senan, E.M., et al.: Diagnosis of chronic kidney disease using effective classification algorithms and recursive feature elimination techniques. J. Healthc. Eng. **2021** (2021). https://doi.org/10.1155/2021/1004767
20. Lamba, R., Gulati, T., Alharbi, H.F., Jain, A.: A hybrid system for Parkinson's disease diagnosis using machine learning techniques. Int J Speech Technol **25**(3), 583–593 (2022). https://doi.org/10.1007/s10772-021-09837-9
21. Friedman, N., Geiger, D., Provan, G., Langley, P., Smyth, P.: Bayesian Network Classifiers. Kluwer Academic Publishers (1997)
22. Hoq, M., Uddin, M.N., Park, S.B.: Vocal feature extraction-based artificial intelligent model for parkinson's disease detection. Diagnostics **11**(6) (2021). https://doi.org/10.3390/diagnostics11061076
23. Freund, Y., Schapire, R.E.: Journal of Computer and System Sciences s SS1504. Journal of Computer and System Sciences (1997)
24. Haq, A.U., et al.: Feature Selection Based on L1-Norm Support Vector Machine and Effective Recognition System for Parkinson's Disease Using Voice Recordings. IEEE Access **7**, 37718–37734 (2019). https://doi.org/10.1109/ACCESS.2019.2906350
25. Al-Husban, A., Abdulridha, M.M., Mohamad, A.A.H., Ibrahim, A.M.: Biocomposite's multiple uses for a new approach in the diagnosis of parkinson's disease using a machine learning algorithm. Adsorption Science and Technology **2022** (2022). https://doi.org/10.1155/2022/6159392
26. Tallapureddy, G., Radha, D.: Analysis of ensemble of machine learning algorithms for detection of parkinson's disease. In: Proceedings - International Conference on Applied Artificial Intelligence and Computing, ICAAIC 2022, pp. 354–361 (2022). https://doi.org/10.1109/ICAAIC53929.2022.9793048

Classification of EEG Signals for Epilepsy Detection Using PCA Analysis

Moushmi Kar(✉)

Pt. S. N. Shukla University, Shahdol, M.P., India
moushmikar@gmail.com

Abstract. Epilepsy is a disease related to the abnormal seizures caused by the central nervous system. The epileptic seizure is occurred due to malfunctioning of the electrophysiological system of the brain. The collection of EEG data is become a tedious and time consuming process so it is difficult to detect of epileptic seizures by visual observation of patient's EEG. It needs an expert to detect epileptic activity by analyzing the entire length of recorded EEG data. In this study, we propose a method for classifying EEG signals for the detection of epilepsy using principal component analysis (PCA) and machine learning techniques. The proposed method involves pre-processing the EEG signals to reduce noise and artifacts, followed by PCA to extract the most relevant features of the signals. It reduces the dimensionality of data and it is linear combinations of the original features. Then the features obtained from the PCA components are feed to machine learning classifier. The most commonly used classifiers are support vector machine or naive bayes, random forest to classify the epileptic brain signals. In this work, the accuracy, sensitivity obtained with PCA reduction and without PCA reduction is compared. The dataset consisted of EEG recordings from multiple electrodes, with a total of over 1000 signals. We used a kfold cross-validation and data partition (33%) method to estimate the performance of the classifier. We have used K-Nearest Neighbors (kNN), Support Vector Machine (SVM), and Navie bayes algorithms by exploiting the PCA feature reduction technique in the dataset to envisage epilepsy, and the performance of classifiers are analyzed with or without using PCA technique. The study revealed that the SVM classifier demonstrated superior performance with a classification accuracy of 96.77% when utilizing the holdout technique. The proposed model exhibited higher measures of accomplishment, including sensitivity (99.34%), specificity (94.2%), and AUC (99%), compared to other classifiers when holdout cross validation was used.

1 Introduction

The brain activity is measured either by using non-invasively EEG signals or invasively using ECOG signals. These signals have been used to identify and analyzes the neurological disorders and abnormalities of the brain. For the sensitive functional test, the most important advantage is the use of EEG signals [1, 2]. The microscopic view of brain is more sophisticated, containing many types of connection involving variety of chemicals in brain. A small group of active neurons record electrical activity on the head [3].

B. K. Singh et al. (Eds.): ICBEST 2023, CCIS 2003, pp. 204–219, 2024.
https://doi.org/10.1007/978-3-031-54547-4_16

1.1 Epilepsy

According to WHO that around the 50 million people of the world having a common brain disease that is Epilepsy. Epilepsy affected one in 100 people in the world [4]. Most of Epilepsy disorder have seen in the earlyhood or adulthood. Few cases also observe at above age of 65.The people affected by this disease die early than the normal person [5].

The cortex is stimulated by epileptic circumstances, and high-voltage pulses (up to 1000 V) that are known as "spikes" or "spike and wave" occur [6].

Chronic brain disease referred to as epilepsy has been brought on by a sudden abnormality of brain nerve cells. Due to incorrect neuronal firing, it causes aberrant electrical activity in the human brain neurons [7–11]. The complex and distinctive traumatic symptoms of seizures include unbalanced bodily movements, jerking body parts, loss of consciousness, stroke, and disruption of cognitive processes [8, 10].

For cases involving two classes, classification accuracies of 99.07%, 91.875%, 97.55%, 90%, and 96.60% can be achieved by combining ANN with the wavelet transform [10], weighted permutation entropy [11], PCA [12], Stockwell transform [13], and time-frequency domain features [14]. The aim of the work is to find the best algorithm for the classification of seizures by applying a PCA (Principal Components Analysis) dimensional reduction technique to the data.

In this paper, we used k-Nearest Neighbors (kNN), Support Vector Machine (SVM) and Navie Bayes algorithms to predict epilepsy using PCA feature reduction technique on the dataset, and the performance of the classifiers is analyzed by PCA and without the PCA technique. The paper reports the best possible classifier among kNN, SVM and Navie Bayes using PCA techniques along with the comparing results with and without the PCA.

2 Related Work

So far different research group have studied in search of suitable model for precise detection as well as prediction of epileptic seizures using EEG signals. Some of recent works have reported very excellent results in this area.

Guerrero, Parada, and Espitia (2021) proposed classification for epilepsy detection using fast fourier analysis and principal component analysis [15]. In this article, author used logistic regression for epileptic classification with a reduction from 5 to 4 dimensions as well as from 8 to 7. The results show that the PCA-enhanced machine learning model achieves high accuracy in identifying epileptic patients, outperforming other techniques, without PCA are 0.560, 0.690, precision and recall respectively; meanwhile, the values obtained using PCA are: precision 0.734, recall 0.787, and F1 score 0.77. The method proposed by Ashwani et al. was based on key point LBP [16]. They present their own approach, which involves feature extraction using Discrete Wavelet Transform (DWT) and classification using Artificial Neural Networks (ANN). The proposed method is evaluated using EEG signals from the CHB-MIT Scalp EEG database, and the results show that it achieves high accuracy in detecting seizures. The authors suggest that their approach has the potential to be used as a real-time seizure detection system for epilepsy patients. The real time automatic detection of seizures with EEG signals has been reported by Lasitha and Khan (2017) [17]. The functions used in these studies were the harmonic wavelet transform and the fractal dimension. Both FD and HWPT power features in all EEG channels at each epoch are positioned according to spatial information when the electrodes are placed on the skull. The classifier used in this study was an importance vector machine because it is effective in classifying sparse but high- dimensional datasets. The author reported a classification accuracy of 99.8% using short-term dataset B and a sensitivity of 96%. 0.1/h average false detection rate and 1.89 s average detection delay. Sharmila and Geethanjali in 2016 have proposed their own method, which involves using Discrete Wavelet Transform (DWT) for feature extraction and classification using both Naive Bayes and k-NN classifiers. The proposed method is evaluated using EEG signals from the CHB-MIT Scalp EEG database, and the results show that it achieves high accuracy in detecting seizures. The paper provides valuable insights into the current state of research on seizure detection and highlights the need for further investigation in this field [18]. Ghosh-Dastidar et al., 2008 is applied principal component analysis (PCA) to EEG signals [19]. The authors use a combination of Principal Component Analysis (PCA) and Cosine Radial Basis Function Neural Network (CRBFNN) to enhance the performance of the system. They proposed a new wavelet-chaos-neural network methodology that shows 96.6% accuracy.

Senger and Tetzlaff have proposed a comparative study of three seizure prediction methods [20]. The first method involves using an eigen value-based principal component analysis (PCA) for pre-processing the EEG signal, followed by a nonlinear convolution neural network (CNN) for seizure prediction. The second method uses a linear signal prediction approach, followed by a level-crossing behaviour analysis. The third method combines the first two methods. The results showed that the combined method (PCA pre-processing + nonlinear CNN + linear prediction + level-crossing behaviour analysis) outperformed the other two methods in terms of sensitivity, specificity, and false prediction rate. The authors concluded that the combined method may have the potential for clinical applications in the prediction of epileptic seizures. Shiao et al. (2017) have applied SVM classifier to predict epileptic seizure [21]. Post Processing is also done in EEG signal because the preictal and interictal data is unbalanced. In the current work, a novel feature extraction procedure has been proposed for the Epilepsy classification utilizing the property of the principal component analysis (PCA). In our proposal, PCA has been used on the selected EEG data channels to reduce information similarity instead of their extracted features. All the EEG channels were divided into several major groups and then we applied the PCA to reduce their dimensions. Since PCA converts and reduces the dimensions remaining the maximum variations from all the signals, it also preserves the required frequency content. Pravez and Paul (2017) propose a new approach for predicting epileptic seizures using global and local features extracted from EEG signals [22]. The authors introduce the concept of undulated features, which are derived by applying wavelet transform to the original EEG signal and then computing the power spectral density. These undulated features capture both global and local information in the signal and are used as input to a support vector machine (SVM) classifier for seizure prediction. The proposed method achieves high accuracy in predicting seizures, with a sensitivity of 92.5% and a specificity of 96.1%. Wang and Luy (2015) proposed the feature extraction method in which frequency and amplitude on epoch basis. In this paper SVM classifier are used and 90% sensitivity is achieved [23].

The objectives of this work are to use the different classifier to get the best classification accuracy of utilizing the proposed feature extraction method. The significant features were selected from the results of PCA feature extraction methods and these features were used to classify the Epilepsy by some well-known classifiers: Support Vector Machine (SVM) [24, 25], k-Nearest Neighbor (kNN) [25, 26] and Navie Bayes [27]. The proposed work are suggested that the classification accuracy found by this feature extraction methods are very effective to classify the epileptic patients.

3 Material and Methodology

This paper proposed the classification of EEG signal for epilepsy diagnosis using PCA feature extraction technique based on machine learning paradigm is shown in Fig. 1.

3.1 Data Acquisition

We used a sample of the data from the University of Bonns Epilepsy Center. For more details, see Andrzejak et al., but we only provide a brief explanation in this part (2001) [1].

The entire collection of data is made up of five sets (A through E), each of which has 100 single channel EEG segments. After being visually checked for artifacts, these sections were removed from the recordings. Using the same 128-channel amplifier arrangement and an average common reference, all EEG signals were captured. The data were constantly written onto the disc of a data acquisition computer system at a sampling rate of 173.61 Hz following 12 bit analogue-to-digital conversion.

Each file has a recording of cerebral activity of 23.6 s. The 4097 data values are being sampled. The value of the recordings is shown at each data point.

The 178-dimensional input vector's category is contained in the value y, where y is one of 1, 2, 3, or 4. All recording input vectors are described as follows: 5-The patients eye was open and closed when they recorded the data. 4-Eyes closed signifies that the patient with eyes closed while they were recording the EEG data. 3-The location of the brain tumour was found. The region that contained the tumours was captured by them. These were all seizure-free participants who belonged to the classes 2, 3, and 4. Seizure activity was only being recorded in the first class. X1, X2, ... X 178.

3.2 Feature Extraction

The reduction of amount of attributes is the purpose of feature extraction phase, by creating new ones from the datasets current ones. This new reduced dataset has most of the information and features from the original datasets [28]. In our study the dataset has 178 features, the training time will long if we used all the features, so in our paper to reduce the feature we apply principle component analysis (PCA).

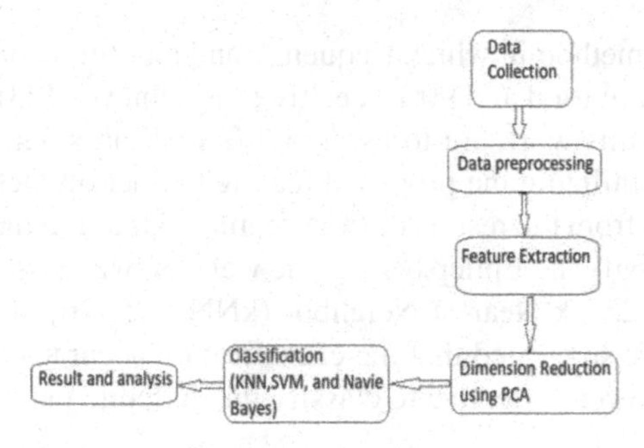

Fig. 1. Proposed model

3.2.1 Principal Components Analysis (PCA)

The principal component analysis (PCA) as a technique for reducing the dimensionality of data while preserving the most important information. They state that PCA involves projecting high-dimensional data onto a lower- dimensional subspace while retaining the key pieces of data that have the highest variance and removing the nonessential parts that have the least variance. The purpose of PCA is to identify a set of input characteristics that can accurately describe the original data's distribution while minimizing reconstruction errors. PCA is an unsupervised learning technique that considers only diversity and is insensitive to data labels [29].

3.3 Classification

The most important part of the seizure detection system is the classification block. The job of classification techniques is to learn a function that associates each set of attributes with a label [30]. The classification modelling was explained as follows. An input data set is used to create a classifier. The model that best fits the relationship between the features set and class label of the input data is the model that each technique uses. The input data needs to be fit into the model. Predicting the class labels of records is also well done. In this study, a group of classifier because multiple classifier systems are more precise and strong then single classifier. To classify the epileptic and non epileptic EEG signal, we use three popular classifiers namely Naive Bayes, kNN and SVM classifier have been utilized in our study explained below. All the work has been done in MATLAB software.

Algorithms Used for Classification

a) Naive Bayes classifiers: Despite being straightforward and linear, Naive Bayes classifiers are extremely effective. Naive Bayes classifier based on conditional probability of bayes theorems. Because it uses every attribute in the dataset and they are all independent of one another, this method is referred to as naïve in practice; the independence assumption is frequently violated. More sophisticated alternatives cannot perform better for small data sizes [31, 32].

b) Support Vector Machine (SVM): In SVM, a hyper plane is constructed to separate the data points into different classes with the largest possible margin, where the margin is the distance between the hyper plane and the closest data points from each class. Support vector are those vector which lies on the margin or on the wrong side of the hyper plane. The biggest possible margin is the distance between the closest data points to each class and the hyper plane. There are materials that are on the wrong side of the hyper plane [33].

c) k-Nearest Neighbors (kNN): In studies on classification, kNN is one of the supervised learning methods that is often utilised [34]. kNN is a learning algorithm based on computing the sample distance, and the distance is evaluated on all samples in the new data, each time the algorithm encounters a new data sample. Following this calculation, class labels are determined by locating the k closest neighbours from known data samples, comparing them to instances in the new example's schooling data, and examining their similarities [35].

3.4 Performance Evaluation

In this section, we focus on a few techniques for assessing the effectiveness of the classifier. Partitioning the dataset for training and testing in order to evaluate the performance of a classifier model is standard procedure. There are various protocols for partitioning data; in this research, we used the holdout method and cross-validation, which are described below.

Holdout Method: In this method, the data set divided into two disjoints sets, called the training and test datasets. Two-thirds of the data should be designated as training datasets, and the final third should be designated as test datasets [36].

Cross-Validation: A statistical technique called cross-validation is used to evaluate the effectiveness and precision of a forecast model. The procedure entails partitioning the dataset into numerous subsets, or "folds," before training and evaluating the model on various fold combinations. The model is then evaluated on the final fold after being trained on k-1 folds. Each fold acts as the test set once during this procedure, which are repeated k times. This paper used 5 fold and 10 fold techniques using following procedure.

Performance measure: Estimation of a performance of classifier are used different measure namely Accuracy, Sensitivity, Specificity, Precision, Area under the receiver operating characteristic curve (AUC), Matthew's correlation coefficient (MCC). These performance measures are explained as:

Overall accuracy: It is defined as the total percentage of test cases Classified correctly samples

$$Accuracy = \frac{TP + TN}{TP + FN + TN + FP} \times 100$$

Sensitivity: Sensitivity is also called true positive rate.

$$Sensitivity(\%) = \frac{TP}{TP + FN} \times 100$$

Specificity: It is also called true negative rate.

$$Specificity(\%) = \frac{TN}{TN + FP} \times 100$$

Area under receiver operating characteristic curve (AUC): It represents a common measure of sensitivity and specificity.

$$AUC(\%) = \frac{1}{2}\left(\frac{TP}{TP + FN} + \frac{TN}{TN = FP}\right)$$

Here, the letters TP, FP, FN, and TN, respectively, stand for true positive, false positive, false negative, and true negative results. The number of seizure samples that were properly identified as being seizures is represented by TP, while the number of samples that were misclassified as non-seizures is represented by FN. Similar to this, TN denotes the number of samples that were properly classified as seizure while FP denotes the number of samples that were incorrectly classified as seizure.

4 Result

This section analyses the results of using different classifiers with PCA feature and without PCA feature reduction techniques to classify the seizure data set, as well as the performance of the classifications using PCA and without using PCA feature reduction. After applied PCA feature selection techniques and SVM classifier with three data partition method 33% hold out, 5 fold and 10 fold we obtained all most same result with the all 178 features. Among these three partition techniques we got best result with 33% hold out techniques with SVM classifier.

Tables 1, 2 and 3 shows the classification performance of different classifier results without PCA reduction with 33% hold out and 5 fold and 10 fold cross validation techniques i.e. when all the 178 features were used for training and testing of classifier model. It is found that the SVM classifier models achieving a better classification accuracy of 96.77% for holdout technique. The other performance measures i.e. sensitivity, specificity, and AUC for proposed model are found to be 99.34%, 94.2%, 99% respectively, for holdout cross validation which are higher than that of other classifier. Figure 2 shows the ROC and AUC curve for the 33% hold out cross validation using SVM classifier.

Tables 4, 5 and 6 shows the classification performance of different classifier model when PCA based feature selection technique is used. In this feature selection technique, we chose top 20 features for evaluation of classifiers because by using top 20 features, we were able to achieve classification accuracy better (97.43% for holdout, 96.93% for 5 fold, 96.93% for 10 fold) to that using all 178 features. In this feature selection technique again SVM outperformed among other classifier. Figure 3 shows the ROC and AUC curve of SVM classifier with 33% hold out cross validation techniques.

Tables 7, 8 and 9 shows the classification performance of different classifier model when PCA based feature selection technique is used. In this feature selection technique, we chose top 10 features for evaluation of classifiers because by using top 10 features, we were able to achieve classification accuracy better (97.43% for holdout, 96.93% for 5 fold, 96.93% for 10 fold) to that using all 178 features. In this feature selection technique again SVM outperformed among other classifier. Figure 4 shows the ROC and AUC curve of SVM classifier with 33% hold out cross validation techniques.

Table 1. Comparison of the different Classification Techniques using without PCA feature reduction 33% holdout

Classification Techniques	Accuracy	Sensitivity	Specificity
kNN (k = 5)	85.05	70.62	99.47
Cosine_kNN	84.85	82.21	87.48
Cubic_kNN	84.58	69.96	99.21
Linear SVM	61.33	22.79	99.87
Quadratic SVM	90.58	82.87	98.29
Cubic SVM	92.62	87.09	98.16
Fine Gaussian SVM	**96.77**	**99.34**	**94.20**
Medium Gaussian SVM	95.32	93.81	96.84
Coarse Gaussian SVM	84.32	69.43	99.21
Gaussian Naïve Bayes	93.87	90.65	97.10

Table 2. Comparison of the different Classification Techniques using without PCA feature 5 fold cross validation

Classification Techniques	Accuracy	Sensitivity	Specificity
kNN (k = 5)	85.37	71.22	99.52
Cosine_kNN	84.87	81.13	88.61
Cubic_kNN	85.07	70.83	99.30
Linear SVM	93.80	90.26	97.35
Quadratic SVM	60.37	20.91	99.83
Cubic SVM	90.59	82.43	98.74
Fine Gaussian SVM	93.50	88.57	98.43
Medium Gaussian SVM	**96.50**	**98.96**	**94.04**
Coarse Gaussian SVM	95.89	94.43	97.35
Gaussian Naïve Bayes	84.67	70.13	99.22

Table 3. Comparison of the different Classification Techniques using without PCA feature 10 fold cross validation

Classification Techniques	Accuracy	Sensitivity	Specificity
kNN (k = 5)	85.90	72.48	99.31
Cosine_kNN	85.33	81.83	88.83
Cubic_kNN	85.78	72.26	99.30
Linear SVM	93.78	90.13	97.43
Quadratic SVM	60.37	20.91	99.83
Cubic SVM	91.39	83.96	98.83
Fine Gaussian SVM	93.83	89.26	98.39
Medium Gaussian SVM	**96.70**	**99.09**	**94.30**
Coarse Gaussian SVM	96.09	94.83	97.35
Gaussian Naïve Bayes	85.15	71.22	99.09

Table 4. Comparison of the different Classification Techniques using top 20 PCA feature 33% holdout

Classification Techniques	Accuracy	Sensitivity	Specificity
kNN (k = 5)	88.27	76.81	99.74
Cosine_kNN	79.05	74.18	83.93
Cubic_kNN	88.27	76.94	99.60
Linear SVM	96.38	94.60	98.16
Quadratic SVM	60.01	20.29	99.74
Cubic SVM	91.96	85.11	98.81
Fine Gaussian SVM	95.59	92.36	98.81
Medium Gaussian SVM	**97.43**	**98.81**	**96.05**
Coarse Gaussian SVM	**96.25**	**94.47**	**98.02**
Gaussian Naïve Bayes	84.32	69.04	99.60

Table 5. Comparison of the different Classification Techniques using top 20 PCA feature 10 fold holdout

Classification Techniques	Accuracy	Sensitivity	Specificity
kNN (k = 5)	88.74	78.22	99.26
Cosine_kNN	79.30	75.48	83.13
Cubic_kNN	89.00	78.78	99.22
Linear SVM	96.00	94.57	97.43
Quadratic SVM	59.43	19.22	99.65
Cubic SVM	93.26	88.11	98.39
Fine Gaussian SVM	95.15	92.22	98.09
Medium Gaussian SVM	**96.93**	**98.74**	**95.13**
Coarse Gaussian SVM	**95.72**	**94.17**	**97.26**
Gaussian Naïve Bayes	84.76	70.61	98.91

Table 6. Comparison of the different Classification Techniques using top 20 PCA feature 5 fold holdout

Classification Techniques	Accuracy	Sensitivity	Specificity
kNN (k = 5)	88.74	78.22	99.26
Cosine_kNN	79.30	75.48	83.13
Cubic_kNN	89.00	78.78	99.22
Linear SVM	96.00	94.57	97.43
Quadratic SVM	59.43	19.22	99.65
Cubic SVM	93.26	88.11	98.39
Fine Gaussian SVM	95.15	92.22	98.09
Medium Gaussian SVM	**96.93**	**98.74**	**95.13**
Coarse Gaussian SVM	95.72	94.17	97.26
Gaussian Naïve Bayes	84.76	70.61	98.91

Table 7. Comparison of the different Classification Techniques using top 10 PCA feature 33% holdout

Classification Techniques	Accuracy	Sensitivity	Specificity
kNN (k = 5)	91.30	84.19	98.42
Cosine_kNN	67.65	70.88	64.43
Cubic_kNN	91.70	84.58	98.81
Linear SVM	95.59	94.33	96.84
Quadratic SVM	57.11	14.62	99.60
Cubic SVM	93.94	90.12	97.76
Fine Gaussian SVM	95.32	97.25	93.54
Medium Gaussian SVM	**95.98**	**96.44**	**95.52**
Coarse Gaussian SVM	94.93	92.89	96.97
Gaussian Naïve Bayes	81.55	64.03	99.08

Table 8. Comparison of the different Classification Techniques using top 10 PCA feature 10 fold holdout

Classification Techniques	Accuracy	Sensitivity	Specificity
kNN (k = 5)	93.17	87.48	98.87
Cosine_kNN	68.96	70.57	67.35
Cubic_kNN	93.13	87.39	98.87
Linear SVM	95.67	93.70	97.65
Quadratic SVM	56.87	13.74	100.00
Cubic SVM	94.43	90.65	98.22
Fine Gaussian SVM	95.48	92.91	98.04
Medium Gaussian SVM	**96.59**	**96.70**	**96.48**
Coarse Gaussian SVM	95.35	93.39	97.30
Gaussian Naïve Bayes	83.87	68.39	99.35

Table 9. Comparison of the different Classification Techniques using top 10 PCA feature 5 fold holdout

Classification Techniques	Accuracy	Sensitivity	Specificity
kNN (k = 5)	92.91	87.09	98.74
Cosine_kNN	69.43	71.43	67.43
Cubic_kNN	92.74	86.65	98.83
Linear SVM	95.70	93.70	97.70
Quadratic SVM	56.93	14.00	99.87
Cubic SVM	94.50	90.91	98.09
Fine Gaussian SVM	94.93	91.87	98.00
Medium Gaussian SVM	**96.54**	**96.74**	**96.35**
Coarse Gaussian SVM	95.07	92.78	97.35
Gaussian Naïve Bayes	83.52	67.70	99.35

Table 10. Accuracy Comparison of Classification Techniques using PCA and without using PCA with top 20 and 10 fold hold out cross validation techniques

Classification Techniques	Accuracy using without PCA	Accuracy using with PCA Top 20
kNN (k = 5)	85.90	88.74
Cosine_kNN	85.33	79.30
Cubic_kNN	85.78	89.00
Linear SVM	93.78	96.00
Quadratic SVM	60.37	59.43
Cubic SVM	91.39	93.26
Fine Gaussian SVM	93.83	95.15
Medium Gaussian SVM	**96.70**	**96.93**
Coarse Gaussian SVM	**96.09**	**95.72**
Gaussian Naïve Bayes	85.15	84.76

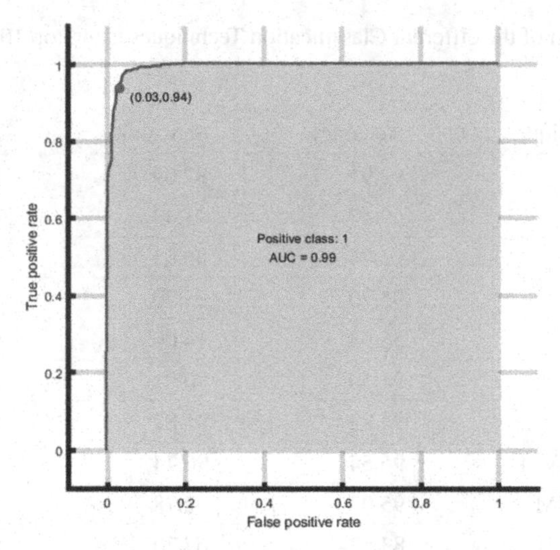

Fig. 2. The ROC and AUC curve for the 33% hold out cross validation without PCA feature selection with SVM classifier.

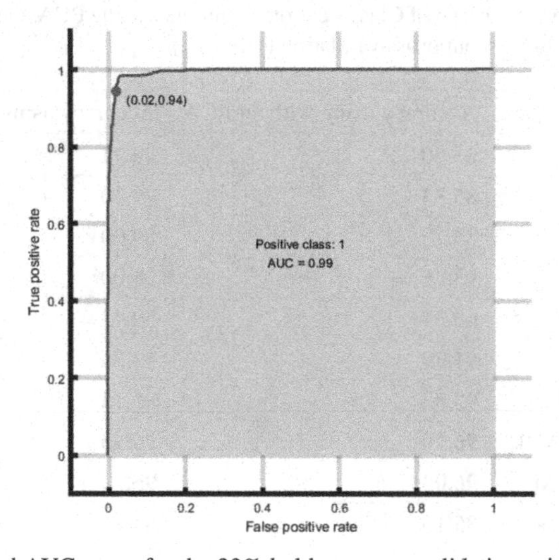

Fig. 3. The ROC and AUC curve for the 33% hold out cross validation using top 20 features of PCA with SVM classifier.

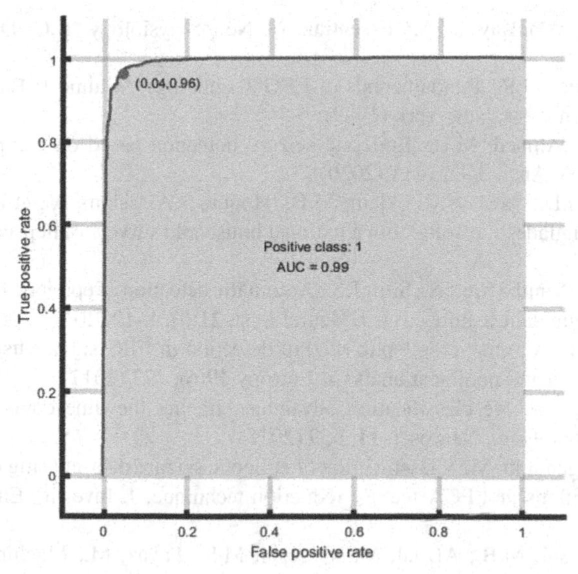

Fig. 4. The ROC and AUC curve for the 33% hold out cross validation using top 10 features of PCA with SVM classifier.

5 Conclusion

One of the most serious causes of death worldwide is epileptic seizures. It is more crucial than ever to receive an accurate and prompt diagnosis because this condition has a significant impact on how the patients live their daily lives.

The main aim of the paper was to determine the optimal classification algorithm for epileptic seizures using the principal components analysis (PCA) method for feature reduction in the dataset. The goal was to identify the most effective technique for accurately classifying seizures. To obtain the results, we used the PCA feature reduction technique to forecast epilepsy. The K-Nearest Neighbors (kNN), Support Vector Machine (SVM), and Navie Bayes algorithms have been applied and the performance of the classifiers both with and without the PCA method was evaluated.

As reported in Table 10, the Gaussian SVM achieves the greatest results when applying the PCA feature reduction in the dataset, with an accuracy of 96.7%, top 20 and 10 fold cross validation, and short computational times (training time and test time). Again, a nearly identical result using the top 10 features and 10-fold cross-validation methods with PCA feature reduction approaches was obtained.

It is important to note that applying PCA feature reduction techniques allows for the same accuracy to be obtained with fewer features overall.

References

1. Andrzejak, R.G., et al.: Indications of nonlinear deterministic and finite-dimensional structures in time series of brain electrical activity: dependence on recording region and brain state. Phys. Rev. E **64**, 061907 (2001)

2. Atwood, H.L., MacKay, W.A.: Essentials of Neurophysiology. B.C. Decker, Hamilton, Canada (1989)
3. Tyner, F.S., Knott, J.R.: Fundamentals of EEG Technology, Volume 1: Basic Concepts and Methods. Raven Press, New York (1989)
4. Ahammed, K., Ahmed, M.U.: Epileptic seizure detection based on complexity feature of EEG. J. Biomed. Anal. **3**(1), 1–11 (2020)
5. Mohammad, Q.D., Saha, N.C., Alam, M.B., Hoque, S.A., Islam, A., et al.: Prevalence of epilepsy in Bangladesh: results from a national household survey. Epilepsia Open **5**(4), 526–536 (2020)
6. Acharya, U.R., Vinitha Sree, S., Suri, J.S.: Automatic detection of epileptic EEG signals using higher order cumulant features. Int. J. Neural Syst. **21**(5), 1–12 (2011)
7. Wang, L., et al.: Automatic epileptic seizure detection in EEG signals using multi-domain feature extraction and nonlinear analysis. Entropy **19**(6), 222 (2017)
8. Yan, T., et al.: Positive classification advantage: tracing the time course based on brain oscillation. Front. Hum. Neurosci. **11**, 659 (2018)
9. Nahzat, S., Yaganoglu, M.: Classification of epileptic seizure dataset using different machine learning algorithms and PCA feature reduction technique. J. Investig. Eng. Technol. **4**(2), 47–60 (2021)
10. Aayesha, Qureshi, M.B., Afzaal, M., Qureshi, M.S., Fayaz, M.: Machine learning-based EEG signals classification model for epileptic seizure detection. Multimed. Tools Appl. **80**, 17849–17877 (2021)
11. Tawfik, N.S., Youssef, S.M., Kholief, M.: A hybrid automated detection of epileptic seizures in EEG records. Comput. Electr. Eng. **53**, 177–190 (2016)
12. Shankar, R.S., Raminaidu, C.H., Raju, V.V.S., Rajanikanth, J.: Detection of epilepsy based on EEG signals using PCA with ANN model. J. Phys.: Conf. Ser. **2070**, 012145 (2021)
13. Baykara, M., Abdulrahman, A.: Seizure detection based on adaptive feature extraction by applying extreme learning machines. Traitement du Signal **38**(2), 331–340 (2021)
14. Srinivasan, V., Eswaran, C., Sriraam, N.: Artificial neural network based epileptic detection using time-domain and frequency-domain features. J. Med. Syst. **29**(6), 647–660 (2005)
15. Guerrero, M.C., Parada, J.S., Espitia, H.E.: Principal components analysis of EEG signals for epileptic patient identification. Computation **9**, 133 (2021). https://doi.org/10.3390/computation9120133
16. Jaiswal, A.K., Banka, H.: Epileptic seizure detection in EEG signal using machine learning techniques. Australas. Phys. Eng. Sci. Med. **41**, 81–94 (2017)
17. Vidyaratne, L.S., Iftekharuddin, K.M.: Real-time epileptic seizure detection using EEG. IEEE Trans. Neural Syst. Rehab. Eng. **25**(11), 2146–2156 (2017)
18. Sharmila, A., Geethanjali, P.: DWT based detection of epileptic seizure from EEG signals using Naive Bayes and k-NN classifiers. IEEE Access **4**, 7716–7727 (2016)
19. Ghosh-Dastidar, S., Adeli, H., IEEE, Dadmehr, N.: Principal component analysis-enhanced cosine radial basis function neural network for robust epilepsy and seizure detection. IEEE Trans. Biomed. Eng. **55**(2), 512–518 (2008)
20. Senger, V., Tetzlaff, R.: New signal processing methods for the development of seizure warning devices in epilepsy. IEEE Trans. Circ. Syst. **63**, 1549–8328 (2016)
21. Shiao, H.T., et al.: SVM-based system for prediction of epileptic seizures from IEEG signal. IEEE Trans. Biomed. Eng. **64**(5), 1011–1022 (2017)
22. Parvez, M.Z., Paul, M.: Seizure prediction using undulated global and local features. IEEE Trans. Biomed. Eng. **64**(1), 208–217 (2017). https://doi.org/10.1109/TBME.2016.2553131
23. Wang, N., Lyu, M.R.: Extracting and selecting distinctive EEG features for efficient epileptic seizure prediction. IEEE J. Biomed. Health Inform. **19**(5), 1648–1659 (2015). https://doi.org/10.1109/JBHI.2014.2358640

24. Vidal, R., Ma, Y., Sastry, S.S.: Generalized Principal Component Analysis. Springer, New York (2016). https://doi.org/10.1007/978-0-387-87811-9
25. Raschka, S., Mirjalili, V.: Python Machine Learning: Machine Learning and Deep Learning with Python, Scikit-Learn, and Tensor Flow 2, 3rd edn. Packt Publishing, Birmingham, UK (2019)
26. Wang, Z., Na, J., Zheng, B.: An improved kNN classifier for epilepsy diagnosis. IEEE Access **8**, 100022–100030 (2020)
27. Rish, I.: An empirical study of the Naive Bayes classifier. In: IJCAI 2001 Workshop on Empirical Methods in Artificial Intelligence, pp. 41–46 (2001)
28. Ippolito, P.P.: Feature Extraction Techniques - Towards Data Science (2019). https://towardsdatascience.com/feature-extraction-techniques-d619b56e31be. Accessed 27 Dec 2020
29. Qiu, J., Wang, H., Lu, J., Zhang, B., Du, K.L.: Neural network implementations for PCA and its extensions. ISRN Artif. Intell. **2012**, 1–19 (2012)
30. Smart, O., Chen, M.: Semi-automated patient-specific scalp EEG seizure detection with unsupervised machine learning. In: IEEE Conference on Computational Intelligence in Bioinformatics and Computational Biology (CIBCB), pp. 1–7. IEEE (2015)
31. Fergus, P., Hignett, D., Hussain, A., Al-Jumeily, D., Abdel-Aziz, K.: Automatic epileptic seizure detection using scalp EEG and advanced artificial intelligence techniques. Bio. Med. Res. Int. **2015**, Article ID 986736 (2015)
32. Bandarabadi, M., Teixeira, C.A., Rasekhi, J., Dourado, A.: Epileptic seizure prediction using relative spectral power features. Clin. Neurophysiol. **126**(2), 237–248 (2015)
33. Rodrigues, J.D.C., Filho, P.P.R., Peixoto, E., Kumar, A., de Albuquerque, V.H.C.: Classification of EEG signals to detect alcoholism using machine learning techniques. Pattern Recogn. Lett. **125**, 140–149 (2019)
34. Yağanoğlu, M., Köse, C.: Real-time detection of important sounds with a wearable vibration based device for hearing-impaired people. Electronics **7**(4), 50 (2018)
35. Mitchell, T.M.: Does machine learning really work? AI Mag. **18**(3), 11 (1997)
36. Kohavi, R.: A study of cross-validation and bootstrap for accuracy estimation and model selection. In: IJCAI (1995)

White Blood Cell Classification Using Deep Transfer Learning

Ramineni Sharvani, Bhawana Sahu, and Pradeep Singh(✉)

National Institute of Technology, Raipur, India
psingh.cs@nitrr.ac.in

Abstract. The count and the classification of WBC in the human body represent the state of the Immune System of a person and indicate the potential risks one can face. If there is any drastic change in the count of WBC from its standard, it implies that the human body is being affected by some disease. The WBC Classification serves as a good guideline in the identification process of deadly diseases. But the manual classification process is very slow and efficiency is proportional to the skills of expert doctors. Hence Computer-Based Automatic algorithms for will play a prominent role in the classification of WBC. In this paper, we proposed a transfer learning-based model for the unmanned classification of White Blood Cells based on blood smear images. The proposed transfer learning model is developed using various deep learning models such as VGG-16, Efficient Net, ResNet-50, Inception, and Xception. We have used the dataset Blood Cell Images from Kaggle which consists of around 12,500 images with sub-types Neutrophils, Lymphocytes, Basophils, Eosinophils, and Monocytes. Comparative analysis of the performance of each transfer learning-based model is performed. The transfer learning model developed using VGG-19 has achieved the highest accuracy of 92% among all compared models.

Keywords: WBC Classification · Deep Learning · CNN · Efficient Net and VGG-19

1 Introduction

In the veins of a healthy person, the circulating blood is about 7% of the body weight [1]. A single drop of blood contains many organic and inorganic elements that are very helpful in identifying the health status of patients by performing different blood tests. Peripheral blood cell examination has a major role in diagnosing many severe diseases like leukemia, malaria, anemia, etc. The major significance of blood in the human body is that it provides crucial information about the health status of a person. Blood Cells include Red Blood Cells (RBCs), White Blood Cells (WBCs), and platelets [2]. 45% of the blood cells are composed of red blood cells and 1% of WBC and platelets [1]. The Immune System of a healthy person is a complex network that consists of organs, tissues, and cells which helps the person to fight against fatal diseases. White Blood Cells (WBC) commonly known as leukocytes, plays the most important part in the human immune

B. K. Singh et al. (Eds.): ICBEST 2023, CCIS 2003, pp. 220–232, 2024.
https://doi.org/10.1007/978-3-031-54547-4_17

system. They protect the human from different kind of diseases, infections, and external invaders [2]. As WBCs play a major part in the human immune system, it is called Immune Cell [3]. Even though WBCs are colorless, hematologists can use special stains to view these cells with the help of a microscope [4].

"By analyzing the count of White Blood Cell and the ratio of its subtypes, the health status of a person can be known, and it helps in the diagnosis process of some diseases like leukemia, malaria, etc. There are five sub-types under WBCs [5, 6]. They are:

- Neutrophils (50–70%)
- Lymphocytes (25–30%)
- Monocytes (3–9%)
- Eosinophils (0–5%)
- Basophils (0–1%)"

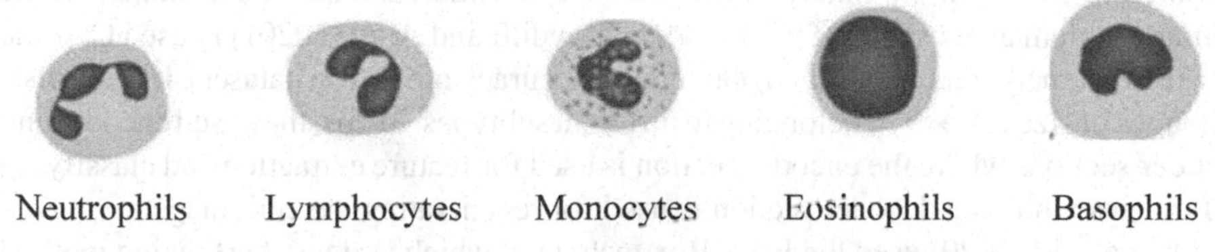

Fig. 1. Sub Types of White Blood Cell

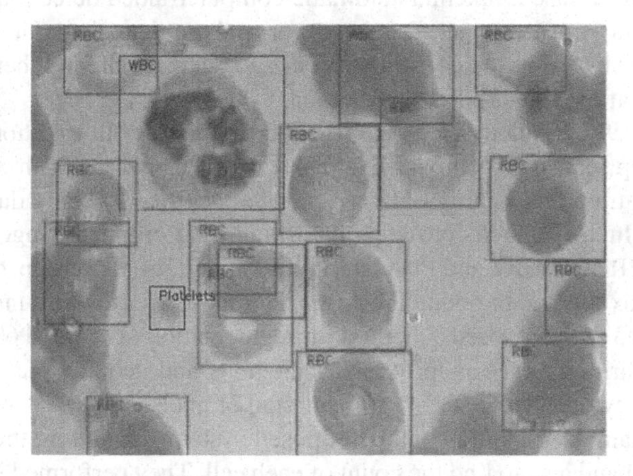

Fig. 2. The image of a human blood cell [7]

Many traditional WBC classification methods are very slow and not much efficient. The accuracy of those manual methods depends on the experience levels of doctors. Hence there is a larger requirement for the development of computer applications. Figure 1 shows the different subtypes of the White Blood cell and Fig. 2 shows image of human blood cell. The objective of this paper is to classify different types of white blood cells as white blood cells are the cells of immune system that helps our body to protect from foreign particles. The examination of immune system is also performed with the help of WBC count. Hence, we need fast method that can tell us the count of different types of WBC-cells like Neutrophils, Lymphocyte, Monocytes, Eosinophils and Basophils. So, in this paper an efficient transfer learning based model for WBC Classification is developed.

The remaining of this paper is organized as follows; Sect. 2 discusses the previous literature work mentioned by various researchers. Section 3 explains the methodology proposed and different models developed. Section 4 summarizes our result of the experiment; Sect. 5 is the conclusion of this paper.

2 Literature Review

The identification of blood-based ailment often involves diagnosing and distinguishing the patient blood samples. Automated Applications of detecting and classifying White Blood Cells have prominent clinical importance. For analyzing the algorithms which are suitable for obtaining maximum results, we considered some of the studies in the similar problem area using different models. Baydilli and Atila (2020) [1] used Capsule networks learned training data and obtained an accuracy of 96% on dataset LISC consists of images of size 720 × 576 belonging to different subtypes. In this, they used encoder and decoder sections where the encoder section is used for feature extraction and classifying cell images while the decoder section is used for regenerating the cell image.

Di et al. [8] (2020) used the Edge Box technique which is state of art region method and tested the model on dataset ALL-IDB (Acute Lymphoblastic Leukaemia Image Dataset) used for disease Leukemia automatic computer-aided detection and also used MP-IDB (Malaria Parasite Image Database) for malaria detection. They modified the edge box method by using the constraint of size to decrease the number of candidates bounding boxes after edge proposal evaluation.

Hegde et al. [9] classified WBC into its subtypes using both traditional Image processing and deep learning methods. Evaluated neural networks with an accuracy of 99.8% and with transfer learning obtained an accuracy of 99% on a dataset collected from 1159 PBS Images which consisting of around 1418 cropped images. These blood cell images have RBC, WBC, and Platelets. In this work, they also cropped actual images using a bound box method to obtain the target region used for the training of the model. Sholeh [10] (2013) has proposed a method for segmentation of original 60 blood images by combining various image preprocessing methods and acquired accuracy 92% for segmentation of cytoplasm, 89% for segmentation of nucleus.

Sinha and Ramakrishnan [11] have proposed system for finding the class of each White Blood cell and also to find the count of each cell. They performed image segmentation, extraction of features, and classifying cells. They obtained an accuracy of 94% using Neural Networks and 94% with SVM classifier on a dataset with 115 blood smear images collected from various patients.

As segmentation of WBC plays a major part in classification into its sub-types - Banik, et al. [2] proposed a k-means algorithm and color space conversion based nucleus segmentation methods. After the segmentation procedure, they proposed a new CNN model by combining first and last convolution layers and giving the input to the layer. They used dropout to overcome the overfitting problem and used the model on different public datasets like BCCD and achieved an accuracy of 96%.

Liang et al. 2018 [3] proposed model by combining CNN and RNN model to produce a framework which can deepen the layers which are used for complex classification. They also applied transfer learning weights which are pre-trained using the ImageNet dataset and used custom loss function to train and execute the model faster.

Wibawa [12] proposed a deep learning solution for WBC classification and compared the results with the other models like SVM RBF, KNN, MLP on the dataset available on Kaggle with 1100 images. The proposed DL model achieved maximum accuracy with around 99%.

3 Proposed Methodology

3.1 Transfer Learning

Transfer learning is a well-known technique in which pre-trained models are used. Transfer learning is a popular choice in deep learning where the one pre-trained models are considered as starting level for another model. The conventional machine learning technologies have been achieved a lot of success, but they worked well under certain assumptions like the data used for training and test belongs to the same feature space and distribution, when the distribution differs it's quite expensive and time-consuming to collect the data and train the model. In such a scenario, transfer learning works well. Transfer learning is useful where data can change with time or where data are uncertain like the problem of sugar beet and volunteer potato classification [13]. Figure 3 demonstrates the setup of transfer learning in terms of domain and tasks. In case of traditional Machine learning model is trained with the data from the same domain for the prediction of the desired task. Whereas in transfer learning the training of the model is performed in one domain task and applied on different task for prediction as shown in the Fig. 3.

More specifically, transfer learning could be summarized as: source domain Ds & learning task Ts, similarly a target domain Dt & learning task Tt, the goal of transfer learning is to improve the learning of target predictive function ft(.) in Dt by the use of knowledge in Ds and Ts, where Ds $\neq$ Dt or Ts $\neq$ Tt [14]. Here the training of the models is performed on Image Net trained model. The model is used for the transfer learning by freezing the convolution layers and updating the fully connected layers by providing the input of segmented blood cells.

In this paper, we have applied different pre-trained models like VGGNet, ResNet-50, Inception, Xception and EfficientNet-B0. These models are used as a starting point then we have provided WBC data to the train the model. Figure 4 illustrates how our proposed model utilizes the transfer learning for model building.

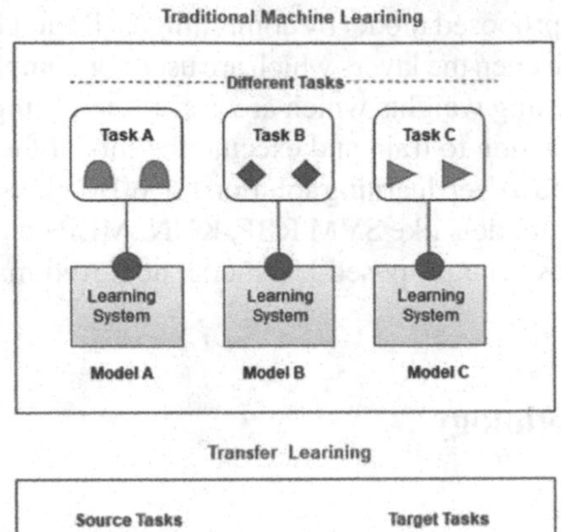

Fig. 3. Differences between Traditional and Transfer Learning Architecture

Table 1. Baseline Network of the proposed model (EfficientNet-B0) [15]

Stage		Kernels size	Resolutions	Channels
1	Conv 3 × 3	3 × 3	224 × 224	32
2	MBConv1	3 × 3	112 × 112	16
3	MBConv6	3 × 3	112 × 112	24
4	MBConv6	5 × 5	56 × 56	40
5	MBConv6	3 × 3	28 × 28	80
6	MBConv6	5 × 5	28 × 28	112
7	MBConv6	5 × 5	14 × 14	192
8	MBConv6	3 × 3	7 × 7	320
9	Conv&Pooling&FC	1 × 1	7 × 7	1280

3.2 Convolutional Neural Network

A convolutional neural network [16] is a deep learning model that has been emulated from the observation of the biological process. It is a type of feed-forward artificial neural network in which the pattern of connectivity between the neurons present in the network, is inspired by the animal visual cortex. The proposed architecture goes through

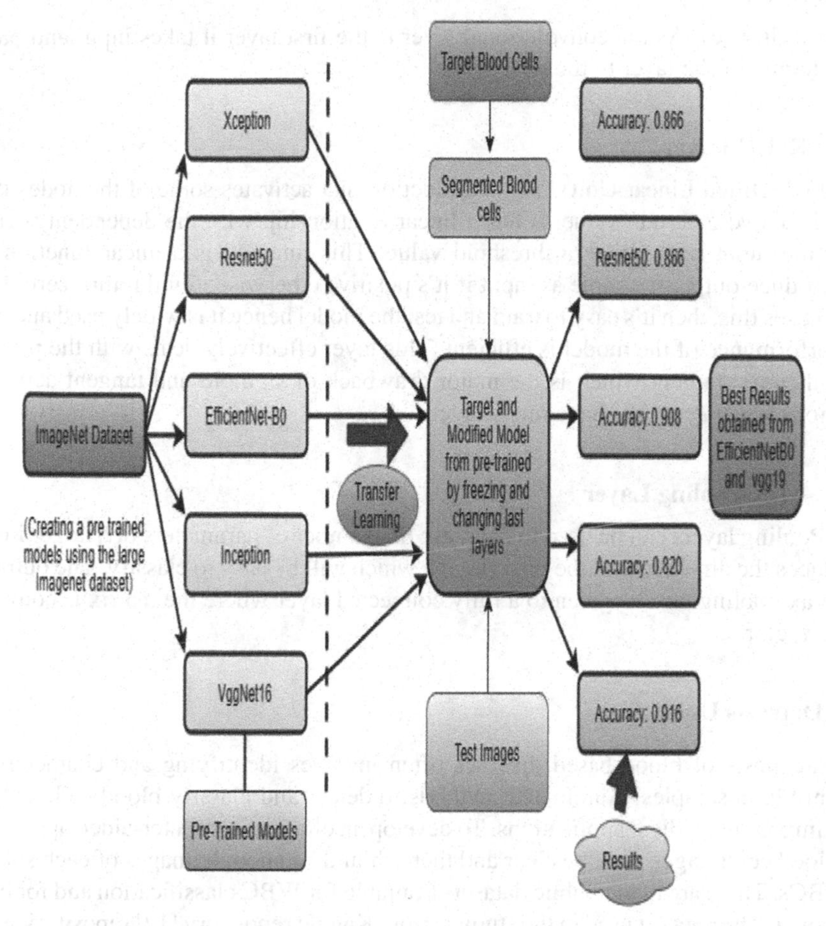

Fig. 4. Pipeline of the proposed model using Transfer Learning

the image processing through different layers. CNN has four layers: Convolutional Layer, ReLU layer, max-pooing layer, and fully connected layer. The input image is given to the initial convolutional layer and the output is fed to the ReLU layer. The output from the relu layer is fed to the max-pooling layer as input where the unwanted pixels will be removed. CNN has different architecture like LeNet, ResNet, GoogleNet, Inception V3, Conv-Net.

3.2.1 Convolutional Layer

The convolutional layer is the first layer of CNN. The convolutional layer is the main building block of a convolutional neural network [16].The input has a shape (image's number) × (width of image) × (height of image) × (depth of image). The layer is used to extract the feature from an input image by applying a filter or a kernel. In this layer, we multiply the image pixel matrix with a filter matrix which is called a feature map. The input has a shape (image's number) × (width of image) × (height of image) ×

(depth of image). As the convolutional layer is the first layer it takes input and passes the outcome of the layer to the next layer.

3.2.2 ReLU Layer

ReLU (Rectified Linear Unit) [17] is a function that activates some of the nodes if the input is above a certain value. It has a linear relationship with the dependent variable when the input rises above a threshold value. This function is a linear function that will produce output the same as input if it's positive otherwise it will return zero. If the model uses this, then it's easy to train and test the model hence it is widely used and often the performance of the model is efficient. This layer effectively deals with the problem of vanishing gradient which is the major drawback of sigmoid and tangent activation functions when networks have many layers.

3.2.3 Max-Pooling Layer

Max Pooling layers can be used to decrease the number of parameters of a larger image. It reduces the dimension of the input image which will be easy to classify. The output of this max pooling will be given to a fully connected layer where the matrix is converted to the vector.

3.3 Datasets Used

The diagnosis of blood-based diseases often involves identifying and characterizing patient blood samples. Automated methods to detect and classify blood cell subtypes have important medical applications. To develop an efficient computer-aided application the blood cell images must be clear and there should be enough images of each subtype of WBCs. There are many public datasets available for WBC classification and for blood cell count. The data set used in the study is from Kaggle repository [12] repository which consists of 12500 augmented blood cell images in the format of JPEG along with the CSV file which has labels of each image with its type. There are four subtypes of WBC images available in this dataset. Around 3000 images are there for each subtype. The cell subtypes available are Lymphocytes, Monocytes, Neutrophil, Eosinophil where the dataset is grouped into four different subfolders each of the above types. This repository also contains around 410 images which are pre-augmented specifically present in the folder dataset-master." Each subfolder has a particular type of WBC. In this dataset we have samples for training and testing separated. Figure 5 is the demonstration of images from the Kaggle dataset.

We have developed our model on the Kaggle repository dataset where there is enough blood images are available with an almost equal ratio of each subtype of WBC. So, we have chosen the Kaggle dataset for the proposed project.

We have used the condition that the dimension of image present in dataset should be equal. Splitting the datasets is performed by dividing the dataset into training and testing. Since we have used 12500 images, we have divided the dataset into 80% and 20%. For this task we are using test_train_split from model_selection library from scikit.

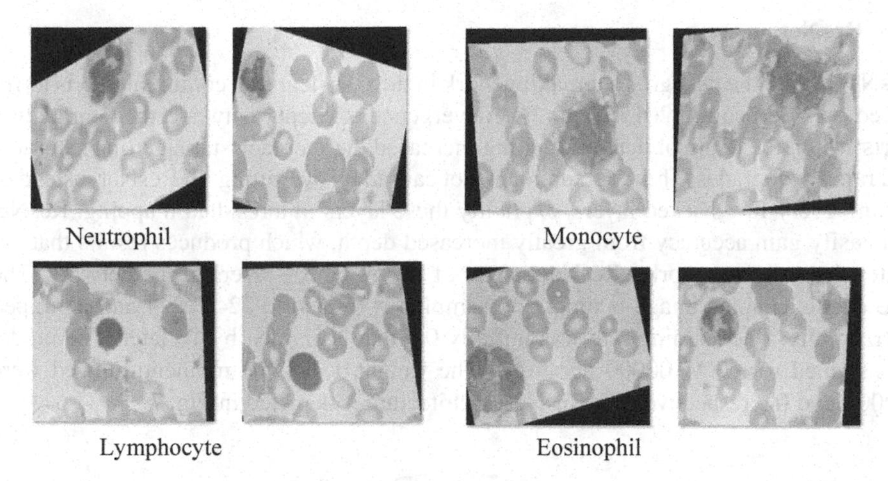

Fig. 5. Images from Kaggle dataset

3.4 VGG Net

The VGG [18] network architecture is another deep CNN model that was introduced by Simonyan and Zisserman in 2014. The network is famous for its simplicity, they use filters of size 3 * 3 stacked on top of each other in increasing depth. The max-pooling handles the reducing volume. Two fully connected layers are used at the end of the architecture each of size 4096 nodes and then followed by a SoftMax classifier. VGG-16 and VGG-19 are two widely used VGGNet networks each having 16- and 19-layers network respectively. These are considered very deep network. Unlike Alex Net, VGG used small receptive fields (3 * 3 with stride 1). It has three ReLU units and hence the decision function is more discriminative. Because the use of 3 * 3 filter parameters is fewer. Due to fewer results, VGG has large numbers of weight layers and hence improved performance. The top-1 error of VGG is 25.5% and top-5 error of 8.0% on ImageNet data. The architecture of VGGNet is shown in Fig. 6.

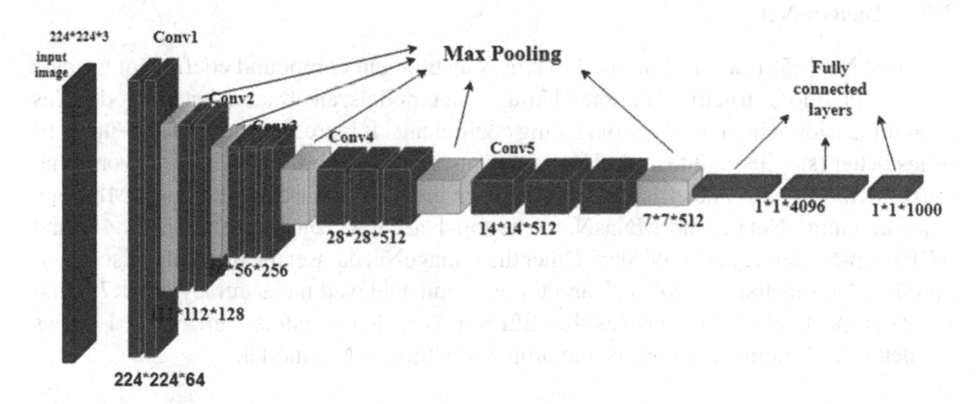

Fig. 6. VGGNet-16 Model

3.5 ResNet

ResNet [19] is the next groundbreaking work in the deep learning community. It is introduced by Microsoft in 2015. With the convergence of deeper network, the degradation starts: when the depth of network getting increased the accuracy starts getting saturated and rapidly degrades. This degradation is not caused by overfitting. In ResNet instead of hoping every few stacked layers, explicitly these layers fit a residual mapping. ResNet can easily gain accuracy from greatly increased depth, which produces results that are better than other networks. ResNet can be of 34, 50 or 101 layers of the network. The size of the sampled image is randomly sampled is resized to 224 * 224 and is flipped horizontally. The learning rate was initially 0.1 that is divided by 10 later. The model was trained for 60 * 100000 iterations. The weight decay and momentum used were 0.0001 and 0.9 respectively. A ResNet Architecture is depicted in Fig. 7.

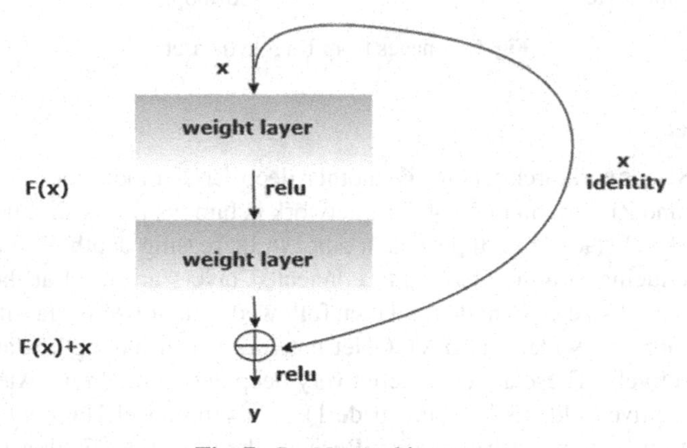

Fig. 7. ResNet Architectures

3.6 Efficient Net

Efficient Net [15] is a novel method that uses an efficient compound coefficient to scale up CNN in a more structured manner. Unlike other models, an efficient net method scales each dimension with a fixed set of scaling coefficients. It increases the accuracy up to 10 times better (smaller and faster). The method uses the AutoML MNAS framework, that makes both accuracy and efficiency better. Then the resultant architecture uses MBConv just like mobileNetV2 and MNasNet. The top-1 accuracy on ImageNet is 84.4% and 97.1% top-5 accuracy are recorded. Other than ImageNet dataset the method is also tested on other datasets like CIFAR-100 and Flowers and achieved the accuracy of 91.7% and 98.8% respectively, which proves that Efficient Net also transfers well. Table 1 shows the details of baseline network of the proposed Efficient Net model.

3.7 Inception Net

The traditional CNN are also prone to overfitting. The Inception Net [20] was designed for fast computation and a deeper network to reduce computational expense. The main idea of inception is to increase the number of filters in each block of the module used. The different sizes of filters usually used are 1×1, 3×3, and 5×5 succeeded by max-pooling. Multiple filters are implemented at the same level; hence the network becomes wider and wider. To make the inception model less expensive, the filter of 1×1 is added before the 3×3 and 5×5 layers and after max-pooling. Since the 1×1 convolutions are much cheaper than 5×5 convolutions. Figure 8 is the illustration of the Inception module with dimension reductions.

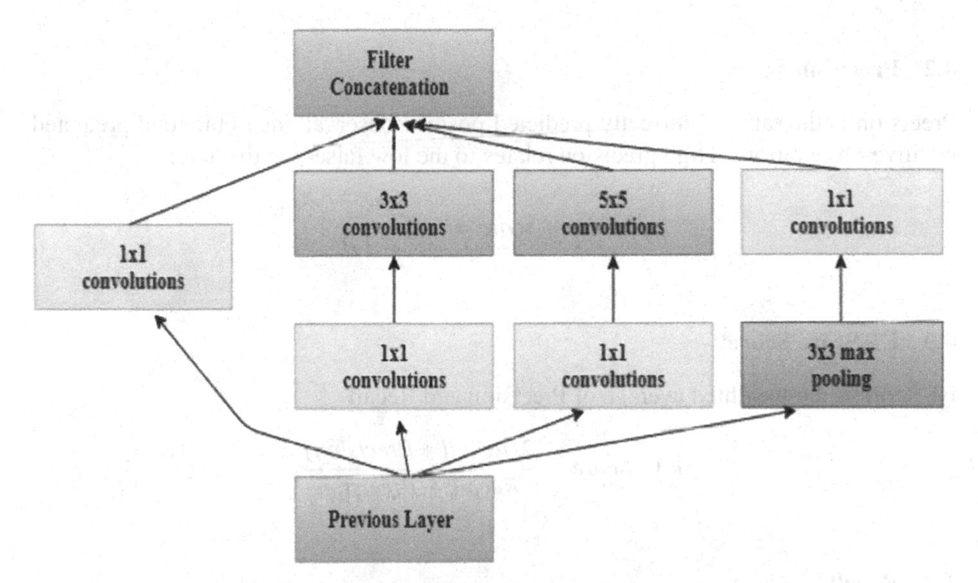

Fig. 8. Inception module with dimension reductions

4 Result and Performance Evolution

The following are the results of each model/method used. The model used to compare the result are VGG-19 [18], Efficient Net [15], Xception, Inception Net [20] and ResNet-50. We have divided the dataset into two parts one for the training of the model and another for testing. Out of 12500 blood cell images, 9957 is treated as training and validation data while rest 2487 images are used for testing the model.

While training the model, again separate the training dataset into two, one for training and others for validation of the model. The training dataset is used for training the model using labelled data and to validate the model validation dataset is used. The ratio of training and validation data is 1:4 i.e. 80% of training data is used to train the model and 20% of the training data was used for validation. Hence, there are two types of accuracy, training accuracy and validation accuracy. A model works efficiently when there is a negligible difference between training accuracy and validation accuracy. Following performance measures are used for evaluation of the model.

4.1 Accuracy

Accuracy is the most intuitive performance measure and it is simply a ratio of correctly predicted observations.

$$Accuracy = \frac{TP + TN}{TP + FP + FN + TN}$$

4.2 Precision Score

Precision is the ratio of correctly predicted positive observations to the total predicted positive observations. High precision relates to the low false-positive rate.

$$Precision\ Score = \frac{TP}{TP + FP}$$

4.3 F1-Score Score

F1 Score is the weighted average of Precision and Recall.

$$F1{-}Score = \frac{2(Recall * Precision)}{Recall + Precision}$$

4.4 Recall Score

Recall is the ratio of correctly predicted positive observations to all observations in the actual class. It is also called as specificity.

$$Recall = \frac{TP}{TP + FN}$$

The Table 2 shown below summarizes accuracy, precision, recall, and F1 score we have achieved for each model. It shows that for a given dataset of blood cells VGG-19 and Efficient gave almost the same accuracy.

Figure 9 shows the training and validation accuracy curve of the Efficient Net model. As shown in the figure, at initial the epoch the training and validation accuracy is low but as the no of epochs increased the accuracy also gets increases. Here, the training and validation accuracy has a very minute difference that shows that there is no case of underfitting or overfitting. The highest validation accuracy achieved 0.9952 at epoch number 45.

Table 2. WBC Classification measure of different Deep learning models

Model	Accuracy	Precision	Recall	F1-score
VGG-19	**0.916**	**0.858**	**0.833**	**0.857**
Efficient Net	0.908	0.843	0.814	0.815
ResNet-50	**0.866**	**0.755**	**0.729**	**0.730**
Xception	0.866	0.760	0.722	0.727
Inception	0.820	0.725	0.6614	0.615

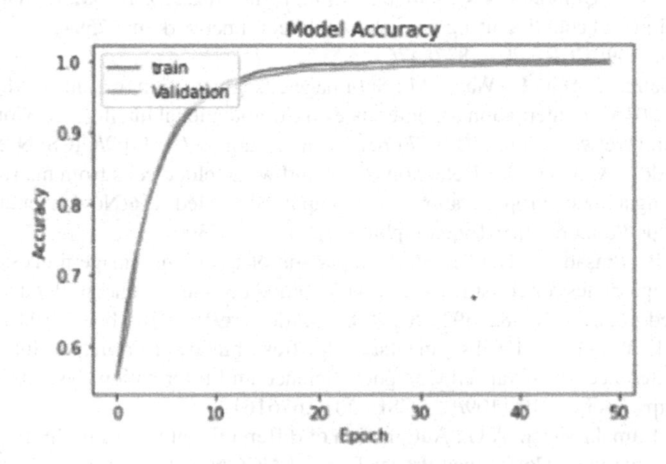

Fig. 9. Training and validation accuracy curve of Efficient Net

5 Conclusion

The complex structure of medical blood cell images makes the diagnosis and analysis of diseases very difficult to image processing tasks. Also, the medical images are not clear and not enough for efficient model development. Hence there is major usage of efficient automatic deep transfer learning-based model which makes the detection process faster and efficient Also it is useful in clinical processes and treatment. In this paper, we focused on developing deep learning models that are pre-trained on Image Net. The models used in the study are VGG-16, Efficient Net, ResNet-50, Inception, and Xception. In order to achieve higher efficiency in dealing with Image processing and WBC Classification, training of pre trained models with WBC data and fine tuning of the trained model is performed. Efficient Net and VGG-19 have achieved higher accuracy in comparison with other convolution models in WBC classification into its subtypes.

References

1. Baydilli, Y.Y., Atila, Ü.: Classification of white blood cells using capsule networks. Computer. Med. Imag. Graph. **80**, 101699 (2020). https://doi.org/10.1016/j.compmedimag.2020.101699

2. Banik, P.P., Saha, R., Kim, K.: An automatic nucleus segmentation and CNN model based classification method of white blood cell. Expert Syst. Appl. **149**, 113211 (2020). https://doi.org/10.1016/j.eswa.2020.113211
3. Liang, G., Hong, H., Xie, W., Zheng, L.: Combining convolutional neural network with recursive neural network for blood cell image classification. IEEE Access **6**, 36188–36197 (2018). https://doi.org/10.1109/ACCESS.2018.2846685
4. Vogado, L.H.S., Veras, R.M.S., Araujo, F.H.D., Silva, R.R.V., Aires, R.T.: Leukemia diagnosis in blood slides using transfer learning in CNNs and SVM for classification. Eng. Appl. Artif. Intell. **72**(October 2017), 415–422 (2018). https://doi.org/10.1016/j.engappai.2018.04.024
5. Yampri, P., Pintavirooj, C., Daochai, S., Teartulakarn, S.: White blood cell classification based on the combination of eigen cell and parametric feature detection (2006)
6. Macawile, M.J., Quiñones, V.V., Ballado, A., Jr., Dela Cruz, J., Caya, M.V.: White blood cell classification and counting using convolutional neural network, pp. 259–263 (2018). https://doi.org/10.1109/ICCRE.2018.8376476
7. Chu, R., Zeng, X., Han, L., Wang, M.: Subimage cosegmentation in a single white blood cell image. In: 2015 7th International Conference on Computational Intelligence, Communication Systems and Networks, pp. 152–157 (2015). https://doi.org/10.1109/CICSyN.2015.36
8. Di, C., Loddo, A., Putzu, L.: Detection of red and white blood cells from microscopic blood images using a region proposal approach. Comput. Biol. Med. **116**(November 2019), 103530 (2020). https://doi.org/10.1016/j.compbiomed.2019.103530
9. Hegde, R.B., Prasad, K., Hebbar, H.: Comparison of traditional image processing and deep learning approaches for classification of white blood cells in peripheral blood smear images. Integr. Med. Res. **39**(2), 382–392 (2019). https://doi.org/10.1016/j.bbe.2019.01.005
10. Sholeh, F.I.: White blood cell segmentation for fresh blood smear images. In: 2013 International Conference on Advanced Computer Science and Information Systems, pp. 425–429 (2013). https://doi.org/10.1109/ICACSIS.2013.6761613
11. Sinha, N., Ramakrishnan, A.G.: Automation of differential blood count. In: IEEE Region 10 Annual International Conference Proceedings/TENCON, vol. 3, no. i, pp. 547–551 (2003). https://doi.org/10.1109/tencon.2003.1273221
12. Wibawa, M.S.: A comparison study between deep learning and conventional machine learning on white blood cells classification. In: 2018 International Conference on Orange Technologies, pp. 1–6 (2018)
13. Suh, H.K., IJsselmuiden, J., Hofstee, J.W., van Henten, E.J.: Transfer learning for the classification of sugar beet and volunteer potato under field conditions. Biosyst. Eng. **174**, 50–65 (2018). https://doi.org/10.1016/j.biosystemseng.2018.06.017
14. Pan, S.J., Yang, Q.: A survey on transfer learning. IEEE Trans. Knowl. Data Eng. **22**(10), 1345–1359 (2010). https://doi.org/10.1109/TKDE.2009.191
15. Le, Q.V., Tan, M.: EfficientNet: improving accuracy and efficiency through AutoML and model scaling. Google AI arXiv Preprint 2019 (2019)
16. Gu, J., et al.: Recent advances in convolutional neural networks. Pattern Recognit. **77**, 354–377 (2018). https://doi.org/10.1016/j.patcog.2017.10.013
17. Agarap, A.F.: Deep learning using rectified linear units (ReLU). arXiv Preprint arXiv:1803.08375, no. 1, pp. 2–8 (2018)
18. Simonyan, K., Zisserman, A.: Very deep convolutional network for large-scale image recognition. In: Conference Paper, ICLR 2015 (2015)
19. He, K., Zhang, X., Ren, S., Sun, J.: Deep residual learning for image recognition. In: Proceedings of the IEEE Computer Society Conference on Computer Vision and Pattern Recognition, vol. 2016-Decem, pp. 770–778 (2016). https://doi.org/10.1109/CVPR.2016.90
20. Szegedy, C., et al.: Going deeper with convolutions. In: Proceedings of IEEE Conference on Computer Vision and Pattern Recognition, pp. 1–9 (2015)

A Novel Feature Selection Algorithm for the Detection of Obstructive Sleep Apnea by Using Heart Rate Variability and ECG Derived Respiratory Analysis

Aditya Prasad Padhy[1], Prateek Pratyasha[2](✉), Saurabh Gupta[2], Kumaresh Pal[1], and Sandeep Mishra[3]

[1] Department of Electrical Engineering, Arka Jain University, Jamshedpur, India
[2] Department of Biomedical Engineering, National Institute of Technology, Raipur, India
prateekpratyasha94@gmail.com
[3] Department of Electronics and Communication Engineering, Dronacharya Group of Institutions, Greater Noida, India

Abstract. Obstructive Sleep Apnea (OSA) is a commonly known sleeping disorder whose undiagnosed and untreated condition can be fatal to cause cardiac fibrillation, arrhythmia and stroke. The recent study aims for a novel computer based methodology for the automated detection of OSA by considering two biosignals from electrocardiogram (ECG) such as Heart Rate Variability (HRV) and ECG-derived Respiratory (EDR). The input signal is retrieved from a publicly available database by using Polysomnography (PSG). Then, the collected data is filtered and pre-process to acquire HRV and EDR parameters from the ECG channel. Three supervised feature selection algorithms, namely Pearson's Correlation Coefficient (FC), ReliefF and Mutual Information Gain Maximization (MIGM) to obtain the optimal features and to balance the feature dimensionality. The input signals are fed into an ensemble learning algorithm for evaluating the effectiveness of proposed feature selection algorithms. The results are compared on the performance matrices and it is observed that MIGM algorithm provides the most optimal features compared two other two selection techniques. Also, the model is evaluated for temporal features and frequency domain features derived from HRV and EDR signals accordingly. The feature domains are also compared depending on their model performance. The comparison states that temporal features are better in accuracy as compared to frequency domain features, however, combination of both the features have improved the validation significantly. Therefore, our proposed methodology acts in a promising way to use the feature selection algorithms and multi-modal analysis for the accurate detection of OSA.

Keywords: Obstructive Sleep Apnea (OSA) · Heart Rate Variability (HRV) · ECG-derived Respiratory (EDR) · Pearson's Correlation Coefficient (FC) · ReliefF · Mutual Information Gain Maximization (MIGM) · eXtreme Gradient Boost (XGBoost)

B. K. Singh et al. (Eds.): ICBEST 2023, CCIS 2003, pp. 233–244, 2024.
https://doi.org/10.1007/978-3-031-54547-4_18

1 Introduction

Sleep is always important for the physical and mental health of an individual. However, sleeping disorders often disturbs the quality of sleep, out of which Sleep Apnea Syndrome (SAS) is a relevant one [1]. SAS is characterized by intermittent of breathing during sleep time. Apnea is of two types: Central Sleep Apnea (CSA) [2] and Obstructive Sleep Apnea (OSA) [3]. Patients with CSA condition often experience cessation of breathing temporarily when the brain lacks in sending proper signals to the muscle nerves [2]. This disorder is less common. But, OSA is a riskier disorder as there is a partial or complete chocking of air at the upper airways due to collapse of throat muscles [3]. Approximately 15% of male and 6% of female with the age group of 30–70 years are suffering from this breath-related sleeping disorder worldwide. The symptoms of OSA for a prolonged time may lead to poor sleeping, morning sickness and day time sleepiness. As a consequence of long term sleep deprivation, the patient may undergo through depression, hypertension, cardiac fibrillation or stroke [4]. Hence, an automated detection and diagnosis methodology is always appreciated. OSA is evaluated with the parameter of Apnea –Hypoapnea Index (AHI) [5] and counted as number of Apneic events per sleep hour. Basically, the score of AHI between 5 and 15 is considered with mild symptoms of OSA while the score between 15 and 30 is taken as a moderate one. However, the score of AHI above 30 is fatal and termed to be severe condition.

Typical invasive methods for the diagnosis and detection of OSA involve the surgery of upper airways, CPAP therapy and weight loss sessions. But the most reliable and approached way is the non-invasive one [6]. There are two non-invasive methods for the detection of OSA: (1) The gold standard method by using Polysomnography (PSG), (2) Computer based automated screening method. The traditional PSG device is used in sleep monitoring laboratories to record the full night sleeping data and Apneic events of patients [7]. Multi-modal channel based devices such as Electrocardiogram (ECG), Electro-encephalogram (EEG), Electro-Myogram (EMG), Electro-oculogram (EOG), nasal airflow, and a microphone attached to a probe. These channels are operated to record the cardiac activities, sleeping pattern, sympathetic characteristics, eyeball movement, body position, breathing pattern, respiratory events and snoring sound respectively. The data recorded in the sleep laboratory with PSG is of high-dimensional, time-consuming and expensive method. Besides that, the device is operated manually by a technician, hence possibility of error while recording, monitoring and storing the database is high. To mitigate this situation, a computer based screening technique is highly in demand for the automated detection of OSA.

Cardiac activities of OSA patient change rapidly between the sleeping stages of wakefulness and Non-Rapid Eye Movement (NREM) [8]. This change is characterized by the sympathetic inhibition from Autonomic Nervous System (ANS) that can decrease the heart rate and blood pressure level during NREM sleep. On the other hand, the heart rate and blood pressure increase during wakefulness of the patient when sympathetic dominance occurs along with sympathetic inhibition. Therefore, only ECG recording is enough to determine the parameters influencing Apneic events. At each ECG data, two different parametric signals namely, Heart Rate Variability (HRV) and ECG-derived Respiratory (EDR) have retrieved [9].

Other existing works have also used various Machine Learning techniques for the identification and severity classification of OSA by using ECG. De Chazal et al. [10] have conducted statistical analysis on HRV and EDR parameters after extracting them from ECG signal. Many authors have used single-lead ECG channels for the detection of OSA which is a reliable and convenient way to use. Shen et al. [11] have used RR intervals from a single-lead ECG channel for the detection of OSA by using multi-scale deep network. Yilmaz et al. [12] also used RR intervals as the statistic features and design the models with classifiers namely, Quadratic Discriminant Analysis (QDA), Support Vector Machine (SVM) and k-Nearest Neighbor (kNN). Sulistyo et al. [13] recorded sleep apnea annotations and HRV features from ECG data to identify the severity of OSA with the help of Artificial Neural Network (ANN), kNN and SVM linear models. From the literature study, some major conclusions have drawn. Firstly, Complex dimensionality of features may cause over fitting to the data. Secondly, retrieving relevant information from the feature set is highly required from the feature pattern recognition. To deal with these shortcomings, feature selection techniques must be included to the data.

The state-of-art on automated OSA detection commonly deals with single modal analysis due to their ease in processing and less computation time. Our recent study uses sleep apnea ECG database from publicly available sources PhysioNet. Each of the signals is obtained from 1 min of segmentation. Then, both HRV and EDR are fed to three different feature selection algorithms such as Pearson's Correlation Coefficient (FC), ReliefF and Mutual Information Gain Maximization (MIGM). Time-domain as well as frequency-domain features are involved in the selection process to obtain the most optimal robust features. Then, the input is fed to an ensembled learning algorithm involving eXtreme Gradient Boost (XGBoost) for evaluating the effectiveness of proposed feature selection algorithms.

The rest of the sections are arranged as follows: Sect. 2 describes the proposed methodology including all the selection techniques and classifiers design in details. Section 3 demonstrates the results obtained from the proposed techniques as well as it narrate a discussion comparing the obtained results with some existing works. Finally, Sect. 4 concludes the entire work with some future scope.

2 Proposed Methodology

The entire work is sub-sectioned as: HRV and EDR parameters pre-processing, application of three feature selection algorithms to reduce the dimensionality of features as well as to obtain an optimal feature set and finally work validation by approaching an ensembled learning algorithm. The framework of our proposed methodology is demonstrated in Fig. 1.

2.1 Description of Dataset

In this work, the database is collected from a publicly available open source called PhysioNet sleep apnea ECG database [14]. The recordings are conducted by using a single-lead ECG channel. Total 70 recordings have collected, out of which 35 recordings are taken as they are the annotated data only. The duration of total recording is 8 h. All

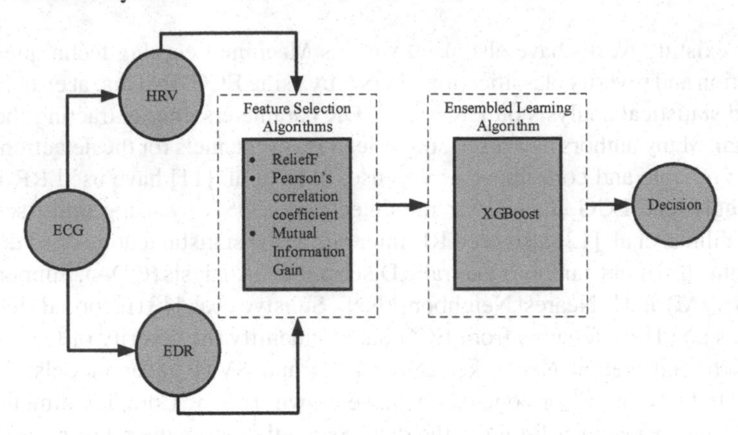

Fig. 1. Framework of Proposed Methodology

the signals are sampled at 100 Hz frequency, 16 bit resolution. The annotation is counted for each minute of Apneic events for which the Apneic events are annotated as 'A' and non-Apneic events are annotated as 'N'. A sampled ECG signal at 1 min segmentation is shown in Fig. 2.

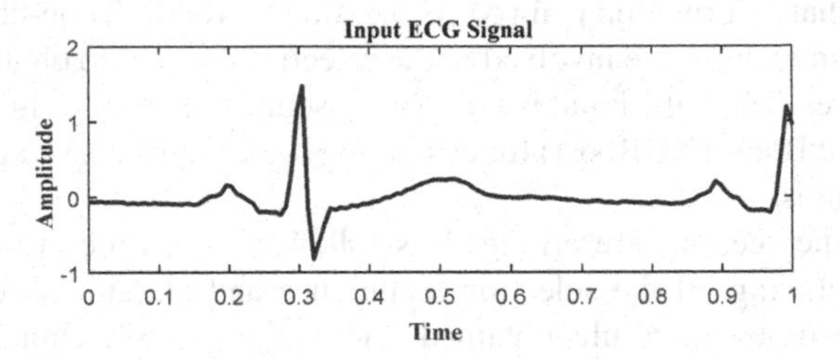

Fig. 2. ECG signal at 1 min of segmentation

2.2 Preprocessing

The first step of ECG signal pre-processing is based on data segmentation. Here, the collected ECG signals are segmented into 1 min duration. Total 2486 number of frames are used for the process of segmentation. The next step is to derive HRV and EDR parameters from each ECG frame. The derivation begins with R-peak detection by using Pan & Tompkins's Algorithm [15]. After the detection of R-peak amplitude, RR intervals have determined which is the ultimate HRV signal. Likewise, for the determination of EDR signal, each of the ECG frame is passed through a band-pass filter at 0.5 Hz and 45 Hz cut off frequency. Then, wavelet analysis is performed to extract EDR signal from the ECG frame [16]. The EDR and HRV signal derived from one ECG frame is depicted in Fig. 3.

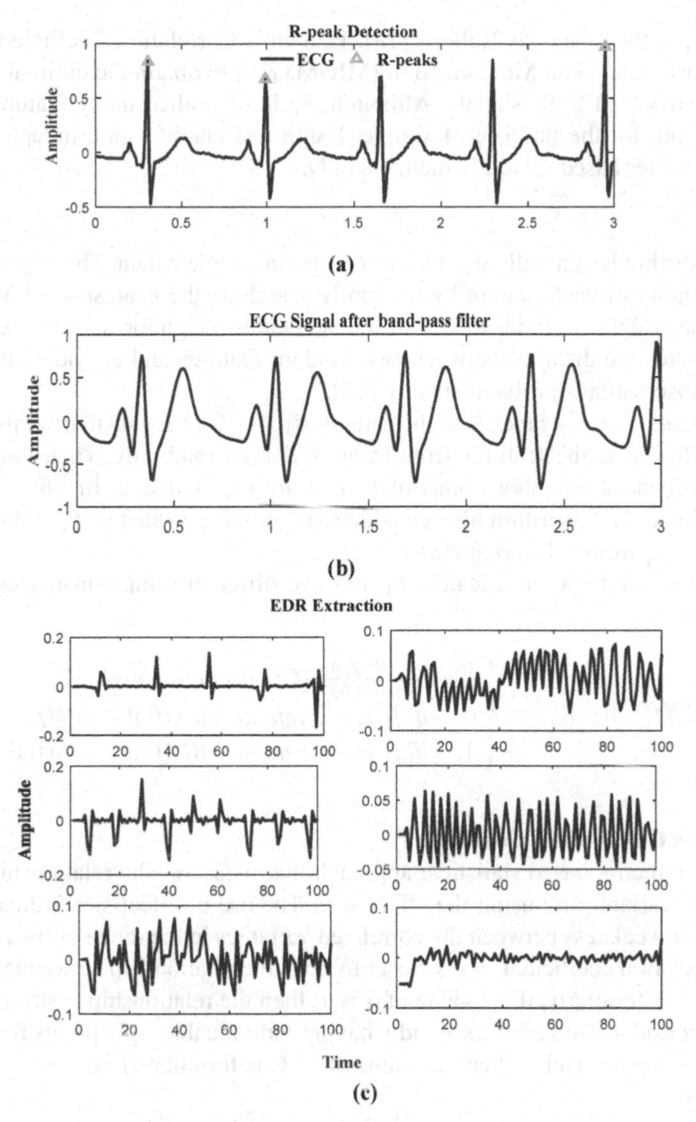

Fig. 3. **(a)** R-peak detection to extract HRV signal, **(b)** ECG signal after Band-pass filter and **(c)** EDR extraction by wavelet analysis

2.3 Feature Selection Algorithms

Feature selection plays an important role for dimensionality reduction, robust feature set and pattern recognition. Moreover, it removes the redundant or irrelevant features to reduce the computation complexity; thereby the performance of classification and detection algorithms would be improved [17]. Many feature selection techniques are also applied when multi-modal features exist in the data dimensionality and there is no certain method regarding combine them. In this study, we have applied three feature selection algorithms such as ReliefF [18], Pearson's Correlation Coefficient [19] and Mutual Information Gain Maximization (MIGM) [20] to obtain the optimal feature set from both HRV and EDR signals. Although we have studied many feature selection algorithms, but for the purpose of simple design and classification independency we have chosen filter based selection methods only.

A. ReliefF

ReliefF algorithm is generally used for noisy and incomplete data. This algorithm evaluates the quality of each feature by randomly searching the nearest neighbor features having some weight. In order to search for the nearest neighbor features, Relief algorithm computes the distance between two random features and considers the missing values of those features probabilistically [18].

For instance, a class label F with features $\{f_1, f_2 \cdots f_k\}$ is taken a training dataset. A sample R_i selects the features from training dataset randomly. Then the algorithm searches for nearest neighbor points of R_i that are mentioned as $\{n_1, n_2 \cdots n_p\}$ from the same class. The algorithm also considers the missing values of R_i and denoted as $\{m_1, m_2 \cdots m_p\}$ from a different class.

The difference between a feature f_1 and two different sample instances R_1, R_2 is given as:

$$diff(f_1, \mathrm{R}_1, \mathrm{R}_2) = \begin{cases} \frac{|R_1(f_1)|-|R_2(f_1)|}{\max(f_1)-\min(f_1)}, \; if\ f_1\ is\ continuous \\ 0, \quad if\ f_1\ is\ discrete\ and\ |R_1(f_1)| = |R_2(f_1)| \\ 1, \quad if\ f_1\ is\ discrete\ and\ |R_1(f_1)| \neq |R_2(f_1)| \end{cases} \tag{1}$$

B. Pearson's Correlation Coefficient

PCC is a parametric based statistical approach that measures the relationship between two or more variables by using the effect size. The use of effect size indicates degree of strength or weakness between the correlated variables in the range of [0–1] [19]. The more a correlation coefficient (α) is closer to zero; the relationship between variables is getting weaker. Similarly, if the value of α is 1, then the relationship is strongest.

Let us consider two variables x and y having individual sample points indexed with k as x_k and y_k respectively. Then, the value of PCC is formulated as:

$$\alpha_{xy} = \frac{1}{m-1}\sum_{k=1}^{m}\left(\frac{x_k - \overline{x}}{\delta_x}\right)\cdot\left(\frac{y_k - \overline{y}}{\delta_y}\right) \tag{2}$$

where δ_x and δ_y are standard deviation for x and y respectively.

The expression for δ_y δ_x is given in equation as:

$$\delta_x = \sqrt{\frac{1}{m-1}\sum_{k=1}^{m}(x_k - \overline{x})^2} \tag{3}$$

$$\delta_y = \sqrt{\frac{1}{m-1}\sum_{k=1}^{m}(y_k - \overline{y})^2} \tag{4}$$

$\overline{y}$ and $\overline{x}$ are the mean formulated as:

$$\overline{y} = \frac{1}{m}\sum_{k=1}^{m} y_k \tag{5}$$

$$\overline{x} = \frac{1}{m}\sum_{k=1}^{m} x_k. \tag{6}$$

However, to understand the purpose of correlation, a correlation coefficient threshold (T) is required after calculating the correlation between two or more different dependent features. When the threshold value attains the utmost result as T $\geq$ 0.95, then the most correlated features are selected. Later on, all the correlation coefficients are calculated with independent variables by reducing the number of dependent features and tuning the threshold value. After that, the features with less significance are automatically deleted to solve the problem of dimensionality reduction and the selected features with strong linear relationship ensure strong independence among variables.

C. Mutual Information Gain

Mutual Information Gain Maximization (MIGM) is also known as Mutual Information Maximization (MIM) is a rank based feature selection algorithm [20]. The concept of MIGM algorithm is based on information theory that depends on feature-class correlation as well as feature-feature correlation. The algorithm first gathers mutual information between a feature and its corresponding class which can be formulated as:

$$I(C, f_k) = \begin{cases} H(f_k) - H(f_k|C) \\ H(C) - H(C|f_k) \\ H(f_k) + H(C) - H(C, f_k) \end{cases} \tag{7}$$

where $H(f_k)$ and $H(C)$ are the entropy of individual feature and class output respectively. $H(f_k \mid C)$ and $H(C \mid f_k)$ are the conditional entropy. $H(C, f_k)$ is denoted as joint entropy.

The same way is formulated to determine feature-dependent information also. All the features are ranked in descending order according to their MIGM values and then based on the priority level the optimal features are selected. The chance of feature redundancy is avoided in this algorithm as both feature-class and feature-feature based information are considered. However, the feature must be of same feature subset $(f_k, f_l) \in F$.

Hence, the MIGM based feature selection formula can be obtained as:

$$MIG = I(C, f_k) - \delta \sum\nolimits_{(f_k, f_l) \in F} I(f_k, f_l) \tag{8}$$

Here the parameter δ is known as redundancy coefficient. The changes in value of δ can affect the entire selection process.

2.4 Design of Ensembled Learning Algorithm

In this study, XGBoost is considered as the ensembled learning algorithm which is a gradient based detection technique [21]. It combines multiple linear classification algorithms on the basis of boosted learning ideas. Although many hyper-parameters are involved in XGBoost, we have focused only 'learning rate', and 'gamma' hyper-parameters in order to avoid over-fitting as well as to increase the model performance. The characteristics of hyper-parameters are mentioned in Table 1.

Table 1. Characteristics of Hyper-parameters

Hyper-parameters	Description	Parameter Range
Learning rate (L)	Reduced step-size used for model update	(0.05–0.3)
Gamma (γ)	Minimum loss value for partition	(0–0.2)

To train the proposed learning algorithm, we have split the data into training and test samples at the ratio of 80:20. Then, the training samples are used to train the model by using XGBoost algorithm. The hyper-parameters are constantly adjusted to build an optimal XGBoost model and tuned by using 10-fold cross validation. To exploit the performance, obtained optimal model is tested on the test samples.

3 Results and Discussion

As the objective of the paper is to classify the OSA and healthy subjects, it is essential to validate the performance of optimal features implemented by our proposed feature selection algorithms. The detection of OSA has performed by using temporal features and time-domain features of both HRV and EDR signals as inputs. Features from each of the signals underpasses through three supervised feature selection algorithm and then performance of each optimal features have conducted by applying an ensembled learning classifier as shown in Tables 2, 3, 4, 5, 6 and 7.

Table 2. Frequency domain feature analysis of HRV signal

methods	Performance Validation		
	Accuracy (in %)	Sensitivity (in %)	Specificity (in %)
ReliefF + XGBoost	62.8	64.18	65.71
PCC + XGBoost	68.91	69.89	70
MIGM + XGBoost	75.14	76.81	75.12

From Tables 2 and 3, we observed that temporal features of HRV are used as input and results maximum accuracy for MIGM based XGBoost model. Using the temporal features as input, the classification accuracy is 75.14% where as using frequency domain features as input, the accuracy increases to 78.61% with the same classifier. Therefore, it is quite evident that frequency domain features of HRV signal shows poor classification accuracy, sensitivity and specificity as compared to the temporal features.

Table 3. Temporal feature analysis of HRV signal

methods	Performance Validation		
	Accuracy (in %)	Sensitivity (in %)	Specificity (in %)
ReliefF + XGBoost	67.55	68.17	69.39
PCC + XGBoost	71.09	72.91	72.64
MIGM + XGBoost	78.61	80.09	80.78

Table 4. Frequency domain feature analysis of EDR signal

methods	Performance Validation		
	Accuracy (in %)	Sensitivity (in %)	Specificity (in %)
ReliefF + XGBoost	63.88	64	64.04
PCC + XGBoost	68.34	69.8	70.94
MIGM + XGBoost	76.7	77.15	77.95

Table 5. Temporal feature analysis of EDR signal

Methods	Performance Validation		
	Accuracy (in %)	Sensitivity (in %)	Specificity (in %)
ReliefF + XGBoost	68.64	69.36	70.24
PCC + XGBoost	72.46	73.95	73.52
MIGM + XGBoost	78.64	80.19	80.9

Tables 4 and 5 shows the model performance by considering frequency domain features and temporal features of EDR signal respectively. Alike Tables 2 and 3, here also MIGM has proved as a better selection algorithm and improved the accuracy to 76.7% for frequency domain features and 78.64% for temporal features.

Table 6. Frequency domain feature analysis of HRV and EDR signal as a combination

methods	Performance Validation		
	Accuracy (in %)	Sensitivity (in %)	Specificity (in %)
ReliefF + XGBoost	80.67	82.41	82.75
PCC + XGBoost	84.32	86.47	84.08
MIGM + XGBoost	91.56	92.75	92.27

Table 7. Temporal feature analysis of both HRV and EDR signal as a combination

methods	Performance Validation		
	Accuracy (in %)	Sensitivity (in %)	Specificity (in %)
ReliefF + XGBoost	81.68	84.66	83.86
PCC + XGBoost	87.65	88.20	88.85
MIGM + XGBoost	94.62	96.21	95.73

Extending our experiment, frequency domain features of both HRV and EDR signals are combined which proved as an improved methodology to increase the classification performance as shown in Tables 6 and 7. It is evident that MIGM based XGBoost algorithm has performed better than other two selection algorithm with an accuracy of 91.56%, sensitivity of 92.75% and specificity of 92.27%. In the similar manner, the same technique is applied on the temporal features of both HRV and EDR signal. The results are again outperformed at 94.62% accuracy 96.21% sensitivity and 95.73% specificity. Hence, it is proved that features of only HRV signal or only EDR signal shows poor performance compared to the similar features fused in a feature level. Automated detection of OSA from HRV and EDR parameters are the prior objective of this work. We have focused on HRV signal because the signal has the ability to capture information during autonomic dysfunction, especially for the detection of cardiac dysfunction related sleep disorders. Also, HRV signal shows higher mean value during Apneic events. This is the reason for using HRV analysis for OSA detection. In addition to that, EDR signals are also shows statistical significance of features.

4 Conclusion

In this work, an automated machine learning methodology was applied for the detection of Obstructive Sleep Apnea (OSA) by using Heart rate Variability (HRV) and Electrocardiogram Based Respiratory (EDR) analysis. In the first step, raw ECG data was collected from an open source at 1 min of segmentation. Then, the signal is filtered and pre-processed to extract HRV as well as EDR parameters. In the next step, three supervised feature selection algorithms, namely ReliefF, Pearson's correlation coefficient (PCC) and Mutual Information Gain Maximization (MIGM) are applied on each of the signals individually. With temporal features and frequency domain features determined from each of the selection algorithms individually, the optimal features were fed to an ensembled learning algorithm, namely eXtreme Gradient Boosting (XGBoost) as an input. The performance of learning algorithm was better for temporal features in terms of accuracy, sensitivity and specificity as compared to input with frequency domain features. To extend our experiment, we have grouped the temporal features of HRV and EDR signals to one feature set, meanwhile frequency domain features of both HRV and EDR signals are combined to form another feature set. Significantly, the performances of features have been improved. In this way, both feature selection algorithm and feature fusion enhances the performance of model and helps in detecting OSA significantly.

Scope of this work proceeds in two directions. The first one is to increase number of features for a deep analysis. The second one is to involve other channels of PSG to improve the fusion process.

References

1. Mannarino, M.R., Di Filippo, F., Pirro, M.: Obstructive sleep apnea syndrome. Eur. J. Internal Med. **23**, 586–593 (2012)
2. Javaheri, S., Dempsey, J.: Central sleep apnea. Compr. Physiol. **3**, 141–163 (2013)
3. Mehra, R.: Sleep apnea and the heart. Clevel. Clin. J. Med. **86**, 10–18 (2019)
4. Bounhoure, J.-P., Galinier, M., Didier, A., Leophonte, P.: Sleep apnea syndromes and cardiovascular disease. Bulletin de l'Academie nationale de medecine **189**, 445–459 (2005). Discussion 460
5. Asghari, A., Mohammadi, F.: Is apnea-hypopnea index a proper measure for obstructive sleep apnea severity. Med. J. Islam Repub. Iran **27**, 161–162 (2013)
6. Haoyu, L., Jianxing, L., Arunkumar, N., Hussein, A.F., Jaber, M.M.: An IoMT cloud-based real time sleep apnea detection scheme by using the SpO2 estimation supported by heart rate variability. Future Gener. Comput. Syst. **98**, 69–77 (2019)
7. Berry, R.B., Hill, G., Thompson, L., McLaurin, V.: Portable monitoring and autotitration versus polysomnography for the diagnosis and treatment of sleep apnea. Sleep **31**, 1423–1431 (2008)
8. Gonzaga, C., Bertolami, A., Bertolami, M., Amodeo, C., Calhoun, D.: Obstructive sleep apnea, hypertension and cardiovascular diseases. J. Hum. Hypertens. **29**, 705–712 (2015)
9. Rahimi, A., Safari, A., Mohebbi, M.: Sleep stage classification based on ECG-derived respiration and heart rate variability of single-lead ECG signal. In: 2019 26th National and 4th International Iranian Conference on Biomedical Engineering (ICBME), pp. 158–163 (2019)
10. De Chazal, P., Sadr, N.: Sleep apnoea classification using heart rate variability, ECG derived respiration and cardiopulmonary coupling parameters. In: 2016 38th Annual International Conference of the IEEE Engineering in Medicine and Biology Society (EMBC), pp. 3203–3206 (2016)
11. Shen, Q., Qin, H., Wei, K., Liu, G.: Multiscale deep neural network for obstructive sleep apnea detection using RR interval from single-lead ECG signal. IEEE Trans. Instrum. Meas. **70**, 1–13 (2021)
12. Yılmaz, B., Asyalı, M.H., Arıkan, E., Yetkin, S., Özgen, F.: Sleep stage and obstructive apneaic epoch classification using single-lead ECG. Biomed. Eng. Online **9**, 1–14 (2010)
13. Sulistyo, B., Surantha, N., Isa, S.M.: Sleep apnea identification using HRV features of ECG signals. Int. J. Electr. Comput. Eng. (IJECE) **8**, 3940–3948 (2018)
14. Schrader, M., Zywietz, C., Von Einem, V., Widiger, B., Joseph, G.: Detection of sleep apnea in single channel ECGs from the PhysioNet data base. In: Computers in Cardiology 2000, vol. 27 (Cat. 00CH37163), pp. 263–266 (2000)
15. Sathyapriya, L., Murali, L., Manigandan, T.: Analysis and detection R-peak detection using Modified Pan-Tompkins algorithm. In: 2014 IEEE International Conference on Advanced Communications, Control and Computing Technologies, pp. 483–487 (2014)
16. Avcı, C., Delibaşoğlu, İ., Akbaş, A.: Sleep apnea detection using wavelet analysis of ECG derived respiratory signal. In: 2012 International Conference on Biomedical Engineering (ICoBE), pp. 272–275 (2012)
17. Pratyasha, P., Gupta, S., Padhy, A.P.: Recent vogues of artificial intelligence in neuroscience: a systematic review (2022)

18. Fan, H., Xue, L., Song, Y., Li, M.: A repetitive feature selection method based on improved ReliefF for missing data. Appl. Intell. **52**, 1–16 (2022)
19. Chen, P., Li, F., Wu, C.: Research on intrusion detection method based on Pearson correlation coefficient feature selection algorithm. J. Phys.: Conf. Ser. **1757**, 012054 (2021)
20. Wang, X., Guo, B., Shen, Y., Zhou, C., Duan, X.: Input feature selection method based on feature set equivalence and mutual information gain maximization. IEEE Access **7**, 151525–151538 (2019)
21. Chen, T., et al.: XGBoost: extreme gradient boosting. R Package Version 0.4-2, vol. 1, pp. 1–4 (2015)

Early Diagnosis of Parkinson's Disease Based on Spiral and Wave Drawings Using Convolutional Neural Networks and Machine Learning Classifier

S. Saravanan, K. Ramkumar(✉), S. Venkatesh, K. Narasimhan, and K. Adalarasu

School of Electrical and Electronics Engineering, SASTRA Deemed University, Thanjavur 613 401, India
ramkumar@eie.sastra.edu

Abstract. Parkinson's disease (PD) is a neurodegenerative condition caused by dopamine-producing nerve cell loss. However, due to a lack of indicators, early detection of Parkinson's disease is difficult. The goal of this study is to create a system for the early detection of Parkinson's disease (PD) based on hand drawings, using pre-trained CNN models as a feature extractor and Machine Learning (ML) classifiers to differentiate PD from healthy persons. Several pre-trained CNN models, including VGG16, VGG19, ResNet-50, IncepetionV3, Xception, and Mobile Net V2, are used as feature extractors. The retrieved characteristics are fed into the various machine learning classifiers as input. The proposed system VGG16 as a feature extractor and Random Forest (RF) as ML classifier performed much better than existing state-of-the-art pretrained models, with a classification accuracy of 97%. The results of the experiment indicate that the proposed strategy works better on publicly available hand-drawn datasets for the early identification of Parkinson's disease.

Keywords: Parkinson's Disease · Spiral drawing · Wave drawing · Machine Learning

1 Introduction

Parkinson's disease (PD) is the second most common form of neurodegenerative disorder that can be found all over the world. Motor and non-motor symptoms can be used to diagnose Parkinson's disease. PD arises due to the loss of dopamine producing neuron cells in the substantia Niagara portion of the brain. These dopamine substances are responsible for communicating the signal from one part of the brain to the other part [1, 2]. However, the cure for PD has not been found yet. The progression of the disease can be characterized by using two different rating scales such as Movement Disorder Society-Unified Parkinson's Disease Rating Scale (MDS-UPDRS) and Hoehn and Yahr scale. Due to the lack of thorough knowledge of Parkinson's disease, there are no trustworthy early detection tests for PD at present, and many patients do not seek treatment in a timely

B. K. Singh et al. (Eds.): ICBEST 2023, CCIS 2003, pp. 245–255, 2024.
https://doi.org/10.1007/978-3-031-54547-4_19

manner [3]. To get the proper treatment, one must identify the disease at the earlier stage to improve the quality life of PD patient. One of the most difficult challenges is acquiring medical information about certain conditions from a patient with the required size of test sample and its suitability to train the pioneering Machine learning (ML) models. In this growing era, machine learning and deep learning techniques have shown greater interest in the field of healthcare [4]. In this article the proposed method is the diagnosis of Parkinson's disease at the earlier stage using spiral and wave drawings. We have presented various state of art deep learning algorithms using transfer learning to select the features from the spiral and wave drawings and classification was accomplished using ML algorithms.

The remaining part of the article is organized as follows: Sect. 2 introduces the related work of PD prediction. Section 3 describes the dataset in detail and demonstrates the methodology of the proposed work. Section 4 investigates the experimental dataset using various pre-trained models. Section 5 finally concludes the article.

2 Related Works

Artificial intelligence (AI) in healthcare can improve patient outcomes overall, enhance anticipatory care, improve quality of life, and generate more accurate diagnoses and treatment approaches. Naseer et al. [5] examined the PaHaw dataset to distinguish between PD and healthy individuals using fine-tuned ImageNet features through transfer learning and achieved a classification accuracy of 98.28%. Meghakamble et al. [6] used ML classifiers to classify PD and healthy controls using three different spiral drawing tasks, such as spiral in static and dynamic motion and circular motion tasks, and obtained a maximum accuracy of 91.6%. In another study, Drotar et al. [7] applied a feature selection algorithm with a support vector machine as a classifier and

recommended analysing the hand drawings of the patients for PD detection. However, D. Impedovo et al. [8] investigated the extent to which dynamic features of the hand drawing technique could be useful in the early detection of Parkinson's disease and developed a classification architecture based on a variety of classifier algorithms with an ensemble method. Alissa et al. [9] investigated the wire cube and pentagon spiral drawing tasks using convolutional neural networks for the early diagnosis of PD and achieved an accuracy of 93.5%. Manuel Gil Martin et al. [10] analysed the drawing movements of the individuals using convolutional neural networks to discriminate PD patients from healthy individuals and obtained a classification accuracy of 96.5%. Kotsavasiloglou C et al. [11] presented a ML model developed in developing countries to distinguish between healthy and Parkinson's disease patients based on the trajectory of horizontal lines drawn on a tab with a stylus - like pen. Destefano et al. [12] conducted a review of recent research on handwriting modelling approaches for the early detection of neurodegenerative disorders.

3 Materials and Methods

We explored the collection of datasets, data augmentation methodologies, and the proposed strategy for the early detection of Parkinson's disease (PD) using various pre-trained models and machine learning classification algorithms in this portion.

3.1 Dataset Description

Experiments in this study used data acquired from Kaggle. A total of 204 images were compiled from the subjects' drawing tasks. The physicians labelled 102 spiral drawings and 102 wave drawings as PD and healthy out of 204 total images. There were 72 training images and 30 test images in each spiral and wave - balanced dataset. These images contain major, recognised patterns that are essential for distinguishing between PD and health. Figure 1 illustrates examples of spiral and wave drawings.

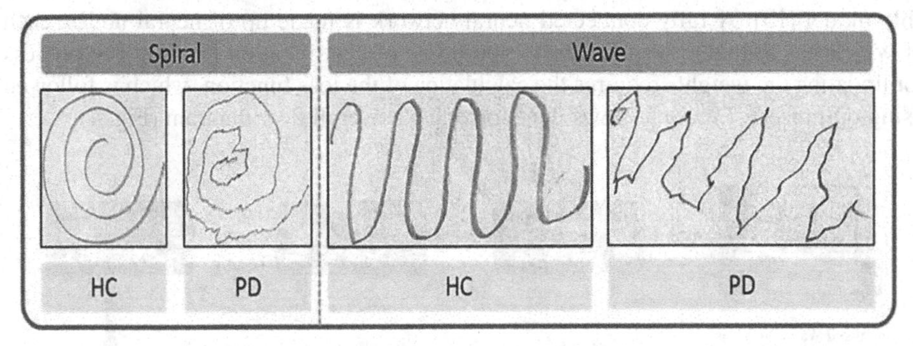

Fig. 1. Sample Images of Spiral and Wave drawings

3.2 Data Augmentation Techniques

Large amounts of data are needed to train good deep learning models, yet industries like healthcare often lack access to such resources. In order to train more accurate deep learning models, a combination of techniques known as "data augmentation" has been proposed [12]. Adding more images to the dataset and making them more diverse throughout training does not necessitate the collection of additional empirical evidence [13]. Over - fitting can also be mitigated with the introduction of generalisation of features by enriching the data [14]. Figure 2 displays some of these enhanced photos.

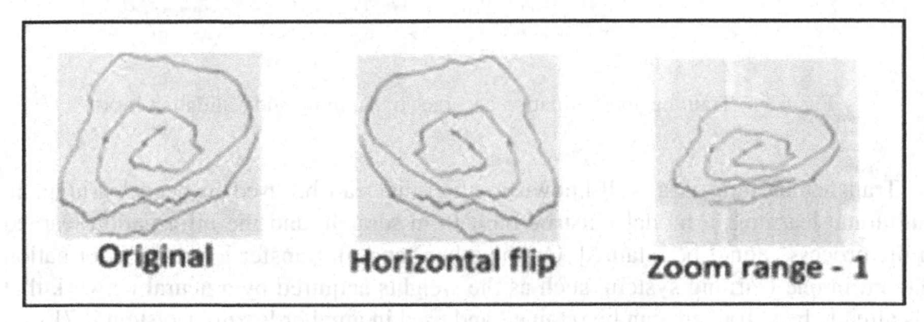

Fig. 2. Sample Augmented Images

4 Experiment and Results

Approach 1: Pre-trained Network as Feature Extractor and Classifier
In this approach 1, the spiral and wave drawings from the Kaggle dataset are pre-processed, and the dataset is divided into a training dataset and a testing dataset in an 80:20 ratio. The pre-trained models were imparted using transfer learning and predicted whether the class would be PD or healthy. In this proposed method, a simple, fully connected neural network layer is built at the end of the feature extraction process. Because the complete model is a pre-trained model, we can obtain an end-to-end trainable model [15]. A fully connected neural network is made up of neural nodes, each of which has trainable weights for the purposes of classification [7, 16]. The process for updating the weights requires the calculation of the loss function, which is followed using optimizers. Figure 3 shows the approach 1's overall flow diagram (Fig. 4).

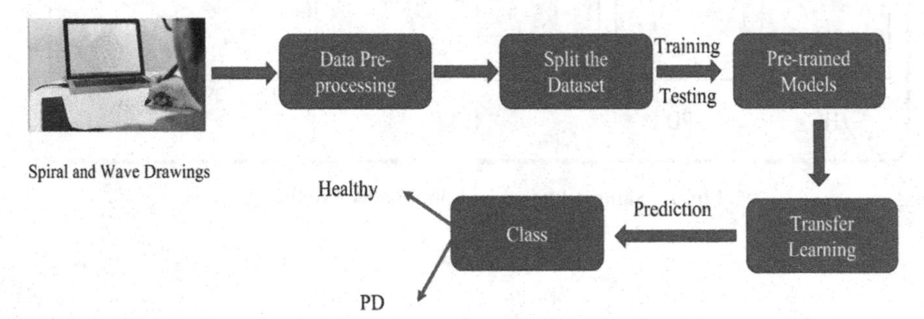

Fig. 3. Overall diagram of Proposed Approach 1

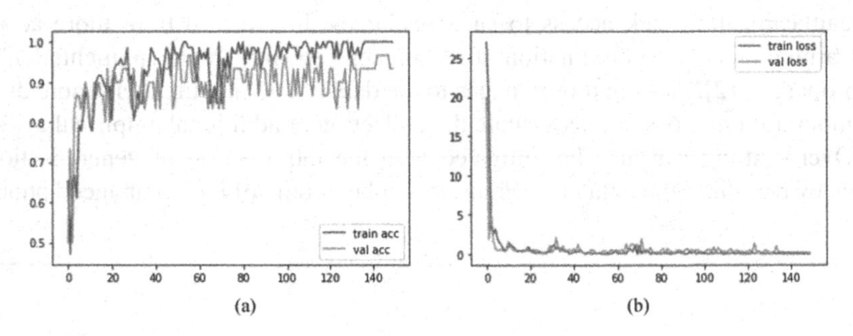

Fig. 4. a) Training and Validation accuracy b) Training and Validation loss

Transfer learning is a well-known method that can be used in deep learning. In traditional learning, a model must be built from scratch, and the information learned in the process cannot be retained. On the other hand, in transfer learning, information learned in one learning system, such as the weights acquired by a neural network that has already been trained, can be retained and used in another learning system [17].

Approach 2: Pre-trained Network as Feature Extractor and Machine learning Classifier

The spiral and wave drawing images are first processed by pre-trained networks that act as feature extractors in the approach 2 that has been proposed, and then the features that have been extracted are provided to the various machine learning classifiers. The classifier model is trained with the assistance of several different machine learning algorithms, including Support Vector Machine, Random Forest, Logistic Regression, Decision Tree, K-Nearest Neighbor, and Naive Bayes [7]. The overall flow diagram of the proposed approach 2 is represented in Fig. 5.

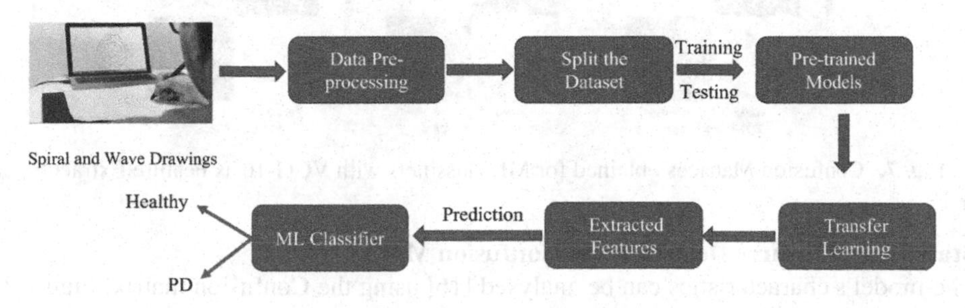

Fig. 5. Overall diagram of Proposed Approach 2

The performance of the PD classification model can be assessed using the confusion matrix. The confusion matrices of various pre-trained models for the early diagnosis of Parkinson's disease are represented in Fig. 6.

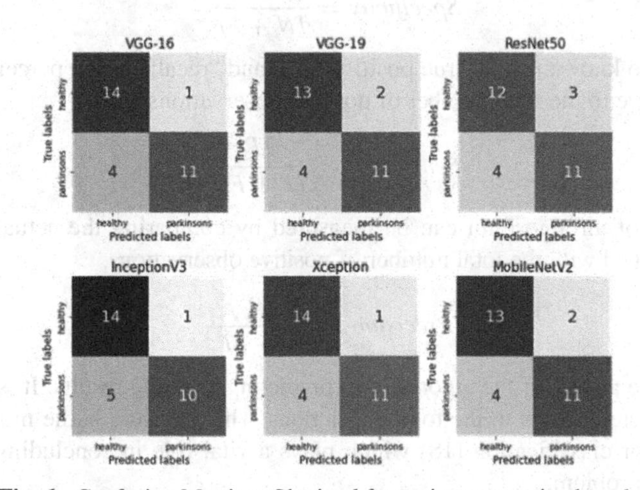

Fig. 6. Confusion Matrices Obtained for various pre-trained models

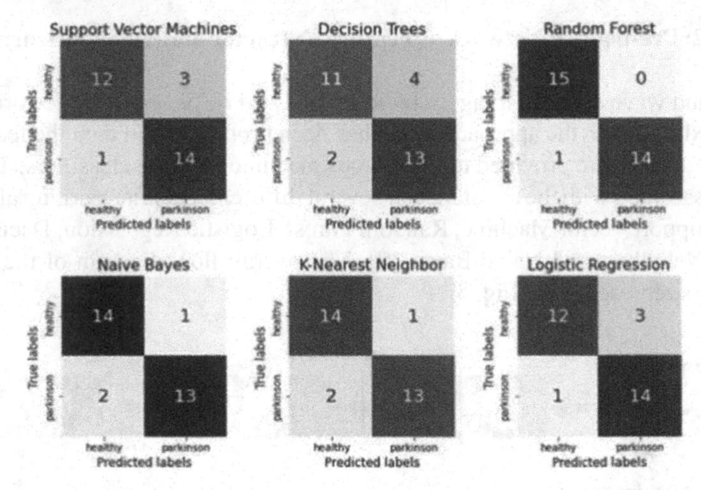

Fig. 7. Confusion Matrices obtained for ML classifiers with VGG-16 as Feature Extractor

Statistical Measures Derived from Confusion Matrix

The model's characteristics can be analysed [16] using the Confusion matrix. Figure 7 Confusion matrices obtained for ML classifiers with VGG-16 as a feature extractor on the matrix by calculating various statistical measures such as specificity, sensitivity, precision, accuracy, F1score, and so on.

Accurately predicting the genuine negatives among the negative observations is the goal of specificity, often known as the true negative rate.

$$Specificity = \frac{TN}{TN + FP} \tag{1}$$

Sensitivity, also known as the "true positive rate" and "recall," is the percentage of true positives relative to the total number of positive observations.

$$Sensitivity = \frac{TP}{TP + FN} \tag{2}$$

The accuracy of an algorithm can be measured by comparing the actual number of positives obtained with the total number of positive observations.

$$Precision = \frac{TP}{TP + FP} \tag{3}$$

Accuracy is the ability of the algorithm in predicting accurate results. It is the ratio of the truly predicted values to the total predictions. This measure is the most important measure for our classification [18] which plays a vital role in concluding to the best model for our problem.

$$Accuracy = \frac{TP + TN}{TP + TN + FP + FN} \tag{4}$$

The F1-score is the harmonic mean (weighted) of the model's precision and recall, and it is used to describe the model's overall accuracy by making use of both positive and negative predicted values. When the data is severely imbalanced, the F1 score should be high on the model's test data in order to balance the classification model using recall and precision. This provides a more trustworthy performance metric than accuracy and precision.

$$F_1\ score = 2 \times \frac{Precision \times Recall}{Precision + Recall} \tag{5}$$

Table 1. Comparing the performances of various pre-trained models

Pre-trained Models	Accuracy	Precision	Recall	F1-score
VGG-16	0.8333	0.9166	0.7333	0.8147
VGG-19	0.8000	0.8461	0.7333	0.7856
ResNet50	0.7666	0.7857	0.7333	0.7585
InceptionV3	0.8000	0.9090	0.6666	0.7691
Xception	0.8333	0.9166	0.7333	0.8147
MobileNetV2	0.8000	0.8461	0.7333	0.7856

Table 2. Comparing the performances of different machine learning classifiers

Machine learning classifiers	Accuracy	Precision	Recall	F1-score
Support Vector Machines	0.8667	0.8235	0.9333	0.8750
Decision Trees	0.8000	0.7647	0.8667	0.8125
Random Forest	0.9667	1.0000	0.9333	0.9655
Naïve Bayes	0.9000	0.9286	0.8667	0.8966
K-Nearest Neighbour	0.9000	0.9286	0.8667	0.8966
Logistic Regression	0.8667	0.8235	0.9333	0.8750

The performance evaluation metrics obtained from the various pre-trained models are displayed in Table 1. The VGG-16 and Xception models perform exceptionally well in distinguishing between PD and healthy people, with an accuracy of 83.33% (Table 2).

The receiver operating characteristics curve of ML classifiers is shown in Fig. 8a. The ROC curve is a diagnostic tool that is often used in any binary classification where various thresholds are assumed for a model to discriminate between Parkinson's disease and health, where the true positive rate is plotted against the false positive rate [19]. In general, the greater the area under the ROC curve, the better a classifier performs for a given task [20]. The precision-recall curve presented in Fig. 8b aids in determining whether the model worked successfully. In our approach, the random forest algorithm surpasses conventional machine learning classifiers.

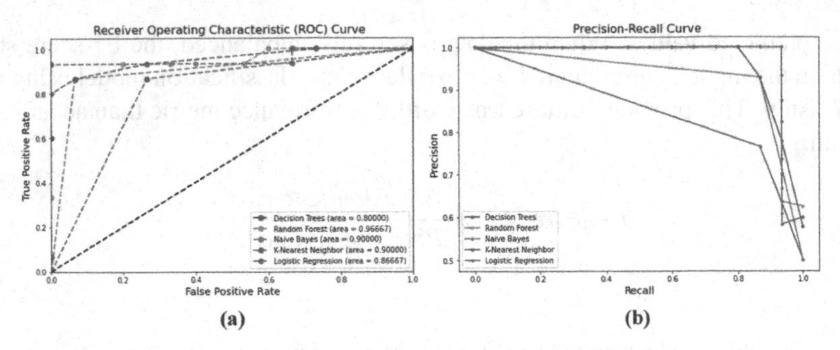

Fig. 8. (a) Receiver Operating Characteristics Curve (b) Precision-Recall Curve

Approach-I

Table 3. Performance of Various pre-trained models for wave dataset

Model	Remarks	Training Accuracy	Validation Accuracy
VGG – 16	Adam, 15ep	0.87	0.90
VGG – 19	Adam, 15ep	0.83	0.90
ResNet50	Adam, 15ep	0.68	0.53
ResNet101	Adam, 15ep	0.61	0.70
InceptionV3	Adam, 15ep	0.91	0.90
Xception	Adam, 15ep	0.91	0.83
DenseNet169	Adam, 15ep	0.91	0.86
EfficientNetV2L	Adam, 15ep	0.47	0.50
MobileNetV2	Adam, 15ep	0.94	0.86

Table 4. Performance of Various pre-trained models for spiral dataset

Model	Remarks	Training Accuracy	Validation Accuracy
VGG – 16	Adam, 15ep	0.97	0.80
VGG – 19	Adam, 15ep	0.93	0.80
ResNet50	Adam, 15ep	0.73	0.73
ResNet101	Adam, 15ep	0.59	0.56
InceptionV3	Adam, 15ep	0.88	0.80
Xception	Adam, 15ep	0.90	0.83
DenseNet169	Adam, 15ep	0.97	0.80

(continued)

Table 4. *(continued)*

Model	Remarks	Training Accuracy	Validation Accuracy
EfficientNetV2L	Adam, 15ep	0.47	0.50
MobileNetV2	Adam, 15ep	0.93	0.80

From Tables 3 and 4 various pre-trained models are trained the spiral and wave dataset independently.

Approach-II

When it comes to the early prediction of Parkinson's disease (PD), various pre-trained models such as feature extractors and machine learning classifiers are utilized. Both the spiral and wave datasets, as well as the combined dataset, are evaluated separately and together. Tables 5 and 6 make it very clear that the VGG16 networks perform exceptionally well as feature extractors, whereas Table 5 demonstrates that Random Forest dominates as a classifier. The findings of the early PD prediction utilizing aggregated datasets are presented in Table 7.

Table 5. Performance of various ML classifier models for wave dataset

S. No	Feature Extractor	Highest Accuracy	Corresponding ML Classifier
1	VGG 16	0.9667	RF
2	VGG 19	0.9333	RF
3	InceptionV3	0.8667	SVM, LR
4	Xception	0.9000	SVM
5	Densenet169	0.9000	SVM
6	MobileNetV2	0.9333	SVM, LR
7	ResNet50	0.7667	RF
8	ResNet101	0.8667	LR

Table 6. Performance of various ML classifier models for spiral dataset

S. No	Feature Extractor	Highest Accuracy	Corresponding ML Classifier
1	VGG 16	0.8667	SVM, LR
2	VGG 19	0.8333	RF, DT
3	InceptionV3	0.8667	NB
4	Xception	0.8333	LR
5	Densenet169	0.9333	SVM, LR
6	MobileNetV2	0.8667	RF
7	ResNet50	0.8333	SVM, LR
8	ResNet101	0.9000	SVM, LR

Table 7. Performance of various ML classifier models for combined dataset

S. No	Feature Extractor	Highest Accuracy	Corresponding ML Classifier
1	VGG 16	0.8667	KNN, LR
2	VGG 19	0.8667	LR
3	InceptionV3	0.8500	SVM, LR
4	Xception	0.9000	SVM, LR
5	Densenet169	0.8833	SVM
6	MobileNetV2	0.8667	SVM
7	ResNet50	0.7666	RF
8	ResNet101	0.8667	LR

5 Conclusion

In this paper, we conduct a comprehensive analysis of the use of pre-trained networks as a feature extractor on spiral and wave drawings for PD prediction and classification by employing a variety of ML classifiers. This research shows that when a dataset of 102 wave drawings is processed with deep learning for feature extraction and machine learning for classification, the results are superior for classifying PD patients and healthy participants. When compared to prior work on spiral drawing datasets, the proposed method's 97% accuracy and 96.67% area under the curve (AUC) achieved by using a VGG-16 network as the feature extractor and a Random Forest classifier stand out. In addition, we programmed a system that can quickly determine whether a given wave drawing image represents Parkinson's disease or a healthy control, saving valuable time that could be better spent elsewhere in a hospital or clinic, where preventative or curative measures could have a greater impact on patients' lives.

References

1. Pereira, C.R., et al.: Handwritten dynamics assessment through convolutional neural networks: an application to Parkinson's disease identification. Artif. Intell. Med.. Intell. Med. **87**, 67–77 (2018)
2. Saravanan, S., et al.: A systematic review of Artificial Intelligence (AI) based approaches for the diagnosis of Parkinson's disease. Arch. Comput. Methods Eng. **29**, 1–15 (2022)
3. Tang, Q.: Early diagnosis of Parkinson's disease using machine learning. Chongqing University of Posts and Telecommunications (2020)
4. Shinde, S., et al.: Predictive markers for Parkinson's disease using deep neural nets on neuromelanin sensitive MRI. Neuro Image: Clin. **22**, 101748 (2019)
5. Naseer, A., et al.: Refining Parkinson's neurological disorder identification through deep transfer learning. Neural Comput. Appl.Comput. Appl. **32**(3), 839–854 (2020)
6. Kamble, M., Shrivastava, P., Jain, M.: Digitized spiral drawing classification for Parkinson's disease diagnosis. Meas.: Sens. **16**, 100047 (2021)
7. Drotár, P., et al.: Decision support framework for Parkinson's disease based on novel handwriting markers. IEEE Trans. Neural Syst. Rehab. Eng. **23**(3), 508–516 (2014)
8. Impedovo, D., Pirlo, G., Vessio, G.: Dynamic handwriting analysis for supporting earlier Parkinson's disease diagnosis. Information **9**(10), 247 (2018)
9. Alissa, M., Lones, M.A., Cosgrove, J., et al.: Parkinson's disease diagnosis using convolutional neural networks and figure-copying tasks. Neural Comput. Applic. **34**, 1433–1453 (2022)
10. Gil-Martín, M., Montero, J.M., San-Segundo, R.: Parkinson's disease detection from drawing movements using convolutional neural networks. Electronics **8**(8), 907 (2019)
11. Kotsavasiloglou, C., Kostikis, N., Hristu-Varsakelis, D., Arnaoutoglou, M.: Machine learning-based classification of simple drawing movements in Parkinson's disease. Biomed. Signal Process. Control **31**, 174–180 (2017)
12. De Stefano, C., Fontanella, F., Impedovo, D., et al.: Handwriting analysis to support neurodegenerative diseases diagnosis: a review. Pattern Recognit. Lett. **121**, 37–45 (2019)
13. San Luciano, M., et al.: Digitized spiral drawing: a possible biomarker for early Parkinson's disease. PLoS ONE **11**(10), e0162799 (2016)
14. Shorten, C., Khoshgoftaar, T.M.: A survey on image data augmentation for deep learning. J. Big Data **6**, 60 (2019)
15. Gazda, M., Hireš, M., Drotár, P.: Multiple-fine-tuned convolutional neural networks for Parkinson's disease diagnosis from offline handwriting. IEEE Trans. Systems Man Cybern.: Syst. **52**(1), 78–89 (2022)
16. Bazgir, O., Frounchi, J., Habibi, S.A.H., Palma, L., Pierleoni, P.: A neural network system for diagnosis and assessment of tremor in Parkinson disease patients. In: 22nd Iranian Conference on Biomedical Engineering (2015)
17. Schwarz, J., et al.: Depression in Parkinson's disease. J. Neurol. **258**(2), 336–338 (2011)
18. Adams, W.R.: High-accuracy detection of early Parkinson's disease using multiple characteristics of finger movement while typing. PLoS ONE **12**, e0188226 (2017)
19. Peter, D., et al.: Analysis of in-air movement in handwriting: a novel marker for Parkinson's disease. Comput. Methods Progr. Biomed. 117(3), 405–411 (December 2014). August 2014
20. Challa, K.N.R., Pagolu, V.S., Panda, G., Majhi, B.: An improved approach for prediction of Parkinson's disease using machine learning techniques (2016)
21. Kamran, I., Naz, S., Razzak, I., Imran, M.: Handwriting dynamics assessment using deep neural network for early identification of Parkinson's disease. Future Gener. Comput. Syst. **117**, 234–244 (2021)

Analysis of Quarantine Norms and Their Healthcare Benefits for Covid-19

Himanshu Jain(✉) and Arvind Kumar Sinha

National Institute of Technology Raipur, Raipur, Chhattisgarh 492001, India
itsjain2905@gmail.com

Abstract. The corona virus, which causes COVID-19 disease, is constantly changing its genetic characteristics, and new variants of the virus are expected to occur as the virus spreads. New variants may become more difficult to stop. Numerous variants of the corona virus that causes COVID-19 are being tracked globally during this pandemic. To break the spread of this virus, people around the globe must practice some prevention methods; one of the most effective methods is to quarantine the infected population. This paper represents the effect of quarantine in India by analysing the mathematical model in the applicable timeframe. This model shows the spread of infection in India with contemporary norms of both restrictions of home quarantine and quarantine facilities provided by the government. We validate the model with the actual data, and statistics calculates the model's accuracy is 91.4% with actual known data. By applying the rough set method to the known real data, we observe that the rough set supports the statistical interference of the data from the mathematical model. Also, this paper shows that the number of active cases in India decreases by applying stricter norms to both home quarantine and quarantine facilities provided by the government.

Keywords: Quarantine · COVID-19 · modelling · rough set

1 Introduction

The novel corona virus is a highly contagious new virus that arose in late 2019 in Wuhan, China, quickly spread throughout the world. COVID-19 is a respiratory disease caused by the corona virus [1]. Like flu, the infection triggers symptoms like cough, fever, and difficulty breathing due to its ability to spread quickly and high fatality rate in more severe cases. WHO announced it as a global pandemic [2]. According to the most recent update on September 30th 2022, there have been 61,47,17,423 confirmed corona virus cases worldwide [3]. In India on October 02, 2022, India had 3375 new corona virus cases, bringing the total number of COVID-19 infections to 4,45,94,487 across the country [3]. This includes 18 further deaths, bringing the total number to 5,28,673 [3]. It is a well-known fact that a virus continuously changes itself via mutation. When the virus mutates and shows a new characteristic compared to the previous version, this mutation is called a variant of the original virus [4]. Over 44 million infected cases of COVID-19 in India have been reported by September 14th, 2022 [3]. Various countries like the USA,

B. K. Singh et al. (Eds.): ICBEST 2023, CCIS 2003, pp. 256–269, 2024.
https://doi.org/10.1007/978-3-031-54547-4_20

United Kingdom, Italy, Germany, Brazil, and India faced most cases in early 2020 and early 2021 caused by two different mutants. After that, the virus infection rate started to decrease in various countries due to strict norms like quarantine and contact tracing of exposed people as WHO and doctors across the world have advised that people can protect themselves by frequently washing hands and avoiding physical contact with people [5]. On November 26th, 2021, a new, more infectious variant of coronavirus named omicron was found in South Africa with more mutation rate [6]. The highly contagious omicron variant has spread worldwide in a brief period. Even though 63.2% of the world population has taken at least a single dose of COVID-19 [7], India is also one of the highly infected countries by the omicron variant.

1.1 Effect of New Variant Omicron in India

The COVID cases and death count stayed low in India in the second half of 2021, giving us a renewed sense of comfort and complacency. However, a large part of the world started experiencing a viral blizzard in November–December of 2021, with the latest Omicron variant defeating the shield of protection from past infection and vaccination [8]. Even in India the people were starting to getting infected by this Omicron variant as two men from Karnataka had tested positive for the Omicron variant of the coronavirus on 2nd of December 2021. They were the country's first cases of the new variant of concern [9]. The numbers started rising sharply in the last few days of December. And in the first two week of January, most states and cities across the country have been reporting an unprecedented surge in new cases. On January 5th 2022, a major populated city in India, Mumbai reported more than 15,000 new cases in 24 h, moving past the peak of 11,163 during the second wave [10]. The omicron variant is indeed leading to many re-infections and breakthrough infections, but fortunately, not an overwhelming number of severe cases so far. This is possibly something you expect in a pandemic, as populations acquire hybrid immunity due to immunization and infection. It is hard to compare the Omi-effect with the Delta effect as the two populations they attacked are quite different in terms of the underlying level of immunity. It is difficult to conclude with certainty whether omicron is truly milder than Delta, given these differences that confound their effects [8]. Rather than inciting an Omicron panic, we must exercise caution and use the tools at our disposal to chart a medium course with effective public health measures. When it comes to preventive steps, we all know that timing is everything. By now, we should be convinced of the virus's unseen presence, which is in active circulation but has yet to be recognized generally. Unfortunately, in today's world, travel limits and testing international travelers will not be enough to avoid it. Even before we realize it, the infection has infiltrated the community. So, we must find something which protect us from the virus and should be more effective way to deal with COVID-19.

We use the rough set theory to for analysis and validation of mathematical model. Pawlak developed the rough set theory in 1982 [11]. The rough set is a field of mathematics that connects the problem of data information to the wider idea of deducing hidden patterns in datasets [12]. The rough set theory has a wide range of applications. For instance, data mining [13–15], medicine [16–18], large data processing [19], and a variety of other sectors. The rough set applications are more important now than they were previously, primarily in the investment database or data mining.

1.2 Effect of Quarantine Norms

The novel coronavirus spreads between people via the droplets produced while sneezing, coughing, or talking. The droplets exhaled by infected people are inhaled into other people's lungs causing new infections. When the droplets fall on surfaces, they can cause virus spread if people touch the contaminated surfaces and then their faces with unwashed hands. While saliva and sputum are significant carriers of the virus, recent research also suggests spreading the disease via air [20]. So, it is well known fact that to stop virus from affecting an individual one should be isolated from the other infectious people. Majorly all the countries follow quarantine as the solution for COVID-19 [21]. The recommended duration of quarantine for Covid-19 based on available information by WHO is up to 14 days from the time of exposure. The purpose of quarantine during the current outbreak is to reduce transmission by Separating contacts of COVID-19 patients from community, Monitoring contacts for development of sign and symptoms of COVID-19, and Segregation of COVID-19 suspects, as early as possible from among other quarantined Persons [22].

Quarantine should continue to be encouraged, according to the WHO. Individuals in quarantine are provided with adequate food, water, protection, hygiene, and communication, as well as access to education for children and paid leave or remote work options from jobs; adequate ventilation and infection prevention and control (IPC) measures are implemented and maintained; and the requirements for monitoring the health of quarantined persons can be met during the quarantine period [22]. But after the vaccination process running across the world CDC now recommends quarantine for 5 days followed by strict mask use for an additional 5 days. Alternatively, if a 5-day quarantine is not feasible, it is imperative that an exposed person wear a well-fitting mask at all times when around others for 10 days after exposure [23].

Government of India also provides quarantine centers for travelers in country and till now it is operated for the peoples [24]. Various guidelines by govt of India is announced time by time [25] for the home isolation and govt quarantine. The effect of this measure taken by government is seen by the graph in the previous variants of COVID-19 from the date it all started in India. The lockdown is the state where all the peoples are quarantined by the orders at their home and during 2020 when government employed the nationwide lockdown on 24th March 2020 [26] after that announcement COVID cases were not that much exaggerated in the country. As compared to when there was no lockdown nationwide [27] but there was red zone circled area defined and announced by Indian government in different part of country where infected people found in more density also many people who are symptomatic to the disease were quarantined in home as well as in government provided quarantine center in this time phase different variant named

Delta variant came in summer of 2021 which causes very high number of death cases and the active cases in country seemed to be uncontrollable at that time phase [28]. Now again when India is suffering with the new variant named omicron there no nationwide lockdown in the country right now also quarantine persons are decreasing in numbers in various states [29]. Due to this ease in restriction the chances of being infected is increasing [29]. Despite the fact that a considerable portion of the adult population has been vaccinated, it is uncertain how long vaccination or natural infection immunity lasts. Furthermore, nothing is known about the omicron variation and how it will affect people in countries where the Delta or omicron variant is prevalent. To protect ourselves from the contagious virus, we should observe the 14-day isolation and quarantine regulations [30]. In late 2021 India was still suffering by delta variant but with less number of severe cases but with the arrival of omicron variant since December 2021 India faced both the variant together. In early December the share of omicron variant to the new cases is about 5% [31]. But as the variant is highly contagious the share of omicron variant in daily cases increases to approx. 95% till 2nd march 2022 [31].

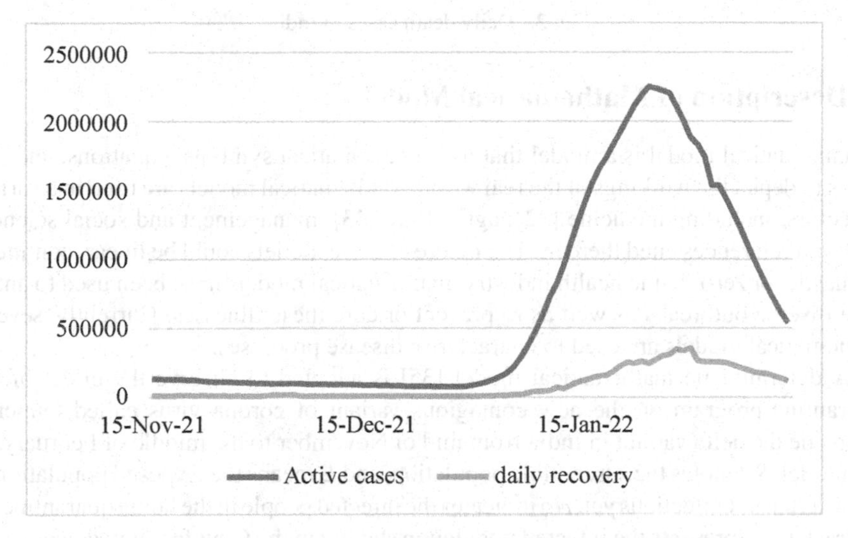

Fig. 1. Daily active cases and daily recovered cases in India.

With both the variant causing daily cases together with increasing share of omicron variant, daily active cases i.e., the total no of daily infectious population and daily recovered cases is shown in Fig. 1. And the graph of daily death case shown in Fig. 2. The data is taken from various official sources from [3].

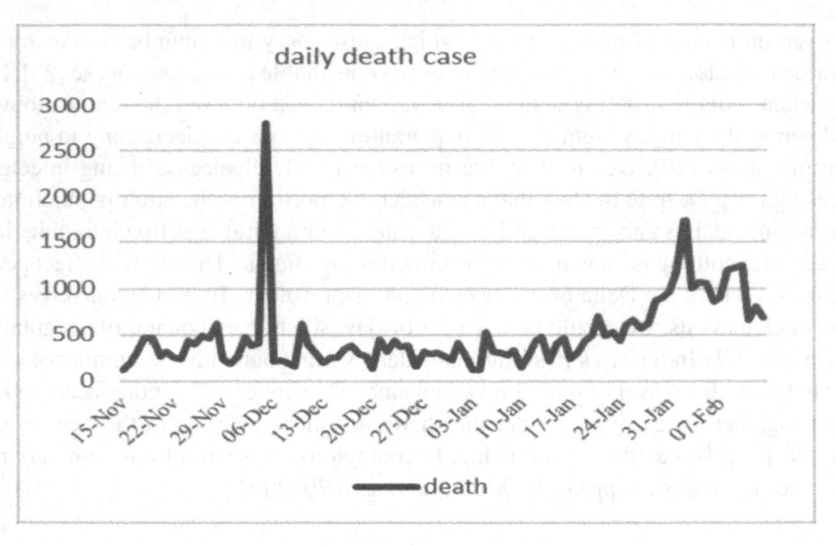

Fig. 2. Daily death cases in India

2 Description of Mathematical Model

A mathematical model is a model that uses mathematical symbols, equations, and formulas to depict the workings of the real world. Mathematical models are used in a variety of sectors, including medicine [32], agriculture [33], management and social sciences [34], and references cited there in. The nature of these models could be linear, nonlinear, stochastic, or zero. In the health industry, mathematical models have been used to anticipate disease outbreaks as well as to prevent or cure these illnesses. Currently, several mathematical models are used to characterise disease processes.

A deterministic mathematical model [35] is adopted to observe the effect of the Quarantine program on the new contagious variant of corona virus called Omicron alongside the delta variant in India from mid of November to the middle of February. In this model, *S* denotes the susceptible population, and *E* means the Exposed population to the virus but not infectious yet. *Hq* indicates the infected people in the home quarantine. In contrast, Gq represents the infected population staying in the Govt facilitated quarantine program. Also, *I* mean the people who are infected and not quarantined or coming from Govt or home quarantine, and R stands for the recovered people. So, the total human population

$$N(t) = S(t) + E(t) + H_q(t) + G_q(t) + I(t) + R(t).$$

The following assumption we have taken as per the recent circumstances of new variant omicron.

(i) Recovered people also treated to be susceptible that means whole population is susceptible whether they are recovered previously by the COVID-19 disease.
(ii) The exposed population is divided into three groups: those who are quarantined at home, those who are quarantined by the government, and those who have been infected. Natural death has also taken a toll on the aforementioned population.

(iii) A portion of the home isolated population becomes infected, while the other becomes recovered. Natural death also has an effect on this group.
(iv) One group of people who were quarantined by the government becomes infected, while the other group recovers. Natural death has a negative impact on this individual.
(v) The recovered class includes a portion of the diseased population. Other people's numbers are decreasing as a result of infection and natural death.

$$\frac{dS}{dt} = A - (q_1 H_q + q_2 G_q)S - \mu_1 S \tag{1}$$

$$\frac{dE}{dt} = (q_1 H_q + q_2 G_q)S - (a_H + a_I + a_G + \mu_1)E \tag{2}$$

$$\frac{dH_q}{dt} = a_H E - h_1 H_q - h_2 H_q - \mu_1 H_q \tag{3}$$

$$\frac{dG_q}{dt} = a_G E - g_1 G_q - g_2 G_q - \mu_1 G_q \tag{4}$$

$$\frac{dI}{dt} = a_I E + h_2 H_q + g_2 G_q - \mu_1 I - \mu_2 I - \theta I \tag{5}$$

$$\frac{dR}{dt} = \theta I - \mu_1 R + h_1 H_q + g_1 G_q \tag{6}$$

With initial condition fitted as per the scenario in 15th November 2021

$$S(0) > 0, \mathrm{E}(0) \geq 0, H_q(0) \geq 0, G_q(0) \geq 0, I(0) \geq 0, R(0) \geq 0;$$

Due to change in scenario in late 2021 the assumption we made and the data we taken are different with respect to model proposed by Pal [35]. The parameters are defined in the Table 1.

Table 1. Specific definition of all the parameters

Parameters	Meaning as per model
A	Recruitment rate of susceptible from changed scenario due to Omicron
q_1	The coefficient of transmission rate from home quarantine to susceptible population
q_2	The coefficient of transmission rate from Govt quarantine to susceptible population
μ_1	The natural death rate of all human epidemiological classes
a_H	The proportion of people who are exposed to Covid-19 who develop mild symptoms (but still have the ability to infect others) and progress to class Hq

(*continued*)

Table 1. (*continued*)

Parameters	Meaning as per model
a_I	The rate at which exposed individuals become infected with Covid 19 and progress to the class I
a_G	the proportion of people who become infected and migrate to class G as a result of their exposure
h_1	The home Quarantined population's recovery rate
h_2	The rate of home quarantined individuals starting to show disease symptoms and migrate to I
g_1	The government quarantined people's recovery rate
g_2	The rate of people migrating to I from Government quarantine as they started to show symptoms
μ_2	The death rate due to disease
θ	The recovery from the infected population

In this following section, we'll run some computer simulations to see if the proposed model is applicable to the recent COVID-19 cases. The simulation is based on data on pandemic infection in India that is currently available. In addition, from a practical standpoint, this numerical simulation is important. We analyzed the proposed Covid-19 system to estimate the model's parameters for India. The major goal is to investigate the effects of two quarantined population parameters, q_1 and q_2, on the pandemic curve using a graphical representation. We examine the behavior of the infected population for 3 months starting on November 15, 2021, by modifying the values of the stated parameters. And a value derived from a variety of sources and matched to current scenario of omicron variant in India. So, we have simulated this model by all the considered parameters taken and fitted as per the situation due to omicron variant alongside delta variant, the taken value and the fitted values are defined in Table 2.

Table 2. Appropriate value of parameters

Parameter	Value taken	Source
A	5000000 day^{-1}	Fitted
q_1	2.7×10^{-10} day^{-1}	Fitted
q_2	1×10^{-10} day^{-1}	Fitted
μ_1	2×10^{-5} day^{-1}	[35]
a_H	0.4 day^{-1}	[35]
a_I	1×10^{-6} day^{-1}	[35]
a_G	0.05 day^{-1}	[35]

(*continued*)

Table 2. (*continued*)

Parameter	Value taken	Source
h_1	0.15 day^{-1}	[35]
h_2	.0028 day^{-1}	[35]
g_1	0.15 day^{-1}	[35]
g_2	2×10^{-3} day^{-1}	[35]
μ_2	0.001 day^{-1}	[35]

The graph of model with fitted parameter is plotted using simulation of model.

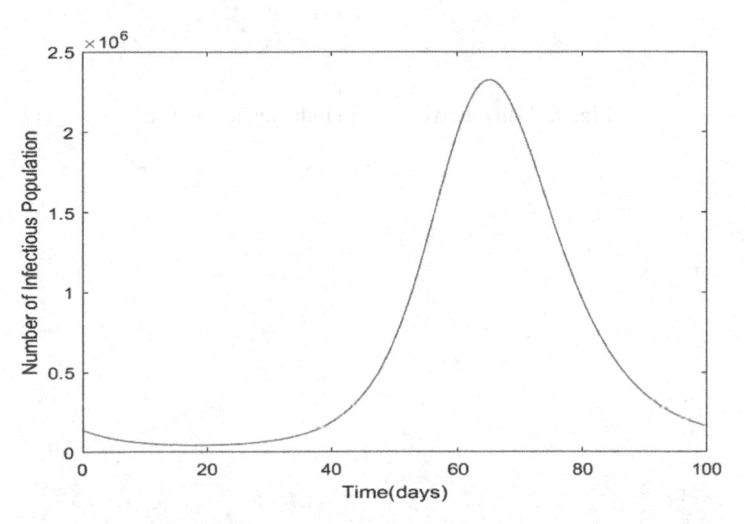

Fig. 3. Daily active cases curve by simulation.

The graph by simulation of mathematical model is shown in Fig. 3 which shows same pattern as the real data of omicron variant, where the graph of real data is shown in Fig. 4. The model's simulated data is then verified by finding error using mean squared error formula and found that the error percentage is 8.6% that means we have 91.4% of accuracy with the real data.

The rough set method is a tool which is used largely for data mining and to find the hidden patterns in data. Here we put all the known data from various sources [3] and using rough set statistics we get the following graph which shows the same pattern as the simulated graph of the mathematical model and with the graph or real data (Fig. 5).

Now, when population from home quarantine as well as government facilized quarantine starts to remains in quarantine strictly the coefficient of transmission rate from home quarantine to susceptible population decreases. That means if we decrease the value of coefficient q_1 and q_2, we are increasing the strictness in the norms of home quarantine and government facilized quarantine peoples, so by using this we tried to

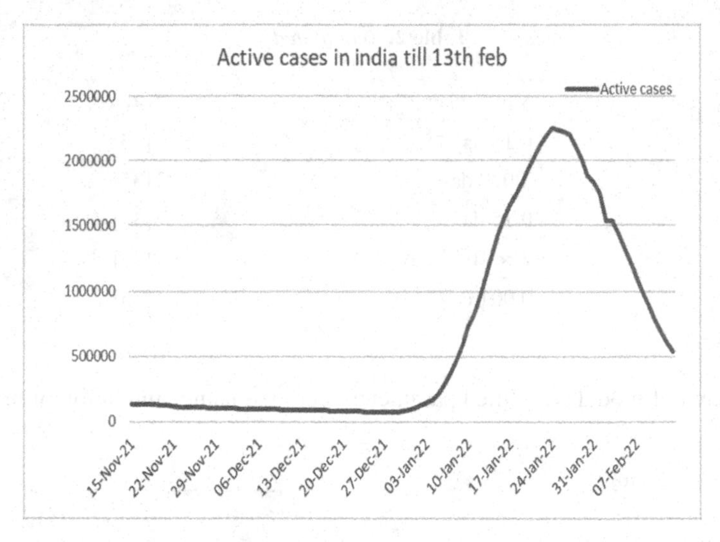

Fig. 4. Daily active cases in India by actual data.

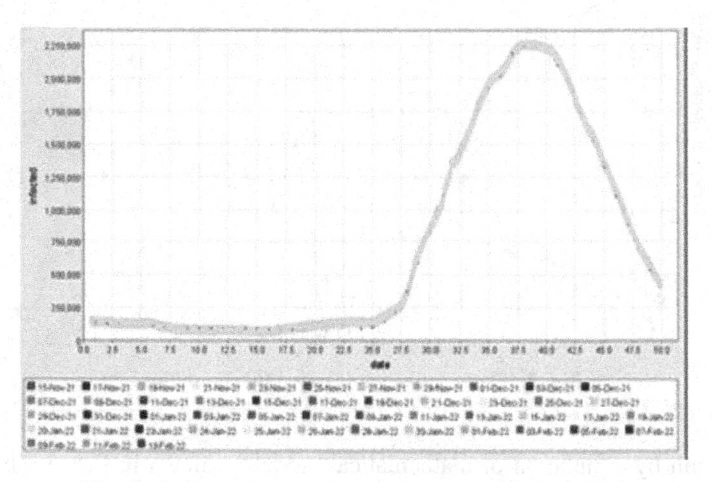

Fig. 5. Daily active case by rough set statistical method.

analyze the situation by simulating the mathematical model and found to have lesser spread of infection if we decrease the value of coefficient q_1 and q_2, we found that the COVID-19 infected cases graph is very flattered as compared to original graph. First, we only reduced the value of coefficient q_1, that means we applied more strictness to home quarantine population which causes the low transmission rate of infection from home quarantine people to susceptible people.

We found that the graph of infected case is flattered. The comparison of both the situation is shown by graph in Fig. 6. Now again if we reduce only the value of coefficient q_2 i.e., we applied more strictness to government facilized quarantine people which cause the low transmission rate of infection from government quarantined people to susceptible

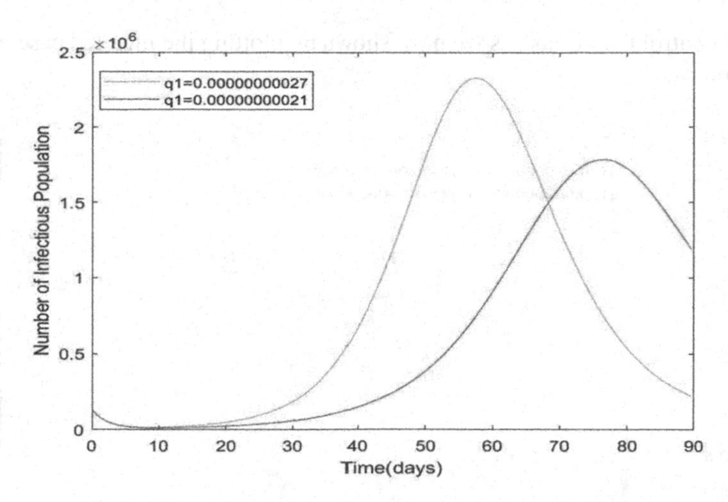

Fig. 6. Effect of increasing the strictness in home quarantine.

peoples. We found that the infection is reduced, Fig. 7 shows the comparison of both the situation and the effect of increase in strictness in quarantine facility provided by government.

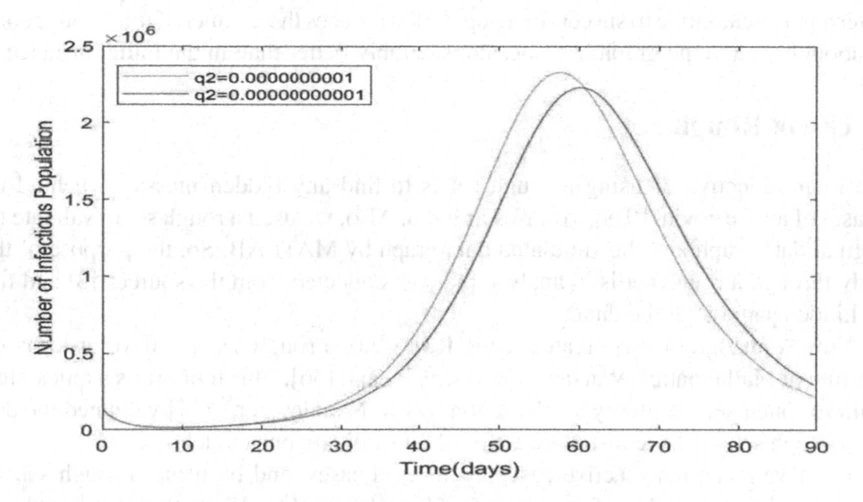

Fig. 7. Effect of increasing the strictness in government quarantine.

If this condition is effective as per model simulation, we tried to show if we apply both the restricted condition together the simulated graph should be flatter as compared with the less strict norms applied to the population so, we simulated model with both strict restrictions together i.e., more strict home quarantine and more strict government quarantine is applied in India then the graph in Fig. 8 is the simulated graph is obtained with both the value of is decreased. So, by simulation, we found that the quarantine can effectively control the disease as we have shown by plotting the infected case with both the situation.

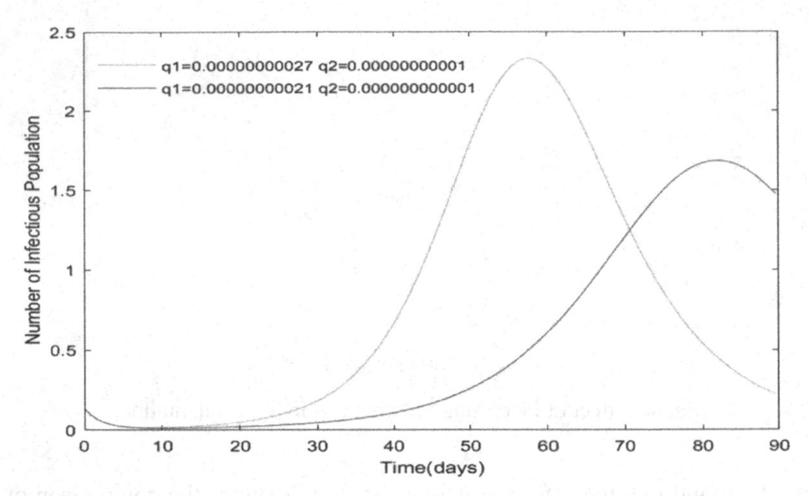

Fig. 8. Effect of strictly applying both government and home quarantine.

As shown in Fig. 8, by lowering the transmission rate from home quarantine and government quarantine to susceptible population reduces the number of infectious people by about half, and the graph becomes considerably flatter than in the initial instance.

3 Use of Rough Set

The main objective of using a rough set is to find any hidden messages behind the dataset of all the covid-19 active cases in India. Also, we used a rough set to validate the original data graph and the simulated data graph by MATLAB. So, the purpose of this study through a rough set is to analyse the data collected from the sources [3] and find the hidden patterns in the data.

Now A histogram was created using RSES 2.0, a rough set tool developed by the Institute of Mathematics, Warsaw University, Poland [36]. This tool offers a quick algorithm to rough set the theory's calculation. As N. Namdev et al. [37] validated the data using rough set, we have also used RSES 2.0 to validate our model.

We have taken daily active cases as infected cases, and by using a rough set, we found that the mean value of the attribute infected is 654664.473 with a standard deviation 752,638, taking characteristics ranging from the minimum value of 75456 on 28th December 2021 to maximum value of 2249335 on 24th January 2022 [3]. The histogram of daily active cases of COVID-19 in India is shown in Fig. 9, which states that the distribution of active cases of COVID19 in India is not as uniform as shown in the histogram in Fig. 9.

We find the mean value 839556.901 and standard deviation 851621 of simulated curve, which lies in the acceptable region. Therefore, the rough set's statistical interference gave the model validation.

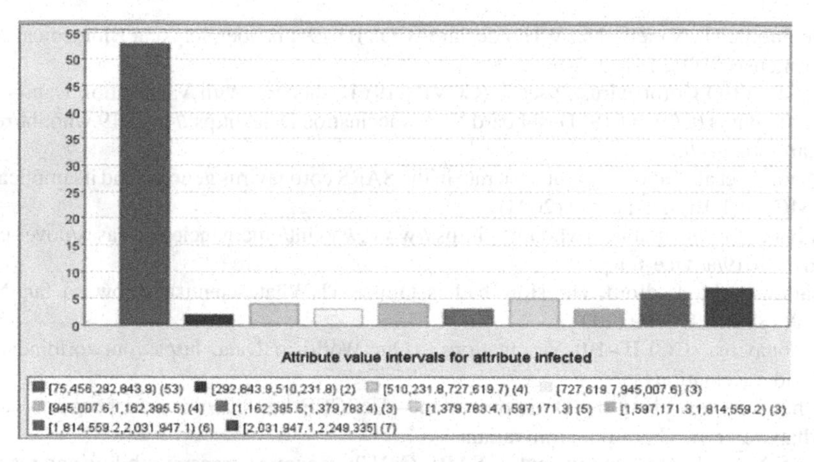

Fig. 9. The histogram of daily active cases.

4 Conclusion

This paper represented the role of quarantine norms in India, which applies to the current variant of corona virus that causes a faster spread of infection. We analyzed a mathematical model showing the effect of the home quarantine and government facilized quarantine in India. We faced the new variant of omicron and dealt with some norms and regulations. We have taken the applicable timeframe from where the Omicron variant started up to its daily infection declined sharply in India. We used the statistical method to find an error in the simulated data of the mathematical model and found the mathematical model to be accurate by 91.4% for presenting the impact of quarantine norms in the country. We showed that with the stricter norms of home quarantine and government facilized quarantine, the cases might reduce to half of their current value with the help of the mathematical model. Finally, we used the rough set method to analyze the recent COVID-19 cases from simulated data of active patients and validated the model by statistical inference.

The paper focused on corona virus transmission in India through public and administration efforts and current norms and regulations. This analysis can be applied to the actual situation in the country and can be effective in dealing better with the future variants of corona virus. The model supports imposing further limitation measures to eradicate the expanse of Covid-19.

Acknowledgement. The authors are grateful to the National Institute of Technology in Raipur (Chhattisgarh), India, for providing the facilities, space, and opportunity to work.

References

1. Liu, Y.C., Kuo, R.L., Shih, S.R.: COVID-19: the first documented coronavirus pandemic in history. Biomed. J. **43**(4), 328–333 (2020)

2. Cucinotta, D., Vanelli, M.: WHO declares COVID-19 a pandemic. Acta Biomedica: Atenei Parmensis **91**(1), 157 (2020)
3. India: WHO Coronavirus Disease (COVID-19) Dashboard With Vaccination Data—WHO Coronavirus (COVID-19) Dashboard With Vaccination Data. https://covid19.who.int/region/searo/country/in
4. Zhao, Z., et al.: Moderate mutation rate in the SARS coronavirus genome and its implications. BMC Evol. Biol. **4**(1), 1–9 (2004)
5. Advice for the public (who.int). https://www.who.int/emergencies/diseases/novel-coronavirus-2019/advice-for-public
6. Callaway, E., Ledford, H.: How bad is Omicron? What scientists know so far. Nature **600**(7888), 197–199 (2021)
7. Coronavirus (COVID-19) Vaccinations - Our World in Data. https://ourworldindata.org/covid-vaccinations
8. What we know about the Omicron variant—UNICEF. https://www.unicef.org/coronavirus/what-we-know-about-omicron-variant
9. Araf, Y., et al.: Omicron variant of SARS-CoV-2: genomics, transmissibility, and responses to current COVID-19 vaccines. J. Med. Virol. **94**, 1825–1832 (2022)
10. COVID-19 Dashboard by Government of Maharashtra. https://www.covid19maharashtragov.in/mh-covid/dashboard-?trenddis=Mumbai
11. Pawlak, Z.: Rough sets. Int. J. Comput. Inf. Sci. **11**, 341–356 (1982)
12. Polkowski, L.: Rough Sets. Mathematical Foundations. Springer, Berlin (2002). https://doi.org/10.1007/978-3-7908-1776-8
13. Chen, H., Li, T., Luo, C., Horng, S., Wang, G.: A decision-theoretic rough set approach for dynamic data mining. IEEE Trans. Fuzzy Syst. **23**(6), 1958–1970 (2015)
14. Grzymala-Busse, J.: Rough set theory with applications to data mining. In: Gh. Negoita, M., Reusch, B. (eds.) Real World Applications of Computational Intelligence. STUDFUZZ, vol. 179, pp. 221–244. Springer, Heidelberg (2005). https://doi.org/10.1007/11364160_7
15. Kumar, M., Yadav, N.: Fuzzy rough sets and its application in data mining field. Adv. Comput. Sci. Inf. Technol. **2**(3), 237–240 (2015)
16. Paszek, P., Wakulicz–Deja, A.: Applying rough set theory to medical diagnosing. In: Kryszkiewicz, M., Peters, J.F., Rybinski, H., Skowron, A. (eds.) RSEISP 2007. LNCS, vol. 4585, pp. 427–435. Springer, Heidelberg (2007). https://doi.org/10.1007/978-3-540-73451-2_45
17. Durairaj, M., Sathyavathi, T.: Applying rough set theory for medical informatics data analysis. Int. J. Sci. Res. Comput. Sci. Eng. **1**(5), 1–8 (2013)
18. Øhrn, A.: Discernibility and rough sets in medicine: tools and applications. Norwegian University of Science and Technology Trondheim, Doctor of Engineering Thesis 999:133 (1999)
19. Slezak, D.: Rough Sets and few-objects-many-attributes problem: the case study of analysis of gene expression data sets. In: Frontiers in the Convergence of Bioscience and Information Technologies, pp. 437–442 (2007)
20. Scientific Brief: SARS-CoV-2 Transmission—CDC. https://www.cdc.gov/coronavirus/2019-ncov/science/science-briefs/sars-cov-2-transmission.html
21. Nussbaumer, B., et al.: Quarantine alone or in combination with other public health measures to control COVID-19: a rapid review. Cochrane Database Syst. Rev. (9), Article no. CD013574 (2020)
22. Considerations for quarantine of contacts of COVID-19 cases (who.int). https://www.who.int/publications/i/item/WHO-2019-nCoV-IHR-Quarantine-2021.1
23. CDC Updates and Shortens Recommended Isolation and Quarantine Period for General Population—CDC Online Newsroom—CDC. https://www.cdc.gov/media/releases/2021/s1227-isolation-quarantine-guidance.html

24. Chaurasiya, P., Pandey, P., Rajak, U., Dhakar, K., Verma, M., Verma, T.: Epidemic and challenges of coronavirus disease-2019 (COVID-19): India response (2020). SSRN 3569665
25. Guidelines for Quarantine facilities (mohfw.gov.in). https://www.mohfw.gov.in/pdf/905426 53311584546120quartineguidelines.pdf
26. Amerta, G., Arora, B., Gupta, R., Anoop, S., Misra, A.: Effects of nationwide lockdown during COVID-19 epidemic on lifestyle and other medical issues of patients with type 2 diabetes in north India. Diab. Metab. Syndr.: Clin. Res. Rev. **14**(5), 917–920 (2020)
27. Chauhan, A., Singh, R.P.: Decline in PM2.5 concentrations over major cities around the world associated with COVID-19. Environ. Res. **187**, 109634 (2020)
28. Roy, I.: Exit Strategy from COVID-19: Vaccination and Alternate Solution. In: Rojas, I., Castillo-Secilla, D., Herrera, L.J., Pomares, H. (eds.) BIOMESIP 2021. LNCS, vol. 12940, pp. 444–459. Springer, Cham (2021). https://doi.org/10.1007/978-3-030-88163-4_38
29. Balasankar, G., Al Jumai, A., Fong, K.N.K., Prasad, P., Kumar, M.S., Kai-Yu, T.R.: Impact of coronavirus disease 2019 (COVID-19) outbreak quarantine, isolation, and lockdown policies on mental health and suicide. Front. Psych. **12**, 565190 (2021)
30. Coronavirus Quarantine: CDC cuts down isolation and quarantine time. https://www.cdc.gov/media/releases/2021/s1227-isolation-quarantine-guidance.html
31. COVID-19 Data Explorer - Our World in Data. https://ourworldindata.org/explorers/coronavirus-data-explorer
32. Pack, A., Murray-Smith, D.: Mathematical models and their applications in medicine. Scott. Med. J. **17**(12), 12401–12409 (1972)
33. Dourado-Neto, D., Teruel, D., Reichardt, K., Nielsen, D., Frizzone, J., Bacchi, O.: Principles of crop modeling and simulation: I. uses of mathematical models in agricultural science. Scientia Agricola **55**(SPE), 46–50 (1998)
34. Pokhariyal, G., Rodrigues, A.: An accurate epidemiological model. Appl. Math. Comput. **53**(1), 1–12 (1993)
35. Pal, D., Ghosh, D., Santra, P.K., Mahapatra, G.S.: Mathematical modeling and analysis of Covid-19 infection spreads with restricted optimal treatment of disease incidence. Biomath **10**(1), 1–20 (2021)
36. Bazan, J.G., Szczuka, M.: The rough set exploration system. In: Peters, J.F., Skowron, A. (eds.) Transactions on Rough Sets III. LNCS, vol. 3400, pp. 37–56. Springer, Heidelberg (2005). https://doi.org/10.1007/11427834_2
37. Sinha, A.K., Namdev, N., Shende, P.: Mathematical modeling of the outbreak of COVID-19. Netw. Model. Anal. Health Inform. Bioinform. **11**(1), 1–9 (2022)

A Deep Learning-Based Sentiment Classification Approach for Detecting Suicidal Ideation on Social Media Posts

Pabbisetty Sai Venkata Tarun Kumar, Dilip Singh Sisodia(✉), and Rahul Shrivastava

Department of Computer Science and Engineering, National Institute of Technology Raipur, Raipur, India
{dssisodia.cs,rshrivastava.phd2018.cs}@nitrr.ac.in

Abstract. The identification of suicidal ideation in a person is known as Suicide Ideation Detection (SID). Nowadays, people share their thoughts and sentiments through social platforms like Facebook, Twitter, and Reddit rather than face-to-face. Recently various machine learning and deep learning models have been applied to the SID task. However, these methods lack in learning the user's intent through social media text by analyzing the semantic meaning and context of the textual post. Additionally, recent methods have employed Convolutional Neural Networks (CNN) and Long Short-Term Memory (LSTM)-based deep learning models for learning users' suicidal intent. But these methods still lack in investigating the user's social media interaction. Hence this work first analyzes the semantics and context of the text and obtains the polarity score of the user's social media posts. Next, this study analyzes the social media interaction between users by incorporating the attention mechanism. We apply Convolutional Neural Networks (CNN), Long Short-Term Memory (LSTM), and LSTM with an attention mechanism to classify the suicidal and non-suicidal intent of the users. We used the dataset from Kaggle, the Suicide and Depression Detection dataset that contains the posts from Reddit (a social media platform). The necessary textual data pre-processing steps, such as removing accent characters, and emojis, fixing the word length, spelling corrections, and lemmatization, are applied. The CNN, LSTM, and LSTM with an Attention mechanism are applied with the Sigmoid function as the activation function and binary cross entropy (BCE) as the loss function. We implemented four convolution layers with different filter sizes in CNN Model, two hidden layers in LSTM Model, and one hidden layer in LSTM along with an attention layer. Precision, accuracy, f1-score, and recall are used as evaluation measures in order to assess the suggested model's competence. The proposed model is compared with the baseline models of other authors. The experimental and evaluation findings indicate the superior performance of the proposed LSTM model, which achieves the highest accuracy of the 93%.

Keywords: Suicide Ideation Detection · Deep Learning models · CNN · LSTM · Word2Vec Embeddings

B. K. Singh et al. (Eds.): ICBEST 2023, CCIS 2003, pp. 270–283, 2024.
https://doi.org/10.1007/978-3-031-54547-4_21

1 Introduction

In today's world, suicide is a serious problem. Suicidal Ideation Detection (SID) discusses the clinical strategies utilizing deep learning for automatic detection based on online social content and social work or other specialists interacting with the targeted persons [1]. Suicidal ideation detection (SID) determines whether a person has suicidal thoughts or ideas by analyzing tabular data on a person or text that a person submitted.

Numerous programs have been created to prevent suicide. Even still, some are reluctant to disclose their identities and are not willing to ask for assistance because they believe it will bring them into disrepute in society [2]. Social media platforms allow for the anonymous communication of oneself while avoiding direct interaction with others [3, 4]. This work seeks to make use of a massive amount of data to locate texts with suicidal ideation and aid in the early diagnosis of suicide [5] by Utilizing machine learning and deep learning techniques.

More than 98% of teenagers use the internet to stay in touch with their friends and family [6, 7]. Youths are more prone to talk about suicidal ideas online since they frequently use social media and digital technology. Consequently, developing state-of-the-art Natural Language Processing (NLP) techniques-based big data systems for extracting suicidal intent from social media data is crucial [2, 8].

The classical Machine Learning models (ML) for classification can be used by changing their hyperparameters. However, to successfully use machine learning to predict whether a post has suicidal intent.

We also need to evaluate whether the person has suicidal intent, which is very difficult to do with classical models such as Logistic Regression, SVM, and Nave Bayes [9]. To completely solve the problem, we need to go for other classification algorithms. Deep Neural Networks like CNN and LSTM can be used to solve the problem completely since they consider the meaning of the sentence and then classify the post. CNN and LSTM use hidden layers to extract the useful part and use them for further layers and then predict whether the post-suicidal intent post or not [10] by applying an attention mechanism [11].

In the previous models, they haven't analyzed the semantics of the sentiment and they haven't considered the user's social media interactions with other users to classify the textual content as suicidal or non-suicidal. Hence in our model, we are going to analyze the semantics and sentiment and the user's social media interactions. The purpose of this research is to detect suicidal intent in social media posts. This study has the following key contributions:

- This study learns the user's intent by learning the textual social media post's lexical significance, semantics, and context.
- This work analyses the social interaction between users by incorporating the attention mechanism.
- We further classify the user's suicidal and non-suicidal intent using CNN, LSTM, and LSTM with an attention mechanism.
- This study validates the model performance using accuracy, precision, recall, and F-1 measure, and evaluation findings are also compared with baseline methods.

The remaining part of the article proceeds as follows: Sect. 2 reviews the literature, Sect. 3 details the comprehensive methodology used for this study, and Sect. 4 presents the results of simulations and evaluations and the discussions and implications of the findings. Finally, the conclusion in Sect. 5 outlines the findings and highlights the study's implications.

2 Related Work

The use of social media has been rapidly expanding in recent years. While some social media sites are forums where users can publish updates about their everyday life and private lives, other social media sites are private messaging services. People have a strong desire to submit their updates every day, and they continue to use them, and their addiction has reached an extreme level [2].

One of the most serious mental health concerns is suicidality. Early detection might be extremely helpful for the individual who is afflicted. Instead of speaking with a person directly, people use internet channels to express themselves [12, 13]. Therefore, having their content checked may be advantageous for the individual. To determine a person's mental state, they used supervised learning to monitor their online material while keeping an eye on their language choices and topic descriptions. It is also considered whether the post's thoughts are favorable or negative. They have employed two neural networks, and four [14–16] supervised classifiers to detect suicidal ideation. Additionally, they have compared six classifiers using topic characteristics, word embedding, statistical, syntactical, and linguistic feature extractions [17, 18].

In recent years, several strategies for combating suicidal ideation have been developed, most of which are based on machine learning and deep learning. The study of SID in Social Media Forums is studied using Deep Learning [9], n-gram analysis, the LSTM-CNN model, and other machine learning approaches. The suicide and Depression Detection Dataset collated the data from Reddit social media, including suicidal and non-suicidal messages. This method employs hyper-parameter tuning via 10-fold cross-validation and can spot suicidal ideation in social media. The proposed model combines LSTM and CNN characteristics to categorize text data. A word embedding layer can generate fixed-length vectors, a dropout layer to prevent over-fitting [19, 20], an LSTM layer for extraction of features, a max-pooling, a flattening layer, and a Soft-Max function to complete the neural network comprise the design. Metrics such as Precision, F1-Score, Accuracy, and Recall are used to evaluate the baseline model [2]. Orthodox suicidal ideation entails interaction between medical personnel and patients. Doctors utilize a series of questions to assess patients' mental health, and the responses reflect the patient's feelings and emotions [17]. To detect suicidal ideation [19, 21, 22] devised a method employing number of fusion, and extracted a language feature set. They presented a rudimentary feature-set-based classification strategy in order to achieve their desired outcomes. Support Vector Machine (SVM) and Decision tree are two classification algorithms that are employed. They beat the data-driven method in terms of performance when employing the MFWF methodology [4]. Table 1 compares the different models using specific column attributes.

Table 1. Literature Review

S.No	Model	Analyzing semantics of sentiment	Investigating user's social interaction
1	SID in Social Media Forums [2]	Yes	No
2	SID using ML methods [1]	No	No
3	Suicidal Behaviour Detection on Twitter using NN [17]	No	No
4	Investigating SID in Twitter [8]	No	Yes
5	SID on social media Using Social Interaction and Post Content [12]	No	Yes
6	Proposed Model	Yes	Yes

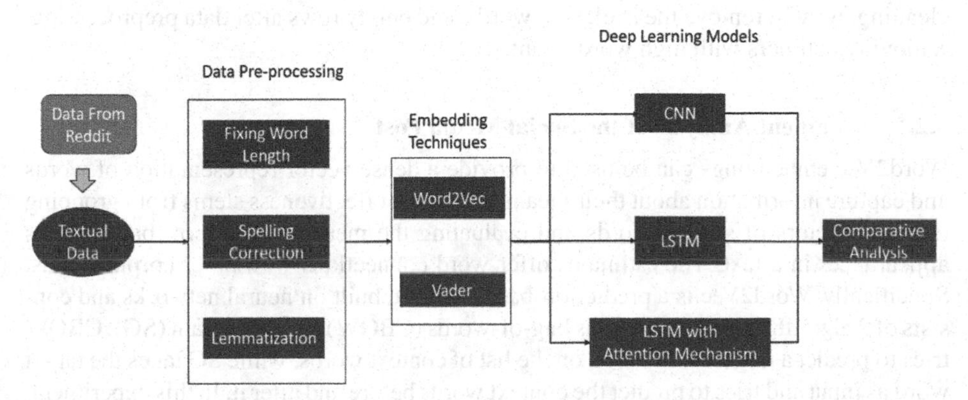

Fig. 1. Architecture of the Suicidal Ideation Detection

3 Methodology

This chapter contains details about the proposed architecture, the steps taken in pre-processing the dataset, various methods used to classify the dataset, and the evaluation metrics.

3.1 Proposed Architecture

The process flow of our project is represented in the form of a diagram in Fig. 1. Because social media data is typically less structured, it requires more specialized preparation and cleaning techniques. Data preprocessing is needed to transform the text data into appropriate formats for the accompanying model building. The data is divided into a training set (80%) and a test set (20%), respectively. The training dataset is used for fitting the model, and the model accuracy is checked on the testing dataset.

We collected the dataset from Kaggle, Suicide and Depression Detection dataset. This dataset has information regarding the posts from Reddit (a social media platform). First, we will do data preprocessing on the dataset and use word2vec word embeddings and Vader classification. Then we will implement a machine learning model (logistic regression) and deep learning models (Convolutional Neural Networks (CNN) and Long Short-Term Memory (LSTM)) along with an attention mechanism for the classification of the text to check whether the post contains suicidal intent or not. Finally, a comparative analysis is conducted. The architecture of the proposed model is shown in Fig. 1.

3.2 Data Preprocessing

Because social media data is typically less structured, it requires more specialized preparation and cleaning techniques. Text processing involves the following methods: removing the accented letters, converting the sentence into lowercase, fixing the word length, expanding short forms, removing emojis, special characters, URLs, spelling correction, and lemmatization. Lemmatization is converting the words to their root word [23]. In data cleaning, we will remove the irrelevant words, and empty rows after data preprocessing, removing outliners with high word count.

3.2.1 Sentiment Analysis of the Social Media Post

Word2Vec embeddings can be used to provide a dense vector representation of words and capture information about their meaning [24]. Its effectiveness stems from grouping together vectors of similar words and evaluating the meaning of a term based on its appearances in a text. The estimates offer word connections with other corpus words. Specifically, Word2Vec is a prediction-based method built on neural networks and consists of 2 algorithms: (1) continuous bag-of-words (CBOW) and skip-gram (SG). CBOW tries to predict a target word based on the list of context words, while SG takes the target word as input and tries to predict the context words before and after it. In this experiment, we will use a continuous bag of words [25, 26].

To train the Word2Vec embeddings, we have set a minimum count of 2, an embedding dimension of 300, and a context window of 10. The minimum count means that the model will ignore words with one occurrence to prevent overfitting the trained embedding [27]. Valence Aware Dictionary and sEntiment Reasoner, also known as Vader, is a rule-based sentiment analysis tool. VADER makes use of a variety of. A sentiment lexicon is a set of words or other lexical items that are assigned good or negative emotional connotations. VADER gives us not just the emotional weight of each Score but also its positivity and negativity [28].

3.3 Classification Using Deep Learning Model: Convolution Neural Network (CNN)

This study attempts to categorize text data; the order of words affects how a sentence is interpreted. We used a number of deep learning methods to analyze the sequences in our training data because the Logistic Regression model had previously failed to capture this feature.

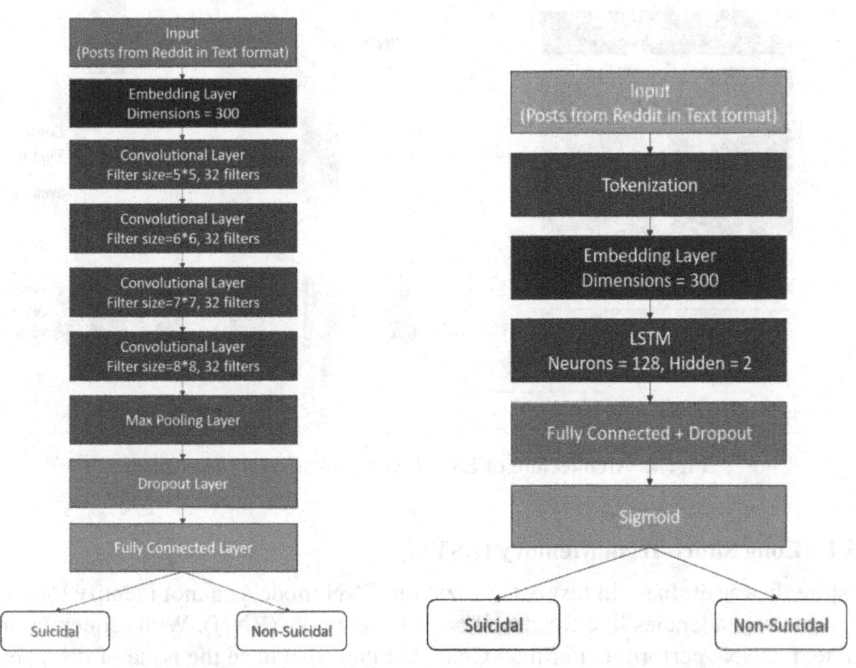

Fig. 2. Model Architecture of CNN **Fig. 3.** Model Architecture of LSTM

As it may produce respectable prediction accuracy using fewer computer resources, the Convolutional Neural Network (CNN) was suggested as an effective method of classifying text input [29]. Figure 2 presents the model architecture for CNN. Our CNN model architecture consists of the following layers: an embedding layer, four convolutional layers, a pooling layer, a dropout layer, and a fully connected layer, as shown in below Fig. 3. First, we tokenized our text data, converting each word into an integer. The embedding layer converts each word of the phrase in our training data into a feature vector with a given embedding size. Word vectors are represented in a semantically defined space by an embedding layer, with words with similar meanings clustered together and words with different meanings scattered apart. The input data is subjected to convolutional layers to extract contextual information. While later convolutional layers are responsible for capturing essential traits and extracting feelings that may influence categorization, earlier convolutional layers are expected to capture simple contextual information. Because we want our model to focus on the most important portions of the text regardless of word order, we employed the max-pooling method after the convolutional layers. Next, as a sort of regularisation, we incorporated a dropout layer in our CNN model to prevent overfitting. In the last layer, which is a completely connected layer, all input and output neurons are connected. The vector will pass through this layer, and the output will be classified as suicidal or non-suicidal using a sigmoid activation function.

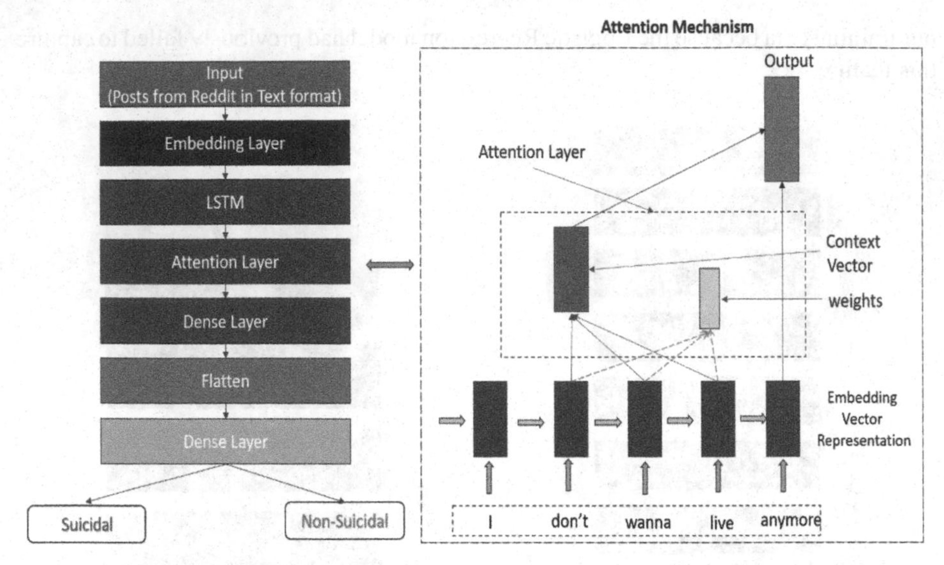

Fig. 4. Architecture of LSTM with an Attention Mechanism

3.3.1 Long Short-Term Memory (LSTM)

Despite their usefulness in text categorization, CNN models cannot identify long range semantic dependencies like Recurrent Neural Networks (RNN). With sequential input, like text, RNNs perform better than CNN, but they also have the issue of disappearing gradients. As a result, we have decided to create an RNN model called a Long Short-Term Memory Network (LSTM) that addresses the drawbacks of RNNs [30]. LSTM can process longer sequences and longer dependencies within a sequence. Figure 3 represents the model architecture of LSTM.

More significantly, LSTM solves the vanishing gradient issue that RNNs have by using a memory cell in the LSTM layer to keep track of values over time. This is accomplished by using a gating mechanism that regulates the flow of information into and out of the cell using input, output, and forget gates. The five layers of our LSTM network architecture are depicted in above Fig. 4 above. The integer word tokens after tokenization are transformed into embedding vectors in the first layer, which is called the embedding layer. Our LSTM layer, the second layer, regulates the flow of information. The sigmoid activation function, which is present in the fourth layer, determines if the sentence is suicidal or not.

3.3.2 LSTM with Attention Layer

Along with the LSTM, we will add an attention layer to the model. The attention layer is utilized to focus on several significant aspects of the input sequence. With the help of specific activation weights, the layer increases the weight of regions we want to amplify and decreases the weight of regions we want to diminish. The layer uses the output from the LSTM layer and then magnifies or diminishes the value of each feature of the input based on those activation weights [11]. The architecture is depicted below in Fig. 4.

The word2vec embedding and VADER generate the word embedding vectors $V = \{v_1, v_2, \ldots v_n\} \in \mathbb{R}^{1*300}$. The size of the embedding vector is 300. Next, the textual social post represented using the V is analyzed for the user's intent using the LSTM. Equation (1) describes the output of the LSTM.

$$\Theta_i = LSTM\,(v_i, \Theta_{i-1}) \tag{1}$$

where Θ_{i-1} represents the output at state i. Finally, the output of the LSTM is described using $O = \{\Theta_1, \Theta_2, \ldots \Theta_n\}$. Further, the attention mechanism is incorporated to investigate the social media interaction of the user for analyzing the user's suicidal or non-suicidal intent. Equations (2) and (3) describe the attention mechanism steps. We first compute the attention (α) over the vector output of the LSTM with initialized weight (W_i) and bias (B_i). Lastly, the final output of the attention mechanism (O') is obtained using Eq. (3).

$$\alpha = OW_i + B_i \tag{2}$$

$$O' = \alpha^T O \tag{3}$$

Figure 4 illustrates the proposed architecture of LSTM with an attention mechanism. The computational complexity of the LSTM with attention mechanism is evaluated by individually analyzing the LSTM and Attention network's computational complexity. The Time Complexity (TC) of the proposed model is mainly affected by the size of the output layer (Θ), hidden layer (H), and memory units (M). So the TC of the LSTM in the proposed model is the $O(\Theta * H + \Theta * S(M))$ which is the aggregated weight updation in hidden layers and output layers for the specified size of the memory cell ($S(M)$). Similarly, the required time by the attention mechanism is $O(\Theta * H)$ which represents the weight updations in hidden layers and the output layer. So, the cumulative TC of the proposed LSTM with Attention mechanism is $O(\Theta * H + \Theta * S(M)) + O(\Theta * H)$.

3.4 Evaluation Metrics

I want to utilize a variety of evaluation metrics to compare the effectiveness of various algorithms with my suggested text classification approach. The evaluation metrics are Accuracy, Recall, Precision, and F1 Score. Figure 5 shows the pictorial representation of the confusion matrix where TP is True Positive, FP is False Positive, FN is False Negative, and TN is True Negative.

Accuracy: A model's accuracy is expressed as the proportion of properly predicted observations to all observed observations. We may verify the result using Eq. (4).

$$Accuracy = \frac{TP + TN}{TP + FP + FN + TN} \tag{4}$$

Precision: Precision is characterized as the measure of accurately positive observations that a model predicts to all positively anticipated observations. Using Eq. (5) below, we can get the precision.

$$Precision = \frac{TP}{TP + FP} \tag{5}$$

Recall (Sensitivity): Recall is calculated as the proportion of accurately positive observations that a model predicts to all actual positive observations. Equation (6) below can be used to compute the recall.

$$Recall = \frac{TP}{TP + FN} \tag{6}$$

F1 Score: The composite Score of recall and accuracy is known as the F1 Score. Equation (7) may be used to compute the F1 Score.

$$F1\ Score = \frac{2 \times Recall \times Precision}{Recall + Precision} \tag{7}$$

4 Experimental Results and Discussion

In this chapter, we will evaluate the performance of different models we built in classifying the text to check whether it has suicidal intent or not. The three models are developed: Convolutional Neural Networks (CNN) and Long Short-Term Memory (LSTM), and LSTM with an Attention mechanism. In this experiment, we have used symspellpy as the word checker to correct the spellings of the words in the Reddit posts. We used lemmatization to change the words into their root words by considering the usage of the word in the sentence. We implemented the experiment by utilizing various libraries: pandas, NumPy, skearn, genism, torch, TensorFlow, and many more.

We have split the dataset into 8:2 ratios to train and test the data, respectively. The dataset is preprocessed, cleaned, and then sent for the training and testing phase of the experiment.

4.1 Dataset Description

We collected the dataset from Kaggle Suicide and Depression Detection dataset. This dataset has information regarding the posts from Reddit (a social media platform). The dataset mainly consists of 2 columns, "text" and "class." Here the text indicates the content of the social media post, and the class signifies the label of the respective posts. The dataset is a collection of 2,32,074 posts from the "SuicideWatch," "depression," and "teenagers" subreddits of the Reddit platform. SuicideWatch posts were classified as suicidal, while the posts gathered from teenagers were classified as non-suicidal.

4.2 Competing Models

The baseline models for the comparative study are taken from machine learning and deep learning models. The machine learning models are SVM [4], XGBoost [2], and logit (Logistic regression), and the deep learning models are CNN [4] and LSTM [4]. Comparison with baseline models is represented below.

Table 2. Comparison of performance of different Models

Models	Accuracy	Precision	Recall	F1 Score
Convolutional Neural Networks	0.9233	0.9004	0.9116	0.9059
Long Short-Term Memory	0.9240	0.8643	0.9319	0.8968
LSTM with Attention Mechanism	0.9315	0.8724	0.9424	0.9060

4.3 Results Discussion

The below table, Table 2, summarizes the key results of the models. For Logistic Regression, Convolutional Neural Networks, and Long Short-Term Memory, the best results are obtained by using the custom Word2vec Embeddings. The deep learning models outperformed the machine learning model. Since we have taken the accuracy score as our metric, the Long Short-Term Memory (LSTM) has outperformed other models.

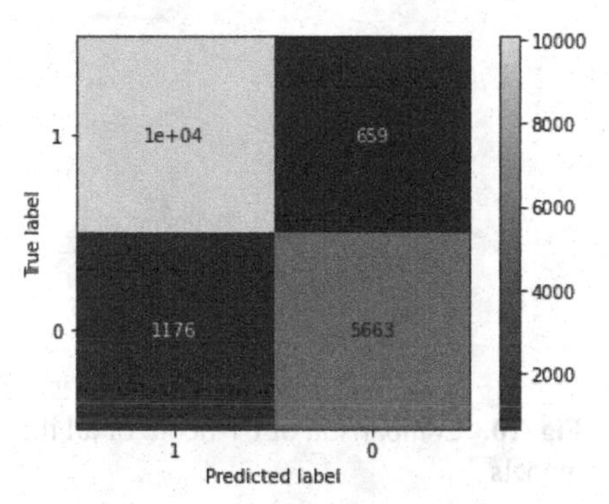

Fig. 5. Confusion Matrix for CNN

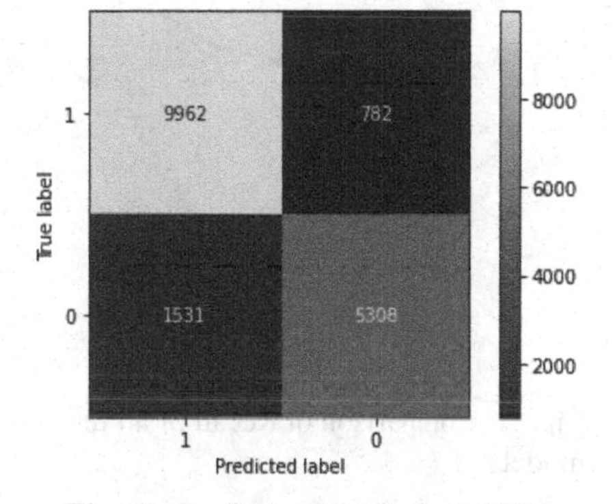

Fig. 6. Confusion Matrix for LSTM

Figure 5 above shows the confusion matrix for CNN Model when we train it using random initialization, and it has a precision of 89% and an F-1 Score of 89%. Figure 6 above shows the confusion matrix for LSTM Model when we train it using random initialization, and it has a precision of 86% and an F-1 Score of 86%. Figure 7 shows the comparison of the accuracy of different models and the proposed model on the Suicide and Depression Detection dataset. The proposed model's accuracy improved more than the baseline model. Figure 8 shows the comparison of the precision of different models and the proposed model on the Suicide and Depression Detection dataset. The precision of proposed models fluctuates since precisions take true positive and true negative into consideration, but we focused mainly on detecting the suicidal ideation, which is positive.

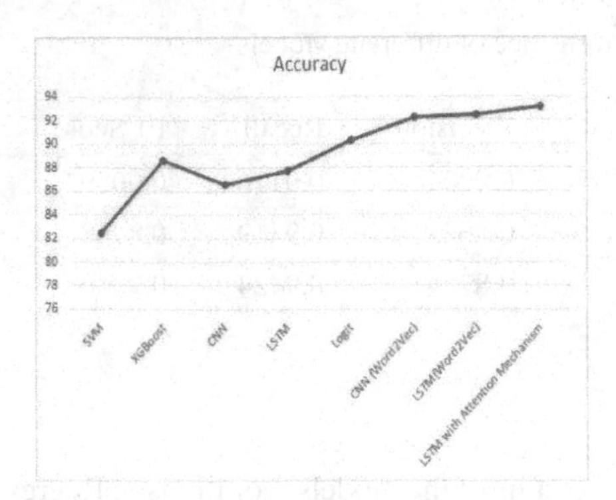

Fig. 7. Comparison of accuracies of all the models

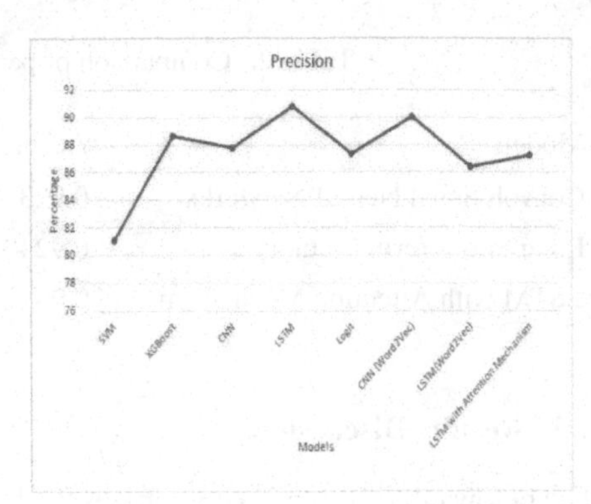

Fig. 8. Comparison of precisions of all the models

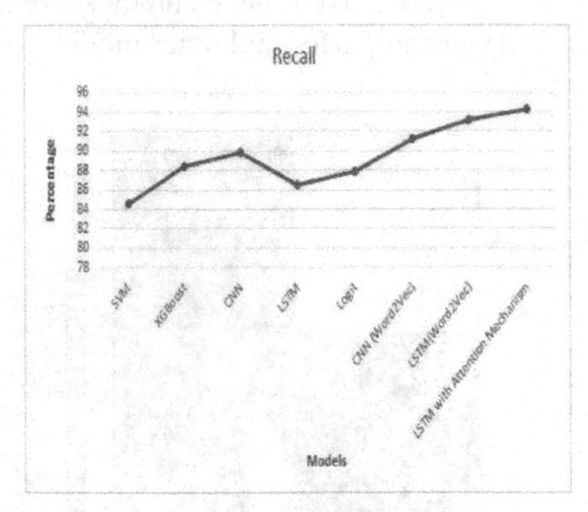

Fig. 9. Comparison of Recall of all the models

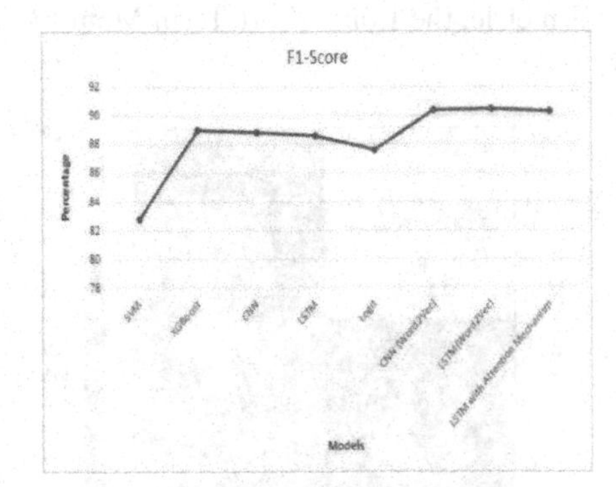

Fig. 10. Comparison of F1-Score of all the models

Figure 9 shows the comparison of recall of different models and the proposed model on the Suicide and Depression Detection dataset. Figure 10 shows the comparison of the f1-score of different models and the proposed model on the Suicide and Depression Detection dataset. Compared to baseline models, our outcomes were better because the F1 Score focuses on positive values (suicidal intent), and our main goal is to detect suicidal intent.

5 Conclusion

In our research, we created a technique for evaluating different Reddit users' mental health. We are going to analyze the semantics and sentiment and the user's social media interactions. This study learns the user's intent by learning the lexical significance, semantics, and context of the textual social media post. This work analyses the social interaction between users by incorporating attention mechanisms. In our research, we created a technique for evaluating different Reddit users' mental health. We are going to analyze the semantics and sentiment and the user's social media interactions.

By identifying suicidal intentions in people's social media interactions, our methodology aims to reduce the probability of suicide in society. This can assist in giving those in need the proper medical care and/or assistance. We may also provide a consciousness class on how to manage mental challenges like stress, worry, and so forth based on the findings of our model. People are more emotional today, which causes massive losses to households, friends, and others in their immediate vicinity. With an accuracy of 93.15%, we categorized the post as suicidal or non-suicidal using an LSTM network with an attention mechanism. When we compared the baseline methods to the method that has been suggested, we can clearly say that the performance has increased substantially. As a part of future work, we would extend the model's capability toward developing the SID with multiple-language support that can analyze the various users with different language backgrounds.

References

1. Ji, S., Pan, S., Li, X., Cambria, E., Long, G., Huang, Z.: Suicidal ideation detection: a review of machine learning methods and applications. IEEE Trans. Comput. Soc. Syst. **8**(1), 214–226 (2021). https://doi.org/10.1109/TCSS.2020.3021467
2. Nikhileswar, K., Vishal, D., Sphoorthi, L., Fathimabi, S.: Suicide ideation detection in social media forums. In: Proceedings - 2nd International Conference on Smart Electronics and Communication, ICOSEC 2021, pp. 1741–1747 (2021). https://doi.org/10.1109/ICOSEC51865.2021.9591887
3. Bhattacharya, D., Karthick, N.S.H., Shahina, A.: Early detection of suicidal tendencies from text data using LSTM. In: 3rd IEEE International Virtual Conference on Innovations in Power and Advanced Computing Technologies, i-PACT 2021 (2021). https://doi.org/10.1109/i-PACT52855.2021.9696630
4. Renjith, S., Abraham, A., Jyothi, S.B., Chandran, L., Thomson, J.: An ensemble deep learning technique for detecting suicidal ideation from posts in social media platforms. J. King Saud Univ. - Comput. Inf. Sci. (2021). https://doi.org/10.1016/j.jksuci.2021.11.010
5. Luo, J., Du, J., Tao, C., Xu, H., Zhang, Y.: Exploring temporal patterns of suicidal behavior on Twitter. In: Proceedings - 2018 IEEE International Conference on Healthcare Informatics Workshops, ICHI-W 2018, July 2018, pp. 55–56 (2018). https://doi.org/10.1109/ICHI-W.2018.00017
6. Gould, M., Jamieson, P., Romer, D.: Media contagion and suicide among the young. Am. Behav. Sci. **46**(9), 1269–1284 (2003). https://doi.org/10.1177/0002764202250670
7. Coppersmith, G., Leary, R., Crutchley, P., Fine, A.: Natural language processing of social media as screening for suicide risk. Biomed. Inform. Insights **10**, 117822261879286 (2018). https://doi.org/10.1177/1178222618792860
8. Sinha, P.P., Mahata, D., Mishra, R., Shah, R.R., Sawhney, R., Liu, H.: #suicidal – a multipronged approach to identify and explore suicidal ideation in twitter. In: International Conference on Information and Knowledge Management, Proceedings, November 2019, pp. 941–950 (2019). https://doi.org/10.1145/3357384.3358060
9. Bhardwaj, T., Gupta, P., Goyal, A., Nagpal, A., Jha, V.: A review on suicidal ideation detection based on machine learning and deep learning techniques. In: 2022 IEEE World AI IoT Congress, AIIoT 2022, pp. 27–31 (2022). https://doi.org/10.1109/AIIoT54504.2022.9817373
10. SCAD Institute of Technology and Institute of Electrical and Electronics Engineers, Proceedings of the International Conference on Intelligent Sustainable Systems (ICISS 2017), pp. 7–8, December 2017 (2017)

11. Galassi, A., Lippi, M., Torroni, P.: Attention in natural language processing. IEEE Trans. Neural Netw. Learn. Syst. **32**(10), 4291–4308 (2021). https://doi.org/10.1109/TNNLS.2020.3019893
12. Ma, Y.: Social media-based suicide risk detection via social interaction and posted content. In: ACM International Conference Proceeding Series, May 2021 (2021). https://doi.org/10.1145/3469213.3470345
13. Xu, X.: Detecting suicide ideation in the online environment: a survey of methods and challenges. IEEE Trans. Comput. Soc. Syst. **9**(3), 679–687 (2022). https://doi.org/10.1109/TCSS.2021.3108976
14. de Choudhury, M., Gamon, M., Counts, S., Horvitz, E.: Predicting depression via social media (2013). www.aaai.org
15. Ahmad, S., Asghar, M.Z., Alotaibi, F.M., Awan, I.: Detection and classification of social media-based extremist affiliations using sentiment analysis techniques. Hum.-Centric Comput. Inf. Sci. **9**(1), 1–23 (2019). https://doi.org/10.1186/s13673-019-0185-6
16. Guntuku, S.C., Yaden, D.B., Kern, M.L., Ungar, L.H., Eichstaedt, J.C.: Detecting depression and mental illness on social media: an integrative review. Curr. Opin. Behav. Sci. **18**, 43–49 (2017). https://doi.org/10.1016/j.cobeha.2017.07.005
17. IEEE Region 10, IEEE Seoul Section, Institute of Electrical and Electronics Engineers. Korea Council, and Institute of Electrical and Electronics Engineers, Proceedings of TENCON 2018: 2018 IEEE Region 10 Conference, Jeju, Korea, 28–31 October 2018 (2018)
18. Chatterjee, M., Samanta, P., Kumar, P., Sarkar, D.: Suicide ideation detection using multiple feature analysis from Twitter data. In: 2022 IEEE Delhi Section Conference, DELCON 2022 (2022). https://doi.org/10.1109/DELCON54057.2022.9753295
19. Huang, S., et al.: Predictive modeling for suicide-related outcomes and risk factors among patients with pain conditions: a systematic review
20. Tadesse, M.M., Lin, H., Xu, B., Yang, L.: Detection of suicide ideation in social media forums using deep learning. Algorithms **13**(1), 7 (2020). https://doi.org/10.3390/a13010007
21. Lalrinmawii, C., Debnath, S.: Analysis of Post centric suicidal expressions and classification on the Social Media Post: Twitter; Analysis of Post centric suicidal expressions and classification on the Social Media Post: Twitter (2020)
22. IEEE Computational Intelligence Society and Institute of Electrical and Electronics Engineers, 2019 IEEE International Conference on Fuzzy Systems: New Orleans, Louisiana, USA, 23–26 June 2019 (2019)
23. Plisson, J., Lavrac, N., Mladenic, D.: A Rule based Approach to Word Lemmatization
24. Mikolov, T., Chen, K., Corrado, G., Dean, J.: Efficient estimation of word representations in vector space (2013). http://arxiv.org/abs/1301.3781
25. Ma, L., Zhang, Y.: Using Word2Vec to process big text data. In: Proceedings - 2015 IEEE International Conference on Big Data, IEEE Big Data 2015, December 2015, pp. 2895–2897 (2015). https://doi.org/10.1109/BigData.2015.7364114
26. Khattak, F.K., Jeblee, S., Pou-Prom, C., Abdalla, M., Meaney, C., Rudzicz, F.: A survey of word embeddings for clinical text. J. Biomed. Inform.: X **4**, 100057 (2019). https://doi.org/10.1016/j.yjbinx.2019.100057
27. Uchida, S., Yoshikawa, T., Furuhashi, T.: Application of output embedding on Word2Vec. In: Proceedings - 2018 Joint 10th International Conference on Soft Computing and Intelligent Systems and 19th International Symposium on Advanced Intelligent Systems, SCIS-ISIS 2018, July 2018, pp. 1433–1436 (2018). https://doi.org/10.1109/SCIS-ISIS.2018.00224
28. Hutto, C.J., Gilbert, E.: VADER: a parsimonious rule-based model for sentiment analysis of social media text (2014). http://sentic.net/

29. Kim, Y.: Convolutional neural networks for sentence classification (2014). http://arxiv.org/abs/1408.5882
30. Smagulova, K., James, A.P.: A survey on LSTM memristive neural network architectures and applications. Eur. Phys. J.: Spec. Top. **228**(10), 2313–2324 (2019). https://doi.org/10.1140/epjst/e2019-900046-x

Bayesian Modeling for a Shape Parameter of Weibull-Lomax Distribution with an Application to Health Data

Mohd Irfan and A. K. Sharma(✉)

Department of Mathematics, National Institute of Technology Raipur, Raipur, India
aksharma.ism@gmail.com

Abstract. In this paper, effective Bayesian estimation procedures have been explored to estimate a shape parameter of the Weibull-Lomax distribution with the aid of classical and Bayesian approaches under informative prior distribution. This approach includes the square error loss function (SELF), precautionary loss function (PLF), quadratic loss function (QLF), general entropy loss function (GELF) and linex loss function (LLF) for optimal estimate of the shape parameter. To validate the theoretical results, an extensive Monte Carlo simulations and a real health data are considered for an illustration of comparative study of classical and Bayesian approach. The mean square error (MSE) are also derived to show the efficiency of the estimators. Further cumsum plots and iterative plots are demonstrated to justify convergence of precision of the estimates.

Keywords: Weibull-Lomax (WL) distribution · Informative prior · Bayesian estimation · Monte Carlo Simulation · Goodness-of-fit

1 Introduction

Bayesian approach has given a lot contribution in past few decades for estimating the unknown model parameters. This approach sparked the development of different approximation techniques of the estimation. In statistical inference, the Bayesian approach commonly depends on selection of prior distribution and the loss functions, where the prior distribution of parameters may be depends on the hyper parameters. In this approach physicians rely heavily on history of the patients like lab blood test, sonography, X-ray, pet-scan, family history etc. The physician would use the previous patient records as prior information and further for computation of decision making prediction can be done by information on posterior distribution.

Although, there is a large literature available on the Bayes estimation including Zellner (1971) [21], Sinha and Howladaer (1993) [13], chung (1995) [5], Ahmad A. Soliman (2000) [14], Wu *et al.* (2006) [19], Soliman and Fahad (2008) [15], Khan *et al.* (2010) [8], Ahmad *et al.* (2013) [1], those who have shown the Bayesian approach for estimation and prediction for different distribution using various loss functions, some of them also discussed life testing and reliability estimates as well. Farahani and Khorram

B. K. Singh et al. (Eds.): ICBEST 2023, CCIS 2003, pp. 284–299, 2024.
https://doi.org/10.1007/978-3-031-54547-4_22

(2014) [7] have derived the Bayes estimators for Weighed exponential distribution under two loss functions.

Tahir *et al.* (2015) [16] initiated a new distribution of four parameters from the extension of Lomax distribution called the Weibull- Lomax (W-L) distribution and derived its numerous statistical properties like moment generating function, mixture representation, quantile function, probability weighted moments, generating function etc. to validate advantages in the effective estimation procedures. It is observed that the W-L distribution of four parameters is highly desirable model in the comparison of Mcdonald - Lomax, beta-Lomax, Kumarswami - Lomax, Exponentiated-Lomax and Lomax model in the estimation of parameter. Further Rady *et al.* (2016) [10] have extended the Lomax distribution and estimates the parameters by classical approaches. Rastogi and Faton (2018) [3] have estimated and derived reliability characteristics of three parameters of Weibull Rayleigh distribution. Yilmaz and Kara (2021) [20] have focused on reliability estimation and the parameter estimation for inverse Weibull model under various loss functions. Boosly (2021) [5] has developed a new weighted composite linex loss function and derived the Bayesian estimation for the Lomax distribution. Usman and Alhaji (2021) [17] have estimates the shape parameter of W-L distribution using the various types of non informative prior distribution. Further Riad *et al.* (2022) [11] have explored some Bayes estimators for Weighted-exponential distribution using non-Bayesian and Bayesian approach with application to health data.

The main goal of this paper is to use informative prior and compute the Bayes estimates of an unknown parameter of W-L distribution under different loss functions, which is highly desirable to health data sets and prompt to estimates clinical health parameters. Therefore it is showed that W-L model is better fit to the health data over the another model on various measure of goodness-of-fit that claim to be more flexible and ethical than traditional model.

2 Motivation

The W-L distribution of four parameter is a very flexible and dominates the many competent distributions. It has monotonic shape of hazard rate function lead to useful in survival analysis of health care data. The ductility and performance of the model have motivated us to estimate the shape parameter of the W-L distribution in Bayesian context under informative prior distribution in the application to health care data.

3 Methodology

This section deals with the method of exploring the classical and Bayes estimators of W-L model under the various loss functions. According to Tahir *et al.* (2015), the probability distribution function (pdf) and cumulative distribution function (cdf) of four parameters Weibull-Lomax (W-L) distribution are given by

$$f(x;\eta,\theta,\lambda,\delta) = \frac{\eta\theta\lambda}{\delta}\left[\left(1+\frac{x}{\delta}\right)^{\theta\lambda-1}\left\{1-\left(1+\frac{x}{\delta}\right)^{-\lambda}\right\}^{\theta-1}\right.$$

$$exp\left[-\eta\left\{\left(1+\frac{x}{\delta}\right)^{\lambda}-1\right\}^{\theta}\right]\Bigg] \tag{3.1}$$

and

$$F(x;\eta,\theta,\lambda,\delta)=1-exp\left[-\eta\left\{\left(1+\frac{x}{\delta}\right)^{\lambda}-1\right\}^{\theta}\right], \tag{3.2}$$

where $x > 0$; $\eta, \theta, \lambda > 0$ are shape parameters and $\delta > 0$ is scale parameter of Weibull-Lomax (W-L) model. The reliability and hazard rate function of W-L distribution are respectively given by

$$S(t)=exp\left[-\eta\left\{\left(1+\frac{t}{\delta}\right)^{\lambda}-1\right\}^{\theta}\right] \tag{3.3}$$

and

$$H(t)=\frac{\eta\theta\lambda}{\delta}\left(1+\frac{t}{\delta}\right)^{\theta\lambda-1}\left[1-\left\{\left(1+\frac{t}{\delta}\right)^{-\lambda}\right\}^{\beta-1}\right], \tag{3.4}$$

$$t > 0, \eta, \theta, \lambda, \delta > 0.$$

, where

The plot of the hazard rate function of the W-L model for various combinations is depicted in Fig. 2 and has the property of increasing, decreasing, and bathtub hazard rate function, which is a nice feature of the model and helpful in fitting the real health data sets.

3.1 Maximum Likelihood Estimation

It is opted the classical approach to estimate the shape parameter η of W-L distribution when θ, λ and δ are known. Let $X = X_1, X_2, \ldots, X_n$ be the random sample drawn from the W-L distribution with probability density function given in the Eq. (3.1), then the likelihood function $L(\eta, \theta, \lambda, \delta|X)$ is defined as the joint density of the random sample as

$$L(\eta,\theta,\lambda,\delta|X)=\prod_{i=1}^{n}\left[\frac{\eta\theta\lambda}{\delta}\left(1+\frac{x_i}{\delta}\right)^{\theta\lambda-1}\left\{1-\left(1+\frac{x_i}{\delta}\right)^{-\lambda}\right\}^{\beta-1}\right.$$

$$\left. exp\left[-\eta\left\{\left(1+\frac{x_i}{\delta}\right)^{\lambda}-1\right\}^{\theta}\right]\right]. \tag{3.5}$$

$$L(\eta,\theta,\lambda,\delta|X)=\left(\frac{\eta\theta\lambda}{\delta}\right)^{n}\left\{\prod_{i=1}^{n}\left(1+\frac{x_i}{\delta}\right)^{\theta\lambda-1}\right\}\left[\prod_{i=1}^{n}\left\{1-\left(1+\frac{x_i}{\delta}\right)^{-\lambda}\right\}^{\theta-1}\right]$$

$$exp\left[-\eta\sum_{i=1}^{n}\left\{\left(1+\frac{x_i}{\delta}\right)^{\lambda}-1\right\}^{\theta}\right]. \tag{3.6}$$

Now the likelihood function with respect to η is expressed as

$$L(\eta|X) \propto \eta^n exp\left[-\eta\sum_{i=1}^{n}\left\{\left(1+\frac{x_i}{\delta}\right)^{\lambda}-1\right\}^{\theta}\right]. \tag{3.7}$$

(See Fig. 1).

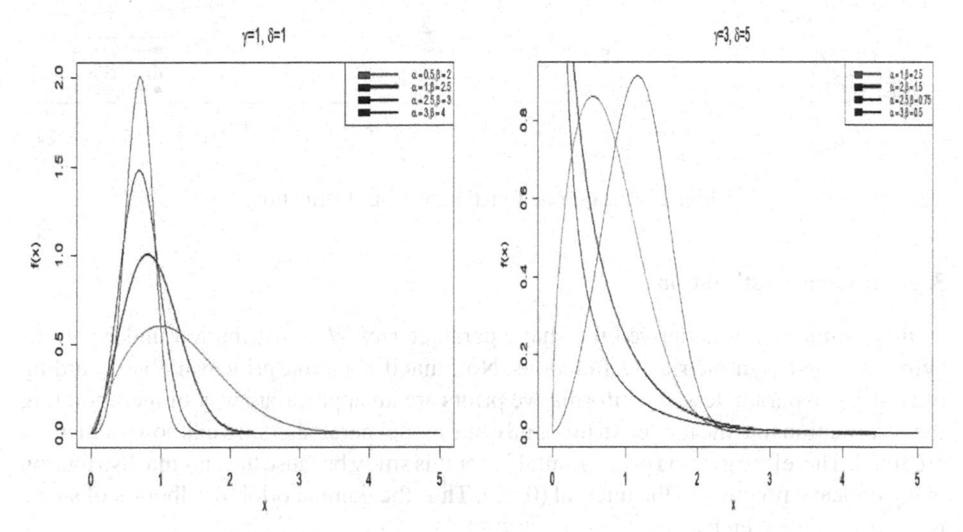

Fig. 1. Probability distribution function (pdf) of W-L distribution.

Let l be the log likelihood function of $L(\eta|X)$, therefore

$$l = nlog(\eta) - \eta\sum_{i=1}^{n}\left\{\left(1+\frac{x_i}{\delta}\right)^{\lambda}-1\right\}^{\theta}. \tag{3.8}$$

On differentiating the Eq. (3.8) partially with respect to η, we have

$$\frac{\partial l}{\partial n} = \frac{n}{\eta} - \sum_{i=1}^{n}\left\{\left(1+\frac{x_i}{\delta}\right)^{\lambda}-1\right\}^{\theta}.$$

On equating the $\frac{\partial l}{\partial n}$ to 0, the MLE of η is obtained as

$$\hat{\eta}_{MLE} = \frac{n}{D}, \tag{3.9}$$

where $D = \sum_{i=1}^{n}\left\{\left(1+\frac{x_i}{\delta}\right)^{\lambda}-1\right\}^{\theta}$.

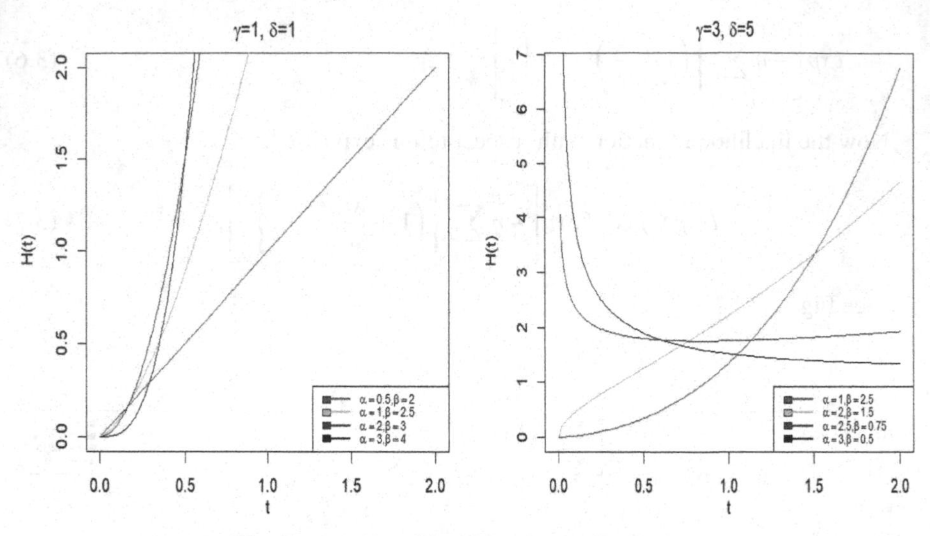

Fig. 2. Hazard rate function of W-L distribution.

3.2 Bayesian Estimation

In this estimation, it is derived the shape parameter of W-L distribution under various symmetric and asymmetric loss functions. Note that if a suitable prior available regarding the unknown parameters, the informative priors are an appropriate way of incorporating the information into the model. In this study one model parameters are unknown and other are fixed. Therefore gamma prior is suitable for this study because the gamma distribution is log concave function in the interval $(0,\infty)$. Thus the gamma prior distribution of shape parameter η is given by

$$g(\eta)=\frac{d^b}{\Gamma b}\eta^{b-1}exp(-d\eta);\ \eta>0;\ b,d>0, \tag{3.10}$$

where b and d are hyper parameters.

Now the posterior distribution is obtained as follows

$$\pi(\eta|X)=K^{-1}\left[\frac{d^b}{\Gamma b}\eta^{b-1}exp(-d\eta)\times\eta^n exp\left[-\eta\left\{\sum_{i=1}^{n}\left(1+\frac{x_i}{\delta}\right)^{\lambda}-1\right\}^{\theta}\right]\right]$$

or,

$$\pi(\eta|X)=K^{-1}\left[\frac{d^b}{\Gamma b}\eta^{b+n-1}exp\{-\eta(d+D)\}\right]. \tag{3.11}$$

where $K=\int_0^\infty\frac{d^b}{\Gamma b}exp\{-\eta(d+D)\}d\eta$, which is known as normalising constant and $D=\sum_{i=1}^{n}\left\{\left(1+\frac{x_i}{\delta}\right)^{\lambda}-1\right\}^{\theta}$.

On solving the Eq. (3.11), we get the posterior density of η for given X as

$$\pi(\eta|X) = \frac{(d+D)^{n+b}\eta^{b+n+-1}exp\{-\eta(d+D)\}}{\Gamma(n+b)}, \tag{3.12}$$

which follows the gamma distribution with parameter (n + b, d + D).

3.2.1 Bayesian Estimation Under Square Error Loss Function (SELF)

One of the most popular symmetric loss function is the square error loss function (SELF) which provide equal penalty for overestimation as well as under estimation. It is well known that the Bayesian estimates of parameter under SELF is the posterior mean. Therefore the Bayes estimator of η under SELF is derived as

$$\hat{\eta}_{BS} = \sqrt{E(\eta|X)}, \tag{3.13}$$

where

$$E(\eta|X) = \int_0^\infty \eta\pi(\eta|X).$$

Using Eq. (3.12), we have

$$\begin{aligned}\hat{\eta}_{BS} &= \int_0^\infty \eta \cdot \frac{(b+D)^{n+a}\eta^{a+n-1}exp\{-\eta(b+D)\}}{\Gamma(n+b)}d\eta \\ &= \frac{n+b}{d+D}.\end{aligned} \tag{3.14}$$

3.2.2 Bayesian Estimation Under Quadratic Loss Function (QLF)

Quadratic loss function (QLF) is an symmetric loss function first addressed by Zellner (1971) [7].

The Bayes estimator of η under QLF is expressed as follows

$$\hat{\eta}_{BQ} = \frac{E(\eta^{-1}|X)}{E(\eta^{-2}|X)}, \tag{3.15}$$

where

$$\begin{aligned}E(\eta^{-1}|X) &= \int_0^\infty \eta^{-1} \cdot \frac{(d+D)^{n+b}\eta^{b+n-1}exp\{-\eta(d+D)\}}{\Gamma(n+b)}d\eta \\ &= \frac{d+D}{n+b-1}\end{aligned} \tag{3.16}$$

and

$$E(\eta^{-2}|X) = \int_0^\infty \eta^{-2} \cdot \frac{(d+D)^{n+a}\eta^{b+n-1}exp\{-\eta(d+D)\}}{\Gamma(n+b)}d\eta$$

$$= \frac{(d+D)^2}{(n+b-1)(n+b-2)}. \tag{3.17}$$

Therefore from the Eqs. (3.15), (3.16) and (3.17), we get

$$\hat{\eta}_{BQ} = \frac{n+b-2}{d+D}. \tag{3.18}$$

3.2.3 Bayesian Estimation of η Under Precautionary Loss Function (PLF)

Precautionary loss function (PLF) is an asymmetric loss function pioneered by Norstrom (1996) [8]. The Bayes estimator of η under PLF is obtained as

$$\hat{\eta}_{BP} = \sqrt{E(\eta^2|X)}. \tag{3.19}$$

Now,

$$E\left(\eta^2|X\right) = \int_0^\infty \eta^2\pi(\eta|X)$$
$$= \frac{(n+b)(n+b+1)}{(d+D)^2}. \tag{3.20}$$

Using the Eqs. (3.19) and (3.20) the Bayes estimator of η under PLF as

$$\hat{\eta}_{BP} = \frac{\sqrt{(n+b)(n+b+1)}}{d+D}. \tag{3.21}$$

3.2.4 Bayesian Estimation of η Under General Entropy Loss Function (GELF)

Entropy loss function is an asymmetric type of loss function which was first introduced by Calabria and Pulcini (1996). The Bayes estimator of η under general entropy loss function (GELF) [9] is expressed as

$$\hat{\eta}_{BE} = [E(\eta^{-p}|X)]^{-\frac{1}{p}}; p \neq 0, \tag{3.22}$$

where

$$E(\eta^{-p}|X) = \int_0^\infty \eta^{-p}\pi(\eta|X)d\eta$$
$$= \frac{(d+D)^{n+b}}{\Gamma(n+b)} \int_0^\infty \eta^{n+b-p-1} exp\{-\eta(d+D)\}d\eta. \tag{3.23}$$

Therefore from the Eqs. (3.22) and (3.23), the Bayes estimator of η under GELF as

$$\hat{\eta}_{BE} = \frac{1}{d+D}\left[\frac{\Gamma(n+b)}{\Gamma(n+b-p)}\right]^{\frac{1}{p}}; p \neq 0. \tag{3.24}$$

3.2.5 Bayesian Estimation of η Under LINEX Loss Function (LLF)

One of the very plausible type of asymmetric loss function is the LLF was first suggested by Varian (1975) [18]. Therefore, the Bayes estimates of η under LLF is obtained as follows

$$\hat{\eta}_{BL} = -\frac{1}{c} ln[E(e^{-c\eta}|X)];\ c \neq 0, \tag{3.25}$$

provided $E(e^{-c\eta}|X)$ exists and finite, where

$$E(e^{-c\eta}|X) = \frac{(d+D)^{(n+b)}}{\Gamma(n+b)} \int_0^{\infty} \eta^{n+b-1} exp\{-\eta(d+c+D)\} d\eta. \tag{3.26}$$

Using the Eqs. (3.25) and (3.26), the Bayes estimator of η under LLF is

$$\hat{\eta}_{BL} = \left(\frac{n+b}{c}\right) ln\left\{1 + \frac{c}{d+D}\right\};\ c \neq 0. \tag{3.27}$$

4 Monte Carlo Simulation

A Monte Carlo simulation is conducted in R software to check relative procession of estimates of the considered parameter. The following steps are considered for the simulation study to estimate the parameter η as:

(i) Select the sample size of $n \in (20, 40, 60, 80, 100)$ from the W-L distribution with shape parameter $\eta = 0.75$, 1, and 1.25.
(ii) Assume the hyper parameters $b = 0.75$, 2, 5 and $d = 1,2,4$ in the Eq. (3.19).
(iii) Take the constant $c = 2$, 2.5 in the Eq. (3.25).
(iv) For given sample size n, $\theta = 2$, $\lambda = 0.5$ and $\delta = 1$; the sample $x_1, x_1, \ldots, x_n$ are generated from the distribution where U

$$x_i = \delta\left[\left[\left\{\frac{-1}{\eta} ln(1-U)\right\}^{\frac{1}{\theta}} + 1\right]^{\frac{1}{\lambda}} - 1\right]$$

follows the uniform distribution with interval [0,1].
(v) The MLE of η i.e. $\hat{\eta}_{MLE}$ are computed from the Eq. (3.9) for different sample size n.
(vi) The Bayes estimators $\hat{\eta}_{BS}$, $\hat{\eta}_{BQ}$, $\hat{\eta}_{BP}$, $\hat{\eta}_{BE}$ and $\hat{\eta}_{BL}$ are calculated from the Eqs. (3.14), (3.18), (3.21), (3.24) and (3.27) respectively.
(vii) On repeating 10,000 times the steps from (iv) to (vi), average estimates and mean square errors of the parameter η are generated, where the MSE formula is given by

$$MSE(\hat{\eta}) = \frac{1}{10000} \sum_{i=1}^{10000} (\hat{\eta}_i - \overline{\eta})^2$$

and

$$\hat{\eta} = \frac{1}{10000} \sum_{i=1}^{10000} \hat{\eta}_i.$$

(viii) The average estimated values of η and its MSEs are depicted in the Table 1 and Table 2 respectively.

Table 1. Estimates of η based on different sample sizes (n) under various loss functions.

η	n	$\hat{\eta}_{MLE}$	$\hat{\eta}_{BS}$	$\hat{\eta}_{BQ}$	$\hat{\eta}_{BP}$	$\hat{\eta}_{BE}$	$\hat{\eta}_{BL}$	$\hat{\eta}_{BL}$
						$p = 1$	$c = 2$	$c = 2.5$
0.75	20	0.7889859	0.7875722	0.710644	0.8101831	0.7492652	0.7598089	0.750391
	40	0.7699615	0.770063	0.7316831	0.7800157	0.7510587	0.7565831	0.7517989
	60	0.7638806	0.7638125	0.7377974	0.7683318	0.7508918	0.7523948	0.7513577
	80	0.7586484	0.7592321	0.7418946	0.7638888	0.7494697	0.7539242	0.750811
	100	0.7572291	0.7580301	0.7419013	0.7611	0.7504682	0.7536123	0.7503627
1	20	1.039957	1.04551	0.9443958	1.067842	0.9958895	0.996679	0.9822398
	40	1.024218	1.023102	0.9721052	1.036808	0.9967	1.000477	0.9922653
	60	1.016652	1.016468	0.9829238	1.024582	0.9987902	0.9981686	0.9943033
	80	1.014743	1.012176	0.9852616	1.018266	0.9991072	0.9980097	0.9966
	100	1.009839	1.010453	0.9906518	1.014839	0.9998735	0.9987476	0.9983634
1.25	20	1.313185	1.3104	1.181513	1.335985	1.238066	1.221668	1.207205
	40	1.279825	1.278961	1.215233	1.292885	1.246496	1.240277	1.228596
	60	1.270769	1.270821	1.229721	1.281476	1.24911	1.24267	1.23628
	80	1.264909	1.263069	1.232261	1.271169	1.247347	1.244651	1.237707
	100	1.264427	1.260013	1.237123	1.267997	1.248754	1.244006	1.2376

Table 2. Mean square error (MSE) of η based on sample sizes (n) under various loss functions.

η	n	$\hat{\eta}_{MLE}$	$\hat{\eta}_{BS}$	$\hat{\eta}_{BQ}$	$\hat{\eta}_{BP}$	$\hat{\eta}_{BG}$	$\hat{\eta}_{BL}$	$\hat{\eta}_{BL}$
						$p = 1$	$c = 2$	$c = 2.5$
0.75	20	0.03698868	0.0345718	0.02886818	0.03886662	0.03065232	0.02916159	**0.02670951**
	40	0.01574955	0.01586111	0.01476284	0.01701481	0.01486323	0.01413834	**0.01377257**
	60	0.01030881	0.01003582	0.009620478	0.0103433	0.009571732	0.009304311	**0.00943541**
	80	0.007478174	0.007177483	0.00715894	0.007718716	**0.00692394**	0.007065366	0.007014149
	100	0.005902779	0.005870313	0.005596353	0.006115135	0.005889749	0.005823276	**0.005562562**
1	20	0.06512978	0.06162494	0.05088617	0.06505584	0.05244382	**0.04704243**	0.04713184

(*continued*)

Table 2. (*continued*)

η	n	$\widehat{\eta}_{MLE}$	$\widehat{\eta}_{BS}$	$\widehat{\eta}_{BQ}$	$\widehat{\eta}_{BP}$	$\widehat{\eta}_{BG}$	$\widehat{\eta}_{BL}$	$\widehat{\eta}_{BL}$
						$p = 1$	$c = 2$	$c = 2.5$
	40	0.02802805	0.02715165	0.02525482	0.0294551	0.9985346	0.02443543	**0.02413868**
	60	0.01785649	0.01765173	0.016718	0.01840823	0.01710871	0.01660043	**0.01616919**
	80	0.01327347	0.01341748	0.01259539	0.01355475	0.01274054	0.01195939	**0.01222951**
	100	0.01079505	0.01048704	0.01005466	0.01045917	0.01027356	**0.009913979**	0.009930365
1.25	20	0.09705324	0.09681593	0.08332093	0.1018142	0.08149138	0.07007211	**0.0671464**
	40	0.04441996	0.04295301	0.03899257	0.04356507	0.04013326	0.03777848	**0.03704435**
	60	0.02835945	0.02744211	0.0266613	0.02997471	0.02588899	0.0253439	**0.02472295**
	80	0.02027306	0.02025714	0.0194942	0.02106445	0.01999755	**0.0185234**	0.01882661
	100	0.01671927	0.01590603	0.01540039	0.01655953	0.01561044	0.01535099	**0.01497322**

5 Real Health Data Analysis

This data set describes the lifetime of relief (in minutes) of 20 patients receiving an analgesic which presented by Gross and Clark (1975) [6].

In order to examine this health data, it is computed the MLEs for the four parameters and goodnessof-fit measures for W-L distribution with some criteria, for instance, the Akaike information criterion (AIC), Bayesian information criterion (BIC), Hannan-Quinn information criterion (HQIC) and Akaike information criterion with correction (AICC). For comparison, some other lifetime models are also presented for goodness of fit, including Lomax distribution (LD), inverse distribution (ILD), Rayleigh Lomax distribution (RLD), Rayleigh distribution (RD) and exponential distribution (ERD).

The summary statistics of the health data set is listed in the following Tables:

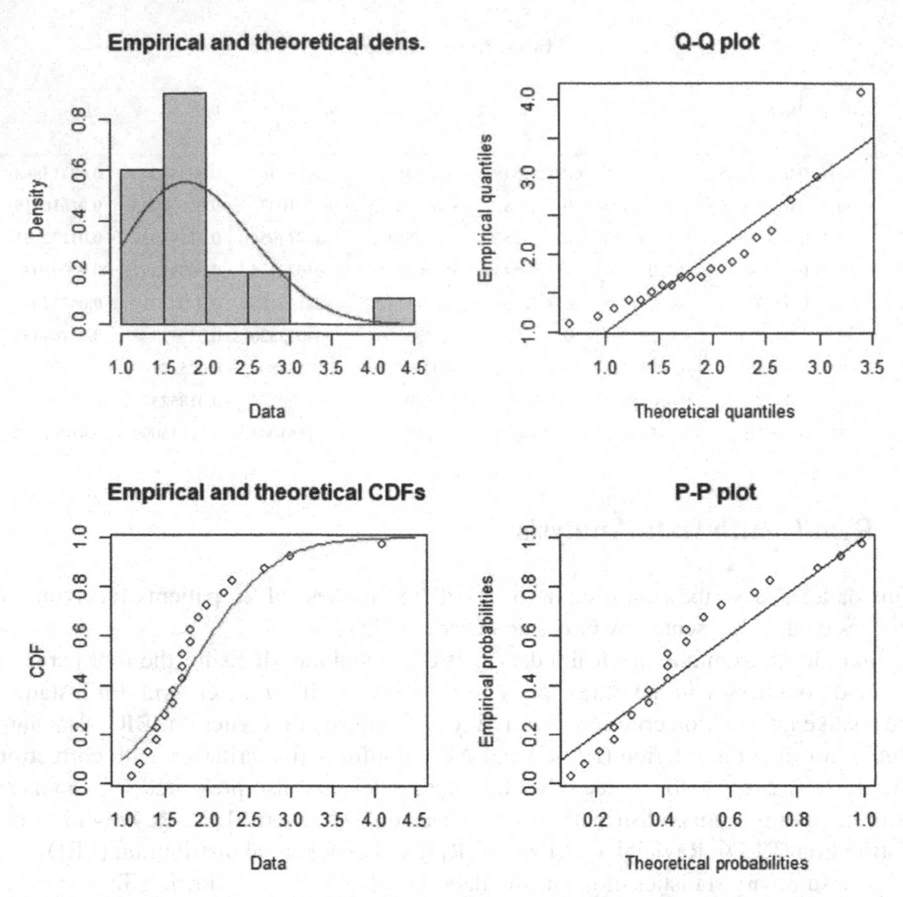

Fig. 3. The estimated density and distribution plots of W-L distribution for the health data set.

Table 3. The goodness of fit measures for the health data set.

Model	-2 log(l)	AIC	BIC	HQIC	AICC
W-LD	**39.268**	**47.268**	**51.251**	**48.046**	**49.935**
LD	66.196	70.196	72.187	70.584	70.902
ILD	65.810	69.810	71.801	70.198	70.516
RLD	41.919	47.919	50.906	48.502	49.419
ED	65.674	69.674	71.666	70.063	70.380

6 Interpretation of Empirical Results

The following interpretation may be read out from Tables 1, 2, 3, 4 and 5 and Figs. 3, 4, 5, 6, 7 and 8:

Table 4. Summary statistics of health data set.

	Est	Std. Error	z value	pr(z)
$\widehat{\eta}$	13.608	0.0176	770.103	<2e-16
$\widehat{\theta}$	5.584	3.863	1.445	0.1484
$\widehat{\lambda}$	0.221	0.1679	1.315	0.188
$\widehat{\delta}$	0.261	0.652	0.400	0.688

Table 5. MLE and Bayesian estimation of α based on real health data set under different loss functions.

	b	d	MLE	SELF	QLF	PLF	GELF	LLF1	LLF2
$\widehat{\eta}$	2	1	2.766	2.673	2.430	2.733	2.551	2.393	2.334
$\widehat{\eta}$	3	2	2.766	2.491	2.275	2.545	2.383	2.255	2.205

(i) The average values and MSEs of the estimators $\widehat{\eta}_{MLE}$, $\widehat{\eta}_{BS}$, $\widehat{\eta}_{BQ}$, $\widehat{\eta}_{BP}$, $\widehat{\eta}_{BG}$ and $\widehat{\eta}_{BL}$ are computed for different sample size n. From the Table 1 and 2 respectively, it is clear that:
  (a) For fixed values of η, the estimates are getting closer to true values of the η with decreasing MSEs as sample size is increasing. This trend desirable in the terms of increased precision of estimates which is highly desirable.
  (b) It is observed that the Bayes estimates of η under the linex loss function for $c = 2.5$ shows the best estimate in term of MSE.
  (c) Table 3 justified that the W-L model is more desirable than the other model based on the measureof goodness-of-fit criteria.
  (d) The summary statistics and different Bayes estimator of the real health data are presented inthe Tables 4 and 5 respectively which is useful for further treatment of the patients.

(ii) Some more interpretation of the estimates of the parameter are shown by the convergence plots, cumsum plots and iterations plots with considerable large sample sizes and large number of iterations in the Figs. 3, 4, 5, 6, 7 and 8 as:
  (a) The empirical and theoretical density, Q-Q plot, empirical and theoretical CDFs, and P-P plotof heath data set are depicted in the Fig. 3 to identify the health status for further process.
  (b) Figures 4 and 5 shows that the convergence plots for $\eta = 1$ and sample size $n = 500$ under maximum likelihood estimation (MLE) and square error loss function (SELF). it is observed that for samll sample size the Bayesian estimation under linex loss function performs better in comparison to classical as well as other loss functions. Similar trend also follow for different values of the parameter η.
  (c) Figures 6, 7 and 8 shows the MCMC cumsum plots and iteration plots for convergence of the parameters η with fixed sample size n $=$ 100 and iterations

N = 10000 under maximum likelihood estimation (MLE), square error loss function (SELF) and linex loss function for efficient further treatment.

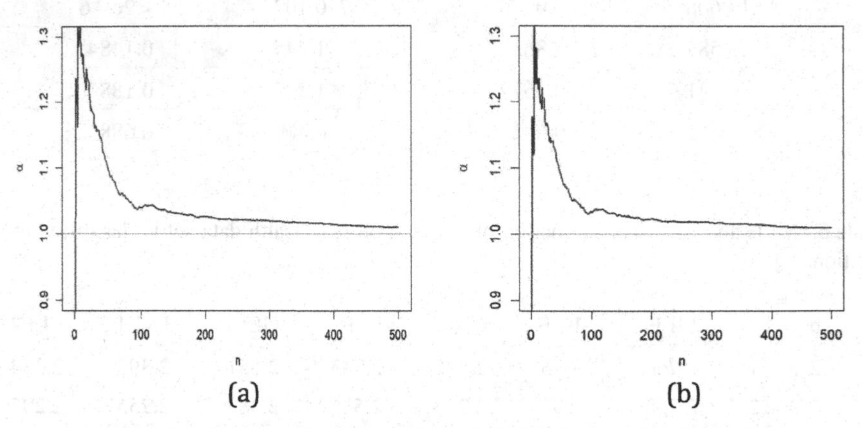

Fig. 4. Convergence plots for $\eta = 1$ under MLE (left) and SELF (right).

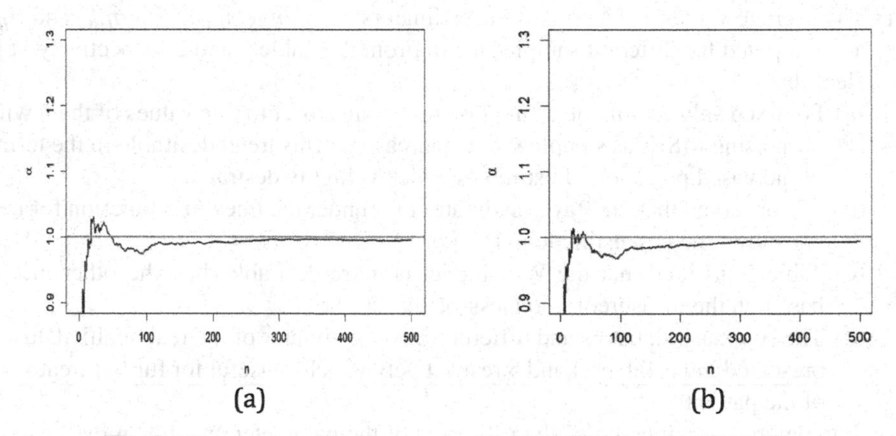

Fig. 5. Convergence plots for $\eta = 1$ under GELF (left) and LLF (right).

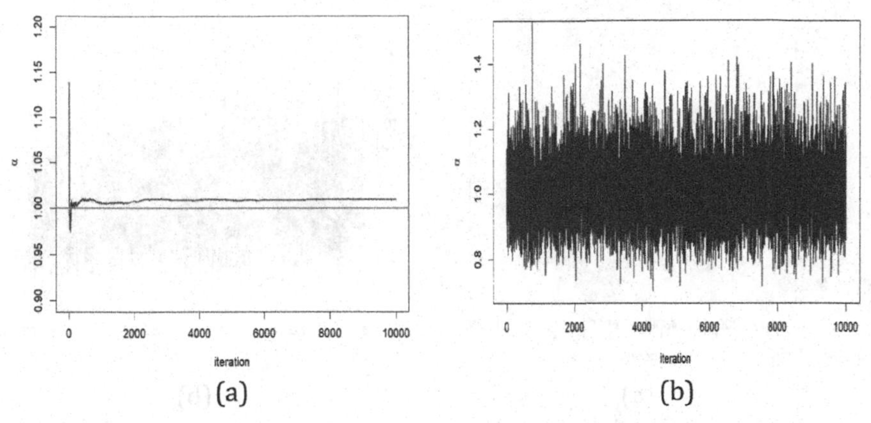

Fig. 6. MCMC cumsum (left) and iteration plots (right) of simulated sample for parameter η under MLE.

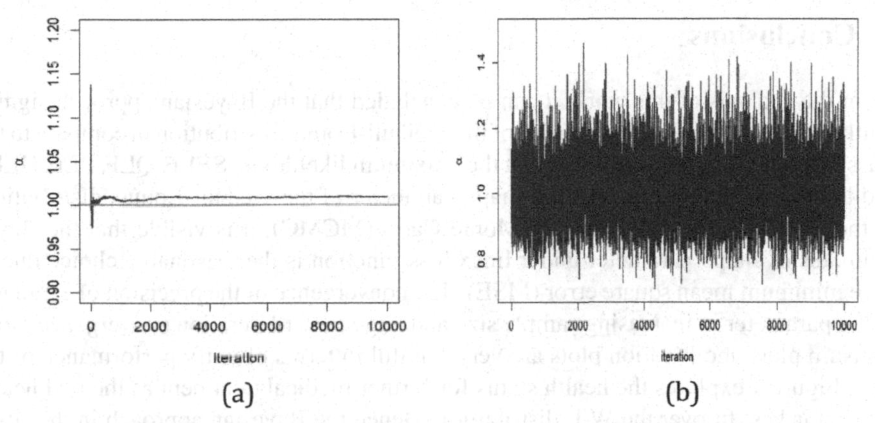

Fig. 7. MCMC cumsum (left) and iteration plots (right) of simulated sample for parameter η under SELF.

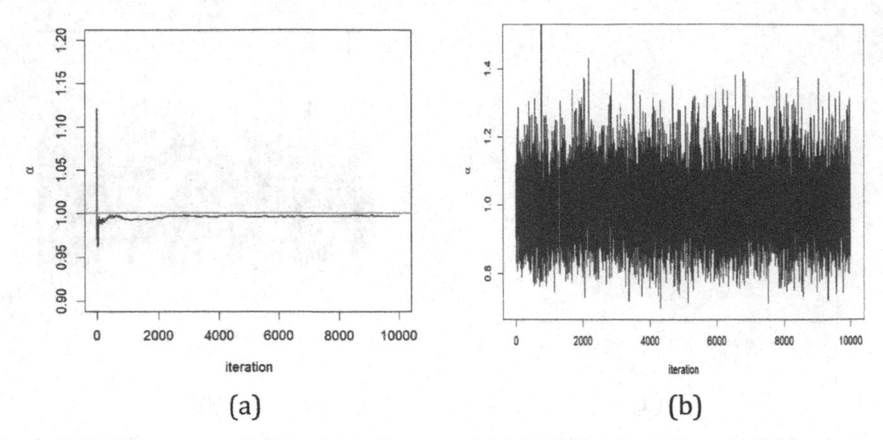

Fig. 8. MCMC cumsum (left) and iteration pots (right) of Simulated Sample for parameter η under LLF.

7 Conclusions

From the analysis of this work, it can be concluded that the Bayesian approach significantly estimates the shape parameter of the Weibull-Lomax distribution in compare to the classical method. It may be noted that the maximum likelihood, SELF, QLF, PLF, GELF and LLF are taken to estimates the shape parameter of the Weibull-Lomax distribution. In the light of simulation study of Monte Carlo (MCMC), it is visible that the Bayes estimates of shape parameter under linex loss function is the reasonable choice due to have minimum mean square error (MSE). The convergence of the precision of estimates of the parameter in increasing sample size and large time of iteration convergence plots, cumsum plots and iteration plots are very faithful in terms of unify performance of the data. Figure 3 explores the health status for further medical treatment as the real health data set is best fit over the W-L distribution. Hence the Bayesian approach in the study is justified the several desirable results. Therefore, the proposed approach for the estimates of shape parameter may be recommended for their application to engineering and medical science.

References

1. Ahamad, A., Ahmad, S.P., Reshi, J.A.: Bayesian analysis of Rayleigh distribution. Int. J. Sci. Res. Publ. **3**(10), 1–9 (2013)
2. Bhat, A.A., Ahmad, S.P.: A new generalization of Rayleigh distribution property and application. Pak. J. Stat. **36**(3), 225–250 (2020)
3. Bossely, A.: E-Bayesian and Bayesian estimation for the Lomax distribution under weighted composite LINEX loss function. Hindawi Comput. Intell. Neurosci. **10**(1) (2021)
4. Calabria, R., Pulcini, G.: On the maximum likelihood and least-squares estimation in the Inverse Weibull distributions. Stat. Appl. **2**(1), 53–66 (1996)
5. Chung, Y.: Estimation of scale parameter from Rayleigh distribution under entropy loss function. Korean J. CAM **2**(1), 33–40 (1995)

6. Gross, A.J., Clark, V.A.: Survival Distributions: Reliability Applications in the Biometrical Sciences. Stephen, Wiley, New York (1975)
7. Farahani, M.S.Z., Khorram, E.: Bayesian statistical inference for weighted exponential distribution. Commun. Stat. Simul. Comput. **43**(6), 1362–1384 (2014)
8. Khan, H.M.R., Provost, S.B., Singh, A.: Predictive inference from a two parameter Rayleigh life model given a doubly censored sample. Commun. Stat. Theory Methods **39**, 1237–1246 (2010)
9. Norstrom, J.G.: The use of precautionary loss functions in risk analysis. IEEE Trans. Son Reliab. **45**(3), 400–403 (1996)
10. Rady, EH.A., Hassanein, W.A., Elhaddad, T.A.: The power Lomax distribution with an application to bladder cancer data. Springer Plus (2016). https://doi.org/10.1186/s40064-016 3464-y
11. Riad, F.H., Hussam, E., Gemeay, A.M., Aldallal, R.A., Afify, A.Z.: Classical and Bayesian inference of the weighted-exponential distribution with an application to insurance data. Math. Biosci. Eng. **19**(7), 6551–6581 (2022)
12. Rastogi, M., Merovcib, F.: Bayesian estimation for parameters and reliability characteristic of the Weibull Rayleigh distribution. J. King Saud Univ. Sci. **30**(4), 472–478 (2018)
13. Sinha, S.K., Howlader, H.A.: Credible and HPD interval of the parameter of and Rayleigh distribution. IEEE Trans. Reliab. **32**(2), 217–220 (1993)
14. Soliman, A.A.: Comparison of linex and quadratic Bayes estimators for the Rayleigh distribution. Commun. Stat. Theory Method **291**(1), 95–107 (2000)
15. Soliman, A.S., Fahad, M.A.: Bayesian inference using record values from Rayleigh model with application. Eur. J. Oper. Res. **185**, 659–672 (2008)
16. Tahir, H., Cordeiro, G.M., Mansoor, M., Zubair, M.: The Weibull-Lomax distribution: properties and applications. Hacettepe J. Math. Stat. **44**(2), 455–474 (2015)
17. Usman, M., Alhaji, B.: Bayesian estimation of a shape parameter of a Weibull-Lomax distribution using non-informative priors. Kasu J. Math. Sci. **2**(1), 101–115 (2021)
18. Varian, H.R.: A bayesian approach to real estate assessment. In: Savage, J.S., Fienberg, E., Arnold, Z., (eds.), Studies in Bayesian Econometrics and Statistics in Honor of Leonard, pp. 195–208. Amsterdam, North Holland (1975)
19. Wu, S., Chen, D., Chen, T.S.: Bayesian inference for Rayleigh distribution under progressive censored sample. Appl. Stoch. Models Bus. Ind. **22**, 269–179 (2006)
20. Yilmaz, A., Kara, M.: Reliability estimation and parameter estimation for inverse Weibull distribution under different loss functions. Kuwait J. Sci. **49**, 1–24 (2021)
21. Zellner, A.:: An Introduction to Bayesian Inference in Econometrics. Wiley, New York (1971). Probability Reprinted in 1987, Melbourne, FL: Krieger

Foot Strike Patterns, Anthropometric Parameters and Somatotypes in Optimizing Multi-sprint Sports Performance

Mohor Chattopadhyay[1], Amit Kumar Singh[2,4], and M. Marieswaran[3](✉)

[1] Department of Sports Biomechanics, School of Sports Sciences, Central University of Rajasthan, Rajasthan, India
[2] Department of Biomedical Engineering, VSB Engineering College, Karur, Tamil Nadu, India
[3] Department of Biomedical Engineering, National Institute of Technology Raipur, Chhattisgarh, India
mmarieswaran.bme@nitrr.ac.in
[4] Department of Biomedical Engineering, Vignan's Foundation for Science Technology & Research, Vadlamudi, Guntur, Andhra Pradesh, India

Abstract. Performances in multi-sprint sports may depend on various attributes like ability to run fast, ability to pass/hit/kick/throw the ball, reaching the ball and maneuvering with the ball. The effect of foot-strike patterns on the ability to run fast and chances of injury occurrence has been evaluated in recent studies on runners, but no study has reported its effect on multi-sprint sports. In this study, the anthropometric and anaerobic parameters have been obtained among athletes with different foot strike patterns to estimate the relationship between suitable foot strike pattern and sports performances. Twenty-six males between the age group of 18–25 years who were free from injury took part in the study. Running-based anaerobic sprint test (RAST) was performed by the participants after checking their oxygen saturation and pulse rate (beats per minute) with a warm-up session. The anthropometric measurements: standing height, body composition, skinfolds (triceps, medial calf, subscapular and supraspinale), breadths (biepicondylar femur and humerus) and girths (arm and calf) were measured. The study has shown that most participants (50%) followed the rear-foot strike pattern during RAST. Significant differences in maximum and aver-age power output in RAST were obtained among the fore-foot strike (FFS) and mid-foot strike (MFS) runners compared with the rear-foot strike (RFS) runners. In the continual six sprints of the RAST, the power outputs were found to be FFS < MFS < RFS. Significant differences were found in fatigue index (FI) and anaerobic capacity (AC) of the MFS and FFS compared with RFS participants. This study has observed a significantly low fatigue index among RFS compared to FFS. This study concludes a higher dependency of the kinesiology and kinanthropometric parameters on multi-sprint sports performances. However, further studies are required to optimise sprinting in multi-sprint sports on different shoed and surface conditions, along with more match specific conditions.

Keywords: Foot strike patterns · fatigue · body composition · RAST · somatotype · kinanthropometric

B. K. Singh et al. (Eds.): ICBEST 2023, CCIS 2003, pp. 300–311, 2024.
https://doi.org/10.1007/978-3-031-54547-4_23

1 Introduction

Anaerobic capacity is a vital parameter to monitor the ability of a sports person to perform anaerobic activities. Activities involving high exertion of force in less time can be classified as anaerobic activities. Multi sprint sports like volleyball, football, hockey, cricket requires a sports person to suddenly increase his speed of running. To evaluate a sports person's anaerobic capacity in such condition many methods are widely followed–Running based anaerobic sprint test, wingate test, maximum accumulated oxygen deficit, critical power concept [1–4]. RAST has been a on field, cheap and reliable method to measure fatigue index and anaerobic capacity when compared to other methods [5]. Wingate test needs a cycle ergometer, treadmill, lab space to measure the anaerobic capacity. Also, participants often complain dizziness, nausea post completion of wingate test. Foot strike patterns can be classified into rear foot strike, mid foot strike and forefoot strike. The rear foot strike (RFS) was defined as the landing where the heel contacts the ground during initial contact prior to ball of the foot, a heel-toe run. The midfoot strike (MFS) is defined as a simultaneous landing of both the heel and ball of the foot. In forefoot strike (FFS), the ball of the foot contacts the ground before the heel, a toe-heel-toe run [6]. In this study anaerobic capacity, fatigue index and somatotypes of male University students (non-elite sports persons) were obtained and reported. The effect on foot strike patterns on these parameters has been analyzed.

2 Methods

26 male students from Central University of Rajasthan of the age group 18 to 25 (22.77 1.34) years were selected as the participants for the study. Subjects who are into physical fitness or sports activities for at least five years (playing any or some of the following games regularly: volleyball, football, basketball, badminton, handball, boxing, cricket and running) were recruited as participants. Informed consent was obtained from the participants regarding their willingness to participate in the study, fitness, and health status. The subjects were explained about all the data collection procedure. Data collection was performed over a period of three days for each subject. First, the participants were introduced to the study on the first day. The RAST, anthropometric tests and flexibility tests were performed on the second and third days. Participants were instructed to come in sportswear and the running shoes regularly used by them while running. They were instructed to come at least after two and half hours of food intake. Pulse rate and oxygen saturation were observed using Oxy-Watch® (ChoiceMMed™) before data collection. Participants were allowed for the test only if their oxygen saturation and pulse rate were in the range of 97% and 72 10 bpm.

400-m oval Running track of the football ground, Central University of Rajasthan, was selected as the location of the test. A 45-m straight-line track was marked for the test, with a recovery section of 10 m was marked (5 m on each side). The 35 m straight-lined track was selected for the test. Two cones (VIXEN®) at each end of 35 m of running track were placed (as described in Fig. 1.) Ground measurement was done by two 13 mm wide fibreglass measuring tapes of 30 m (FREEMANS TOP-LINE).

He participants were asked to stand at one end of the 35-m marked track and start with a maximal sprint on the 'go' command. 10 se post completion of the first sprint,

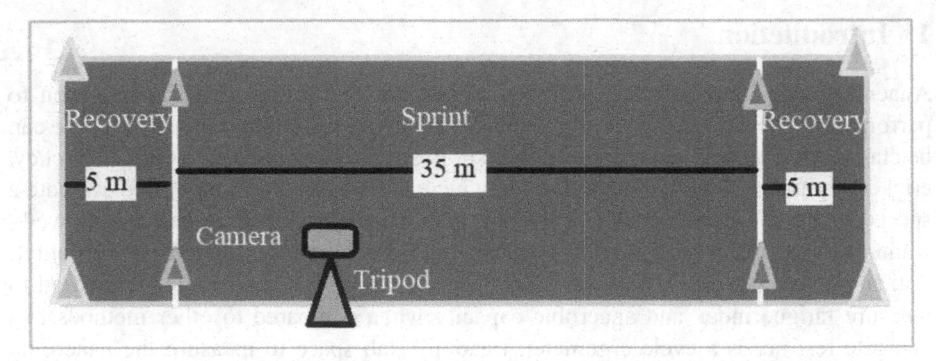

Fig. 1. Track Preparation for the Running-Based Anaerobic Sprint Test (RAST) and set up video capturing.

the next sprint was started on the observer's instruction, from the other end of the 35 m track. The procedure was continued until six sprints were completed. The time taken for the sprints along with bodyweight (obtained later while collecting the anthropometric data on day 3) were used to compute the anaerobic capacity, maximum power output, average power output, and fatigue index (FI) [7]. For all six sprints (n = 6), the power output was obtained from the Eq. (1):

$$PO_i(\,watt\,) = \frac{body\,mass\,(kg) \times distance\,(\,meter\,)^2}{S_i(\,second\,)^3} \tag{1}$$

where PO_i is the power output by a participant of the ith sprint (i = 1, 2,...,6) and distance is 35 m, which has been covered in each sprint. S_i is the time taken for completion of ith sprint in seconds. Maximum or peak power output (PO_{max}) and minimum power output (PO_{min}) are obtained as the largest and smallest value of the PO_i's respectively [8]. The *AVG PO* is the average value of power outputs of the 6 sprints. *RPP* is the relative peak power output obtained by Eq. (2). RPP is a fair comparison among the participants of different body weights.

$$RPP = \frac{Peak\ Power\ Output}{Body\ Mass\ (kg)} \tag{2}$$

Fatigue index (FI) and anaerobic capacity (AC) were also obtained from the Eqs. (3) and (4). FI was estimated to represent the rate of declination of power output in the performance of the participant. The lower FI indicates a better ability of the participant to maintain the performance and vice versa. Anaerobic capacity is the amount of total work completed during the whole RAST duration [8].

$$FI = \frac{PO_{\max} - PO_{\min}}{\sum_{i=1}^{6} S_i} \tag{3}$$

$$AC = \sum\nolimits_{i=1}^{6} PO_i \tag{4}$$

A tripod stand at the height of 10 in. with a mobile camera (Redmi Note 5) mounted on it was set to capture the video of foot strike pattern in the sagittal plane, as described in Fig. 1.

2.1 Foot Strike Classification

The foot-strike classification defined by Lieberman [6] is used in this study. With the highest % of appearance in recorded steps of RAST, the strike pattern was considered the dominant strike pattern for the participant for his sprinting. Video data collected was analysed using Kinovea-0.8.15 software in slow motion (30%) to extract the foot strike pattern. For each runner, approximately 4 to 7 foot strikes were captured while video recording. A sample of screenshots of video data for each foot strike is provided in Fig. 2. Using Kinovea-0.8.15, five sequences of images were obtained covering 0.7 to 0.8 seconds of the clips. Foot-strike classification was per- formed for each participant by visual observation from the videos. Foot strike pattern analysis from 2D videos has been accepted as a validated procedure [9].

2.2 Sit and Reach Flexibility Test

The standard sit and reach flexibility test, with the foot line at 10 cm were used. The participants performed the test without warming up. Each participant was asked to perform the test three times with a rest period of 30 s between each assessment, following the traditional method [10]. The best score (range 0 cm to 30 cm) of all three attempts were counted for the data analysis.

2.3 Obtaining the Anthropometric Parameters

Height of the participants was measured using stadiometer. Body weight and body compositions (body water, muscle mass, fat mass) were measured using Salter Body Analyser & Scale based on Body Impedance Analysis (BIA). Following the somatotype measuring protocols given by Heath and Carter in 2002, four skinfolds, two girths and two breadths were measured to obtain the somatotype of the participants [11]. Somatotyping is performed to evaluate body composition and shape. The procedure provides a three-numbered rating, which represents endomorphy, mesomorphy and ectomorphy, respectively [11]. Endomorphy is defined as the relative fatness of an individual whereas, mesomorphy represents the relative musculoskeletal robust- ness, and ectomorphy is defined as the relative linearity of a physique. A total of ten anthropometric dimensions are required for the estimation of somatotypes of which height and body mass were obtained previously. The remaining eight measurements of four skinfolds (triceps, medial calf, subscapular and supraspinale), breadths of tow bones-biepicondylar humerus and femur, girths of arm (tensed and flexed) and calf were obtained from right sides of the samples.

Digital sliding calliper, flexible fibreglass tape measure, and a skinfold calliper were used to obtain the bi-epicondylar humerus/femur breadth, arm/calf girth and skinfold thickness.

Skinfold calliper, SH5020 (SAEHAN®), has been used to obtain the skinfolds of the participants. The equations which were used to obtain the somatotype (endomorphy, mesomorphy and ectomorphy) are:

$$Endomorphy = -0.7182 + 0.1451\left(X^{1}\right) - 0.00068\left(X^{2}\right) + 0.0000014\left(X^{3}\right) \quad (5)$$

Fig. 2. Examples of fore-foot, mid-foot and rear-foot strike pattern while performing RAST, 5 frames (from left to right) From top to bottom: Fore-foot strike (right foot in frame), Mid-foot strike (right foot in frame), and Rear-foot strike (left foot in frame).

$$Mesomorphy = 0.858 \times HB + 0.601 \times FB + 0.188 \times CAG + 0.161 \times CCG - 0.131 \times H + 4.5 \quad (6)$$

$$X = (Subscapular\ sk\inf old + Triceps\ sk\inf old + Supraspinale\ sk\inf old) \times H. \quad (7)$$

where H is the height, HB is the biepicondylar humerus breadth, FB is the biepicondylar femur breadth, CAG is the corrected arm girth and CCG is the corrected calf girth of the participant, in Eqs. (8) and (9).

$$CAG = flexed\ arm\ girth - \frac{triceps\ skinfold}{10} \quad (8)$$

$$CCG = flexed\ calf\ girth - \frac{calf\ skinfold}{10} \quad (9)$$

To estimate ectomorphy three different equations were used with respect to height – weight ratio (HWR), where the HWR is calculated by Eq. (10).

$$HWR = \frac{height\ (cm)}{\sqrt[3]{body\ weight\ (kg)}} \quad (10)$$

$$\text{If } HWR \geq 40.75, \text{ then, } Ectomorphy = 0.732 \times HWR - 28.58 \quad (11)$$

$$\text{If } 38.25 \leq \text{HWR} \leq 40.75, \text{ then, } Ectomorphy = 0.463 \times HWR - 17.63 \quad (12)$$

$$\text{If HWR } \leq 38.25\text{, then, } Ectomorphy = 0.1 \quad (13)$$

To estimate the somatotypes among the three groups, the individual and group mean of the 3–point score (endomorphy–mesomorphy–ectomorphy) were evaluated and rounded off to the nearest integer value to plot in the Heath–Carter somato-type chart. Somatopoint (S) is a point in 3D space determined from the three somato-type scores, endomorphy, mesomorphy and ectomorphy, respectively, which can be plotted as a triad of x, y, z coordinates for the three components.

The density (D) of the participants were obtained using Womerseley Eq. (14). The Siri Eq. (15) was used to estimate body fat % (F) using density and the four skinfold measurements as input variables (triceps, medial calf, subscapular and supraspinale skinfolds) [12].

$$D = 1.1631 - 0.0632 \times \log_{10}^{\sum 4SF} \quad (14)$$

$$F = \left(\frac{495}{D}\right) - 450 \quad (15)$$

2.4 Statistical Analysis of Data

To verify the significance of the differences in the values among the three groups of the participants FFS, MFS and RFS, Kruskal–Wallis Test was performed using SPSS for K-independent samples (K = 3). The significant difference in Kruskal–Wallis test implies the difference among all the three groups. To identify the significant differences between any two of the three groups, Mann–Whitney U test, was performed for two independent samples.

3 Results and Discussions

The numbers of participants classified as forefoot strikers, midfoot strikers and rearfoot strikers are provided in Table 1. Participants were found to be RFS (50 %), followed by FFS (27 %) and MFS (23 %). Mean values of maximum Power Output, minimum Power Output and average Power Output for each set of runners are provided in Table 2. Power Output values were higher for forefoot strikers followed by mid-foot strikers and rear foot strikers. Results of Kruskal Wallis H Test (Table 2) indicates that there is significant difference ($p < 0.05$) among the 3 groups of foot strikers for maximum Power Output and average Power Output. Table 3 provides the results of Mann Whitney U Test (pairwise comparison) performed for the maximum Power Output and average Power Output of 3 classified groups. Significant differences ($p < 0.05$) were observed between maximum and average Power Output values of forefoot strikers and rear foot strikers. Variation of maximum and average Power Output for forefoot, midfoot and rear foot strikers is illustrated in Fig. 3(a). Power Output of sprint sequences is averaged for each group of classified runners and are provided in Table 4. Figure 3(b) illustrates the variation in Power Output of each sprint sequence for the classified groups of runners.

Table 1. Participants Classified based on Foot Strike Patterns-Fore-Foot Strike (FFS), Mid-Foot Strike (MFS) and Rear-Foot Strike (RFS)

Groups	Number of participants	%
Total Participants	26	100
Forefoot strikers	7	27
Midfoot strikers	6	23
Rearfoot strikers	13	50

Table 2. Results of Kruskal Wallis test for RAST parameters of runners with different foot strike patterns

Power Output PO (Mean ± SD in Watts)	Foot Strike Patterns			Kruskal Wallis H Test	
	FFS	MFS	RFS	ψ^2	*p* value
Maximum Power Output	684.70 ± 239.56	603.79 ± 128.83	463.52 ± 113.49	6.24	**0.04**
Minimum Power Output	402.44 ± 168.35	309.84 ± 45.04	284.18 ± 50.42	4.21	0.12
Average Power Output	533.95 ± 192.20	436.45 ± 50.11	367.39 ± 83.67	6.77	**0.03**

Table 3. Pairwise comparison (Mann Whitney U test) for Maximum Power Output and Average Power Output

Maximum Power	FFS	MFS	RFS
FFS	NA		
MFS	0.31	NA	
RFS	**0.02**	0.05	NA
Average Power	FFS	MFS	RFS
FFS	NA		
MFS	0.31	NA	
RFS	**0.02**	0.05	NA

Power Output is expected to decrease drastically from first sprint sequence to sixth sprint sequence for forefoot runners as these participants spend higher amount of energy for running and their anaerobic capacity will not help them to maintain a constant Power Output. Power Output values for midfoot and rear foot strikers are found to decrease steadily and not drastically as in the case of forefoot strikers.

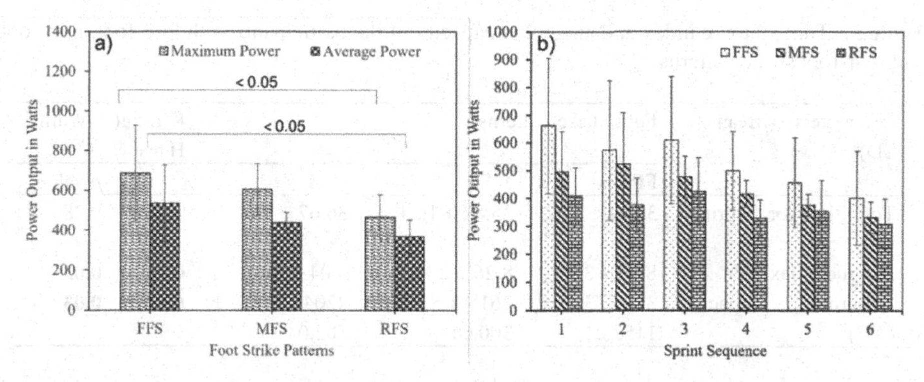

Fig. 3. Variation of maximum and average power for different foot strike patterns–FFS, MFS & RFS (a), Graphical Representation of the Power Outputs of RAST among the participants with FFS, MFS and RFS (b)

Table 4. Power Outputs of each of the six sprints of the participants with FFS, MFS and RFS

Power Output (Mean ± SD in Watts)	FFS	MFS	RFS
PO1	661.85 ± 244.22	495.21 ± 144.32	409.68 ± 100.02
PO2	574.34 ± 250.57	522.65 ± 184.93	375.93 ± 96.30
PO3	609.16 ± 229.06	476.36 ± 74.37	424.59 ± 121.73
PO4	499.04 ± 140.53	416.39 ± 47.39	330.41 ± 64.60
PO5	456.89 ± 160.97	376.78 ± 38.28	354.06 ± 110.57
PO6	402.44 ± 168.35	331.28 ± 55.62	309.70 ± 88.99

The time taken to complete six sprints, their fatigue index and anaerobic capacity for participants with different foot strike patterns is provided in Table 5. Kruskal Wallis H Test indicates that there is significant difference ($p < 0.05$) among the 3 groups of foot strikers for fatigue index and anaerobic capacity. Table 6 provides the results of Mann Whitney U Test (pairwise comparison) performed for the fatigue index and anaerobic capacity of 3 set of runners. Significant differences ($p < 0.05$) were ob- served between fatigue index values of rear foot strikers when compared with front foot and mid foot strikers. Likewise, significant difference ($p < 0.05$) was observed between anaerobic capacity values of rear foot strikers when compared with front foot strikers. The variation of fatigue index and anaerobic capacity as a function of foot strike patterns in provided in Fig. 4 (a) & (b).

The mean value of measured anthropometric parameters of the three sets of runners are provided in Table 7. Somatotype of each set of runners is obtained from the measured anthropometric values. Forefoot strikers and midfoot runners are reported to have a somatotype score of 352 (endomorphic mesomorph–mesomorphy is dominant and endomorphy is greater than ectomorphy). Rear foot runners are reported to have a somatotype score of 443 (central type–no component differs from the other two by

Table 5. Time, fatigue index and anaerobic capacity of the participants with fore-foot, mid-foot and rear-foot strike patterns

Parameters (Mean ± SD)	Foot Strike Patterns			Kruskal Wallis H test	
	FFS	MFS	RFS		*p* value
Total time for 6 sprints (in s)	34.41 ± 3.42	35.82 ± 1.24	36.67 ± 2.61	2.583	0.28
Fatigue Index (FI)	8.49 ± 3.36	8.26 ± 2.99	5.04 ± 2.42	6.70	**0.03**
Anaerobic Capacity (AC)	3203.73 ± 1153.21	2618.68 ± 300.64	2204.37 ± 502.04	6.77	**0.03**

Table 6. Pairwise comparison (Mann Whitney U test) for fatigue index and anaerobic capacity

Maximum Power	FFS	MFS	RFS
FFS	NA		
MFS	0.77	NA	
RFS	**0.04**	**0.02**	NA
Average Power	**FFS**	**MFS**	**RFS**
FFS	NA		
MFS	0.31	NA	
RFS	**0.02**	0.05	NA

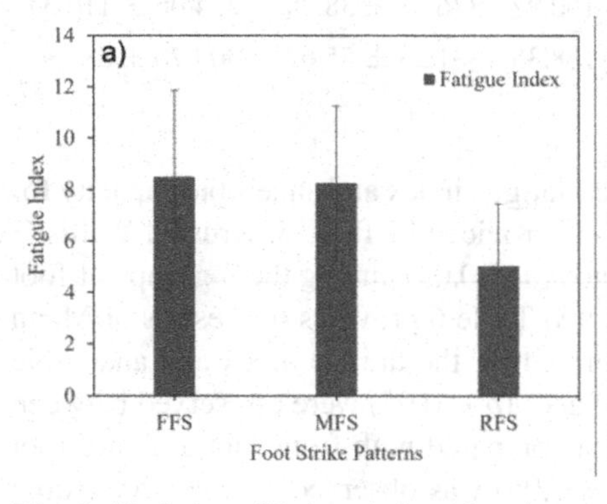

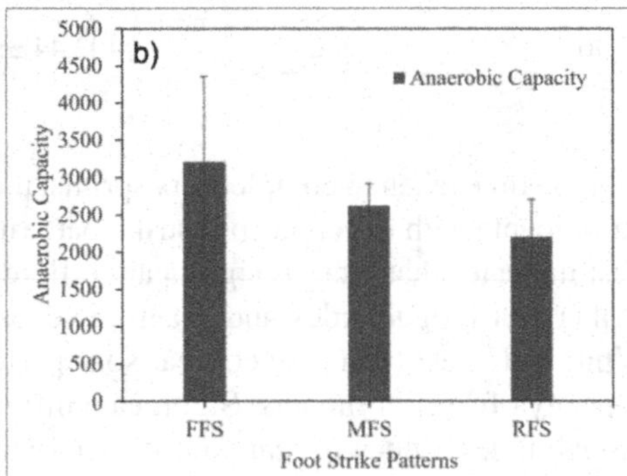

Fig. 4. Variation of fatigue index (a) and anaerobic capacity for various foot strike patterns–FFS, MFS & RFS

more than 1 unit). Forefoot runners and midfoot runners have better musculo-skeletal composition than the rear foot runners.

Research groups have reported that rear foot strike is the most commonly observed foot strike pattern (>90%) among recreational and distant runners [13, 14]. In this study, participants involved in multi-sprint sports have reported rear-foot strike in 50% of cases, but not as high as the previous studies (>90%). This observation may imply

Table 7. Results of Kruskal Wallis test for anthropometric parameters of runners with different foot strike patterns

Parameters	Foot Strike Patterns			Kruskal Wallis H Test	
	FFS	MFS	RFS	ψ^2	*p* value
Medial Calf Skin fold (mm)	6.86 ± 2.42	7.17 ± 1.07	7.87 ± 2.38	0.99	0.60
Biepicondylar Femur Breadth (cm)	8.68 ± 0.22	8.33 ± 0.42	8.37 ± 0.41	2.59	0.27
Corrected Calf Girth (cm)	38.50 ± 1.02	38.36 ± 1.79	37.33 ± 1.80	2.48	0.28
Body Muscle %	44.97 ± 1.43	45.55 ± 1.83	45.27 ± 2.23	0.18	0.91
Body Fat %	15.90 ± 3.04	15.75 ± 2.02	16.66 ± 2.10	0.43	0.80
Sit and Reach Test (cm)	21.29 ± 5.39	24.17 ± 4.91	21.73 ± 5.36	1.08	0.58

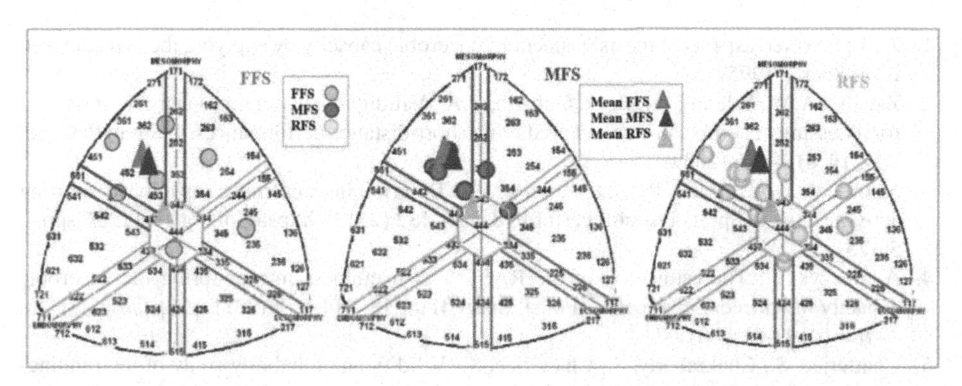

Fig. 5. Somatotype Plot (Heath-Cater) among FFS (352), MFS (352) and RFS (443)

that dynamics involved in a multi-sprints sports have a role to play in the determining foot strike pattern while running. Moreover, the running foot strike patterns are of 6 short distance sprinting (35 m each, 10-s recovery), which is more related to the multi-sprint sports [8].

Average power output in RAST among soccer players were found to be 693.80 in a study among 33 soccer players and which was found high to low depending on playing positions forward (735.94 ± 168.3), defence (676.40 ± 183.4) and midfield (669.04 ± 144.7) respectively [15]. These reported values are closer to the average power output of the FFS participants (533.95 ± 192.20) of this study which is the highest among all the three groups.

4 Conclusion

Fatigue index and anaerobic capacity has been obtained and reported for runners with different foot strike patterns using RAST method. Somatotypes of participants with different foot strike patterns is also obtained from anthropometric parameters of participants and reported. Fatigue index and anaerobic capacity of participants has been observed to be directly linked with foot strike patterns. Significant differences have been observed between the fatigue index and anaerobic capacity of forefoot strikers and rear foot strikers. The number of participants in this study is very less. Identifying the onset of fatigue from EMG signal of the calf muscle/quadriceps/hamstrings obtained during RAST procedure for runners with different foot strike patterns will help us to validate this study. Correlation studies can be performed by measuring blood lactate concentration to quantify the degree of fatigue developed for runners with different foot strike patterns.

References

1. Z., J.: Selected aspects of the assessment of anaerobic capacity by applying the Wingate test. Biol. Sport. (1995)
2. Zagatto, A.M., Wladimir R. Beck, Gobatto, C.A.: Validity of the running anaerobic sprint test for assessing anaerobic power and predicting short-distance performances. J. Strength Cond. Res. **10**, 610–616 (2009)
3. Noordhof, D.A., Skiba, P.F., De Koning, J.J.: Determining anaerobic capacity in sporting activities. Int. J. Sports Physiol. Perform. **8**, 475–482 (2013). https://doi.org/10.1123/ijspp.8.5.475
4. Adamczyk, J.: The estimation of the RAST Test usefulness in monitoring the anaerobic capacity of sprinters in athletics. Pol. J. Sport Tour. **18**, 214–218 (2011). https://doi.org/10.2478/v10197-011-0017-3
5. Chatterjee, S., Chakraborty, S., Chatterjee, S.: Validity and reliability study of the running-based anaerobic sprint test for evaluating anaerobic power performance as compared to Wingate test in Indian male track and field sprinters. Eur. J. Phys. Educ. Sport Sci. **8**, 38–50 (2022). https://doi.org/10.46827/ejpe.v8i4.4327
6. Lieberman, D.E.: What we can learn about running from barefoot running: an evolutionary medical perspective. Exerc. Sport Sci. Rev. **40**, 63–72 (2012). https://doi.org/10.1097/JES.0b013e31824ab210
7. Queiroga, M.R.,Cavazzotto, T.G., Ferreira, S.A.: Validity of the RAST for evaluating anaerobic power performance as compared to Wingate test in cycling athletes. Motriz. J. Phys. Educ. (2013).https://doi.org/10.1590/S1980-65742013000400005
8. Erformances, S.H.I.P., Agatto, A.L.M.Z., Eck, W.L.R.B., Obatto, C.L.A.G.: Validity of the running anaerobic sprint test for accessing anerobic power and predicting short distance performances. J. Strength Cond. Res. 1820–1827 (2009)
9. de Oliveira, F.C.L., Fredette, A., Echeverría, S.O., Batcho, C.S., Roy, J.S.: Validity and reliability of 2-Dimensional video-based assessment to analyze foot strike pattern and step rate during running: a systematic review. Sports Health **11**, 409–415 (2019). https://doi.org/10.1177/1941738119844795
10. Wells, K.F., Dillon, E.K.: The sit and reach—a test of back and leg flexibility. Res. Q. Am. Assoc. Heal. Phys. Educ. Recreat. **23**, 115–118 (1952). https://doi.org/10.1080/10671188.1952.10761965

11. JEL Carter: Part 1: The Heath-Carter Anthropometric Somatotype. Instr. Man. 1–25 (2002)
12. Durnin, B.Y.J.V.G.A., Womersley, J.: Body fat assesed from total body density and its estimation from skinfold thickness : measurements on 481 men and women aged from 16 to 72 years. Br. J. Nutr. **32**, 77–97 (1973)
13. Almeida, M.O. De, Saragiotto, B.T., Yamato, T.P., Lopes, A.D.: Is the rear-foot pattern the most frequently foot strike pattern among recreational shod distance runners? Phys. Ther. Sport. (2014).https://doi.org/10.1016/j.ptsp.2014.02.005
14. Kasmer, M.E., Liu, X., Roberts, K.G., Valadao, J.M., Incorporated, O.: Foot- strike pattern and performance in a marathon Mark. Int J Sport. Physiol Perform. **8**, 286–292 (2013)
15. Çetinkaya, E., Tanır, H., Çelebi, B.: Comparison of agility, sprint, anaerobic power and aerobic capacities of soccer players by playing positions. J. Educ. Train. Stud. **6**, 184–190 (2018). https://doi.org/10.11114/jets.v6i9.3560

Minimal Surface Based Hybrid Cellular Material Model for Use in Healthcare Engineering

Ajay Kumar Sahu[1], Aman Khandwe[1], Jitendra Kumar[1], Nitesh Kumar Singh[1], Rati Verma[2], and Nishant Kumar Singh[1](✉)

[1] Department of Biomedical Engineering, National Institute of Technology (NIT), Raipur, India
nksingh.bme@gmail.com

[2] Biomechanics Lab, School of Biomedical Engineering, Indian Institute of Technology, Varanasi, India

Abstract. Triply Periodic Minimal Surface (TPMS) is distinctive structure that supports numerous cellular activities and encapsulating trabecular bonemimicking hyperboloidal topography. They have emerged as a prospective choice in reconstructive orthopedic surgery. The hybrid design of these structures is an effective approach to constructing functional implant structures. In this study, we have used TPMS structures to develop the hybrid models with given transition boundaries to investigate the multi-morphology cellular material model for its effectiveness as scaffold use. Sigmoid function was employed to connect more than two TPMS structures, with a special shape at the transition boundary region. Furthermore, numerical simulation was performed to investigate the mechanical performance on hybrid model of IWP (I-graph-wrapped package curved surface) and Diamond (D) (both are the family of TPMS unit cell libraries) with accurately controlled porosity of 50%, 60%, 70% and 80%. The numerical results showed that the proposed hybrid scaffold architecture has the potential to be advantageous for accurately controlling the spatial porosity distribution to tailor the specific bone attributes while maintaining the benefit of the TPMS-based hybrid unit cell libraries.

Keywords: hybridization · metamaterial · material design · scaffold · TPMS · porosity · porous structure

1 Introduction

Bones are one of the most vital components of the human body as they provide shape, mechanical support, and protection to the body and also helps in facilitating movement apart from that they also help in the homeostasis of the body and serves as the home for the bone marrow [1–4]. Daily, bones are subjected to various compressive loads, and the ability of bone tissues to tolerate this load decreases, as the load increases with time, and bone fracture occurs when these forces exceed the tolerability of the bone tissue [3]. There are several kinds of bone fractures that can be categorized by a variety of characteristics, such as the shape and pattern of fragmented pieces, which can be transverse, oblique,

B. K. Singh et al. (Eds.): ICBEST 2023, CCIS 2003, pp. 312–325, 2024.
https://doi.org/10.1007/978-3-031-54547-4_24

spiral, or comminute. There are three types of fractures based on their etiology: traumatic, fatigue, and pathological, and then there are closed and open fractures. Compression fracture, gunshot fracture, green stick fracture, and avulsion fracture are some of the other types [1, 3]. Bone implants and scaffolds are common orthopedic treatments used to fix and stabilize injured body parts and responsible for the beginning of the healing process. Artificial materials, such as bone replacements, may be used to make bone implants including metal alloys and polymers [5, 6]. Despite recent interest in these materials, they still need to fulfil a number of biological and mechanical criteria. Due to the excellent biocompatibility, mechanical strength, and resistance to corrosion qualities, medical-grade titanium alloy (Ti6AL4V) is often employed in clinical applications and the same is used to do the numerical analysis in this study [5–7].

A scaffold's primary purpose is to act as a three-dimensional (3D) template for tissue development and cell organization, hence removing the limitations of both autologous and allogeneic bone transplantations [8]. Currently, the use of light weight implants to repair segmental bone abnormalities is receiving good attention from researchers. Intentional porosity generation with the use of porous TPMS structures into the scaffolds can be a promising candidate for lightweight and superior mechanical properties which can handle the benefits of reducing the stress shielding, enhancing bone growth, and reducing relative micro-motion between the implant and surrounding tissue. According to recent studies, TPMS is a flexible way of creating biomorphic porous scaffold designs. The selection of TPMS was made possible by the fact that its overall porous structure can be precisely defined by mathematical expressions, and that some of its fundamental characteristics, such as porosity and volume-specific surface area, can be directly manipulated by the variables of the function expression. Additionally, the surface of TPMS is extremely smooth, the structure is highly inter- connected, and there are no sharp twists or connecting points of the lattice porous structure. Due to their continuity throughout space, periodicity in three separate directions, and bonelike characteristics, these lattice-type structures most closely resemble the nature of bone [9–13]. Distinctive sheet-based metamaterials TPMS cellular structures can outperform strut-based structures in terms of mechanical and other engineering aspects. In terms of mechanical properties, these sheet-based cellular structures are superior to strut-based structures due to extensive nodal connectivity that have shown bending as the predominant source of deformation. [14–18]. The continuous and smooth shells of TPMS sheet structures allow larger surface areas and continuous interior channels. TPMS structures can have greater specific stiffness and strength when compared to stretch-dominated porous structures [17, 19, 20]. Furthermore, TPMS offers the advantage of developing scaffolds for specific requirements by varying the porous morphologies high/low in accordance with the application requirements. Also, with the proper use of parameter and the placement of TPMS structure one can accurately developed the hybrid multi-morphological structure with precisely controlled volume fraction, porosity and other mechanical parameters and replicate the tissues microstructure [21–23].

There are several types of TPMS family among them four types of TPMS structures namely Primitive, Gyroid, I-WP, and D exhibit quiet similar morphology with bone tissues. According to existing literature, D structures exhibit an intense stretching-dominated behavior as compared to the Gibson and Ashby group of structures [17]. IWP is another family of TPMS structure where the letter 'I' stands for Body-Centered Cubic (BCC) structure and the 'WP' is an acronym to 'wrapped package,' because of the finite portions of the IWP structure resemble string arrangements on a simple wrapped package. Researchers have proposed several design approaches, such as varying strut diameter through its length, in relation to studies on the mechanical characteristics of hybrid scaffolds and have demonstrated their superiority using quasistatic mechanical characterization [24, 25]. Another efficient method for developing hybrid graded structures has been suggested: integrating two or more types of unit cells with distinct morphologies. In this situation, it is possible to modify the morphological characteristics independently and create a transition strategy that is more versatile. In this study, D and IWP structure''s trigonometric function was used for developing regular unit cells and sigmoid function was used for hybrid development of scaffolds unit cell libraries. Finally, numerical simulation was performed for hybrid model of D and IWP of different porosity ranging from 50%–80% to study the bulk properties and deformation behavior to ascertain the suitability of these hybrid models for hard tissue application.

2 Materials and Methods

2.1 Development of TPMS Cellular Structure

Figure 1 shows two porous TPMS unit cell libraries; the D and IWP surfaces are considered for hybridization in this study. These porous structures can be described by the following functions based on our prior research.

IWP-TPMS structure:

$$2.^{*}\left(\cos(2.^{*}\ \mathrm{pi}.^{*}x)^{*}\cos(2.^{*}\ \mathrm{pi}.^{*}y)+\cos(2.^{*}\ \mathrm{pi}.^{*}y)^{*}\cos(2.^{*}\ \mathrm{pi}.^{*}z)+\cos(2.^{*}\ \mathrm{pi}.^{*}z)*\cos(2.^{*}\ \mathrm{pi}.^{*}x)\right)-\left(\cos(4.^{*}\ \mathrm{pi}.^{*}x)+\cos(4.^{*}\ \mathrm{pi}.^{*}y)+\cos(4.^{*}\ \mathrm{pi}.^{*}z)\right)+C\cdot \tag{1}$$

D-TPMS structure:

$$\cos(2.*\ \mathrm{pi}.^{*}x).*\cos(2.*\ \mathrm{pi}.^{*}y).*\cos(2.*\ \mathrm{pi}.*z)-\sin(2.*\ \mathrm{pi}.^{*}x).*\sin(2.*\ \mathrm{pi}.^{*}y).*\sin(2.*\ \mathrm{pi}.^{*}z)+C \tag{2}$$

where 'C' is the level set constant used to control the porosity of the structures [26].

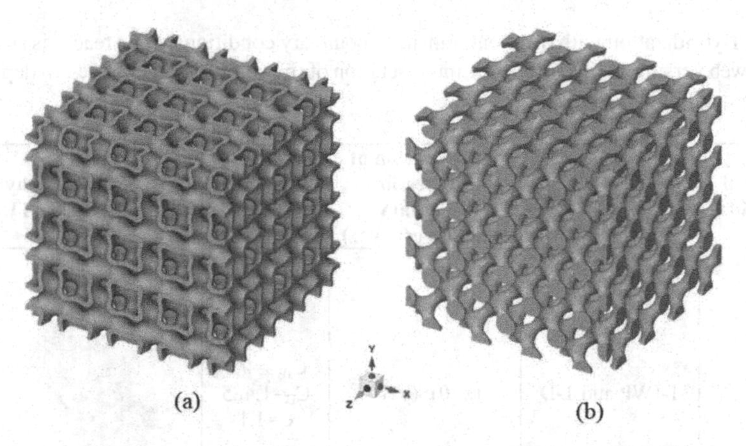

Fig. 1. Representation of TPMS based unit cell libraries (a) IWP (b) D

2.2 Implementation of Sigmoid Function and Construction of Hybrid TPMS Unit Cell Libraries

Hybrid of TPMS structures can be obtained by using the specific sigmoid function which can give a smooth transition between the two morphologies. Mathematically, the whole scaffold structure is defined by:

$$f_h(x, y, z) = s(x, y, z)f_1(x, y, z) + (1 - s(x, y, z))f_2(x, y, z) \tag{3}$$

Equation (3) gives the hybrid unit cell of two morphologies, where (f1) and (f2) are the primary morphologies functions and s (x, y, z) is a weight function that governs the transition between two morphologies. It provides the monotonic variation from 0 to 1 [27, 28].

$$s(x, y, z) = \frac{1}{1 + e^{\alpha g}(x, y, z)} \tag{4}$$

where g (x, y, z) = 0 defines the intersection boundary of the two TPMS structure and it is a continuous structure and is the gradient control parameter which provides a smooth and steady variation of structures at the transition boundary. To generate hybrid models, we have used fixed value of $\alpha = -15$ for mismatched structure hybridization. By modifying the intersection boundary function g (x, y, z) as listed in Table 1, we have generated seven different hybridized models by using four distinct TPMS morphologies namely; D, IWP, Gyroid (G), Primitive (P) by employing sigmoid function as given in Eq. 4 to demonstrate the effectiveness of the custom code developed in MATLAB® (MathWorks, Natick, MA). Several notations were used to describe hybrid structure and its parameters which are listed at the bottom of Table 1.

Table 1. Hybridization with different transition boundary conditions. (The reader is suggested to the refer web version of this article for interpretation of the color references used to depict the 3D hybrid model in this table.)

Name of hybrid	Number and types of TPMS Structure used	Modification of intersection boundary function g(x, y, z)	Level set constant 'C'	Developed 3D hybrid model in MATLAB
DI_{Cross}	1-IWP and 1-D	(x+0)×(y+0)	C_{IWP}=0.1 C_D=1.485 C=1.1	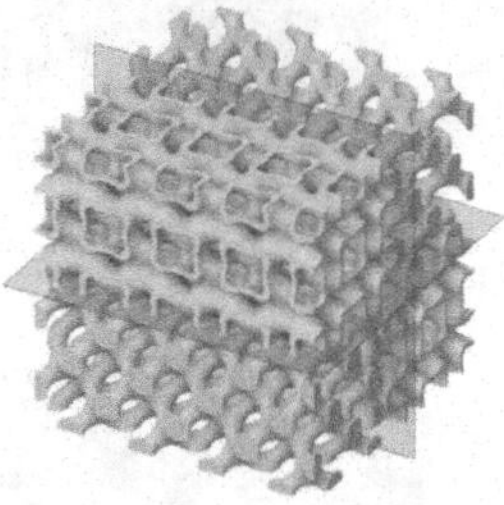
$DI_{parallel}$	1-IWP and 1-D	x+0	C_{IWP}=0.1 C_D=1.485 C=1.072	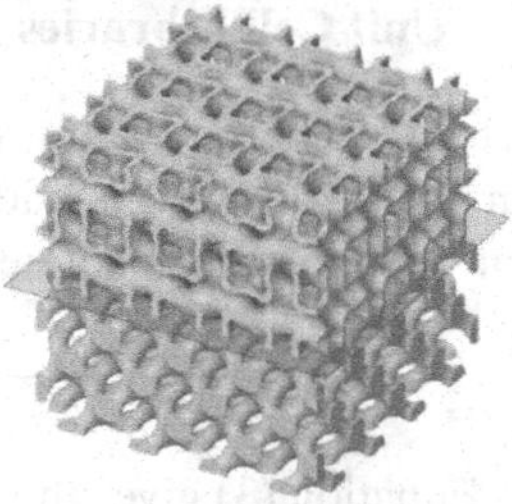
$DIGP_{Cross}$	1-IWP, 1-D, 1-Primitive and 1-Gyroid	(x+0)×(y+0) and (z+0)	C_{IWP}=0.1 C_D=0.6 C_{DI}=0.3 C_G=0.6 C_P=0.6 C_{GP}=0.4 C=1	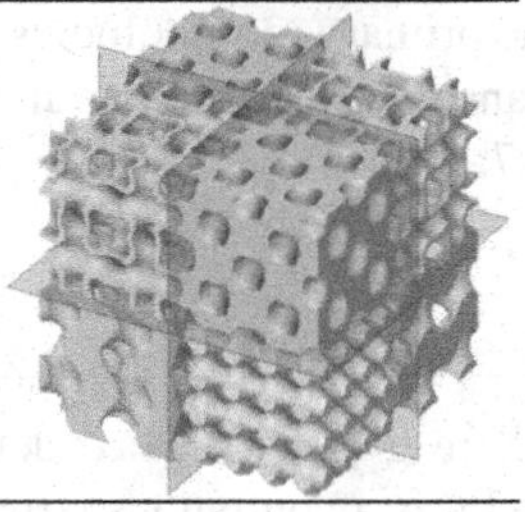

(continued)

Table 1. (*continued*)

$DIGP_{Parallel}$	1-IWP, 1-D, 1-Primitive and 1-Gyroid	(x+0) and (y+0)	C_{IWP}=0.1 C_D=0.6 C_{DI}=0.3 C_G=0.6 C_P=0.6 C_{GP}=0.4 C=1	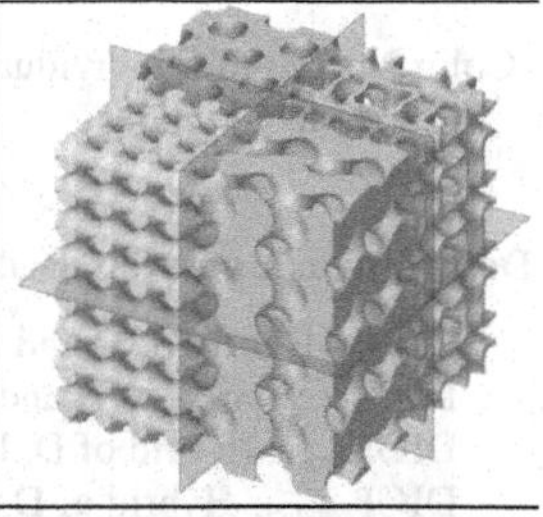
$DIGP_{Cross2\rho}$	1-IWP, 1-D, 1-Primitive and 1-Gyroid with 2 different densities of each structure	(x+0), (y+0) and (z+0)	C_{IWP1}=0.1, C_{IWP2}=0.2 C_{D1}=0.6, C_{D2}=0.9 C_{G1}=0.6, C_{G2}=0.6 C_{P1}=0.9, C_{P2}=0.2, C_1=0.2, C_2=0.4 C=1.08	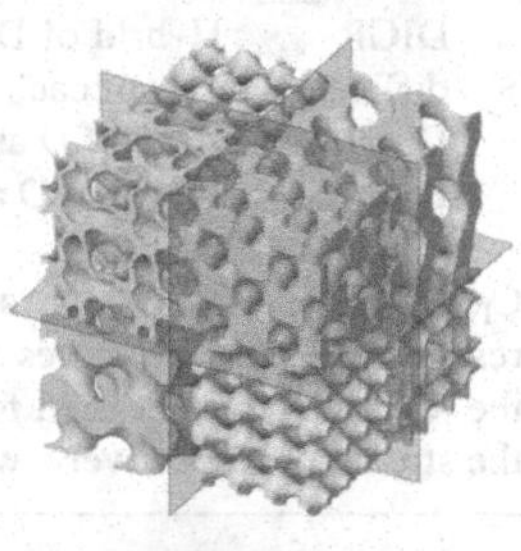
$DI_{Spherical}$	1-IWP and 1-D	$(x^2+0)+(y^2+0)+(z^2+0)-R^2$ Where R is the radius of boundary.	C_{IWP}=0.1 C_D=1.485 C=1.2	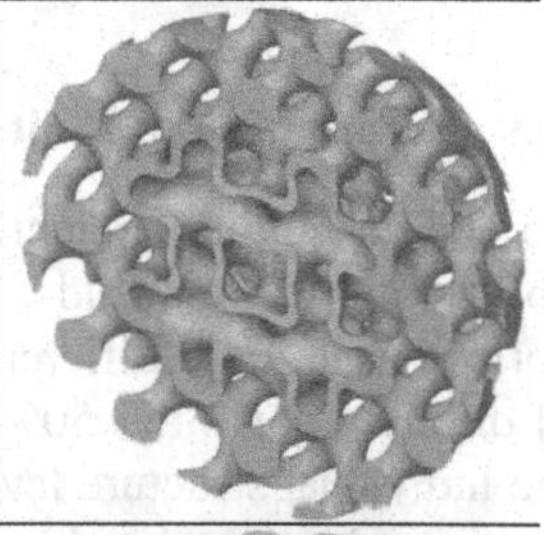
$DI_{Cylindrical}$	1-IWP and 1-D	$(x^2+0)+(y^2+0)-R^2$ Where R is the radius of boundary.	C_{IWP}=0.1 C_D=1.485 C=1.1	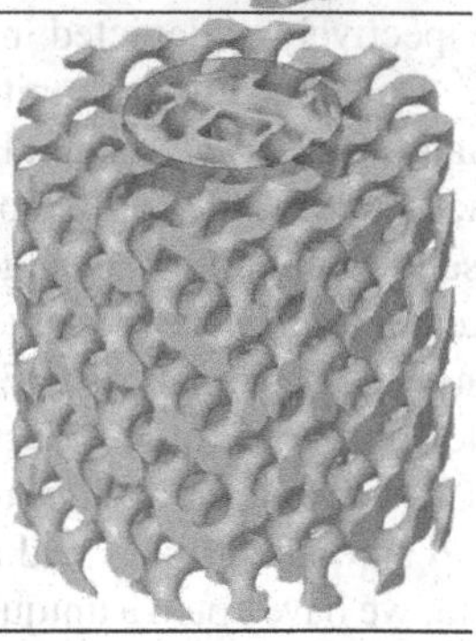

(*continued*)

Table 1. *(continued)*

Color Notation for individual TPMS Strucures	IWP	D	G	P

Description of notaton used for hybridized structures:

DI_{Cross} : Hybrid of D and IWP with cross transition boundary
$DI_{parallel}$: Hybrid of D and IWP with parallel transition boundary
$DIGP_{Cross}$: Hybrid of D, IWP, Gyroid and primitive with cross transition boundary
$DIGP_{parallel}$: Hybrid of D, IWP, Gyroid and primitive with parallel transition boundary
$DIGP_{Cross2p}$: Hybrid of D, IWP, Gyroid and primitive with cross transition boundary with 2 different density of each structures
$DI_{Spherical}$: Hybrid of D and IWP with spherical transition boundary
$DI_{Cylindrical}$: Hybrid of D and IWP with cylindrical transition boundary

C_{IWP},C_D,C_P,C_G,C_{DI},C_{GP}: C stands for level set constant and subscript IWP, D,P,G stands for respective TPMS structures and DI, GP stands for dual TPMS hybrid structures and C_1 and C_2 are the level set constant used for the $DIGP_{Cross2p}$ in which both denote the different density set of the structures respectively, whereas C is the overall levelset constant of respective structure.

2.3 Preparation of Finite Element Model of Hybrid TPMS Scaffold

To demonstrate the potential of hybridization of regular TPMS structures for its suitability as bone scaffolds, fused unit cell $DI_{parallel}$ was used to study its mechanical performance. Four different unit cell of the size 2mm x 2mm x 2mm were developed of different porosities 50%, 60%, 70% and 80% respectively by appropriately varying the individual structure level set constant 'C_D' and 'C_{IWP}' and combined constant 'C' respectively as depicted in Fig. 2.

Developed hybrid unit cells were exported in STL format through STL write command. Since STL triangulations cannot be used directly for FEM purposes due to the fact that STL files only represent "bucket of faces" and not solid models, it is difficult to use them for data exchange purposes such as STEP and IGES. Also, object boundaries are commonly too irregular, it is difficult to create effective meshes that can be used for finite element analysis (FEA). The elements must have a unique shape, optimal size, and the appropriate quality factor for the precise calculation of governing equations used in FEA software in background. Hence, it is essential to simplify the boundaries of STL into smooth surfaces and convert the scaffold into finite element/solid model. To this end, we have opted a unique method to convert these complex hybrid TPMS unit cell into finite element models. Initially we used six planes to split and omit each facet present on the face of rectangular shaped unit cell. The remaining part of triangulated surfaces of unit cells which represent highly curved surfaces were carefully patched into rectangular smooth surface by the method described in our previous research [28]. The patched model face boundaries (of which facets were removed) were closed by smooth surfaces in PTC Creo Parametric 2.0 and finally obtained solid model of hybrid TPMS scaffold. The procedure was likewise used to generate all four solid models of hybrid TPMS. The faces which were missing or if any gap between generated patches is there, they

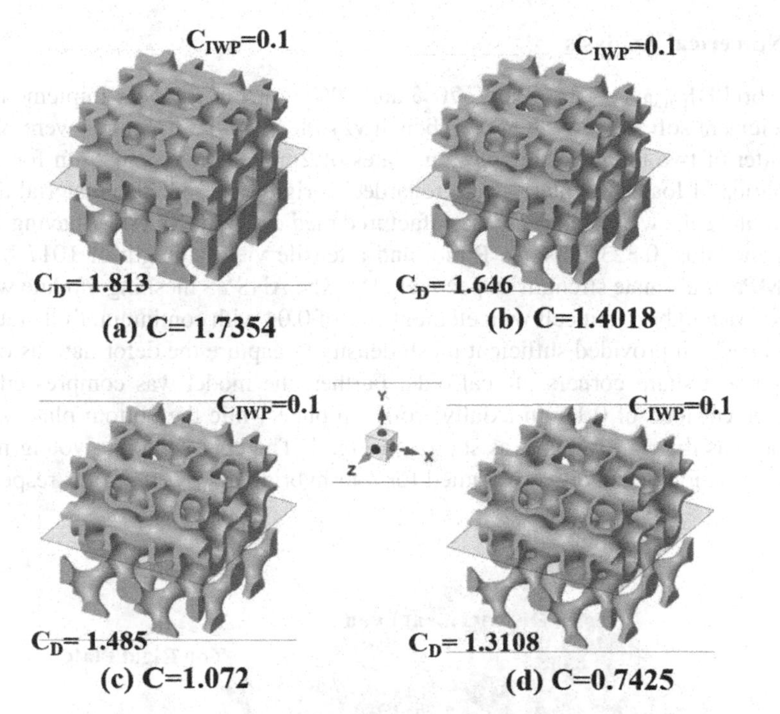

Fig. 2. Hybrid DIparallel structure with two TPMS unit cell IWP at top and D at the Bottom of transition boundary respectively. (a) Porosity 50% (b) Porosity 60% (c) Porosity 70% (d) Porosity 80%.

were carefully checked and rebuilt by repair tool present in ANSYS SpaceClaim. The complete sequence of events to convert the STL model into solid model is represented in Fig. 3.

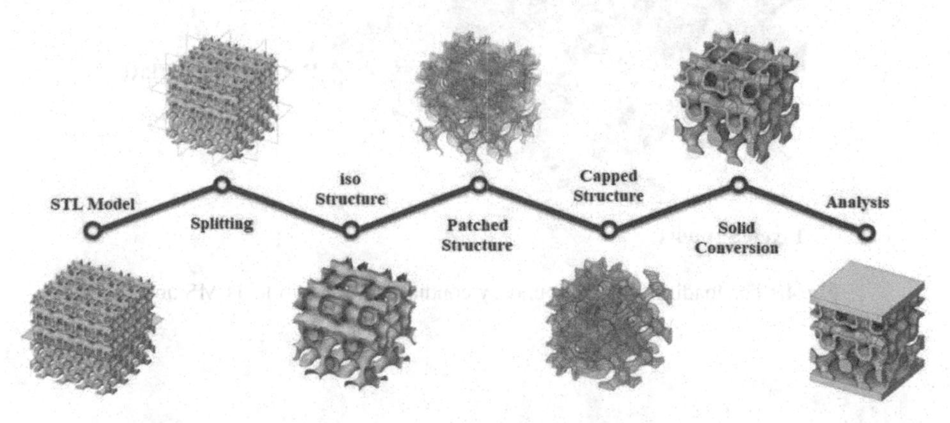

Fig. 3. Complete sequence for finite element model development

2.4 Numerical Analysis

The hybrid $DI_{parallel}$ of 50%, 60%, 70% and 80% respectively were implemented into finite element software ANSYS Workbench v21.0R2. All the structures were placed in the center of two parallel compression plates of 2mm × 2mm × 0.2mm for uniform distribution of loads, the plates were regarded as rigid material. The material assigned to each unit cells were additively manufactured medical grade Ti6Al4V having 107GPa young modulus, 0.323 Poison's Ratio, and a tensile yield strength of 1017 MPa and 1096 MPa of ultimate strength respectively [5]. The ANSYS meshing module was used to mesh each hybrid unit cell with element size of 0.04 with continuum finite tetrahedra elements which provided sufficient mesh density to capture the deformations occurred at edges and sharp corners of scaffolds. Further, the model was compressed with a displacement load of 0.1 mm axially from top plate while the bottom plate was kept stationary as the rigid support as shown in Fig. 4. The corresponding young modulus and deformation behavior were studied for four hybrid $DI_{parallel}$ scaffolds respectively.

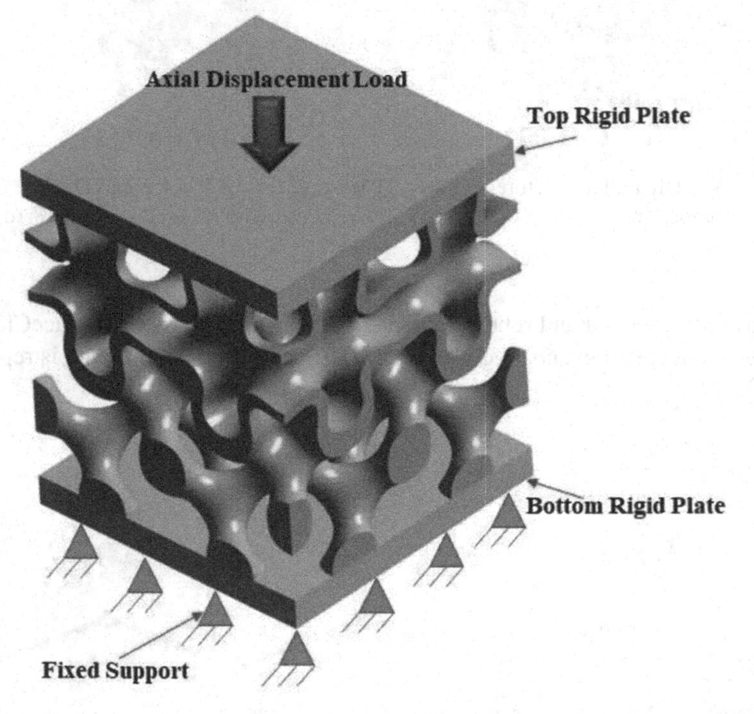

Fig. 4. The loading and the boundary condition of the hybrid TPMS unit cell

3 Results and Discussion

A set of four hybrid structures of D and IWP with parallel transition boundary, having different porosities ranging from 50% to 80%, were used to investigate the mechanical characteristics. The mechanical behavior of these different hybrid scaffolds was determined by performing axial displacement loading of 0.1 mm. The viability of TPMS porous hybrid morphology use as orthopedic implant was also taken into account. Engineering stress-strain curve for the hybrid model were plotted in Fig. 5. The stress- strain curve has two sections: linear section which shows the elastic property and nonlinear which shows the plastic property of the hybrid structures. The linear relation of the stress-strain curve was used to calculate the effective young's modulus of hybrid morphologies. The hybrid structural modulus was determined by using the linear elastic response of the load displacement relationship by using Eq. 5.

$$\text{Young Modulus} = \frac{\text{Force Reaction} \times \text{Initial Length of Structure}}{\text{Top surface area of the Structure} \times \text{Shortening of the Structure}} \quad (5)$$

The yield compressive strength of each hybrid structure was determined by 0.2% offset strain method, the yield strength was evaluated at the point where the offset line intersects the stress strain curve [5]. The calculated young's modulus and yield compressive strength of hybrid porous structure having porosities ranging from 50% to 80% respectively, are listed in Table 2.

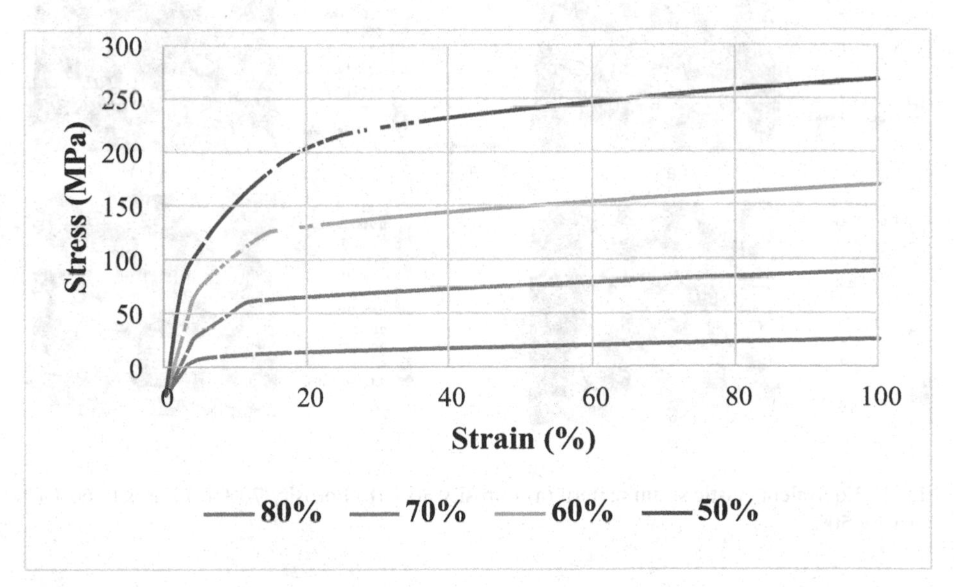

Fig. 5. Stress-strain curve for the hybrid morphology (a) Porosity 80% (b) Porosity 70% (c) Porosity 60% (d) Porosity 50%

Table 2. Calculated effective mechanical properties of hybrid DIparallel structures on different porosities.

Mechanical Property Parameter	Porosity 50%	Porosity 60%	Porosity 70%	Porosity 80%
Young's Modulus (in GPa)	17.909	9.871	6.368	3.721
Yield Strength (in MPa)	130	107.5	60	33.5

The effective young modulus of the hybrid morphologies of D and IWP for different porosities is observed in the range of 3.721–17.909 GPa and the effective compressive strength lied in the range of 33.5–130 MPa. It is obvious from the results that increasing the porosity from 50% to 80% leads to a decrease in modulus and compressive strength of the hybrid structures by 13.26% and 9.49% respectively. Figure 6 shows the von mises strain pattern of the $DI_{parallel}$ hybrid structure of different porosity. It is evident from the strain pattern that at the central region of strut, higher concentrations of strains were noted, whereas it decreases as we move to the border region of the hybrid structure.

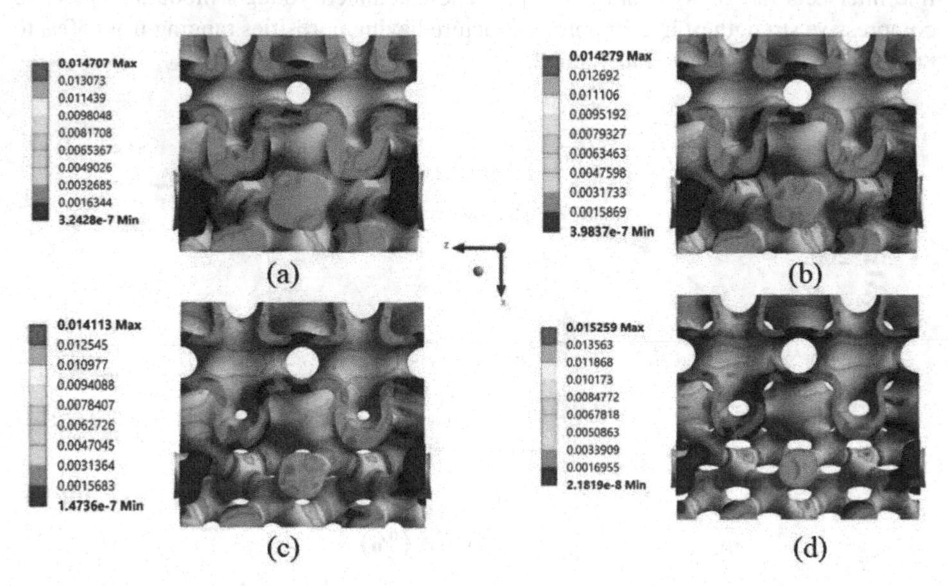

Fig. 6. Equivalent elastic strain pattern (a) Porosity 80% (b) Porosity 70% (c) Porosity 60% (d) Porosity 50%

Axial compression test simulation of hybrid structures yielded values of young modulus between 3.7 GPa to 17.9 GPa, comparable to human bone experimental data. Young bone has a much lower elastic modulus than solid titanium scaffolds, with a range of 3–30GPa for cortical bones and 0.02-1GPa for trabecular bone [29]. However, modulus mismatch between implants and its neighboring bone tissue gives rise to stress shielding phenomena which is undesirable [30, 31]. Observing the values of young modulus which agrees well with the elastic modulus of human bone, the findings of the numerical analysis suggests that the young modulus of the proposed structure is consistent for its application as bone scaffold [29]. It has been seen that by lowering the elastic moduli that matches the host bone, the suggested TPMS hybrid shape may reduce the stress shielding effect [32–34]. Furthermore, modulus of hybrid scaffold can be uniquely matched by tuning the exact modulus by altering the morphological parameters of the hybrid equations, an advanced approach for patient specific implant development.

The hybrid morphology of TPMS proposed in this study can be a suitable candidate for bone scaffold where the properties can be uniquely matched with the host bone, during the bone defect treatment, bone scaffolds must exactly form the defective component, offer substantial mechanical support to withstand on physiological loading also stimulate tissue regeneration simultaneously. These requirements contradict design and manufacturing objectives of the scaffolds. Since, dense scaffolds improve mechanical performance, whereas porous scaffolds improve cell transport phenomena. Hence, these scenarios can be fulfilled by using hybrid porous scaffolds, keeping the dense structure in the outer cortical layer and the lower density structure in the inner part of the hybrid by carefully varying the level set constant of the hybrid morphology. Another suitable application of hybrid morphology can be on the long bones where the proposed structure can be utilized to generate the implant to support the higher density and load bearing property of the epiphysis and lower density stretching property of the diaphysis. From the results comparison between IWP and D structures, it is found that IWP has a greater young's modulus at a particular density with greater damping ratio making it an optimum choice for the load bearing areas with D structure having greater stretching property with lower young's modulus at a particular density that can be utilized for the shaft of the femur [5, 17, 24]. Therefore, the hybridization can have very promising application in case of the hard tissue reconstructions when specific variations are desired in replacement devices in accordance with host bones.

4 Conclusion

Bone is a vital part of our body structure, and it is very important that the implants we propose should have similar properties and mechanical strength as that of the hard tissues. TPMS structure have shown potential to fulfill the design requirement of bone tissue replacement. The implicit function and sigmoid function employed in this study can be uniquely selected and level set constant can be tuned to obtain the hybrid optimal scaffold design to fulfill the subject specific and location specific requirements. As the implicit function and sigmoid function can facilitate designing a scaffold as per the specific need and favoring the location of the scaffold with the desired mechanical properties. Also, the careful transition between the structures governed by the sigmoid function the hybrid gives smooth curvature of transition making it optimum for enhancing tissue integration properties of scaffolds. Recursive sigmoid function uses are also an outstanding feature which provides facility to hybrid more than two TPMS structures as shown in Table 1. In the field of implantology, where multi-morphology implant is required for an individual implant, TPMS hybridization offers the excellent option as substitute over other porous structure hybridization where transition boundary is impossible and can only be used in welded structures. In this study, numerical analysis was performed on the hybrid of D and IWP structures which have shown promising results to mimic human bone with varying mechanical properties.

References

1. Ahmad, O., et al.: Bone injury and fracture healing biology. Biomed. Environ. Sci.: BES **28**(1), 57–71 (2015)
2. Bigham-Sadegh, A., Ahmad, O.: Basic concepts regarding fracture healing and the current options and future directions in managing bone fractures. Int. Wound J. **12.3**, 238–247 (2015)
3. Oryan, A., et al.: Current concerns egarding healing of bone defects. Hard Tissue, vol. 2, no. 2, Open Access Publishing London (2013)
4. Boonen, S., et al.: Aging and bone loss. Osteoporosis Men 207–219 (2010)
5. Kumar, J., et al.: Mechanical property analysis of triply periodic minimal surface inspired porous scaffold for bone applications: a compromise between desired mechanical strength and additive manufacturability. J. Mater. Eng. Perform. 1–13 (2022)
6. Zhao, D., et al.: Poly (lactic-co-glycolic acid)-based composite bone-substitute materials. Bioactive Mater. **6.2**, 346–360 (2021)
7. Wang, S., et al.: Pore functionally graded Ti6Al4V scaffolds for bone tissue engineering application. Mater. Design **168**, 107643 (2019)
8. Limmahakhun, S., Cheng, Y.: Graded cellular bone scaffolds. Scaffolds Tissue Eng.: Mater. Technol. Clin. Appl. 75–94 (2017)
9. Jung, Y., Torquato, S.: Fluid permeabilities of triply periodic minimal surfaces. Phys. Rev. E **72**(5), 056319 (2005)
10. Nikolaos, K., et al.: Mechanical and FEA-assisted characterization of fused filament fabricated triply periodic minimal surface structures. J. Comp. Sci. **5**(2), 58 (2021)
11. Fu, H., Kaewunruen, S.: Experimental and DEM investigation of axially-loaded behaviours of IWP-based structures. Int. J. Mech. Sci. **235**, 107738 (2022)
12. Verma, R., et al.: Low elastic modulus and highly porous triply periodic minimal surfaces architectured implant for orthopedic applications. Proc. Inst. Mech. Eng. Part E: J. Process Mechan. Eng. 09544089221111258 (2022)
13. Yan, C., et al.: Ti–6Al–4V triply periodic minimal surface structures for bone implants fabricated via selective laser melting. J. Mech. Behav. Bio-Med. Mater. **51**, 61–73 (2015)
14. Brush, K.: Finite Element Analysis FEA. Software Quality, 1 (2019). www.techtarget.com/searchsoftwarequality/definition/finite-element-analysis-FEA
15. Hutton, D.V.: Fundamentals of Finite Element Analysis. McGraw-Hill Science Engineering (2003)
16. Novak, N., et al.: Quasi-static and dynamic compressive behaviour of sheet TPMS cellular structures. Comp. Struct. **266**, 113801 (2021)
17. Al-Ketan, O., et al.: Opology-mechanical property relationship of 3D printed strut, skeletal, and sheet based periodic metallic cellular materials. Additive Manufact. **19**, 167–183 (2018)
18. Oraib, A.-K., et al.: He Effect of Architecture on the Mechanical Properties of Cellular structures Based on the IWP Minimal surface. J. Mater. Res. **33**(3), 343–359 (2018)
19. Al-Ketan, O., et al.: Microarchitected stretching-dominated mechanical metamaterials with minimal surface topologies. Adv. Eng. Mater. **20.9**, 1800029 (2018)
20. Kapfer, S.C., et al.: Minimal surface scaffold designs for tissue engineering. Biomater. **32.29**, 6875–6882 (2011)

21. Yoo, D.-J., Kim, K.-H.: An advanced multi-morphology porous scaffold design method using volumetric distance field and beta growth function. Int. J. Precis. Eng. Manuf. **16**(9), 2021–2032 (2015)
22. Fu, H., Kaewunruen, S.: Experimental and DEM investigation of axially-loaded behaviours of IWP-based structures. Int. J. Mech. Sci. **235**, 107738 (2022)
23. Miralbes, R., et al.: Mechanical properties of diamond lattice structures based on main parameters and strain rate. Mech. Adv. Mater. Struct. 1–13 (2022)
24. Ang, N., et al.: Mathematically defined gradient porous materials. Mater. Lett. **173**, 136–140 (2016)
25. Urmeneva, M.A., et al.: Fabrication of multiple-layered gradient cellular metal scaffold via electron beam melting foregmental bone Econstruction. Mater. Design **133**, 195–204. Elsevier BV (2017)
26. Ma, Songhua, et al. "Biological and mechanical property analysis for designed heterogeneous porous scaffolds based on the refined TPMS." Journal of the Mechanical Behavior of Biomedical Materials 107 (2020): 103727
27. Zhang, X.-Y., et al.: Biomechanical influence of structural variation strategies on functionally graded scaffolds constructed with triply periodic minimal surface. Additive Manufac. **32**, 101015 (2020)
28. Verma, A., et al.: Design and analysis of biomedical scaffolds using TPMS-based porous-tructures inspired from additive manufacturing. Coatings **12**(6), 839 (2022)
29. Poltue, T., et al.: Design exploration of 3D-printed triply periodic minimal surface scaffolds for bone implants. Int. J. Mech. Sci. **211**, 106762 (2021)
30. Wang, L., et al.: Mapping porous microstructures to yield desired mechanical properties for application in 3D printed bone scaffolds and orthopaedic implants. Mater. Design **133**, 62–68 (2017)
31. Mullen, L., et al.: Selective laser melting: a regular unit cell approach for the manufacture of porous, titanium, bone ingrowth constructs, suitable for orthopedic applications. J. Biomed. Mater. Res. Part B: Appl. Biomater.: Official J. Soc. Biomater. Japanese Soc. Biomater. Australian Soc. Biomater. Korean Soc. Biomater. **89.2**, 325–334 (2009).Gibson, L.J.: Biomechanics of cellular solids. J. Biomech. **38**(3), 377–399 (2005)
32. Parthasarathy, J., et al.: Mechanical evaluation of porous titanium (Ti6Al4V) structures with electron beam melting (EBM). J. Mech. Behav. Biomed. Mater. **3.3**, 249–259 (2010)
33. Mehboob, H., et al.: Finite element modelling and characterization of 3D cellular microstructures for the design of a cementless biomimetic porous hip stem. Mater. Design **149**, 101–112 (2018)
34. Al- Ketan, O., Rashid, K., Abu, A.-Rub.:MSLattice: a free software for generating uniform and graded lattices based on triply periodic minimal surfaces. Mater. Design Process. Commun. **3.6**, e205 (2021). Yang, N., Quan, Z., Zhang, D., Tian, Y.: Multi-morphology transition hybridization CAD design of minimal surface porous structures for use in tissue engineering. Comput.-Aided Design **56**, 11–21 (2014)

Apnea Controlled CPAP for Obstructuve Sleep Patient

S. Gowthami(✉), R. Boopathi Raaj, R. Muthu Selvam, and P. V. Saraswathy

Bannari Amman Institute of Technology, Sathyamangalam, Erode , Tamilnadu 638401, India
gowthamiiselvarajan@gmail.com

Abtract. In modern medicine, treatment and diagnosis are focused on improving early detection. Sleep-disordered breathing and other upper respiratory ailments are frequently accompanied by snoring, and it's a very common symptom. These symptoms may lead to Obstructive sleep apnea (OSA) which is caused by the blockages of the airway and features a range of symptoms, including snoring, episodes of oxygen desaturation, and arousals from sleep. The Obstructive sleep apnea (OSA) may cause insomnia which makes the person to get difficulty in sleeping and hypersomnia which cause in excessive daytime sleeping time. The main objective of the system is to detect Apnea during sleep as it may cause damage to the body organs. The shortness of breath can be detected by this system which manipulates the breathing pattern of the person and ensures that whether apnea is incorporated. This can be achieved by using suitable low pressure differential sensors which gives values on the inhalation and exhalation process. Compatible software platform can be used to process the output gained from the sensor which in turn activates the CPAP device as per the needs. This helps the soft muscle around the throat region to keep open so the patient can breathe easily. The proposed solution can be worked with patients of old age, children, lung affected and persons with respiratory issues. The system can also be connected via IOT to transmit the data to patient's family or the healthcare center so that the manual assistance can be provided to the patient when needed.

Keywords: Sleep apnea · obstruction · CPAP · respiratory · low pressure differential sensor

1 Introduction

MORE than 60 distinct sleep disorders, categorized by "The International Classification of Sleep Disorders" which has recognized this sleeping issues into seven categories [1]. Breathing while sleep disorders, fall under the second category which includes central sleep apnea disorder, obstructive sleep apnea (OSA), and other disorders associated to sleep such as hypoxemia and hypoventilation. In these disorders, Obstructive Sleep Apnea (OSA) is the most prevalent disease and is characterized by partial or total ventilation during sleep that is impacted by upper airway blockage and repeated collapse [2]. Complete or partial obstruction of the upper airway when sleeping is a sign of the obstructive sleep apnea syndrome (OSA). It is analyzed that the apnea affected range

B. K. Singh et al. (Eds.): ICBEST 2023, CCIS 2003, pp. 326–335, 2024.
https://doi.org/10.1007/978-3-031-54547-4_25

varies from 6%–17% comparing with the apnea-hypopnea index. It is also found the male people are affected more in prevalence when compared with females. This prevalence is also connected to the aging factor among male and females.

Polysomnography is used as the recording method for the disorders when test the signs of apnea during sleep. PSG is considered as accurate in cases with 80%. It is considered to measure the physiological factors which is connected to the deep sleep and wakefulness. The PSG study can reveal disorders such as narcolepsy, REM sleep issues, disorders related to limb movements in a periodic way and other sleep related disorders [2]. The average detection factors of the apnea can be analyzed by evaluating the number of obstructing events in the respiratory pathway during sleep for an hour. In case of apnea conclusion, it is estimated that the sleeping obstructi ng events must be equal to or greater than 15 times/hour. This count may be not be concluded as apnea if one such counting is noted for the first time. The detection and monitoring carried over few times and based upon the results, it can be concluded whether OSA is incorporated for the patient. The count of 5–15 may conclude a mild apnea and 15–30 may conclude moderate apnea [3]. Whereas the count of above 30 can termed as severe apnea in which the person need to kept in medication with appropriate respiratory devices throughout the sleep. The apnea process can be initialized with obstructing, followed by a brief waking that improves the permeability of the upper airways.

PSG offers accurate data, but the procedure is time-consuming and expensive because the patient must frequently be present at a sleep laboratory under the guidance of a trained technician. Hence other methods are used in order to minimize the time and expense. In turn for the therapeutic purpose, CPAP is used in many cases. Millions of adolescents around the world are negatively impacted in terms of their health and behavior. Nasal CPAP was first developed as a sleep apnea syndrome therapy approach by Sullivan and colleagues in 1981 [4]. In particular for OSA, the CPAP equipment is employed as a therapeutic method for sleep apnea. The cornerstone of medical care is a device that delivers continuous positive airway pressure (CPAP) during night. The device draws air via the upper airways through a mask on the nose or mouth in order to keep the neck open. The device isa popular and successful non-surgical treatment for OSA; it has been shown to be useful in reducing OSA patient's levels of daytime sleepiness. The veil, which is coupled to the tubing connections and transmits the air pressure to the patient, is the most crucial component of the CPAP apparatus. Nasal CPAP, as the device with veil is commonly known, covers the nose (nCPAP). However, a nose and mouth veil can be used to apply it. The majority of the machine will normally have a calibration range of (4 to 15 cmH2O), with the ability to go as high as (25 cmH2O). In order to create a feedback loop for accurate pressure monitoring, a CPAP device included a pressure sensor.

The interface, electro-pneumatic, and microprocessor modules were all present in this device. It is easy to construct and suitable for use in low- and middle-income nations. Additionally, it is helpful in the case of a pandemic. Designing a continuous positive airway pressure system with simple components that are easily replaceable when they break down and are readily available in online stores, as well as a continuous positive airway pressure system with an Arduino Uno board to minimize electrical wire connections and because of the Uno performance, which makes it simple to programme it via an IDE environment with available codes for various electronics materials. Developing a simple device which can detect the disorder as well as used for the purpose of therapeutic process will be well appreciated in terms of respiratory issue related disorders which in turn reduces the frequent interaction of the patients with the hospital.

2 Literature Review

PAPER 1: "Compliance with CPAP therapy in patients with the sleep apnea/hypopnoea syndrome" by Heather M Engleman, Sascha E Martin, Neil J Douglas. PMID: 8202884, PMCID: PMC1021157, DOI https://doi.org/10.1136/thx.49.3.263, 1994.

CPAP therapy is obtrusive, and it is not surprising that patients do not use it all night every night. However, the results in this study indicate that CPAP usage is extremely variable between patients and that, on average, CPAP use is less than either five hours or night. One would therefore anticipate that optimal CPAP pressure would only be achieved for 93% of the night in patients using a 20 min delay on one occasion if the machine was switched on for the average 4–7 h/night. It thus seems that these patients are achieving satisfactory CPAP pressures for most of the time during which the CPAP machines are switched on, with the one noticeable exception of the patient whose pressure monitor port cap was displaced.

PAPER 2: "Nasal CPAP or Intubation at Birth for Very Preterm Infants". By Colin J. Morley, M.D., Peter G. Davis, M.D., Lex W. Doyle, M.D., Luc P. Brion, M.D., Jean-Michel Hascoet. N Engl J Med, Feb 14; 358(7):700-8. doi https://doi.org/10.1056/NEJMoa072788. 2008.

Ventilation and oxygen therapy are linked to bronchopulmonary dysplasia. This randomized experiment looked at whether nasal CPAP could be used to treat bronchopulmonary dysplasia in extremely preterm infants instead of intubation and ventilation soon after delivery. In this article 610 new born babies were randomly assigned who were born between 25 and 28 weeks of gestation to CPAP or intubation and ventilation. Surfactant usage is now considered to be standard care for very preterm newborns. When the research got underway, there was mounting observational evidence suggesting extremely preterm newborns receiving early CPAP did not need ventilation or surfactant use.

PAPER 3: "Detection of Sleep Apnea using Pressure Sensor", by Kumari Sneha, International Research Journal of Engineering and Technology, Volume: 03 Issue: 10 | Oct-2016.

The author used MPX100DP pressure sensor to detect the sleep apnea which is integrated with the Matlab software to get the values displayed from the patient nostril via the microcontroller. Bluetooth technology is employed in this idea to assist the wireless communication for data transfer.

PAPER 4: "A Review on Detection and Treatment Methods of Sleep Apnea", by Rajeswari Jayaraj, Jagannath Mohan, Adalarasu Kanagasabai. Journal of Clinical and Diagnostic Research. 2017 Mar, Vol-11(3): VE01-VE03, 2017.

Micro Electrical Mechanical Systems (MEMS) technology is used as the primary component in this article. The sensor converts the value into the signal which will be processed by the time domain signal processing integrated circuits. MEMS lacks in term of sensitivity and repeatability making it not suitable for accurate results.

PAPER 5: "Low-cost, easy-to-build noninvasive pressure support ventilator for under-resourced regions: opensource hardware description, performance and feasibility testing" by Onintza Garmendia, Miguel A. Rodriquez Lazaro, Jorge Otero, Phuong Phan, Alexandrina Stoyanova, Anh Tuan Dinh Xuan, David Gozal, Daniel Navajas, Josep M. Montserrat, Ramon Farre European Respiratory, European Respiratory Society, vol. 55 no. 6, https://doi.org/10.1183/13993003.00846-2020.

This article ensures the usage of ventilator type of devices that is constructed from off-the- shelf materials along with usage of pressure transducers for detecting the respiration rate with Arduino Nano as microcontroller for the condition purpose. The device works efficiently as the commercial devices but the accuracy tends to be in doubt during times.

PAPER 6: "Tele monitor care helps CPAP compliance in patients with obstructive sleep apnea: a systemic review and meta-analysis of randomized controlled trials" by Chongxiang ChenJiaojiao Wan, Lanlan Pang, Yanyan Wang, Gang Ma, and Wei Liao. Volume 11, 2020.

Using TM care is more beneficial than regular care for enhancing CPAP compliance. Additionally, in the sleep unit, HAPT demonstrated non-inferior efficacy in CPAP compliance compared to SPT. Clinicians should think about using more efficient techniques to increase CPAP compliance in OSA patients.

PAPER 7: "A Simplified Design of CPAP Device Construction by Using Arduino NANO for OSA Patients", by Athra'a sabeeh Mikha, Hadeel K. Aljobouri, Article in Design Engineering (Toronto) DOI https://doi.org/10.4053/DE.21.06.13.6174, ISSN: 0011-9342, 2021.

The object is provided with using the pressure sensor MPS20N0040D and identifying the sleep apnea disorder by the help of Arduino Nano microcontroller and then allowing them to get treated with the help of BLDC motor.

PAPER 8: "Randomized Controlled Trial of Variable-Pressure Versus Fixed-Pressure Continuous Positive Airway Pressure (CPAP) Treatment for Patients with Obstructive Sleep Apnea/Hypopnea Syndrome (OSAHS)", by Marjorie Vennelle, RN; Sandra White BSc (Hons); Renata L. Riha, FRCPE; Tom W. Mackay, FRCPE; Heather M. Engleman, PhD; Neil J. Douglas, MD, DSc Department of Sleep Medicine, Royal Infirmary of Edinburgh, Edinburgh, UK, Vol. 33, No. 2, 2022.

All new cases to determine inclusion and exclusion criteria, eligible patients with a diagnosis of obstructive sleep apnea/hypopnea syndrome (OSAHS) were found. Up until 200 individuals were enrolled, successive eligible participants were approached with research material and asked to consent. The study's written informed permission from each participant was obtained before it began, and the local IRB gave its prior approval. On the morning following the CPAP titration night, those who agreed to participate were randomly assigned to the study (Fig. 1).

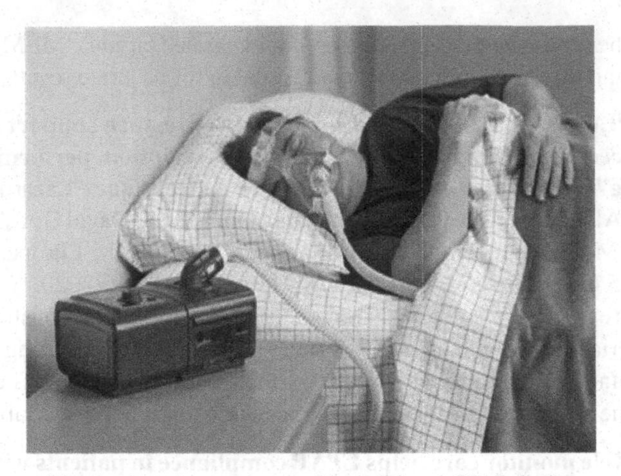

Fig. 1. A patient assisted with traditional CPAP device

3 Materials and Methods

So it is clear that an external device which includes the working of CPAP as well as a machine which can work on the real time values of the patient need to be associated and developed which will possess features of less expensive, reliable, compact, accurate and timer friendly. The primary function of a traditional CPAP device is to shield OSA patient's soft palates from obstruction. The pressure in the alveoli above atmospheric pressure at the conclusion of expiration is known as positive end-expiratory pressure (PEEP). PEEP can be administered with CPAP, which likewise keeps the target pressure constant during the entire breathing cycle, including both inspiration and expiration. Bilevel positive airway pressure (BiPAP), as contrast to CPAP, delivers a different amount of pressure depending on whether the patient is inhaling or exhaling. Expiratory positive airway pressure (EPAP) and inspiratory positive airway pressure (IPAP) are the names for these pressures (EPAP).

In CPAP, no additional pressure is applied over the predetermined amount, and patients must start every breath on their own. The use of CPAP preserves PEEP, has the potential to lessen atelectasis, increases the alveolar surface area, enhances V/Q matching, and consequently enhances oxygenation. Although CPAP alone is frequently insufficient for supporting ventilation, which necessitates extra pressure support during inspiration for non-invasive ventilation, it can also indirectly aid in ventilation.

The current CPAP is made up of a number of components, each of which serves a specific purpose to guarantee that the overall design is compatible. The blower, motor, pressure sensor, and power supply are some of the components. One of the primary components of the project idea is the BLDC motor which works as the remedial measure component in the entire process and the patient would get the required air pressure from this component. Is a brushless motor where the torque is higher compa to the brushed motor. It lacks the brushes but nevertheless exhibits many of the same speed and torque properties as the Brushed DC (BDC) motor. Technically, it is a Permanent Magnet Synchronous Motor (PMSM), but because there are no brushes and it runs on a DC power source through an inverter with a straightforward commutation scheme, it is referred to as a Brushless DC motor. The CPAP device's design structure must be as simple, comfortable, and affordable as well as feasible hence only the primary element of the device is used for these advantages. It is also made to be tiny for portability. In addition to offering services to all patients, regardless of their age, sex, or body weight. This method enables multiple operating modes within the approximate (4–16 cmH2O) range that is compatible with commercial CPAP machines.

4 Electronic Parts

The operation of the various electronic components used to build an electronic circuit, including a pressure sensor, microcontroller, motor, and power source, is covered in this section. The MPS20N0040D pressure sensor model is used as the design's pressure gauge. This sensor has roughly measured pressures from (0 kPa–40 kPa). The amount of wires connecting the sensors to the Arduino Microcontroller was reduced by using a breakout board. It is a small, faultless, open-source building board that uses the ATMEGA328P microcontroller IC. It uses the flash memory with 32 KB of storage and 2 KB each for the bootloader and the processor clock (16 MHz). A brushless 12/24V DC silent centrifugal blower was used as the therapeutic component which is used as the primary component of the commercial CPAP device. It is specifically used with auto CPAP devices to treat OSA patients which provides the air that is transported to the patient with the help of the tube. The blower is powered by a BLDC motor.

A switching power source was used in the hardware design to power the motor drive. In addition to the above listed components, other study materials have been used as test supplements. These components are the CPAP mask and tubing, a simple full-facial, transparent CPAP mask with headgear fastened to the volunteer's face to direct airflow to their mouth and nose. Plastic and silicone are used to make it. Numerous benefits of this mask include its adaptability to different facial shapes, convenience of use when worn (with a headpiece), and ease of cleaning (clean in warm soap suds). The 1.8-meter-long tube that connects the blower and mask on the developed model (Fig. 2).

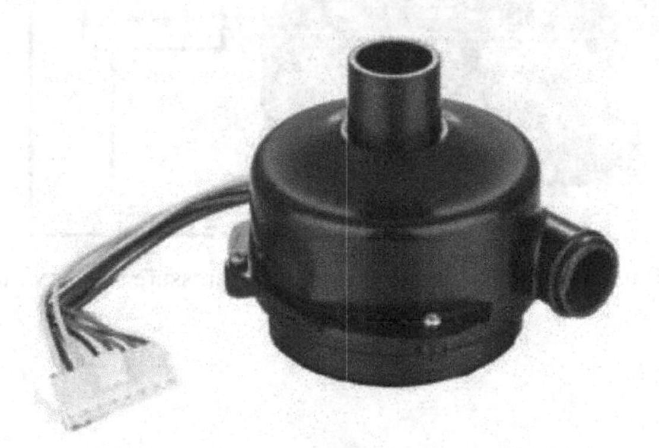

Fig. 2. A model of a BLDC motor

5 Configuration

The Uno board is home to all of the electronic components, each of which was wired in accordance with the appropriate Arduino pin. Powering the blower and motor requires connecting the (12 V) output voltage poles of the power supply to the positive (Vin) of the motor drive board. For direct operation from AC 220 V general electricity, the power supply (L and N) poles are wired to an external plug. For the majority of clinical CPAP applications, the Blower's flow of (110 l/min) at up to (16 cmH2O) under 12V might be adequate. The differential pressure sensor MPS20N0040D is used as the detecting element in this idea where the respiration pressure would be measured and the output of the value will be transported to the Arduino Uno board where the digital value is checked with the condition that is predefined in the board and determines whether apnea is incorporated or not. The power supply is connected via the VIN Module power supply pin of 3.3–5 V whereas the SLC acts as the I2C Clock pin. The Digital output data pin is connected to any one of the digital pins of the Arduino boards for data fetching.

The Arduino Uno and other electronic components are programmed using an IDE programme for the software model of the current system. The application is launched with a fresh Arduino sketch after connecting the Arduino via USB. The Arduino board style (ATMEGA328P old Bootloader) is then selected from the tbar to activate the board on the IDE environment and the Port COM number. Additionally, libraries for the differential pressure sensor were installed. After creating an Arduino sketch was finished, the sketch was then uploaded to the Arduino Uno. The chosen approach was a system design that was contained in a single white box and prepared for a performance test. The electronic components inside the box are secured using screws. The syringe end was also glued to and attached to the tube (Figs. 3 and 4).

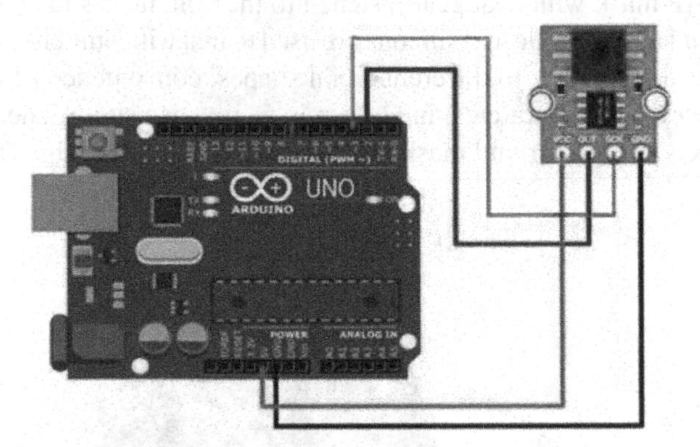

Fig. 3. The illustration of the circuit formation of pressure sensor with the Arduino

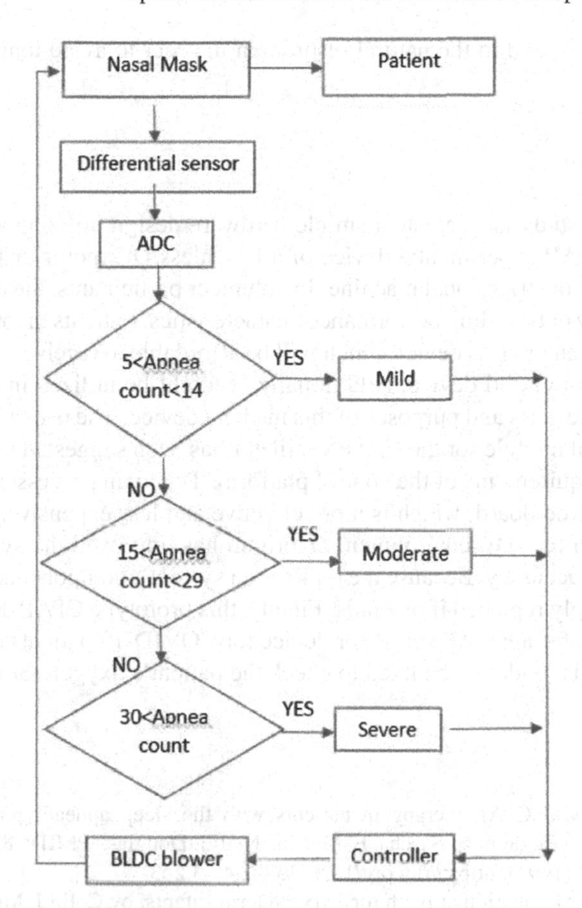

Fig. 4. The schematic representation of the idea

6 Block Diagram of the Idea

7 Results and Discussions

The gadget is turned on when the value from the controller is deflecting from the threshold value that is set up as a criteria in it. The value is compared to the threshold value for the apnea range of 2–6 mmHg when the low- pressure differential sensor detects the pressure in the airway path. The motor driver will be working will the value from the sensor as the input simultaneously. The air gets via the tube and reaches the mask which is placed on the patient's face. To execute the detection, this value would be altered in the Arduino IDE.

Since the air pressure is imported to the respiratory pathway of the patient, the patient can have a comfortable sleeping period during this time. The above device must be set to only assist the patient and the patient should not continuously addict to the device at any stage. SO it is mandatory to maintain a particular amount of time for the activation purpose. It much be activated in terms if regular intervals of time to allow the patient to comfortably respond to the natural respiration in order to avoid long term addiction effects.

8 Conclusion

The goal of this study is to create a simple hardware design utilizing an Arduino Uno board for the CPAP experimental device of a Brushless DC motor that can be used as an alternative for the traditional machine. In volunteer participants, the demonstration is expected to show outstanding performance characteristics. Patients in lower and middle-income nations can use this device which will be affordable to receive breathing support from this simple-to-build device. Additionally, it might be utilized in the lab to teach students about the parts and purposes of this medical device. The use of an Arduino Uno board in a control module for the CPAP ventilator has been suggested in order to satisfy the functional requirements of the control platform. The main processor was decided to be the Arduino Uno board, which is more effective and less expensive. Adoption of an interface through the IDE environment algorithm has improved the system's response time and control accuracy. Because the replication system's components are not built-in, they may be simply replaced if one fails. Finally, this prototype CPAP design may serve as the inspiration for a BIPAP ventilator device for COVID-19 patients in the future. An oximeter sensor is in idea to be fixed to check the patient's oxygen saturation.

References

1. Compliance with CPAP therapy in patients with the sleep apnea/hypopnoea syndrome, by Heather M Engleman, Sascha E Martin, Neil J Douglas. PMID: **8202884**, PMCID: PMC1021157 (1994). https://doi.org/10.1136/thx.49.3.263
2. Nasal CPAP or Intubation at Birth for Very Preterm Infants, by Colin J. Morley, M.D., Peter G. Davis, M.D., Lex W. Doyle, M.D., Luc P. Brion, M.D., Jean-Michel Hascoet. N Engl J Med, Feb 14 **358**(7), 700–8 (2008). https://doi.org/10.1056/NEJMoa072788
3. Sateia, M.: International classification of sleep disorders-third edition: highlights and modifications, wwwwwwwl Chest, vol. 146, no. 5, pp. 1387–1394 (2014)
4. "Trends in CPAP adherence over twenty years of data collection a flattened curve", by Brian W. Rotenber ,Dorian Murariu, Kenny P. Pang
5. "A Review on Detection and Treatment Methods of Sleep Apnea", by Rajeswari Jayaraj, Jagannath Mohan, Adalarasu Kanagasabai
6. A Review of Obstructive Sleep Apnea Detection Approaches", by Fabio Mendonc ¸Sheikh Shanawaz Mostafa, Antonio G. Ravelo-Garc´ıa, Fernando Morgado-Dias
7. Cpap, C., Pressure, P.A.: Non-invasive ventilation device settings : a brief guide, no. April, pp. 1–2 (2013)
8. Garmendia, O., et al.: Low-cost, easy-to-build noninvasive pressure support ventilator for under-resourced regions: open source hardware description, performance and feasibility testing. Orig. Artic. Respir. INTENSIVE CARE **55**, 1–11 (2020)
9. Azagra-Calero, E., Espinar-Escalona, E., Barrera-Mora, J.M., Llamas-Carreras, J.M., Solano-Reina, E.: Obstructive sleep apnea syndrome (OSAS). Review of the literature. Med. Oral Patol. Oral Cir. Bucal **17**(6), e925–e929 (2012). https://doi.org/10.4317/medoral.17706
10. Virk, J.S., Kotecha, B.: When continuous positive airway pressure (CPAP) fails. J. Thorac. Dis. **8**(10), E1112–E1121 (2016). https://doi.org/10.21037/jtd.2016.09.67

11. Giles, T.L., Lasserson, T.J., Smith, B.J., White, J., Wright, J., Cates, C.J.: Continuous positive airways pressure for obstructive sleep apnoea in adults: a Cochrane collaboration review. Cochrane Database Syst. Rev. **1**, 1–15 (2006)
12. Tele monitor care helps CPAP compliance in patients with obstructive sleep apnea: a systemic review and meta- analysis of randomized controlled trials, by Chongxiang ChenJiaojiao Wan, Lanlan Pang, Yanyan Wang, Gang Ma, and Wei Liao,· vol. 11 (2020)
13. A Simplified Design of CPAP Device Construction by Using Arduino NANO for OSA Patients, by Athra'a sabeeh Mikha, Hadeel K. Aljobouri, Article in Design Engineering (Toronto) (2021). https://doi.org/10.4053/DE.21.06.13.6174. ISSN: 0011-9342
14. Randomized Controlled Trial of Variable-Pressure Versus Fixed-Pressure Continuous Positive Airway Pressure (CPAP) Treatment for Patients with Obstructive Sleep Apnea/Hypopnea Syndrome (OSAHS), by Marjorie Vennelle, RN; Sandra White BSc (Hons); Renata L. Riha, FRCPE; Tom W. Mackay, FRCPE; Heather M. Engleman, PhD; Neil J. Douglas, MD, DSc Department of Sleep Medicine, Royal Infirmary of Edinburgh, Edinburgh, UK, vol. 33, no. 2 (2022)

FPGA Implementation of Biological Feature Based Steganography Method

Vijayakumar Peroumal(✉), R. Harishankar, R. Ramesh, and R. Rajashree

School of Electronics Engineering, Vellore Institute of Technology, Chennai, Tamilnadu 600127, India

{Vijaya.kumar,ramesh.r}@vit.ac.in

Abstract. Due to the fast development of information technology, secure communication is important. To ensure secure communication we need to send an encrypted message so that it can be read only by the expected recipient. There is some existing method which is used for secure communication like cryptography and its limitation is adversary knows about your message and it impacts its security level. In order to overcome the above limitation, a proposed DNA steganography method provides a greater level of security and the adversary has no idea about you hiding something. The proposed steganography method consists of DNA computing, RSA algorithm, and LSB embedding. DNA computing is a branch of computing that performs computations using DNA molecules. The RSA algorithm is a frequently used public key cryptographic algorithm for secure communication. Finally, the proposed DNA computing-based steganography was implemented using Xilinx and MATLAB and the RSA and DNA mapping have been done in Verilog and LSB embedding has been done in MATLAB. The RSA and DNA mapping have been simulated using Questa Sim-64 10.6c and implemented on vertex 7 using Xilinx ISE 14.7. The LSB embedding is implemented using MATLAB program in MATLAB R2022a.

Keywords: DNA Computing · ECC · Steganography · Security · Encryption · Cryptography · RSA

1 Introduction

Every individual has the right to privacy. Whether it's the privacy we have in our homes or simply interacting with friends over the internet, we value privacy. To preserve privacy and safe communication, we must send encrypted messages to ensure that only the intended receiver may read them. Cryptography and steganography are two methods for ensuring secure communication. The study of secure communications methods that enable only the sender and targeted receiver of an information to read its data is known as cryptography. Here, information is transformed to cipher data using a secret key, and then both the secret key and cipher data are given to the receiver for decryption. Steganography is the art of hiding a hidden message within a regular message. The steganography and cryptography are look like same thing, but two are different concepts.

B. K. Singh et al. (Eds.): ICBEST 2023, CCIS 2003, pp. 336–348, 2024.
https://doi.org/10.1007/978-3-031-54547-4_26

Both have nearly the same aim at their root, which is to safeguard an information or messages from third parties. They do, however, safeguard the information via a completely different approach. Cryptography converts data into cypher text, which is unreadable without a decryption key. If someone intercepted this encrypted message, they would immediately notice that encryption had been used. Steganography, on the other hand, does not alter the information's format but hides the message's presence. In other words, while sending secret information, steganography is more discrete than cryptography. The disadvantage is that if the presence of a secret is detected, the hidden message is easier to extract. Steganography is used in a wide range of applications, including cyber-security, computer security, homeland security, cyber-forensics, watermarking, and so on. Copyright protection, feature marking, and covert communication are just a few of the uses for digital steganography. Intelligence services, radar systems, army department agencies, healthcare imagery devices, digital synchronization, broadcasting services, advanced data structures, intellectual property applications, remote sensing, companies' safe transmission of hidden information, identification cards, smart IDs, and other fields and applications use steganography. Examiners looking for secret information or signs of secret information in digital media might employ steganography in the cyber-security industry [1].

Digital steganography has now become increasingly realistic and effective as the digital age has progressed, as data is the lifeblood of communication systems. The digital data boom has resulted in significant changes in daily life. As the amount of data exchanged over the communication system grows, data security seems to have become a key concern, necessitating the usage of data integrity and confidentiality to defend against unwanted access and use. As a result, the aspect of data concealment has exploded in popularity. Cyber-forensics, cyber-attacks and cyber-security have all become major concerns for both private businesses and governments. Data concealing has piqued the interest of designers, developers, and assessors for handling sensitive data and establishing covert pathways to hide sensitive data for many years. Data concealing is a technique for encoding secret message into a digital medium in such a way that only the intended receiver may see it. Steganography technology has sparked worldwide attention, and scientific research has skyrocketed in recent decades. This is due to the ability of digital steganography to create covert communication channels. It is feasible to hide data within a digital file without incurring perceptual damage using digital steganography. This could be accomplished by embedding data within a file using a technology that tries to manipulate or transforms the unused bits of digital media in a non-visible manner. Following that, the receiver of the message can also retrieve the secret message from the cover medium. Steganography gives the ability to conceal sensitive data, as well as the difficulty of finding the secret data and the intensity of the secret data's confidentiality. Steganographic techniques have existed for years, but with the arrival of digital technology, they have gained a new form [2].

Steganography has a variety of applications, the most common of which being data security. They also are attracting attention as a result of increased self-awareness about the need to improve security systems, as well as following the global security attack. The use of steganography to hide information inside image files is a subject of significant interest. Given the widespread availability of digital images and the considerable extent of redundancy included in their binary representation, there seems to be a surge in demand for employing images as steganography cover-objects. For better security and working of steganography system. It consists of DNA computing, RSA algorithm and LSB embedding. DNA computing is a branch of computing which performs the computations using DNA molecules and the use of ribonucleic acid (RNA) or recombinant DNA strands for computation is referred to as DNA computing [3]. The RSA algorithm is a frequently used public key cryptographic algorithm for secure communication. The least significant bit is abbreviated as LSB. The concept of LSB embedding is that if we modify a pixel's final bit value, the colour will modify just slightly. For example, black is 0. Modifying the value of 0 to 1 won't necessarily make that much difference because the colour remains black, but lighter. The ultimate purpose of this study is to create an image steganography technique that may be used in data protection and security applications [3]. The goal of this work is to implement the DNA steganography using Xilinx ISE 14.7 suite and MATLAB R2022a. The main contribution of this study is mentioned below:

- Propose a Novel DNA computing based Steganography method to improve the security level.
- Implement the proposed DNA steganography method using Xilinx ISE and MATLAB

2 Literature Survey

Tsu Yang Wu, Xiaoning Fan et al. proposed a DNA computation-based image encryption technique for cloud surveillance cameras. Here, image encryption algorithm is based on hyperchaotic map and DNA coding. It offers a more safe and efficient computations than some recent related research [4]. Mohammed Abbas Fadhil Al-Husainy et al. demonstrated the encryption and decryption of data by using randomized encryption keys, strong substitution, and transposed operations. Finally, it proved that proposed encryption algorithm achieves better performance compared with other encryption techniques and also it obtains a higher avalanche effect value of over 50% [5]. Sun et al. [6] presented a cloud-based antimalware technology and the proposed method achieves reliable and effective security solution to IoT network devices. Ukil et al. [7] proposed the idea of trusted computing by examining the need of embedded security, to develop techniques and answers for rejecting the cyber-attacks. At last, the proposed supplied technologies is used for tamper the proof for embedded systems. Bairagi et al. [8] proposed the three colour image steganography techniques for securing data in an IoT environment. The third and first techniques require three (green, blue and red) channels for data transmission, whereas the second technique requires two (blue and green) channels. With the use of a shared secret key, dynamic positioning were employed to hide data in the deeper levels of the image channels.

Anwar et al. [9], presented a method to protect any image, mainly medical images. They wanted to preserve the confidentiality of electronic health records and also maintaining its authentication and accessibility service to the authorized person. Abdelaziz et al. [10], evaluated the findings of security flaws and danger concerns in smartphone medical apps. These apps can be classified as treatment support, monitoring system, medical data, diagnostic services, communication and awareness and education and coaching and mentoring for healthcare staff, as per the risk factor guidelines.

3 Proposed Biological Feature Based DNA Steganography

The proposed biological based steganography system consists of DNA mapping, RSA algorithm and LSB embedding as shown in Fig. 1. Ron Rivest, Adi Shamir, and Len Adleman of MIT created the first public key technique in 1977. The RivestShamir-Adleman (RSA) public key cryptographic algorithm is now the most generally recognized and developed. The public key technique is establish on the use of two separate keys, one for decryption and the other for encryption. The entire procedure entails computing the remaining after performing large number modular and exponential operations. In most cases, a third party is used to produce a set of public keys and deliver them to the sender and recipient. The value of n should be known by both the sender and the recipient. Only the sender knows the public key e, whereas the recipient knows the private key d. As a result, a third-party produced secret key (d, n) and public key (e, n) are distributed independently to the sender and recipient.

3.1 DNA Mapping Technique

DNA computing is a branch of computing which performs the computations using DNA molecules and the use of ribonucleic acid (RNA) or recombinant DNA strands for computation is referred to as DNA computing. DNA mapping is done using DNA molecule to map the binary value of the data based on ASCII value. As shown in Table 1, if data is 00 it is assigned to 65 which is ASCII value of "A", if data is 01 it is assigned to 84 which is ASCII value of "T", if data is 10 it is assigned to 67 which is ASCII value of "C", if data is 11 it is assigned to 71 which is ASCII value of "G".

Table 1. DNA mapping table

Data	DNA Molecule	ASCII
00	A	65
01	T	84
10	C	67
11	G	71

The DNA mapping is implemented in the following steps:

1. Get the input data.
2. Compare the data from left to right based on DNA mapping table. If the value is matched then the 2bit will be replaced by ASCII value.
3. The ASCII value is converted into binary format.

The 8 bit input is converted into a 32 bit value based on DNA mapping. If n bit input is given, the output will be 4n bit.

3.2 LSB Embedding Technique

The least significant bit is abbreviated as LSB. The concept of LSB embedding is that if we modify a pixel's final bit value, the colour will modify just slightly. For example, black is 0. Modifying the value of 0 to 1 won't necessarily make that much difference because the colour remains black, but lighter. Each pixels in a grayscale picture is expressed by eight bits. The final bit in a pixel is known as the Least Significant bit since its value has only a "1" value influence on the value of pixel. As a solution, this attribute is employed to obscure the image's data. If the last set of bit are considered LSB bits, they will only influence the value of pixel by "3". This support in the storage of more information. The Least Significant Bit (LSB) steganography technique replaces the image's least significant bit with message bits. We encrypted the original information prior to embedding it in the picture to create a more secure because this technique is unsafe for analysis.

While the encryption technique adds to the complexity of the procedure, it also improves the security. This technique is fairly effortless. A few of the least significant bits or all of the bits in a picture are changed with bits of the secret data using this technique. Many methods for data hiding inside multimedia carrier message are based on the LSB embedding technique. LSB technique can even be used in specific data domain, such as embedding a concealed data in the RGB bitmap colour values or the Jpeg file frequencies coefficients. LSB technique can be used on different data forms and types. As an effect, LSB embedding is among the most generally used steganography technique today.

The benefits of using an LSB-based information hiding approach include that it is straightforward to embed the message's bits straight into the LSB of an image pixel, and numerous ways use this mechanism. Because the LSB alteration causes no picture distortion, the generated stego image totally looks like the cover image. The LSB based method allows for a high embedding rate while also recovering the secret information completely and without error. As a result, it is commonly used in image steganography. Based on number of pixels chosen, the quantities of information to be embedded can be variable or fixed in size. The key advantage of this technique is that changing the LSB plane has no effect on the overall image statistics because the amplitude change of the pixel value is limited to ± 1. The least significant bits of the cover media's digital information are employed to hide the information in LSB steganography. LSB substitution is the most basic LSB steganography approach. The final bit of every pixel value is changed in LSB substitute steganography to represent the in- formation that has to be concealed. Assume an 8-bit bitmap grayscale picture in which every pixel is represented by a byte with a grayscale values. The proposed steganography includes RSA algorithm, DNA mapping and LSB embedding and it consists of seven blocks. The blocks are data block, RSA algorithm block, cipher data block, DNA mapping block, LSB embedding block, Cover image block and stego image block. The binary data input will be given to RSA algorithm from data block. In RSA algorithm block by utilize the formula M^e mod n = C, the data is encrypted and converted into cipher data. After the cipher data is obtained it is sent to DNA mapping block. The data in from cipher data block is compared from left to right to DNA mapping table. If data is 00 it is assigned to 65 which is ASCII value of "A", if data is 01 it is assigned to 84 which is ASCII value of "T", if data is 10 it is assigned to 67 which is ASCII value of "C", if data is 11 it is assigned to 71 which is ASCII value of "G" and the ASCII is converted to binary value.

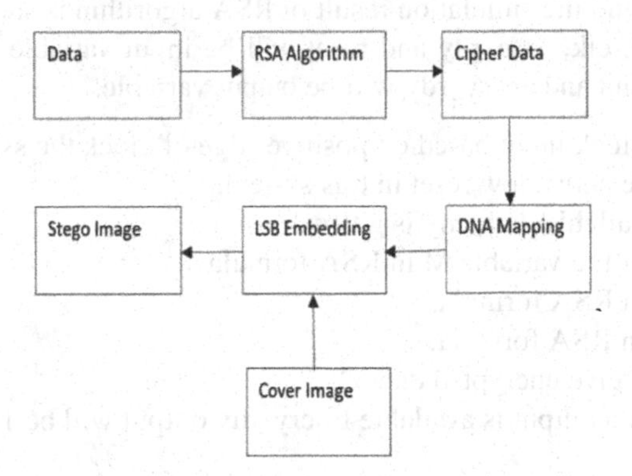

Fig. 1. Proposed DNA steganography

The binary value is given to LSB embedding block from DNA mapping block. In LSB embedding block the pixel value will be converted to binary and LSB of pixel is compared to MSB of message if same, no change value is changed and if value is different, the pixel's LSB is subsequently substituted with the message's MSB. Similarly the LSB of next pixel is changed depend on the message bit. The pixel value of the LSB embedding block is obtained from the cover image after the LSB embedding the stego image will be obtained as shown in Fig. 1.

DNA steganographic Algorithm

Step 1: Get the data value.
Step 2: Compute the value of encrypted value, using $C = M^e \bmod n$.
Step 3: DNA mapping is done on binary value.
Step 4: LSB embedding is done using the binary value from DNA mapping using cover image pixel.
Step 5: Get the stego image after LSB embedding.

4 FPGA Implementation Results and Its Discussions

The results of proposed architecture's implementation are shown in this section. The implementation of RSA and DNA mapping has been done in Verilog and LSB embedding has been done in MATLAB. With the purpose of achieving high security, the architecture was simulated using Questa Sim-64 10.6c and synthesized using Xilinx ISE 14.7. The following images show the RTL schematic, simulation and Technical Schematic of RSA and DNA mapping.

4.1 RSA Cryptographic Algorithm Results

For RSA algorithm to implement C = M^e mod n formula. The block diagram for RSA is shown in Fig. 2 and the simulation result of RSA algorithm is shown in Fig. 3. Here, ku1, ku2, message, clk, data_rdy and reset will be input variable for RSA algorithm block and encry_data and encry_rdy will be output variable.

- clk will be the clock input based on positive edge of clock the system is performed.
- Reset will be the active low reset in this system.
- When data is available data_rdy is given 1.
- Message input is the variable M in RSA formula
- Ku1 input is e in RSA formula.
- Ku2 input is n in RSA formula.
- Encry_data will give encrypted data C.
- When Encry_data output is available Encry_rdy output will be 1

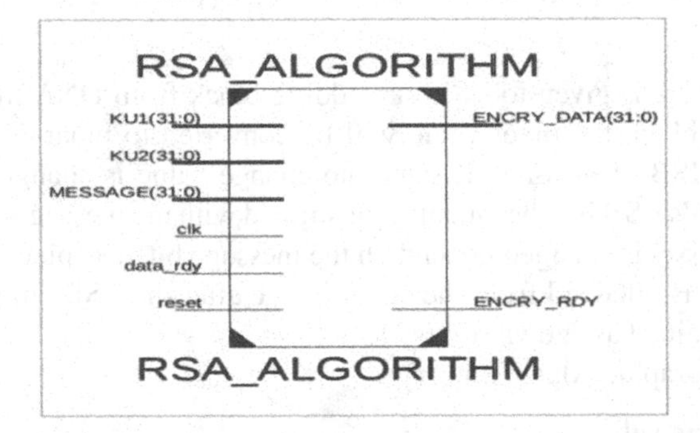

Fig. 2. RSA block diagram

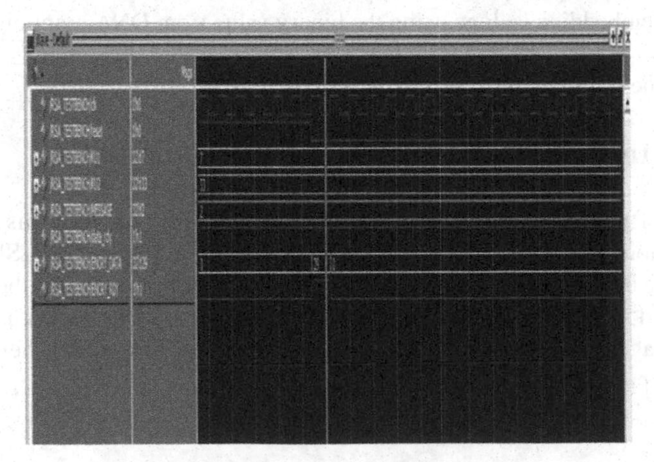

Fig. 3. Simulation result of RSA algorithm

Here, for RSA algorithm m, e and n is used as input. The output variable out will give the result of m^e mod n. For example in simulation result for the m = message = 2, ku1 = e = 7, ku2 = n = 33 and result of encrypted data = 29. Technical Schematic of RSA is shown in Fig. 4 and the RTL schematic of RSA is shown in Fig. 5. Table 2 shows Device utilization summary of RSA and Timing details.

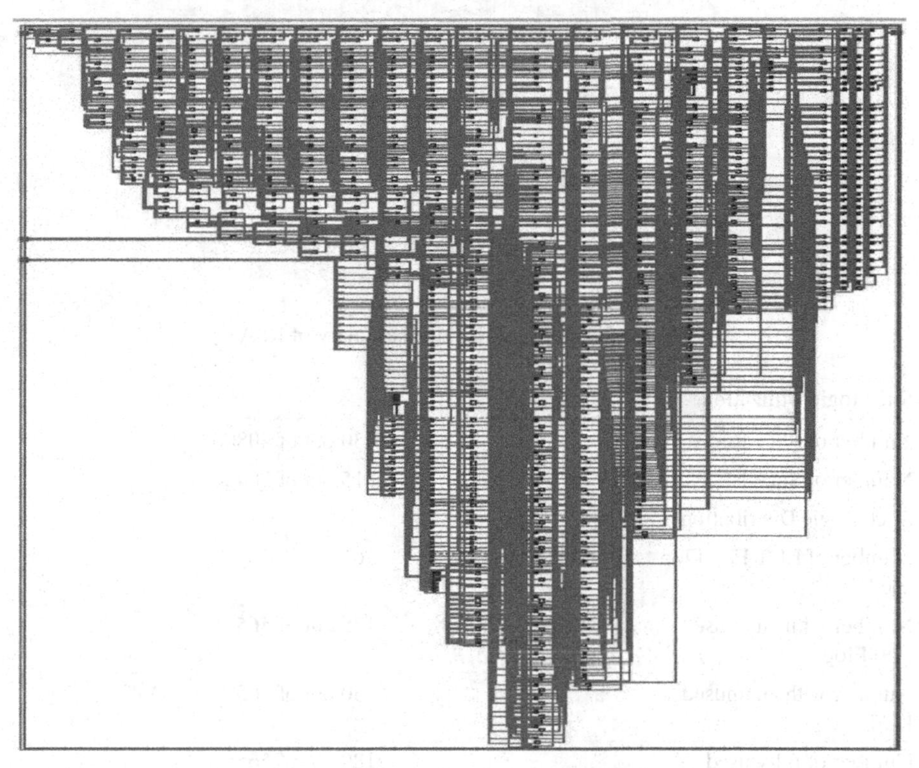

Fig. 4. Technical Schematic of RSA

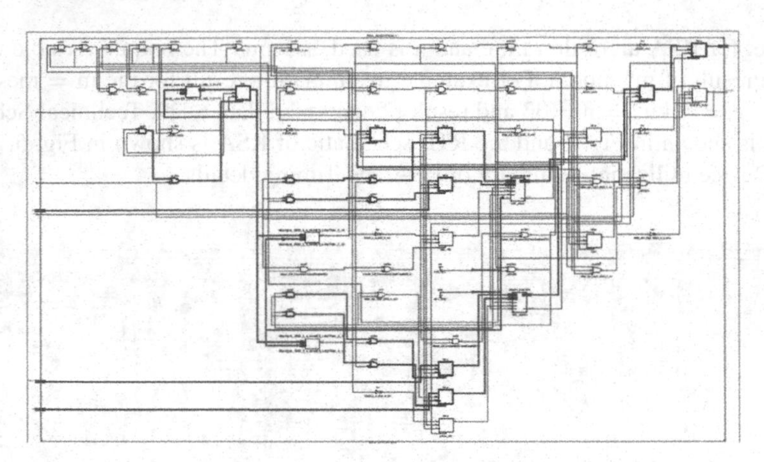

Fig. 5. RTL schematic of RSA

Table 2. Device utilization summary of RSA

Slice logic Utilization	
Number of Slice Registers	430 out of 408000
Number of Slice LUT	315 out of 204000
Slice Logic Distribution	
Number of LUT Flip Flop used	565
Number with an unused Flip Flop	135 out of 565
Number with an unused LUT	250 out of 565
Number of fully used LUT-FF pairs	180 out of 565
Number of unique control set	14
IO Utilization	
Number of IO's	132
Number of Bounded IOBs	132 out of 600
Specific Feature Utilization	
Number of BUFG /BUFGCTRLs	1 out 32
Number of DSP48E1s	3 out of 1120
Total timing for RSA	5.519 ns

4.2 DNA Mapping Technique Results

Device utilization summary and Timing details for DNA mapping is shown in Table 3.

Table 3. Device utilization summary of DNA mapping

Slice logic Utilization		
Number used as Logic	32 out of 204000	
Number of Slice LUT	32 out of 204000	
Slice Logic Distribution		
Number of LUT Flip Flop used	32	
Number with an unused Flip Flop	0 out of 32	
Number with an unused LUT	0 out of 32	
Number of fully used LUT-FF pairs	0 out of 32	
Number of unique control set	0	
IO Utilization		
Number of IO's		160
Number of Bounded IOBs		160 out of 600
Total timing for RSA		0.675 ns

4.3 LSB Embedding Technique Results

LSB embedding is done using MATLAB R2022a. The input to for LSB embedding is grayscale image that is cover image and output will be the stego image obtained after LSB embedding. The steps for LSB embedding encoding are as follows:

1. Transform the picture to grayscale.
2. If necessary, resize the picture.
3. Convert the data to binary code.
4. Make the output picture the same as the input picture.
5. Perform the following steps for every pixel in the picture:
   - Transform the value of pixel to binary.
   - Obtain the next data segment to be integrated.
   - Generate a temp variable.
   - Set temp = 0, if the pixel's LSB and data bit are same.
   - Set temp = 1, if the pixel's LSB and data bit are different.

- The XOR of the data bit and the value of LSB pixel can be used to set the temp.
- Change the pixel values of the output picture to the pixel values of the input picture + temp.

6. Continue to update the output picture until all of the data bits are integra ted.
7. Finally, output picture can be obtained after LSB.

Fig. 6. Cover image

Fig. 7. Stego image

The Cover image is shown in Fig. 6 and the size of the input cover image is 512 × 512, and the data is embedded in the cover image to create the stego image. The output Stego image is shown in Fig. 7. The stego image created after LSB embedding with a size of 512 × 512 is shown above. When we look at the stego image and the cover image that is based on the LSB embedding steganography technique, there would be no much difference. We have implemented the steganography and obtain the results of proposed DNA steganography. The RSA and DNA mapping is implement on vertex 7 using Xilinx ISE and LSB embedding is implemented using MATLAB program in MATLAB R2022a. For RSA and DNA mapping, the synthesis output for time and area is shown in Table 4 and it shows for RSA, the synthesis output for time and area and Table 5 shows the time and area synthesis output for DNA mapping:

Table 4. RSA timing and area table

LUTs Used	Total LUTs Present	Delay(ns)
315	204000	5.519

Table 5. DNA mapping timing and area table

LUTs Used	Total LUTs Present	Delay(ns)
33	204000	0.675

For RSA the total slice LUTs used is 315 out of 204000 and it has a timing delay is 5.519 ns and for DNA mapping the total LUTs used is 33 out of 204000 and it has a timing delay is 0.675 ns.

5 Conclusion

The expansion of communication media necessitates the development of unique security measures, particularly on computer networks. Due to the openness of the internet, there seems to be a possibility that confidential material transferred may be intercepted or manipulated by unwanted observers. As a result, secure data transmission and information concealment have exploded in popularity. Furthermore, the information concealment method can be applied to a wide range of applications, including business, digital forensics, advertising, anti-criminal, and military. Steganography is a hidden communication technique that has attracted a lot of interest. We proposed a novel design and implemented the proposed DNA steganography. In this paper, the focus is not only on the DNA mapping, but also on the stages, such as secret message encryption using RSA and embedding area selection using LSB embedding improve the security. The implementation of RSA and DNA mapping has been done in Verilog and LSB embedding has been done in MATLAB. The RSA and DNA mapping has been simulated using Questa Sim-64 10.6c and implement on vertex 7 using Xilinx ISE 14.7. The LSB embedding is implemented using MATLAB program in MATLAB R2022a.

References

1. Jan, A., Parah, S.A., Hussan, M.: Double layer security using cryptostego techniques: a comprehensive review. Health Technol. **12**, 9–31 (2022). https://doi.org/10.1007/s12553-021-00602-1
2. Biswas, D., et al.: Digital image steganography using dithering technique. Procedia Technol. **4**, 251–255 (2012). https://doi.org/10.1016/j.protcy.2012.05.038

3. Gehani, A., LaBean, T., Reif, J.: DNA-based Cryptography. In: Jonoska, N., Păun, G., Rozenberg, G. (eds.) Aspects of Molecular Computing. Lecture Notes in Computer Science, vol. 2950, pp. 167–188. Springer, Berlin, Heidelberg (2003). https://doi.org/10.1007/978-3-540-24635-0_12
4. Wu, T.Y., et al.: A DNA computation-based image encryption scheme for cloud CCTV systems. IEEE Access **7**, 181434–181443 (2019). Digital Object Identifier. https://doi.org/10.1109/ACCESS.2019.2958491
5. Al-Husainy, M.A.F., Al-Shargabi, B., Aljawarneh, S.: Lightweightcryptography system for IoT devices using DNA. Elsevier, Comput. Electr. Eng. **95**, 107418 (2021)
6. Sun, Y., Lu, Y., Yan, X., Liu, L., Li, L.: Robust secret image sharing scheme against noise in shadow images. IEEE Access **9** (2021)
7. Ukil, A., Sen, J., Koilakonda, S.: Embedded security for Internet of Things. In: Proceedings 2nd National Conference Emerging Trends Applications in Computer Science (NCETACS), pp. 50–55 (2011)
8. Bairagi, A.K., Khondoker, R., Islam, R.: An efficient steganographic approach for protecting communication in the Internet of Things critical infrastructures. Inf. Secur. J. **25**(4–6), 197–212 (2016)
9. Anwar, A.S., Ghany, K.K.A., Mahdy, H.E.: Improvingthe security of images transmission. Int. J. Bio-Med. Inf. eHealth **3**(4), 7–13(2015)
10. Abdelaziz, A., Elhoseny, M., Salama, A.S., Riad, A.M.: A machine learning model for improving healthcare services on cloud computing environment. Measurement Elseveir **119**, 117–128 (2018)

Dual-Band THz Absorption Based Refractive Index Sensor for Bio-Sensing

Durgesh Kumar[✉], Gaurav Varshney, and Pushpa Giri

National Institute of Technology Patna, Patna, India
durgeshk.ph21.ec@nitp.ac.in

Abstract. A dual-band absorber is implemented for terahertz (THz) applications. The absorber provides the dual narrow bands which can be utilized for the refractive index sensing. The absorber provides the resonance peaks at 1.03 and 1.49 THz with the absorption of more than 95% at the peak frequency. The absorber provides the ultra-narrow Full-width-half-maxima of 0.04225 and 0.0541 THz with maximum sensitivity of 48 and 108 GHz/RIU in the lower and upper band, respectively which makes it suitable to be utilized in the application of THz bio-sensing.

Keywords: Absorber · Meta-material · bio-sensing · THz

1 Introduction

The latest advancement in the electromagnetics are working for the development of the planar frequency selective surfaces (FSS) [1–4]. These planar structures are useful in the multiple application like defense, communication, and medical [5]. The absorption based FSS are becoming popular in the THz applications especially as refractive sensor and bio-sensors [6]. These are the passive device which can react in a desired way for the incident electromagnetic wave on it [7, 8]. The basic mechanism of these structures is based on the confined localized surface plasmons at the metal-dielectric interface [9]. The variation in the refractive index of the surrounding medium changes the confinement of the generated localized surface plasmon resonance and hence the impact comes in the form of the variation in the frequency response of the device response [10]. This feature makes them capable of being utilized in the absorption-based sensing applications [11].

In the current time, researchers are working to develop the bio-sensors by considering the THz frequency spectrum [12]. The dimensions of the most of the virus generally remains of the order of micro or nano scale which can provide the resonance in the somewhere in the THz frequency spectrum [12–14]. In general, the signature of the resonance of the virus structures remains in the lower THz range between 0.1 to 10 THz [14–16]. This encourages to serve the community by implementing the planar absorber structures. Furthermore, the implementation of the multiband absorber is always required for their utilization in the multiple applications simultaneously [17].

A survey of the literature shows that the implementation of dual-band absorption with the narrow-band spectrum with ultra-narrow full-width half-maxima (FWHM) is

B. K. Singh et al. (Eds.): ICBEST 2023, CCIS 2003, pp. 349–356, 2024.
https://doi.org/10.1007/978-3-031-54547-4_27

quite difficult [9, 18, 19]. The main performance indices of the sensors are senitivity (S) defined over the variation in the refractive index, quality factor (Q), FWHM, and figure-of-merit (FOM) [19–21]. The proposed dual-band absorber provides the ultra-narrow FWHM with the absorption of more than 95% in each band. The proposed absorber provides the maximum sensitivity of 48 and 108 GHz/RIU with FWHM of 0.0401 and 0.0503 THz in the lower and upper band, respectively. Moreover, the proposed absorber provides the high value of the maximum quality factor as 25.17 and 27.414 in the lower and upper band, respectively. The evaluation of these performance indices confirms the proposed absorber as a good candidate for the THz sensing applications. The proposed absorber can be utilized as a refractive index sensor and for the sensing of the influenza virus in the human blood.

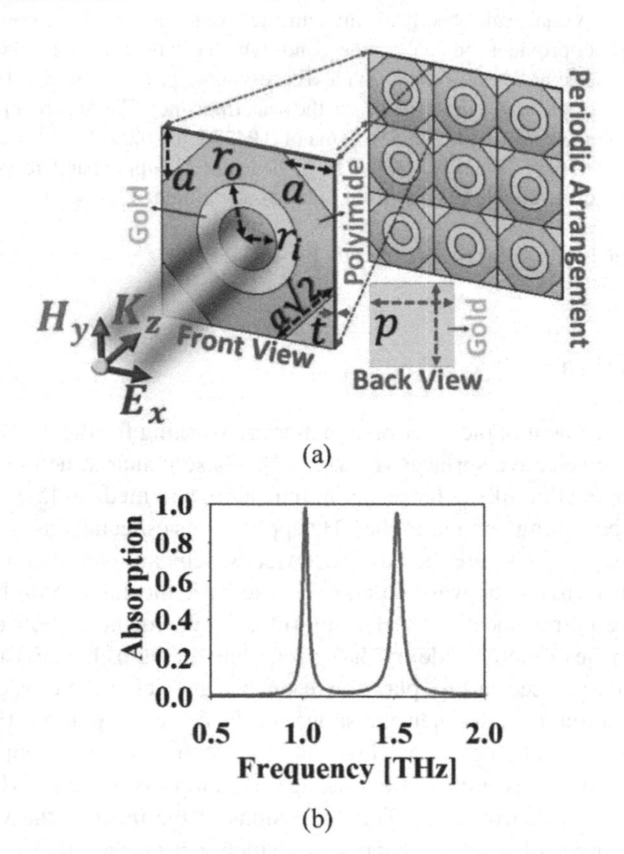

Fig. 1. The geometry of proposed dual-band absorber.

2 Absorber Structure, Evolution and Results

The structure of the proposed absorber contains a substrate of polyimide with the relative permittivity $\epsilon_s = 3.5$. The metal coatings are grown at both the sides of the substrate. The thickness of metal is kept so that it can avoid the transmission of the signal in the forward direction. Gold can be selected as metal for the implementation of resonator at one side of substrate and reflecting plane at another side of the substrate. The metal coating at one side of the substrate is patterned for making the structure of resonator. The micro-fabrication techniques can be utilized for obtaining the geometry of the resonator. Figure 1 shows the structure of the unit cell with the period p of along the horizontal axis. The simulation is performed with the consideration of unit-cells with the periodic arrangement and open boundaries in the direction normal to the absorber surface which is considered as the direction of propagation of incident electromagnetic wave. Figure 1(a) shows the structure of the proposed absorber. The dimensions of the proposed absorber are mentioned in Table 1. This absorber provides the dual-band absorption response with the level of absorption more than 95% as shown in Fig. 1(b). Figure 2 shows the evolution of the proposed absorber structure. This contains two resonators; a ring resonator and a squared resonator. The absorption response of these all the resonators is shown in Fig. 2. The ring resonator provides the resonance at the lower frequency and squared resonator provides the resonance at upper frequency. These both the resonators can be merged to find the dual-band response.

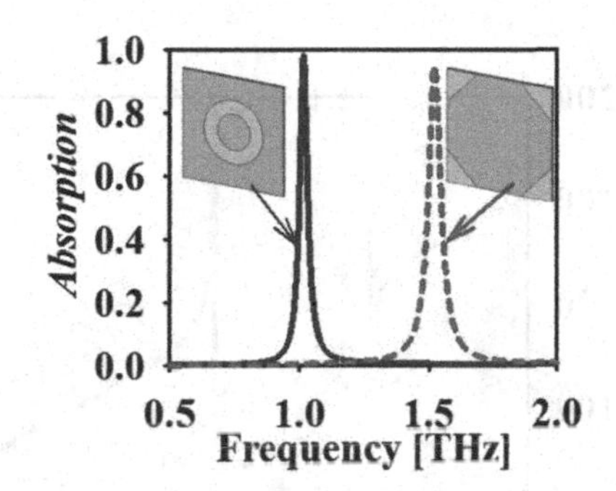

Fig. 2. The absorption response of different resonators used in the structure.

Table 1. The dimensions of the proposed antenna

Dimensions	Value (μm)
p	120
r_i	20
r_o	35
a	35
t	4

The proposed absorber operates over the frequency range of $f = 0.5-3$ THz with two absorption peaks at and. Absorption spectrum evident the absorption of electromagnetic waves with very narrow peaks $f_z = 1.02$ THz and $f_2 = 1.51$ THz. Figure 3 reflects about dual transmission dips at the resonating frequencies and zero transmission over the entire

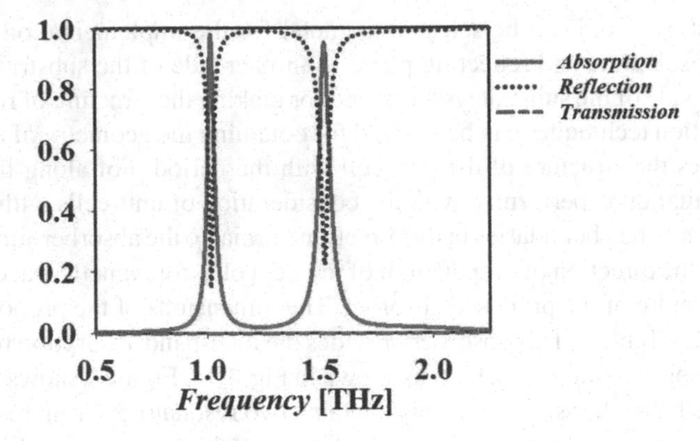

Fig. 3. The absorption, Reflection and Transmission response of proposed absorber

spectrum signifying the perfect absorbing capability of the proposed absorber-based sensor.

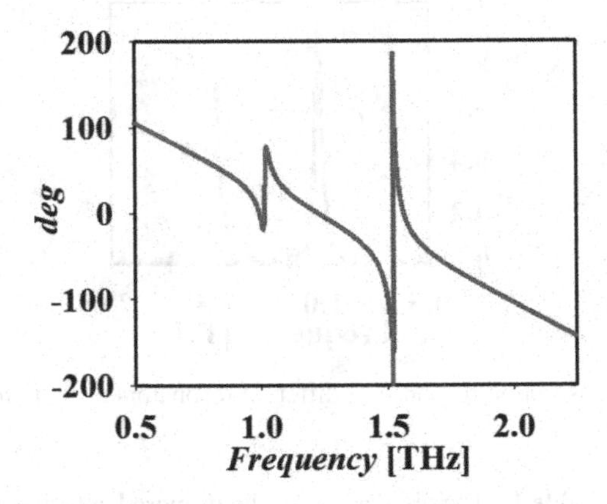

Fig. 4. The phase response of proposed absorber

The phase response of the proposed absorber structure is shown in the Fig. 4. There exist two sharp phase change at the resonating frequency which confirms about the absorption at two different frequencies.

Figure 5 shows the influence of the variation in the azimuthal angle ϕ of the electromagnetic wave on the absorption spectrum. The parameter ϕ is varied with the interval of starting from $\phi = 0° - 90°$. At all the variation of the azimuthal angle ϕ the absorption spectrum suffers no changes and the absorption spectrum is maintained at reflecting the insensititvity the parameter ϕ on the spectrum.

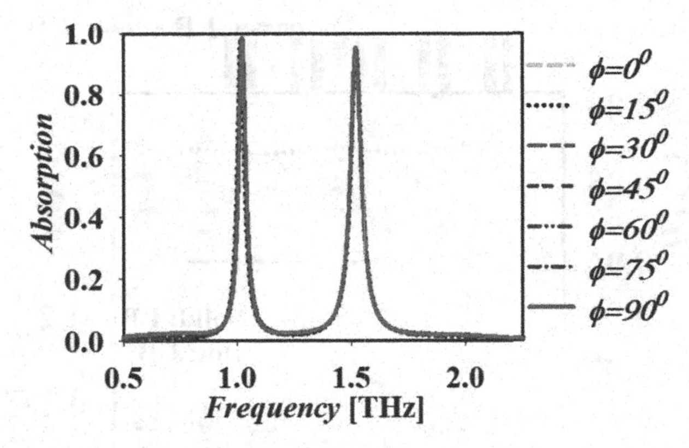

Fig. 5. Absorption response of absorber with different polarization angle

3 Refractive Index Sensing

An analyte of thickness 2 μm is placed at the top of the absorber structure. The thick- ness of the analyte is selected in the way that it can be nearby to the most of the virus cell size at the lower THz frequency [1]. The variation in the refractive index of the surroundings of the metal-dielectric interface-based devices changes their frequency response due to the change in the confined localized surface plasmon resonance. The variation in the frequency response of the proposed absorber is with the variation in the refractive index of the analyte is plotted in Fig. 6. It can be seen in the plot that increment in the refractive index of the analyte results in the red-shift in the absorption peaks.

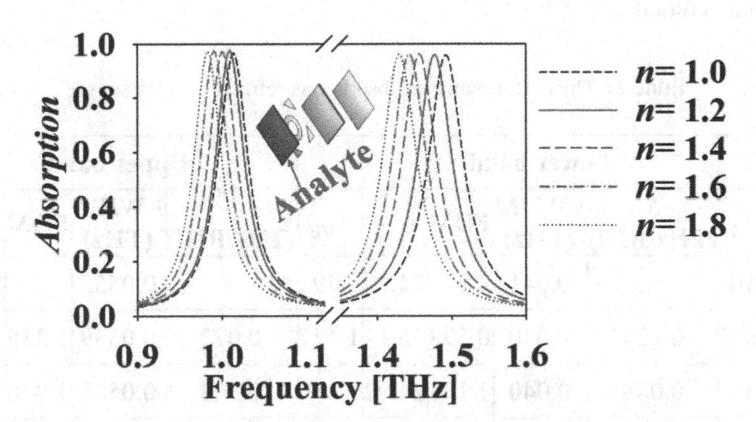

Fig. 6. Absorption response for the variation in the refractive index of analyte.

The performance indices of the proposed absorber are plotted in Fig. 7 for the variation in the refractive index of the analyte medium. It can be seen that absorber provides the ultra-narrow FWHM with high quality factor. Furthermore, the sensitivity of the proposed absorber varies within the range up to 200 GHz/RIU which can be considered

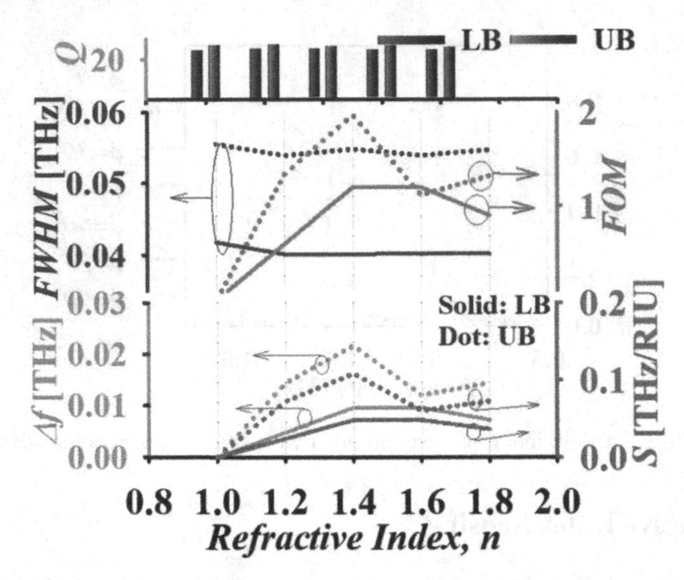

Fig. 7. The performance indices for the variation in the refractive index of analyte.

as a high value of sensitivity over many existing research works available in the literature [1]. The performance of the proposed absorber as refractive index sensor is mentioned in Table 2. The ultra-narrow value of FWHM of the absorption spectrum makes the absorber capable of to be utilized in the sensing of the different viruses like malaria in human blood with its different closely packed variants having the nearby values of the refractive index. Also, the proposed absorber can be utilized in the sensing of glucose in the human blood.

Table 2. The performance of absorber as refractive index sensor

	Lower band					Upper band				
n	f_{LB}	S (THz/RIU)	FWHM (THz)	FOM	Q_{LB}	f_{UB}	S (THz/RIU)	FWHM (THz)	FOM	Q_{UB}
1	1.01		0.041		24.29	1.492		0.0555		26.88
1.2	1.005	0.024	0.040	0.598	25.17	1.477	0.072	0.0539	1.335	27.41
1.4	1	0.048	0.040	1.196	24.92	1.456	0.108	0.0548	1.970	26.56
1.6	0.99	0.048	0.040	1.195	24.66	1.444	0.06	0.0539	1.113	26.79
1.8	0.98	0.036	0.040	0.896	24.49	1.429	0.072	0.0547	1.316	26.13

4 Conclusion

A dual-band THz absorber has been implemented. The absorber has been implemented with the usage of two type of resonators engraved in a single geometry; a ring resonator and a square shaped resonator. Both the resonators provide the resonance at the different frequencies in the lower THz frequency region. The performance of the proposed absorber has been investigated as THz refractive index sensor. A narrow value of FWHM as 0.04225 and 0.0541 with the sensitivity as 48 and 108 GHz/RIU and quality factor as 25.17 and 27.414 has been achieved in the lower and upper band, respectively. The proposed absorber can be utilized as THz refractive index sensor. The ultra-narrow bandwidth of the obtained absorption spectrum can allow it to be utilized in the different bio-sensing application like in the detection of influenza virus and glucose in the human blood.

References

1. Varshney, G., Giri, P.: Bipolar charge trapping for absorption enhancement in a graphene-based ultrathin dual-band terahertz biosensor. Nanoscale Adv. **3**, 5813–5822 (2021). https://doi.org/10.1039/d1na00388g
2. Varshney, G.: Wideband THz absorber: by merging the resonance of dielectric cavity and graphite disk resonator. IEEE Sens. J. **21**(2), 1635–1643 (2020). https://doi.org/10.1109/JSEN.2020.3017454
3. Varshney, G., Rani, N., Pandey, V.S., Yaduvanshi, R.S., Singh, D.: Graphite/graphene disk stack-based metal-free wideband terahertz absorber. J. Opt. Soc. Am. B **38**(1), 1–9 (2021)
4. Soni, A., Giri, P., Varshney, G.: Metal-free super-wideband THz absorber for electromagnetic shielding. Phys. Scr. **96**(125866), 0–11 (2021)
5. Khan, M.S., Varshney, G., Giri, P.: Altering the multimodal resonance in ultrathin silicon ring for tunable THz biosensing. IEEE Trans. Nanobiosci. **20**(4), 488–496 (2021). https://doi.org/10.1109/TNB.2021.3105561
6. Khan, M.S., Giri, P., Varshney, G.: Generating multiple resonances in ultrathin silicon for highly selective THz biosensing. Phys. Scr. **97**(8), 85009 (2022). https://doi.org/10.1088/1402-4896/ac807f
7. Singh, R., Cao, W., Al-Naib, I., Cong, L., Withayachumnankul, W., Zhang, W.: Ultrasensitive terahertz sensing with high- Q Fano resonances in metasurfaces. Appl. Phys. Lett. **105**(17), 171101 (2014). https://doi.org/10.1063/1.4895595
8. Nejat, M., Nozhat, N.: Ultrasensitive THz refractive index sensor based on a controllable perfect MTM absorber. IEEE Sens. J. **19**(22), 10490–10497 (2019). https://doi.org/10.1109/JSEN.2019.2931057
9. Wang, B.X., Huang, W.Q., Wang, L.L.: Ultra-narrow terahertz perfect light absorber based on surface lattice resonance of a sandwich resonator for sensing applications. RSC Adv. **7**(68), 42956–42963 (2017). https://doi.org/10.1039/c7ra08413g
10. Yan, F., et al.: Ultrasensitive tunable terahertz sensor with graphene plasmonic grating. J. Light. Technol. **37**(4), 1103–1112 (2019). https://doi.org/10.1109/JLT.2018.2886412
11. Mohanty, A., Acharya, O.P., Appasani, B., Mohapatra, S.K., Khan, M.S.: Design of a novel terahertz metamaterial absorber for sensing applications. IEEE Sens. J. **21**(20), 22688–22694 (2021). https://doi.org/10.1109/JSEN.2021.3109158
12. Homola, J.: Surface plasmon resonance sensors for detection of chemical and biological species. Chem. Rev. **108**(2), 462–493 (2008). https://doi.org/10.1021/cr068107d

13. Emami Nejad, H., Mir, A., Farmani, A.: Supersensitive and tunable nano-biosensor for cancer detection. IEEE Sens. J. **19**(13), 4874–4881 (2019). https://doi.org/10.1109/JSEN.2019.2899886
14. Keshavarz, A., Vafapour, Z.: Sensing avian influenza viruses using terahertz metamaterial reflector. IEEE Sens. J. **19**(13), 5161–5166 (2019). https://doi.org/10.1109/JSEN.2019.2903731
15. Iravani, S.: Nano- and biosensors for the detection of SARS-CoV-2: challenges and opportunities. Mater. Adv. **1**(9), 3092–3103 (2020). https://doi.org/10.1039/d0ma00702a
16. Yahiaoui, R., Strikwerda, A.C., Jepsen, P.U.: Terahertz plasmonic structure with enhanced sensing capabilities. IEEE Sens. J. **16**(8), 2484–2488 (2016). https://doi.org/10.1109/JSEN.2016.2521708
17. Soni, A.K., Varshney, G.: Multiband generation and absorption enhancement in a graphite-based metal-free absorber Plasmonics (2020).https://doi.org/10.1007/s11468-020-01286-5
18. Ali, F., Aksu, S.: A narrow-band multi-resonant metamaterial in near-IR. Mater. (Basel) **13**(22), 1–12 (2020). https://doi.org/10.3390/ma13225140
19. Hu, D., Meng, T., Wang, H., Ma, Y., Zhu, Q.: Ultra-narrow-band terahertz perfect metamaterial absorber for refractive index sensing application. Results Phys. **19**, 103567 (2020). https://doi.org/10.1016/j.rinp.2020.103567
20. Wang, Z., et al.: Ultrahigh-Q tunable terahertz absorber based on bulk dirac semimetal with surface lattice resonance. Photonics **9**(1) (2022). https://doi.org/10.3390/photonics9010022
21. Luo, S., Zhao, J., Zuo, D., Wang, X.: Perfect narrow band absorber for sensing applications. Opt. Express **24**(9), 9288 (2016). https://doi.org/10.1364/oe.24.009288

A Descriptive Analysis on Various Depression Detection Models of Human Brain: A Review Article

Sweety Singh[1(✉)], Poonam Sheoran[1], and Manoj Duhan[2]

[1] Department of Biomedical Engineering, Deenbandhu Chhotu Ram University of Science and Technology, Murthal, Haryana 131039, India
singhsweety250@gmail.com

[2] Department of Electronics and Communication Engineering, Deenbandhu Chhotu Ram University of Science and Technology, Murthal, Haryana 131039, India

Abstract. It is scientifically proved that mental health affects other organs of human. So, people start giving attention to depression condition of patients. This mild depressive state can be converted into major depressive state if not treated at right time. Many persons are affected by depressed state every year which lead to their death, suicidal thoughts with some health- related problems. EEG tool is proved helpful in diagnosis of depressive state very soon to prevent any worrying symptoms in future. The EEG feature extraction technique is able to differentiate healthy one from depressed subject. This methodology is helpful in finding exact cause behind real mechanisms to find out biomarkers for detection of abnormalities. This comprehensive study is focused on EEG-based algorithms for MDD patients. Different papers are explored extensively to understand the EEG frequency rhythm, band power, various filtration techniques, pre-processing methods, feature extraction and classification techniques. In many papers, more than one classifiers are used to differentiate between best and worst outcomes. The performance of these classifiers is calculated using parameters such as accuracy, sensitivity, specificity, entropy, S.D. (Standard Deviation) and ROC (Receiver Operating Characteristic) curve etc. This study aims at providing concrete ideas regarding possibilities of future research based on this study.

Keywords: MDD (Major Depressive Disorder) · EEG (Electroencephalography) · ROC curve · technique

1 Introduction

EEG is the most significant modality used to understand the different states of human brain. In this progression, we worked to analyze depressive state of brain. Depression disease is related to mental health and can be classified into mild and major depressive states. This disease is affecting more than 350 million people worldwide. The rate of depression is more in female (51%) as compared to males (36%). In this way this disorder is affecting sleeping pattern, eating habits and other general life including physical health

B. K. Singh et al. (Eds.): ICBEST 2023, CCIS 2003, pp. 357–375, 2024.
https://doi.org/10.1007/978-3-031-54547-4_28

of person [1]. This disease is main cause of disability according to report of WHO (World Health Organization) [2]. The main characteristics representing depression are pervasive, persistent, depressed mood, low mood or loss of interest in all activities [3]. This disease is associated with cognitive impairment or decline that result in Alzheimer's disease along with suicidal attacks [4]. The MDD (Major Depressive Disorder) is considered as 2nd largest disorder over the other diseases of world [5]. It is observed that nearly 10 lakh per year suicides is due to MDD, that cover nearly 54% of all suicides [6, 7].

The rate of depression in young college students is noted between 10% to 85% that is quite significant and to be focused upon seriously [8]. The CES-D10 (Centre for Epidemiological Studies Short Depression Scale) is used to detect level of depression (32.2%) in undergraduate students [9]. The DSM (Diagnostic and Statistical Manual of Mental Disorder) is the depression testing standard used by healthcare professionals [10]. There are various types of therapies used to treat disorder related to mental health (depression) [11]. It is observed from results that brain oscillation or frequency band pattern (delta, theta, alpha and beta range) is different for depression patient as compared to healthy person [12]. The EEG technique is low cost, non-invasive method for acquisition of signals at high temporal resolution. The computer aided EEG system is effective in detection of depression [14–16]. The functional connectivity method is used to differentiate between healthy person and mental depressed person [17]. The type of classifier used are SVM (Support Vector Machine), best first, GSW (Greedy Stepwise), GS (Genetic Search), RS (Renve Search), bayes net, KNN (K-Nearest Neighbor), LR (Logistic Regression), RFA (Random Forest Approach) and LDA (Linear Discriminant Analysis) [18–20].

The performance of signal is calculated with the help of statistical parameter such as accuracy, sensitivity, specificity, ROC, bar graph etc. These finding helps to reach a conclusion point to differentiate between healthy person and MDD (Major Depressive Disorder) person. In this way based on the existing research works, we tried to provide a review on technique of identification of depressive state. A summary of depression state, EEG based acquisition, pre-processing stage, feature extraction, classification and performance evaluation are added step by step in each study to understand our analysis in better way. The main aim is to find the best method to characterize depressive symptoms with the help of EEG signal. In the end, main conclusion part is added in every study to highlight the best result. There are lots of studies used for investigation to differentiate between healthy mind and depressed mind. This article is structured in the following manner: EEG signal acquisition section presents a brief discussion on various methods used to pick-up signal from human brain, pre-processing section represents the methods that change the condition of raw EEG signal, feature extraction section illustrates the essential features required to differentiate between various state of human brain, feature classification section represents classifier outcomes, statistical analysis section represents mathematical calculation applied on outcomes, result section gives focus on the results obtained through various methods, research gap with possible solutions section highlights limitations with feasible solution for future research and conclusion section

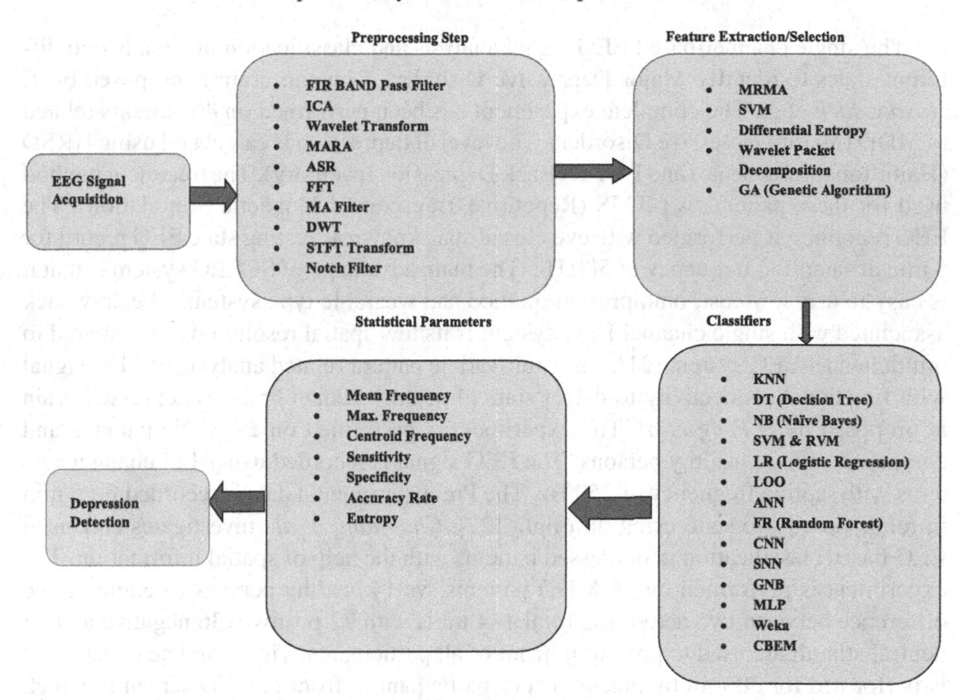

Fig. 1. Block diagram of detection of depression using EEG signal analysis though various methods.

highlights the essence of the present review paper. The Fig. 1 represents the various steps used for detection of depression using EEG signals including signal acquisition, pre-processing, feature extraction, classifiers, statistical methods in the form of block diagram. The detail of comparative study of each section is also explained here in this article:

2 EEG Signal Acquisition

The first step used to identify state of human brain is signal acquisition. EEG signal is taken/ picked up from human brain with the help of electrodes. In this review article, we include many signal acquisition methods and highlighted the best method out of them. From previous studies, it is proved that multichannel EEG signal analysis is very complex process. In multichannel system many electrodes are attached on scalp of human brain that is very irritating process for both patient and operator. So, scientists search a new method known as single channel-based EEG signal acquisition.

The single channel-based EEG signal analysis and classification of signals into different states to identify Major Depressive Disorder of human brain is proposed by *F. Hasanzadeh et al.* The complete experiment has been performed on 46 patients related to MDD (Major Depressive Disorder). The level of depression is calculated using HRSD (Hamilton Rating Scale) and BDI-II (Back Depression Inventory). The treatment method used for these patients is r-TMS (Repetitive Transcranial Magnetic Stimulation). The EEG recording is performed with eye closed state known as resting state EEG record for 5 min at sampling frequency of 500 Hz. The main advantage of SCEEG system is that it is easy to use, low cost, omnipresent method and wearable type system. The drawback associated with single channel EEG system is its low spatial resolution as compared to multichannel EEG system [21]. The multivariate pattern related analysis of EEG signal with functional connectivity to detect state of healthy human brain vs depressed brain is proposed by *H.Peng et al.* The experiment is performed on 27 MDD patients and compared with 28 healthy persons. The EEG signal is recorded using 128 channel systems with sapling frequency of 250 Hz. The Pre-experimental data is recorded for 5 min in relax closed eye state using dim light [22]. *Chaojiang et al.*, investigates enhanced EEG based classification in depressed patients with the help of spatial information. The experiment is performed on 16 MDD patients' vs 14 healthy persons to compare the difference between two activities. Total 144 trails with 72 positive, 36 negative and 36 neutral stimuli are used to present in front of all participants. The complete experiment is performed for 30 min by placing every participant in front of LCD screen in which positive blocks are designed by pressing button "1" during positive stimuli and similarly button "5" is pressed during negative stimuli [23].

The analysis of EEG signal in functional brain network of MDD patient based on some improved EMD (Empirical Mode Decomposition) method and comparing result with resting state EEG signal is proposed by *Xuexiao Shao et al.* The experiment is performed on 30 subjects including equal number of MDD (Major Depressive Diseases) and NC subjects. The experiment is based on use of EMD (Empirical Mode Decomposition) method to get time frequency analysis of EEG signal by decomposing acquired signal into sum of different IMF (Intrinsic Mode Function) values. The main drawback of simple EMD method is mode mixing which has been overcome by EEMD (Ensemble Empirical Mode Decomposition). But, this EEMD method also has some drawbacks such as residual noise produced may create problem and computational cost of method is also high. In this way progression is made by researchers to overcome all these problems by generating improved EMD (Empirical Mode Decomposition) method. In this method WGN (White Gaussian Noise) is mixed in the input signal for better processing of signal [24]. The new signal acquisition method based on taking signal only from forehead area is introduced by *C.T.Lin et al.* This signal analysis method is helpful in headache prevention, proper management of sleep and treatment of depression in better way. As we come to know from previous studies that EEG signal acquisition using many wet electrodes is a difficult and irritating task for patient and a time-consuming process too. To overcome this problem, researchers work by placing electrode only on forehead area.

By comparing our results, we understand that this forehead-based analysis is sufficient to provide relevant features for cognitive assessment of brain in depressed patient. The experiment is performed on 10 healthy persons to measure the sleep pattern. In this way EEG signal of forehead and PSG signals of two frontal-parietal regions are recorded at frequency of 256 Hz. To accomplish this task, a wearable head band system using dry contact-based EEG sensor device with foam cover is used that cover complete forehead by placing around head. The designing of headband is performed in such a way that it is having light weight with miniature EEG data acquisition circuit inbuilt with wireless telemetry system [25].

The brain analysis using traditional method is a difficult task; in order to solve this problem a new network based functional algorithm is searched by researcher to provide analysis and classification of signal in MDD patient. The resting state EEG signal is collected using 64 electrode-based recording for 5 min and this signal is classified into various segments for analysis [26]. A new hybrid EEG network system based on CNN (Convolution Neural Network) to extract EEG features and classification of depressed signal effectively to learn each feature is introduced by *Z.wan et al.* In this method deep learning algorithm is used to classify signal at more accurate rate. The experimental study of CNN (Convolution Neural Network) such as hybrid EEG NET is having two parallel lines to learn EEG signal features to differentiate between MDD patients from control group. This experiment is performed on 35 subjects including 12 from normal control group, 12 from Un-medicated patients and remaining 11 from medicated group patients. The resting state EEG signal of all subjects is calculated for 8 min in dim light relax room [27].

A new EEG based system for detection of mild depression using DE (Differential evolution) method is proposed by *Y. Li et al.* In this study feature selection and classification of signal using algorithm for detection of depression is the main focused point. The experiment is performed on 20 subjects including 10 mild depressed vs 10 healthy person. Total 60 facial expression-based pictures/images are taken to represent stimuli, in which 15 images are related to negative expression to represent sadness. The EEG signal is recorded using 16 electrodes using 128 channel HCGSN (HydroCel Geodesic Sensor Net) systems with sampling frequency of 250 Hz. This experiment is proved helpful because detection of mild depression in accurate way is considered to be most complex task [20]. The new experimental based strategy proved to be helpful in recognition of mild depression using EEG -EM (Electroencephalogram-Eye Movement) related signal is proposed by *Z.zhu et al.* This network will simultaneously record EM (eye movement) and EEG signals during mild depression condition and comparing it with normal control state. The multimodal based fusion method can be implemented in the study to differentiate between mild depression and normal control state. This multimodal system is helpful for doctors to achieve diagnosis and monitoring mild depression. The total experiment is performed using16 electrodes based on 128 electrodes 10/20 [28].

One more approach related to emotional arousal based ASSR (Auditory Steady State Response) examination using 40 Hz frequency is described by *Zhang et al.* The experiment is performed on 24 young subjects having correct eye vision and normal hearing state. The ASSR is accessed using sequence of chirp sound. There are two types of tasks involved in this current experiment, out of that one is EI (Emotion Induction) and other is SS (Sound Stimuli) listening [29]. The improved EMD (Empirical Mode Decomposition) based method for EEG (Electroencephalogram) signal analysis for detection of depression using three electrode systems (FP1, FP2 and FPz) from prefrontal area is proposed by *Jian et al.* The experiment is accomplished by dividing complete data into four sets. The data set 1 for 81 depressed vs 89 healthy, data set 2 for 160 depressed vs 116 healthy, data set 3 for 105 depressed vs 109 healthy subjects and set 4 for 105 depressed vs 70 healthy subjects. The four types of data sets are collected according to affective AS (Auditory Stimulus). All the resting state data is recorded for duration of 40 records in eye closed condition [30]. The automated diagnosis of MDD (Major Depression Disorder) using 3 D CNN (Convolutional Neural Network) with the help of brain effective connectivity using DMN (Default Mode Network) is represented by *D.M Khan et al.* The experiment is performed on 30 MDD patient's vs 30 HC (Healthy Control) group to compare effect of stress on human brain. The resting state EEG signal is recorded for 5 min using 19 channel EEG system based on 10/20 system. The whole task is divided into 2 steps, one is related to training of subjects using 3 D CNN network and in the 2nd step testing of subjects is input to train network to get classification result. The effective connectivity is accomplished in DMN (Default Mode Network). The 19*19*64 matrices are reduced to 6*6*64 matrices with the help of DMN algorithm. This process is followed by 3 D CNN for connectivity of 6*6*64 PDC based EEG signal [31]. The CTD (Coupled Tensor Decomposition) method to identify hypo connectivity and hyper connectivity is introduced by Liu et al., to study Major Depression (MD) in patients. The experiment is performed on 20 MDD vs Healthy Control by giving task of listening music. The 64-electrode based 10/20 system with frequency (fs) of 1000 Hz is used for acquisition of EEG signal from all participants [32]. The multi-modal based approach for detection of depression by fusion of EEG signal and some classifiers to understand change in features of MDD patient is given by *Xiaoweizhang et al.* The complete experiment is performed on 170 subjects including 81 depressed vs 89 Healthy Control with the help of 3 pervasive electrode- based system by placing electrodes on forehead area (FP1, FP2, FPz) for EEG signal acquisition [33]. The role of machine learning technique in detection of stress at various multiple level in mental person is explained by *Ahmad Rauf Subhani et al.* The experiment is performed on 22 subjects. The EEG signals are acquired using 128 channel system with net amp. 300 amplifiers [34]. The optimal channel selection method for EEG based detection of depression using KTA (Kernel Target Alignment) is given by *Jian Shen.* The EEG signal acquisition is performed on 15 depressed vs 20 healthy subjects (age 18 to 55 years). The resting state EEG signals are recorded for 5 min using 64 channel-based electrode system of 10/20 international protocol with sampling frequency of 1000 Hz and comparing it with 128 channel system-based recording of 5 min using sampling frequency of 250 Hz with the help of 24 depressed vs 29 healthy subjects [35].

A new improved classification model for detection of depression using EEG signal data and eye tracking data is given by *Zing et al.* The eye tracking experiment is performed on 20 HC (Healthy Control) vs 28 MDD (Major Depressive Disorder). The EEG signal data experiment is performed using 128 channel EEG systems with net station software and amplifier on 17 HC (Healthy Control) vs 17 MDD (Major Depressive Disorder) patients. The eye tracking device (eye link 1000) is used to monitor the pupil and movement of eye perfectly [36]. A new model known as CBRM (Case-Based Reasoning Model) is designed using 3 (three electrode) EEG system (*Hanshucai et al*). The experiment is conducted on 86 depressed vs 92 normal person having age between 18 to 55 years. The EEG signal is collected in three steps, first step is related to resting state data for 1 min and after this EEG signal is collected with audio stimulus of various emotions [37]. Table 1 presents the overall summary from various reviewed studies including type of study with number of subjects, pre-processing techniques, feature extraction and classification modalities, and final result.

Table 1. Representing comparison between various studies used for complete survey in detection of depression.

Author's Name	Type of study (EEG acquisition and no. of patients)	Pre-processing	Feature extraction	Feature Classification	Results and Conclusions
F. Hasanzadeh et al., [21]	• Single channel EEG system. • 46 MDD patients.	• FIR band pass filter • ICA (Independent Component Analysis) • MARA (Multiple Artifact Rejection Algorithm)	• MRMR (Minimal Redundancy Maximal Relevance) based algorithm.	• KNN (K-Nearest Neighbor) Classifiers	• We get high accuracy rate of 91.67% (training data sets) and 80% (testing data sets) for F8 channel.
H. Peng et al., [22]	• Multivariate pattern related analysis. • 28 Healthy Control vs 27 MDD patient.	• FIR filter • PLI (Phase Lag Index) analysis.	• Binary SVM (Support Vector Machine) method	• KNN (K-Nearest Neighbor) classifier • DT (Decision Tree) and NB (Naïve Bayes) • SVM classifier	• Binary SVM gives accuracy rate of 92.73%. • MDD patient having high low frequency coherence in alpha and theta band
Chaojiang et al., [23]	• EEG based classification using spatial information.	• Ocular correlation algorithm • Band pass filter	• Differential entropy • Wavelet Packet Decomposition • GA (Genetic Algorithm)	• SVM • KNN and LR (Logistic Regression) function	• With TCSP condition, Classification rate is found to be 84% (+ve stimuli) and 85.7% (-ve stimuli). • Change in gamma

(continued)

Table 1. (*continued*)

					frequency during facial expression task of depressed patient.
Xuexiao shao et al., [24]	EEG acquisition with improved EMD. 30 subjects (15 MDD vs NC)	High pass filter ICA, ASR, FIR FFT Matlab R 2016a	-------------	LMT (Logistic Model Trees) algorithm LOO (leave One Out) Cross Validation	We get classification rate of 80%. unctional connectivity of normal group is higher as compared to MDD group
C.T. Lin et al., [25]	EEG signal acquisition using forehead electrode system. 0 healthy persons.	Preamplifier MA filter FIR band pass filter STFT transform FFT	----------------	RVM (Relevance Vector Machine) and SVM (Support Vector Machine) SVM RBF (Support Vector Machine with Radial Basis Function)	SVM RBF accuracy is 73.5±11.2%. Theta wave band power is found to be lower.
Zhang et al.,[26]	Functional network algorithm for analysis of EEG signal. 24 MDD vs Healthy Control	Bandpass filter DWT (Discrete Wavelet Transform) Adaptive noise rejection filter FFT and PLI	----------------	SVM linear kernel KNN (K-Nearest Neighbor) ANN (Artificial Neural Network) FR (Random Forest)	We get highest accuracy rate of 93.16%. MDD group synchronization rate is found to be higher as compared to HC in theta and alpha band.
z. wan et al., [27]	Multichannel EEG acquisition system. 35 subjects (12 NC, 12 un medicated and 11 medicated)	ADC (Analog to Digital converter)	--------------	Deep learning algorithm 5 type CNN (Convolution Neural Network)	Hybrid EEG net CNN model is best (accuracy of 79.08%) as compared to others. Alpha peak is affected in case of depressed patient as compared to NC (Normal Control) and MD(medicated depressed patient).
Y. Li et al., [28]	EEG signal acquisition using 16 electrode system with 10 healthy subjects vs 10 MDD (mild depressive disorder) patients. fferential evolution method	Low and high pass filter Notch filter Fast ICA Matlab R2016a	Hanning filter	KNN DE+ KNN	With DE+ KNN Model, Classification accuracy (97.44%) is highest for beta band as compared to other bands.
Z. Zhu et al.,[29]	Multimodal strategy for depression detection EEG and EM	Lowpass and high pass filter Fast ICA	Hanning filter	Linear SVM RBFSVM (Radial Basis	We get accuracy rate (linear SVM) of 83.42%. Performance of highest frequency

(*continued*)

Table 1. (*continued*)

	(Eye Movement) analysis comparison ▯ 0 mild depressed vs 19 healthy subjects			Function SVM) ▯ GBD tree (Gradient Boosting Decision Tree) ▯ RF (Random Forest) ▯ SNN (Soft Normalization on Neural Networks) ▯ BNMLP (Batch Normalized Multilayer Perceptron)	band (beta and gamma) is best as compared to lowest frequency (delta) band. ▯ Hidden layer fusion method gives better result to identify mild depression.
Zhang et al., [30]	▯ Emotional arousal based ASSR examination. ▯ 4 channel electrode-based system ▯ 28 subjects	▯ Band pass filter ▯ ICA	▯ STFT with Hanning window/ filter	▯	▯ Emotion influences our auditory processing in both valence and arousal condition
Jian shen et al.,[31]	▯ Improved EMD based EEG signal detection ▯ lobes (FP1, FP2 and FPz) ▯ 1 (81 depressed vs 89 HC) ▯ 2 (160 dep. Vs 116 HC) ▯ 105 dep. Vs 109 HC) ▯ 4 (105 dep. Vs 70 HC)	▯ Bandpass filter ▯ 24-bit ADC ▯ DWT ▯ K-kalman filter	▯ Improved EMD	▯ SVM	▯ We get accuracy rate of 83.27%. ▯ Improved EMD method gives better outcomes as compared to FFT and traditional method.
D.M. khan et al., [32]	▯ Detection of depression using 3 DCNN (Convolutional Neural Network) ▯ 30 MDD vs 30 HC	▯ Band pass filter ▯ Amplifiers ▯ Notch filter	-	▯ 3 DCNN (Convolutional Neural Network)	▯ With 3DCNN accuracy rate is 100%.
Liu et al. [33]	▯ f MD (major depression) using CTD (Coupled Tensor Decomposition) ▯ 20 MDD vs Healthy Control	▯ ass, high pass and notch filter ▯ ICA	▯ PCA (Principal Component Analysis)	▯ CTD (Coupled Tensor Decomposition)	▯ Investigation of oscillatory hypo-connectivity and hyper- connectivity network is affected by stimuli of music in depressed patient.
Xiaoweizhang et al., [34]	▯ Multimodal based approach for detection of depression ▯ bjects (81 dep. Vs 89 HC) ▯ rode-based system (FP1, FP2 andFPz)	▯ High pass and low pass filter ▯ DWT ▯ man filtering ▯ ntation and other filtration techniques.	▯ 45-dimensional EEG signal features ▯ voice features	▯ KNN ▯ SVM ▯ GP (Gaussian Processes) ▯ RF (Random Forest) ▯ MLP (Multilayer Perceptron) ▯ GNB	▯ We get highest accuracy rate of 76.41%. ▯ Accuracy rate of vocal signal is better as compared to EEG signal

(*continued*)

Table 1. (*continued*)

				(Gaussian Naïve Bayes)	
Ahmad raufsubhani et al., [35]	▯ Machine learning technique of detection of stress ▯ 22 subjects	▯ Filter for removing DC and notch filter	▯ FFT (Fast Fourier Transform)	▯ LR (Logistic Regression) ▯ SVM (Support Vector Machine) ▯ NB (Naïve Bayes)	▯ We get better outcomes (accuracy of 83.43%) in detecting stress using MLtechniques.
Jian shen [36]	▯ Optimal channel section method for EEG based detection using KTA. ▯ 15 dep. Vs 20 HC (64 channel) ▯ 24 dep. Vs 29 HC (128 channel)	▯ Analog band pass filter ▯ ICA algorithm	----------	▯ SVM ▯ KNN ▯ T (Differentia l Tree)	▯ We get accuracy rate of 80%. ▯ nify importance of left frontal and bilateral temporal region in detection of depression in MDD.
Zing zhu et al., [37]	▯ Detection of dep. Using EEG and eye tracking data ▯ 20 HC vs 28 MDD for EEG ▯ 17 HC vs 17 MDD for eye movement	▯ Band pass filter ▯ Fast ICA algorithm	▯ Hanning filter To get feature of α, β, θ wave bands.	▯ Weka ▯ 10-fold cross validation ▯ Lib SVM ▯ CBEM (Content-BasedEnse mble Method)	▯ highest accuracy rate of 91.18%. ▯ king modality become superior in detection of depression as compared to complex electrode placement of EEG system.
Hanshucai et al., [38]	▯ Detection of depression using 3 electrode system of forehead area. ▯ 86 dep. Vs 92 HC	▯ Notch filter ▯ Wavelet Transform	-------------	▯ SVM ▯ KNN ▯ NB ▯ T (Decision Tree)	▯ We get highest accuracy rate of 91.25%. ▯ tion of depression from Forehead area (using 3 electrode) is easy process as compared to complex more channel EEG system.

3 Pre-processing Phase

This phase is defined after acquisition of EEG signal from patient using different modalities. The acquired signal is known as raw signal having some noise and artifacts in it. This phase mainly includes some filters, amplifiers and some transform techniques to change the conditioning of acquired signal. In this review article, many methods followed by different scientists are discussed step by step.

The FIR band pass filter (1–42 Hz range), ICA (Independent Component Analysis) and MARA (Multiple Artifact Rejection Algorithm) based methods are used to remove several types of artifacts in processing of signal acquired from single channel system [21]. The acquired signal is processed through FIR filter (1–40 Hz range) using MATLAB EEG LAB toolbox to remove noise such as base line drift [22]. The band pass filter (0.05–100 Hz range) is used to process the EEG signal and ocular algorithm is used to remove eye related artefacts. The other method known as spatial method is implemented to separate a multivariate signal into other additive sub components to extract relevant features [23]. The high pass filter (0.5 Hz), FIR filter (0.5–30 Hz), ICA (Independent Component Analysis) and ASR methods are used for processing of signal to remove extra artifacts from raw signal [24]. The FIR band pass filter (1 to 30 Hz range) and MA (Moving Average) filter are used for processing of EEG signal by removing power line interference. The amplification of signal is accomplished using amplifier and other technique known as STFT (Short Time Fourier Transform) is involved in this study for analysis of EEG patterns [25]. The frequency range during preprocessing phase is set between 0.5–50 Hz using band pass filter. The DWT (Discrete Wavelet Transform) and additive noise filters are used to remove several types of artifacts to get desired EEG signal. The power spectrum analysis of EEG signal is performed using FFT (Fast Fourier Transform) technique [26]. The several type of filters such as low pass filter (cut off 70 Hz), high pass filter (cutoff 0.5 Hz), notch filter (remove baseline drift of 50 Hz) and fast ICA (Independent Component Analysis; remove ocular artifacts of 0–16 Hz) are used to remove various type of noise from raw EEG signal [20]. The filtration of signal is done using low pass filter (cutoff frequency 40 Hz) and high pass filter (cutoff frequency 1 Hz); the other type of artifacts such as ocular artifacts (0–16 Hz) and muscle artifacts are removed using waveform techniques and fast ICA (Independent Component Analysis) method [28]. The band pass filter (0.5–200 Hz) and ICA (Independent Component Analysis) method is used to process desired EEG signal by removing artifacts. The analysis of EEG signal is accomplished using ERSP (Event Related Spectral Perturbation) method both in time and frequency domain [29]. The band pass filter is designed using high pass filter (cutoff frequency 1 Hz), low pass filter (cutoff frequency 40 Hz) and other filters such as DWT and K-Kalman filters are used for removing eye movement related artifacts [30]. The band pass filter (0.1–70 Hz) and notch filter (remove 50 Hz) is used to process EEG signal. The amplifier is used to get enhanced desired signal [31].

The band pass filter (1–30 Hz), notch filter (remove 50 Hz) and ICA (Independent Component Analysis) method is used to process desired signal by removing eye related and other artifacts [32]. The band pass filter (1–40 Hz) with DWT (Discrete Wavelet Transform) and Kalman filters are used to process EEG signal after removing eye related artifacts. The acquired signal (using microphone) is processed using segmentation and filtration (range 60–4500 Hz) technique [33]. The signal is preprocessed using filter 0.1 Hz to remove artifacts of DC and 50 Hz frequency are removed using notch filter. The other type of artifacts such as eye blink and muscle related artifacts are removed manually [34]. The ICA (Independent Component Analysis) algorithm is used to remove all type of artifacts (EOG, EMG and ECG related artifacts) from EEG signal acquired from both channel (64 and 128 channel) [35]. The band pass filter (0.5 Hz) along with hanning filter (4–30 Hz) is used to get the frequency range of theta, alpha and beta bands. The other type of artifacts such as ocular artifacts is removed using fast ICA method [36]. The band pass filter (0.5–50 Hz range) and notch filter (remove 50 Hz) are used to get desired EEG signal. The wavelet transform is used to remove some amplitude related artifacts [37].

4 Feature Extraction

All the relevant features are extracted from EEG signal after pre-processing phase for complete analysis of signal to differentiate between healthy mind and depressed mind. Some relevant features are selected from extracted features to get final result. The several type of features such as KFD (Katz Fractal Dimension), CD (Correlation Dimensions), Hjorth parameters, HFD (Higuchi Fractal Dimension), RE (Rényi -Entropy), Power Spectrum and Bi- Spectrum features are extracted using different modalities. The mRMR (Minimal Redundancy Maximal Relevance) based algorithm is selected for selection of features before actual classification process begins [21]. The feature extraction is done using entropy; change in entropy represents the change in human brain during cognitive task. The Wavelet Packet Decomposition (WPD) method is used to extract signal in different frequency bands. The GA (Genetic Algorithm) strategy is used for selection of some relevant features. In this progression roulette wheel selection algorithm is used to calculate the fitness value [23]. The FFT method is used for analysis of signal to extract feature matrix [27]. Hanning filter is used to extract signal of desired frequency range of θ (4–8 Hz), α (8–13 Hz) and β (13–30 Hz) [20]. The EEG data is segmented into different bands and filter using Hanning filter for extraction of relevant features. Total 14 feature (12 non-linear and 2 linear) are calculated [28]. The power spectrum of signal is calculated using STFT (Short Time Fourier Transform) by applying Hanning window [29].

The improved EMD (Empirical Mode Decomposition) method is used for extraction of relevant features from EEG signal [30]. The PCA (Principal Component Analysis) method is used in the study to define various brain regions [32]. In this experimental study 45-dimensional EEG features and 162-dimensional voice features are extracted for analysis of signal [33]. The feature extraction method is accomplished by calculating parameters using ROC value, t- test and Bhattacharya distance [34]. The required features (19 features including 9 linear, 8 non-linear and 2 other features) are calculated using AR model [36]. The 4 linear features are collected from theta band (4–8 Hz), alpha (8–13 Hz), beta (13–30 Hz) and gamma bands (30–50 Hz). The non-linear features such as K-Entropy and Shannon Entropy are calculated to measure the desired outcomes [37].

5 Feature Classification

The extracted features are classified into different zones to understand difference between normal human brain and depressed brain in different frequency bands. The classification process is accomplished using various methods defined by different researchers through this review article. The KNN (K-Nearest Neighbor) classifier is used to understand difference between two groups (responder vs non responder) [21]. The binary SVM (Support Vector Machine) method, KNN (K-Nearest Neighbor), DT (Decision Tree) and NB (Naïve Bayes) classifiers are used to find out best classifier for same experimental study [22]. The machine learning technique such as SVM (Support Vector Machine), KNN and LR (Logistic Regression) function classifiers are employed for classification of signal [23]. The LMT (Logistic Model Tree) algorithm and LOO (Leave One Out) cross validation classifiers are used to evaluate performance [24]. The RVM (Relevance Vector Machine) and SVM (Support Vector Machine) are two methods used to classify signal [25]. The classifiers used in this study are SVM (Support Vector Machine) linear kernel, KNN (K-Nearest Neighbor), ANN (Artificial Neural Network) and FR (Random Forest) for classification of signal in various frequency bands [26].

The total 5 types of CNN (Convolution Neural Network) model such as sgn EEG net, Reg EEG-net, Deep CONN net, Ach CNN and EEG net are used for classification of

EEG signal [27]. The KNN classifier outcomes are compared with DE+ KNN (Differential Entropy + K-Nearest Neighbor) outcomes for both emo-block and neu-block data [20]. The SVM (Support Vector Machine), RBFSVM (Radial Basis Function SVM), GBD tree (Gradient Boosting Decision Tree), RF (Random Forest), SNN (Soft Normalization Neural Networks) and BNMLP (Batch Normalized Multi-layer Perceptron) are different type of classifiers used in this experiment to classify EEG signal [28]. The SVM method is used for detection of depression in less sample size [30]. The k-fold cross validation method is used to evaluate classification rate [31]. The 4th order CPTD (Coupled Tensor Decomposition) method is used to calculate adjacent correlation maps [32]. A reliable depression detection model is formed with various classifier such as KNN (K-Nearest Neighbor), SVM (Support Vector Machine), GP (Gaussian Process), RF (Random Forest) MLP (Multilayer Perceptron) and GNB (Gaussian Naïve Bayes) [33].

The various type of classification models used to compare stress vs control condition are LR (Logistic Regression), SVM (Support Vector Machine) and NB (Naïve Bayes) [34]. The classification of EEG signal is done using SVM (Support Vector Machine) method [35]. The three types of classifiers known as weka, 10-fold cross validation and lib SVM are used to compare classification result of EEG signal. The CBEM (Content Based Ensemble Method) is used for both experiments (eye tracking vs EEG signal) [36]. The four classifiers such as SVM (Support Vector Machine), KNN (K-Nearest Neighbor), NB (Naïve Bayes) and DT (Decision Tree) are taken to compare results to find out best classifier for 3 electrode-based system [37].

6 Statistical Analysis and Results

There are various types of statistical method mentioned in this review article for complete analysis of signal. These statistical parameters are sufficient to reach a final conclusion that defines best method/ modalities to understand the state of human brain. These parameters mainly include Mean, Median, Max Frequency, S.D. (Standard Deviation), Variance, Accuracy, Sensitivity and Specificity. The type of methods used to calculate statistical parameters are ANOVA, Wilcoxon Signed-Rank method, AR model, time-varying Hjorth parameter, ROC value, t-test, Bhattacharya distance, Z-score and MANOVA (Multivariate Analysis of Variance) method. The F8 Channel (single channel) is having high accuracy rate of 91.67% (training data sets) and 80% (testing data sets) as compared to multichannel system accuracy rate of 86.11% (training data sets) and 60% (testing data sets) [21]. The binary SVM (Support Vector Machine) classifier give high accuracy rate of 92.73% whereas DT (Delta Tree) and NB (Naïve Bayes) classifier gives accuracy of 85.46%. These results showed that MDD related patients are having high coherence in low frequency range (alpha and theta band) [22]. In task condition, the classification rate is found to be 84% and 85.7% in case of positive and negative stimuli with the help of TCSP condition; whereas for without TCSP condition rate is found to be 81.7% and 83.2% for both positive and negative stimuli. The leave one out cross validation classifier gives improved results with TCSP condition as compared to without TCSP condition [23]. The effect of classification is observed best for beta band with score of 80%. It is found IMFs waveform decomposition using network metrics analysis (improved EMD) gives stable result [24]. The SVM RBF (Support Vector Machine with Radial Basis Function) has high accuracy rate of 73.5± 11.2% to classify pattern of EEG signal during responder and non-responder group. The theta power frequency taken from front area (fore head) is sufficient to reached a conclusion to differentiate between depressive vs normal mind [25]. The accuracy rate is calculated for all classifiers such as RF, SVM, KNN and ANN with respect to MDD as potential biomarkers. The performance of MDD is evaluated and lowest accuracy rate is found to be 85.62%; whereas highest accuracy rate is 93.16% [26]. The accuracy, sensitivity and specificity rate are found to be 79.08%, 68.78% and 84.45% respectively. These results proved that hybrid EEG Net CNN model is best suited for analysis of brain state/ functioning as compared to another CNN model. The sensitivity rate for Syn EEG Net classifier is found to be highest with value of 83.8% [27]. It is proved that DE (Differential Entropy) based KNN classifier accuracy is higher as compared to simple KNN classifier. The average accuracy rate for emo-block and neu-block data is found to be 94.67% and 97.44% respectively [20]. The accuracy rate for EEG signal is found to be 81.03% and 76.92% for neu-block and emo-block respectively. Similarly, for EM signal, the accuracy rate is found to be highest (80.17%) for SVM classifier for Neu-block. The classification accuracy rate with SVM classifier by fusion beta band with EM in case of Neu block is found to be 82.05%, whereas for GBD tree classifier accuracy rate is found to be 78.46 % by fusion gamma band with EM in case of Emo-block. The classification accuracy rate for linear SVM during hidden fusion method is found to be highest (83.42 %) with S.D. rate of 2.09%. The above results signify/indicates that linear SVM classifier is considered as best classifier for large scale data [28]. It is concluded that enhanced ASSR value is observed in emotion condition as compared to neutral condition. The bar graph of power and three emotions (+ve, -veand neutral) showed that low amplitude of ASSR is observed for neutral condition as compared to +veand -ve emotional state. The emotional experience showed powerful impact for auditory response of person. It is proved that emotion influences our auditory processing in valence and arousal condition/ dimension [29]. The accuracy rate for data set 1 and 2 with improved EMD (Empirical Mode Decomposition) is found to be (83.27%) higher as compared to traditional one (82.65%). The accuracy rate for data set 1, 2, 3 and 4 is found to be 83.27%, 85.19%, 81.98% and 88.07% respectively. It is observed from result that data set 1 and data set 3 exhibits more stress as compared to data set 2 and data set 4 [30]. The classification results using functional connectivity show that 2 DCN network method achieve better accuracy rate as compared to 1 DCN network. The most effective outcome is obtained with 3 DCNN with accuracy rate of 100%. The high level of accuracy is found to be greater than 95% for 15-foldand lower than 95% for 10 folds [31]. The type of algorithm known as FHALS (Fast Hierarchical Alternative Least Square) is used with CTD (Coupled Tensor Decomposition) problems. The result of stimulated data gives average tensor fit with value of 0.864[32]. The graph parameter proved that classification accuracy of vocal signal is better as compared to EEG signal. The SVM classifier gives highest accuracy rate of 74.12% and 72.35% for vocal and EEG signal respectively. The accuracy rate of multi-agent strategy is found to be highest 76.41% [33]. The t-test performance accuracy rate is found to be 94% and 93.9% with NB classifier and SVM classifier respectively. The accuracy level of proposed work is found to be 83.43% and compared with performance of other studies to know the difference while comparing two level of stress [34]. The rate of accuracy is found to be 65.71% and 80% for 58 channel and19 channel system respectively in order to detect level of depression. The highest accuracy rate of m-KTA Chsel method with SVM (Support Vector Machine) classifier is found to be 80% and F1 score with value of 88.89%. These results are highest as compared to KNN and DT classifier with same m-KTA Chsel method. These findings of PET scanning proved that left frontal lobe and bilateral temporal region plays significant role in detection of depression in MDD patient [35]. The CBE method get accuracy rate of (for eye movement) 78.50% ± 4.54% for dynamic condition, whereas 82.50 ± 3.52% for static condition. The highest rate of accuracy is found to be 78.24 ± 3.53 % for integrated data of RFC (Random Forest Classifier). The accuracy rate for CBEM for resting state EEG in dynamic condition is found to be 85 % ± 4.04 %, whereas for LOSO method CBEM accuracy rate is 88.24 % for dynamic model and 91.18 % for static model [36]. The highest accuracy rate is 81.09 % for NB classifier for second segment. The highest recognition rate for resting state is found to be 76.83 % for KNN classifier that identify depression state of patient. The case base reasoning technique achieves highest accuracy (91.25%) as compared to another classifier such as KNN (81.44 %) [37].

7 Research Gaps with Possible Solutions

There are various types of limitations associated with studies reviewed in our work, these research gap can be overcome by possible future solutions described below:

- The number of electrodes and number of subjects taken for experiment is very small, which can be improved by performing experiment at large scale with more electrode system [21].
- The validity of experiment can be improved by performing experiment on dataset of more subjects by measuring level of depression (mild, major and severe depression). We should find out changes inside human brain during depression [22].
- The experiment can be performed with other techniques like deep learning to obtained more improved results [23].
- The EEG signal taken for experiment does not provide functional connectivity of brain accurately so, we should compare analysis with other techniques such as fMRI (Functional Resonance Imaging), PET (Positron Emission Tomography) scan [24].
- Decoding of frontal area activity is very complex process and the EEG system is also limited in spatial resolution. The experiment should be considered with analysis of brain activity with various imaging system [25].
- The information provided in dataset is limited, we should include dataset with functional connectivity showing changes inside the brain. The Potential Biomarkers based information conformation is very complex process [26].
- We can perform experiment on large number of subjects with complete 128 channel EEG system to get more reliable outcomes. We should find out spatial information differentiating MDD (Major Depressive Disorder) patient from Healthy ones including EEG rhythm changes. We can perform experiment with other classifiers for better interpretation of experiment [27].

- We can combine differential evolution operator with other classifiers for better outcomes. The other techniques such as deep learning can be used for analysis of experiment to achieve high classification value [20].
- We can perform experiment on large scale with a greater number of subjects that not only define mild depression but also consider major depression and severe depression states for classification at each level [28].
- We can perform experiment considering more diverse state of emotion by improving selection of method to evaluate effect of various emotions [29].
- The experiment can be performed on large scale with more MDD (Major Depressive Disorder) vs HC (Healthy Control) subjects for clinical applications [31].
- We can perform experiment including anatomical imaging-based MRI (Magnetic Resonance Imaging) system to evaluate more improved accuracy. The experiment can be investigated with various types of music by correlating relationship between neural activity with specific music [32].
- We can perform experiment with other physiological signals such as EMG (Electromyography) and eye movement related signals to understand depressive state in more effective way. The experiment can be considered with other deep learning techniques to get more reliable outcomes [33].
- We can perform experiment on more subjects for validity of mKTAChsel (Kernel-Target-Alignment

  Channel Selection) based system for EEG signal analysis of depressed person [35].

- We can use GUI (Graphic User Interface) technique to build an app that will help in detection of depression using sensors (detect eye movement) and camera-based device that support our mobile phone for easy with less costly continuous diagnosis [36].
- The experiment can be performed with more electrode system for depression detection using case-based reasoning system [37].

8 Conclusions

This study highlighted many points to understand depressive state of human being. In this way, EEG (Electroencephalography) modality proved to be best method in acquisition of signal from human brain. Some studies are related to acquisition of EEG signal using multiple channel system, but single channel EEG also gives significant outcomes in acquisition of signal for MDD (Major Depressive Disorder) patient. In this progression, the forehead EEG signal acquisition method proved to be revolutionary step in acquisition of signal for depressed patient.

The type of filtering techniques mentioned in this survey are FIR (Finite Impulse Response), ICA (Independent Component Analysis), MARA (Multiple Artifact Rejection Algorithm), high pass filter, band pass filter, ASR (Artifact Subspace Reconstruction). The processing of signal is accomplished using MATLAB EEG LAB toolbox. The type of modalities used for feature extraction are KFD (Katz Fractal Dimension), RE (Rényi Entropy), CD (Correlation Dimensions), HFD (Higuchi Fractal Dimension), KNN (K-Nearest Neighbor), SVM (Support Vector Machine), DT (Decision Tree), NB (Naïve Bayes), KRC (Kendall Rank Correlation), Entropy, WPD (Wavelet Packet

Decomposition), LR (Logistic Regression), GA (Genetic Algorithm), DE (Differential Entropy), EMD, improved EMD (Empirical Mode Decomposition), LMT (Logistic Model Tree), LOO (Leave One Out) Cross Validation. The power spectrum of EEG signal is calculated using FFT (Fast Fourier Transform) algorithm and STFT (Short Term Fourier Transform). The statistical analysis of signal is performed using various statistical modalities such as ANOVA (Analysis of Variance), permutation test, chi square, T-test, F measurement etc. The analysis is represented in terms of accuracy, specificity, sensitivity and precision. These analysis and results showed comparison between Healthy Control group, mild depressed and major depressed patient by change in various frequency bands. It is proved from this survey that pre frontal region of brain is mostly affected in depressed patient. Moreover, this survey identifies the research gaps of the existing works and presents the possible solutions which can provide valuable insights for the budding researchers working in this field.

References

1. Depression (NIH Publication No. 15-3561), U.S. Dept. Health Hum. Services, Nat. Inst. Health, U.S. Government Printing Office, Bethesda, MD, USA (2015)
2. World Health Organization. Depression (2019). http://www.who.int/en/news-room/fact-sheets/detail/depression
3. Bear M.F., Connors B.W., Paradiso M.A.: Neuroscience: Exploring the Brain. Lippincot, Williams and Wilkins, Philadelphia, PA, USA (2007)
4. Malik, J., Dahiya, M., Kumari, N.: Brain wave frequency measurement in gamma wave range for accurate and early detection of depression. Int. J. Adv. Res. Innov. **6**(1), 21–24 (2018)
5. Sayers, J.: The world health report 2001-Mental health: new understanding, new hope. Bull. World Health Org. **79**(11), 1085 (2001)
6. Zhang, X., Shen, J., Din, Z.U., Liu J., Wang, G., Hu, B.: Multimodal depression detection: fusion of electroencephalography and paralinguistic behaviors using a novel strategy for classifier ensemble. IEEE J. Biomed. Health Informat. **23**(6), 2265–2275 (2019)
7. Zhang, B., et al.: Ubiquitous depression detection of sleep physiological data by using combination learning and functional networks. IEEE Access **8**, 94220–94235 (2020)
8. Ibrahim, A.K., Kelly, S.J., Adams, C.E., Glazebrook, C.: A systematic review of studies of depression prevalence in university students. J. Psychiatric Res. **47**(3), 391–400 (2013)
9. Peltzer, K.: Depressive symptoms in relation to alcohol and tobacco use in South African University students. Psychol. Rep. **92**(3), 1097–1098 (2003)
10. American Psychiatric Association, Diagnostic and statistical Manual of Mental Disorders (DSM). American Psychiatric Association, Ishington, DC, USA, pp. 143–147(1994)
11. Wasserman D.: Depression (The facts), p.112. Oxford Univ. Press, Oxford, U.K. (2011)
12. Knott, V., Mahoney, C., Kennedy, S., Evans, K.: EEG power frequency asymmetry and coherence in male depression. Psychiatry Res. Neuroimaging Sect. **106**(2), 123–140 (2001)
13. Nystrom, C., Matousek, M., Hallstrom, T.: Relationships between EEG and clinical characteristics in major depressive disorder. Acta Psychi-atrica Scandinavica **73**(4), 390–394 (1986)
14. Liu Y., Zhang H., Chen M., Zhang L.: A boosting-based spatial spectral model for stroke patients' EEG analysis in rehabilitation training. IEEE Trans. Neural Syst. Rehabil. Eng. **24**(1), 169–179 (2016)
15. Ozcan, A.R., Erturk, S.: Seizure prediction in scalp EEG using 3D convolutional neural networks with an image-based approach. IEEE Trans. Neural Syst. Rehabil. Eng. **27**(11), 2284–2293 (2019)

16. Acharya, U.R., Sudarshan, V.K., Adeli, H., Santhosh, J., Koh, J.E.W., Adeli, A.: Computer-aided diagnosis of depression using EEG signals. Eur. Neurol. **73**(5–6), 329–336 (2015)
17. Zhang X., Hu B., Ma X., Xu L.: Resting-state whole –brain functional connectivity networks for MCI classification using L2-regularized logistic regression. IEEE Trans. Nanobiosci. **14**(2), 237–247 (2015)
18. Hosseinifard, B., Moradi, M.H., Rostami, R.: Classifying depression patients and normal subjects using machine learning techniques and nonlinear features from EEG signal. Comput. Methods Programs Biomed. **109**(3), 339–345 (2013)
19. Erguzel, T.T., Ozekes, S., Tan, O., Gultekin, S.: Feature selection and classification of electroencephalographic signals: an artificial neural network and genetic algorithm-based approach. Clin. EEG Neurosci. **46**(4), 321–326 (2015)
20. Li, Y., Hu, B., Zheng, X., Li, X.: EEG-based mild depressive detection using differential evolution. IEEE Access **7**, 7814–7822 (2019)
21. Hasanzadeh, F., Mohebbi, M., Rostami, R.: Single channel EEG classification: a case study on prediction of major depressive disorder treatment outcome. IEEE Access **9**, 3417–3427 (2021)
22. Peng, H., et al.: Multivariate pattern analysis of EEG-based functional connectivity: a study on the identification of depression. IEEE Access **7**, 92630–92641 (2019)
23. Jiang, C., Li, Y., Tang, Y., Guan, C.: Enhancing EEG-based classification of depression patients using spatial information. IEEE Access **29**, 566–575 (2021)
24. Shao, X., et al.: Analysis of functional brain network in MDD based on improved empirical mode decomposition with resting state EEG data. IEEE Access **29**, 1546–1556 (2021)
25. Lin, C.T., et al.: Forehead EEG in support of future feasible personal healthcare solutions: sleep management, headache prevention and depression treatment. IEEE Access **5**, 10612–10621 (2017)
26. Zhang, B., Yan, G., Yang, Z., Su, Y., Wang, J., Lei, T.: Brain functional networks based on resting state EEG data for major depressive disorder analysis and classification. IEEE Access **29**, 215–229 (2021)
27. Wan, Z., Huang, J.J., Zhou, H., Yang, J., Zhong, N.: Hybrid EEGNet: a convolutional neural network for EEG feature learning and depression discrimination. IEEE Access **8**, 30332–30342 (2020)
28. Zhu, J., et al.: Multimodalmild depression recognition based on EEG-EM synchronization acquisition network, vol. 7, pp. 28196–28210 (2019)
29. Zhang, L., Liu, S., Liu, X., Zhang, B., An, X., Ming, D.: Emotional arousal and valence jointly modulate the auditory response: a 40 Hz ASSR study. IEEE Trans. Neural Syst. Rehabil. Eng. **29**, 1150–1157 (2021)
30. Shen, J., Zhang, X., Hu, B., Wang, G., Ding, Z., Hu, B.: An improved empirical mode decomposition of electroencephalogram signals for depression detection. IEEE Trans. Affect. Comput. **3**(1), 262–271 (2022)
31. Khan, D.M., Yahya, N., Kamel, N., Faye, I.: Automated diagnosis of major depressive disorder using brain effective connectivity and 3D convolutional neural network. IEEE Access **9**, 8835–8846 (2021)
32. Liu, W., Wang, X., Xu, J., Chang, Y., Hamalainen, T., Cong, F.: Identifying oscillatory hyper connectivity and hypoconnectivity networks in major depression using coupled tensor decomposition. IEEE Trans. Neural Syst. Rehabil. Eng. **29**, 1895–1904 (2021)
33. Zhang, X., et al.: Multimodal depression detection: fusion of electroencephalography and paralinguistic behaviors using a novel strategy for classifier Ensemble. IEEE J. Biomed. Health Inform. **14**(8) (2015)
34. Subhani, A.R., Mumtaz, W., Saad, M.N.B.M., Kamel, N., Malik, A.S.: Machine learning framework for the detection of mental stress at multiple levels. IEEE Access **5**, 13545–13556 (2017)

35. Shen, J., et al.: An optimal channel selection for EEG-based depression detection via Kernel-Target alignment. IEEE J. Biomed. Health Inform. **25**(7), 2545–2556 (2020)
36. Zhu, J., et al.: An improved classification model for depression detection using EEG and Eye tracking data. IEEE Trans. Nano Biosci. **19**(3), 527–537 (2020)
37. Cai, H., Zhang, X., Zhang, Y., Wang, Z., Hu, B.: A case-based reasoning model for depression on three-electrodes EEG data. IEEE Trans. Affect. Comput. . **11**(3), 383–392 (2020)

Implementing a Calibration System for Demand Pacemaker Using a Web-Based Approach

RaamaNarayanan AnanthaNarayanan, B. Geethanjali(✉), Sona Mariya, and Mahesh Veezhinathan

Department of Bio-Medical Engineering, Sri Sivasubramaniya Nadar College of Engineering, Kalavakkam 603110, India

{raamanarayanan1939,sona1955}@bme.ssn.edu.in, {geethanjalib, maheshv}@ssn.edu.in

Abstract. According to current estimation surveys, 18.1 million premature deaths in 2021 are being directly related to cardio-vascular disorders, accounting for about 30% of all fatalities worldwide. The cardiac electrical conduction problem is corrected by a pacemaker whose pivotal function is to maintain the heart rate consistent. The consistency needs to be maintained and monitored remotely for effective delivery of pacing pulse. This necessitates the calibration of pacemakers in inaccessible areas encompassing the conception of telemetry. Designing and calibrating the synchronous pacemaker becomes mandatory to deliver the required pacing pulse for the effective functioning of the heart. The proposed device was designed and implemented using a low noise and high slew rate amplifier. The heart rate was extracted and used to simulate the pacing pulse required by the developed pacing algorithm. The repeatability and reproducibility of the pacing algorithm was statistically validated ($p = 0.779$) as an evaluation metric to ascertain the relationship between the pre-determined heart rate and the measured heart rate from the proposed design, which concluded that both individual parameters have a strong degree of association. The results confirm that the generation of pacemaker pulse as and when the demand arises from the patient.

1 Introduction

In India, CVDs caused over 28 of 100 fatalities and 14.1% of total of all years with a disability that were disability-adjusted, compared to 15.3% and 6.8%, respectively. The cumulative gross fatality rates for victims from any cause and had an underlying condition abnormally slow or rapid heartbeat—was 63.3 per 1,000 individuals. Most CVD fatalities are experienced in low and middle-income nations [2]. India ranks at the top of the list of countries with the greatest prevalence of acute coronary syndrome.

Cardiovascular disease is a term used to describe ailments that impact the heart or blood arteries. Coronary artery disease, excessive blood pressure, heart attacks, cardiomyopathy, dysrhythmias, pulmonary embolism, stroke, and peripheral vascular disease are the most prevalent forms of cardiovascular illness. The primary diagnostic instrument used to acquire and interpret ECG signals is an electrocardiogram (ECG) [1].

B. K. Singh et al. (Eds.): ICBEST 2023, CCIS 2003, pp. 376–387, 2024.
https://doi.org/10.1007/978-3-031-54547-4_29

As people age, the cardiovascular system starts to deteriorate and becomes more susceptible to arrhythmia. A life-threatening arrhythmia is an irregular heart rhythm. Recurrent life-threatening arrhythmias that afflict the older population include tachycardia, paroxysmal atrial flutter, and ventricular fibrillation [3–5] Arrhythmias slow down the heart's ability to beat regularly and normally. Furthermore, pacemakers can help the heart's chambers beat in unison, allowing the patient's heart to supply blood to the tissues more efficiently.

The heart may beat unusually rapidly, too weakly, or erratically when experiencing an arrhythmia. The electrical system in the heart stimulates each pulse and delivers signals that enable the heart to function [6, 8]. The heart's pace and rhythm are influenced by these electrical impulses. The upper right chamber of the heart, known as the right atrium, possesses the sinus node, which stimulates the heart to function in a regular rhythm [9–11]. The natural pacemaker of the heart is the sinus node. The heart can pump either swiftly, more slowly, or unpredictably when the sinus node is not fully functional. Tachycardia is the serious condition for a rapid heartbeat (greater than 100 beats per minute in adults). Bradycardia is a condition for a stagnating heart rate (just under 60 beats per minute) [38]. A pacemaker's pivotal function is to maintain the heart rate consistent, if the heart's natural pacemaker becomes too slow or as a result of an obstruction in the electrical conduction system [12, 13]. Depending on the type of heart malfunction, either temporary artificial pacing during treatment or permanent electronic pacing for a healthy existence following therapy is necessary [14–16]. As need for effective pacing of pacemaker is necessary for people with cardiac pacemaker the device needs to be calibrated remotely. This study aims to design and implement the simulation model of a demand type synchronous pacemaker which can be used for remotely calibrating the device using web publishing tool.

2 Methodology

The workflow process of the pacemaker design is depicted in Fig. 1. The instrumentation amplifier AD 620, which is unpretentious by drift and has a constant gain over the entire bandwidth range is used to amplify the signal of physiologic electrical activity obtained from the Cardio-Sim simulator due to its typical low amplitude [17–19]. The resultant continuous signal has a mean variation over a specified period as a result of which the signal needs to be detrended. The forward double-density Discrete Wavelet Transform, which segments the signal up into multiple sets, is employed to eliminate noise from the signal [20–22]. Each set has a time series of coefficients which portray the signal's temporal dynamics in the corresponding frequency spectrum [23–25]. The wavelet coefficients are being subjected to a deblurring technique known as soft thresholding over all scale and sub-bands [26–28]. To extricate the R-R intervals, the denoised portion of the signal is exposed to IIR Butterworth frequency selective filtering. A filter which further uses moving mean weight to extract the R-R intervals smoothens the signal generated by this segment [29, 30].

The intended design of the synchronous pacemaker was initially simulated in NI Multisim 14.2 environment. The waveform's frequency and voltage settings were pre-set at 1.4 Hz and 50.46 mv correspondingly. The ECG instrumentation amplifier deployed

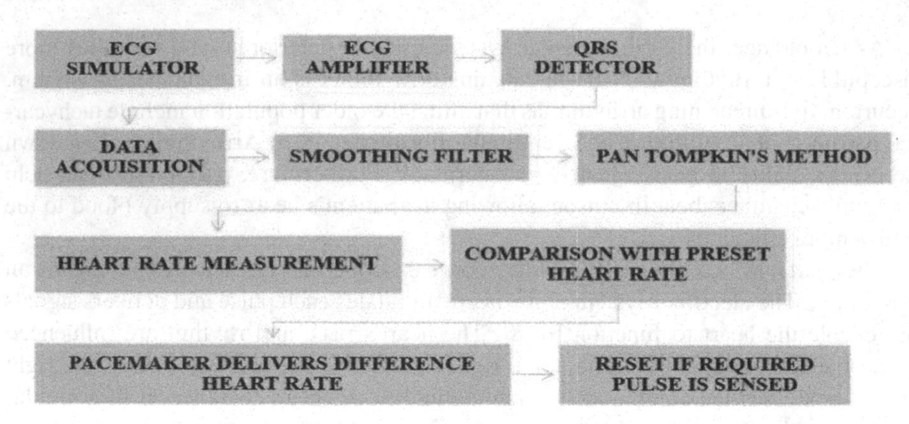

Fig. 1. Workflow of the Proposed Pacemaker Design

is the AD 620, which is acknowledged for its discrete design and great accuracy, with embedded integrated resistors that assist in attaining the optimal common mode rejection ratio in contrast to other ICs [31, 32] (Fig. 2).

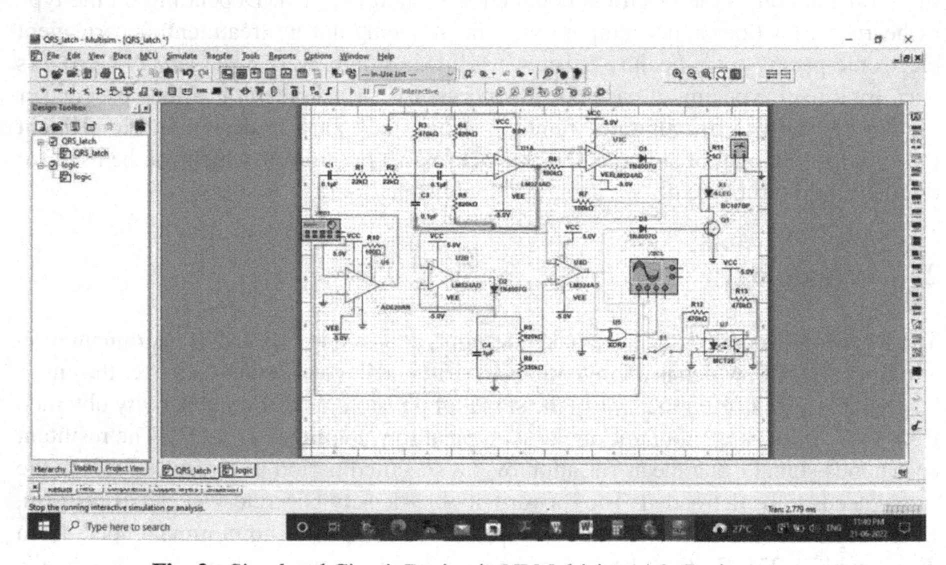

Fig. 2. Simulated Circuit Design in NI Multisim 14.2. Environment

The recorded ECG is processed through a band delimiting filter with a central frequency of 10 Hz and a frequency band of 15 Hz, which is harnessed by the QRS detector to distinguish the QRS complex. This filtered signal is half wave rectified, the peak voltage of the output is stored in the capacitor which is subsequently compared with the output from the filter and a pulse is produced each time the filtered wave exceeds the threshold limit of the preset heart rate which drives a LED This leads to the determination of a new threshold voltage limit by recharging of the capacitor based on the preceding

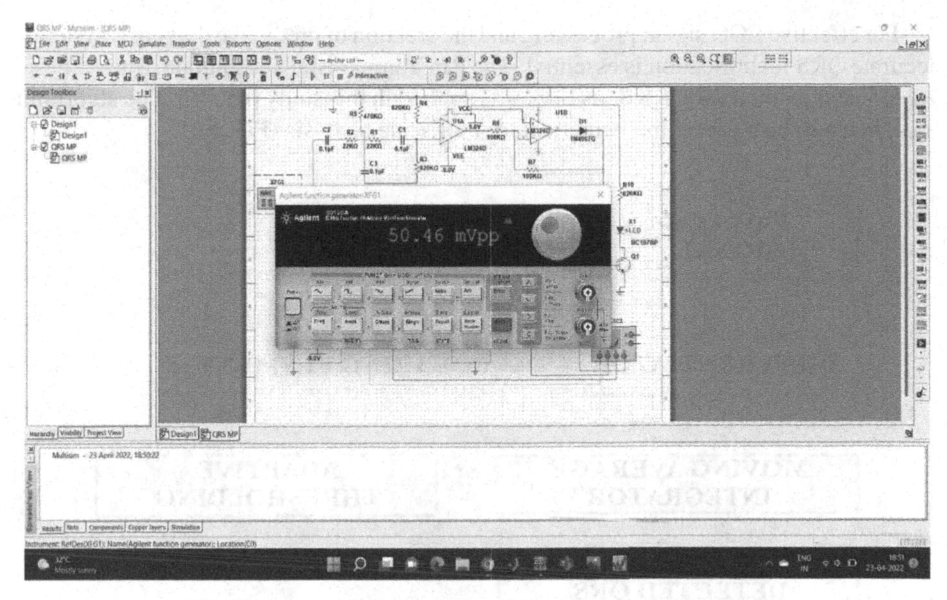

Fig. 3. Input voltage set in the Agilent Function Generator

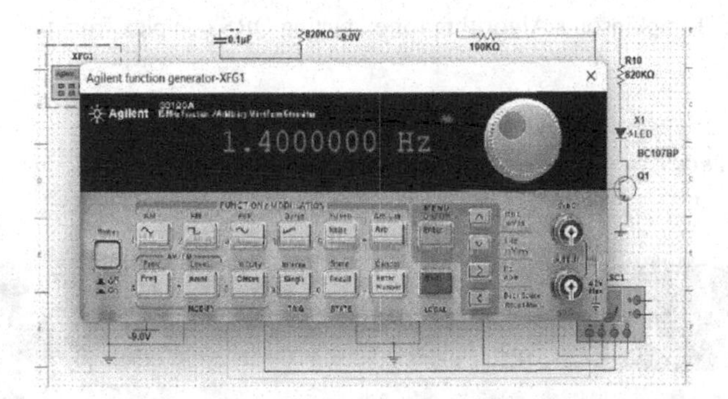

Fig. 4. Frequency Parameters set in the Agilent Function Generator

signals [33–35]. When the preset threshold limit of the pulses was fixed at 40,50 and 55 beats per minute which are considered at the limits of bradycardic condition, the subsequent new threshold limits determined due to recharging of the capacitor was 42.7,53.5 and 59.3 beats per minute (see Fig. 3).

Employing universal logic gates, the acquired mono-pulse was latched with a reference pulse to investigate the operation of the pacemaker design in this study. The inputs of the gate were the digitized signals of the acquired and the reference waves [36, 37, 39]. When both the waves were present, the timing circuit of the pacemaker resets itself (i.e., when both the inputs are high, the output of the gate is zero). When the acquired pulse is absent, the latching pulse compensates for it by delivering the required pulse (i.e., when one of the inputs is low, the output of the gate is high) (see Fig. 4).

For effective ECG signal processing and the creation of predictive diagnosis systems, accurate QRS identification is essential. The Pan Tompkinson's technique, a novel basic yet efficient method for QRS detection, was applied in this study [45, 46]. The algorithm was shown to be a useful mathematical tool for scale analysis (Fig. 5).

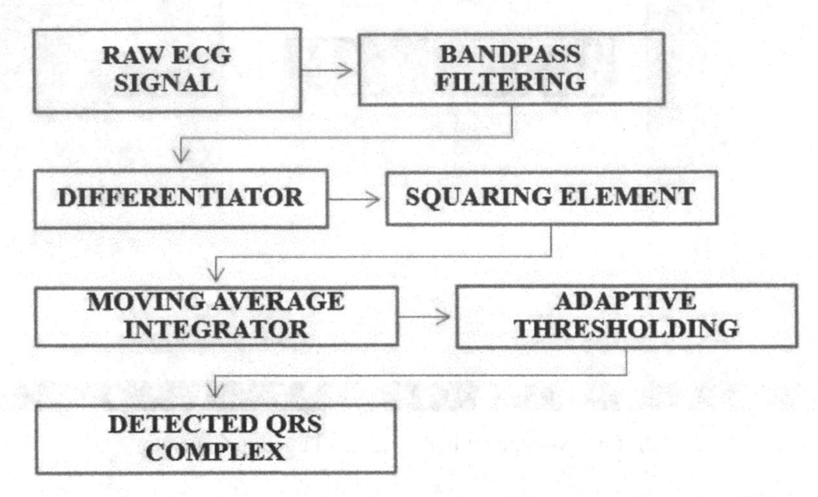

Fig. 5. Pan Tompkinson's Algorithm for extracting QRS complex from the ECG signal

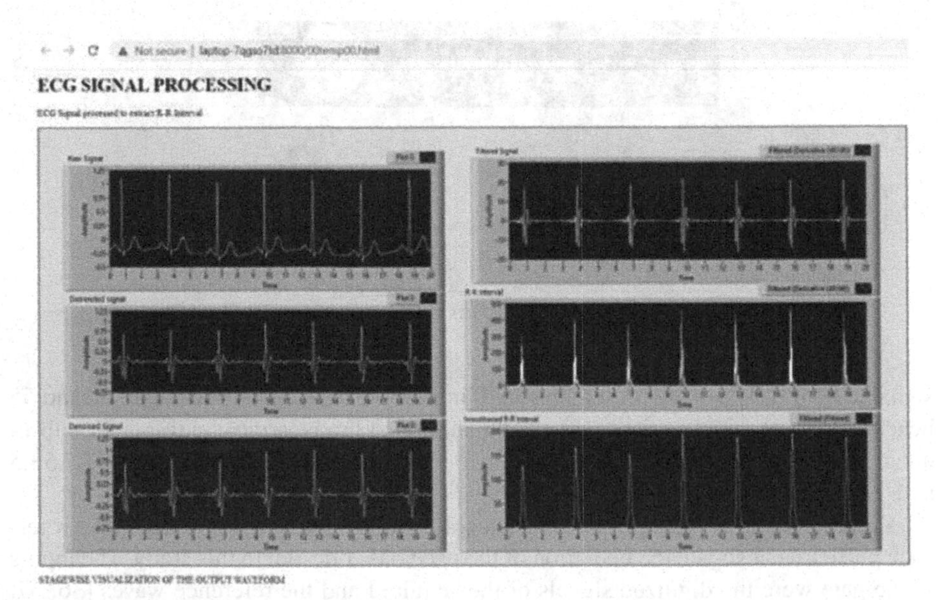

Fig. 6. a) Raw ECG Signal b) Detrended Signal c) Denoised Signal d) R-Wave Differentiated Signal e) R-R Intervals f) Smoothened R-R Intervals

The hardware implementation was initiated after prudently reviewing the in NI Multisim 14.2 simulation results. The ECG signal was acquired using ECG ring electrodes. The

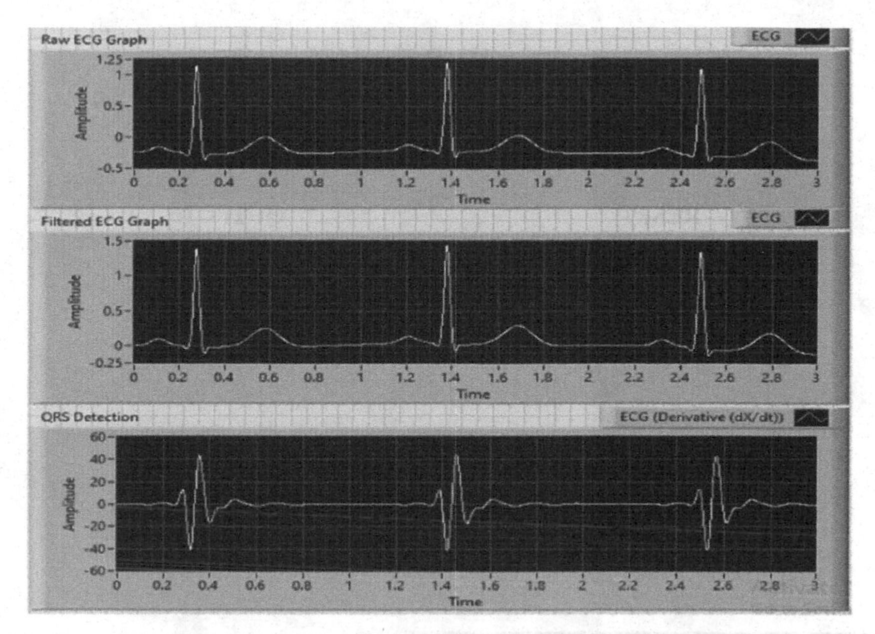

Fig. 7. a) Raw ECG signal b) Filtered ECG Signal c) QRS Interval Discrimination Wave

signal is then amplified using an AD620 amplifier with a suitable gain of 200. This signal is then interfaced into NI LabVIEW 2022 software through a data acquisition system (DAQ Signal Acquisition palette) (see Fig. 6 and 7).

The hypothesized synchronous pacemaker design's timing circuit is illustrated with its operational conditions in Fig. 8, 9 and 10. The components of the circuitry includes electrodes for signal acquisition, amplifier for amplifiying the acquired signal, the reset and pulse deliver circuits which are the operational conditions of the proposed synchronous pacemaker design [40, 41].

If the natural cardiac pulse is sensed, the pacemaker resets itself. The pacemaker delivers the required difference when the natural pulse is not sensed or when it does not match with the pre-set heart rate. The observations prove that the production of pacemaker pulses is in response to patient demand.

2.1 Remote Monitoring

The Web Publishing Tool of LabVIEW 2022 software was utilized to record the results which are presented in Fig. 11. Display panels from a VI are presented as static images on the Internet by the LabVIEW Web Server. This is accomplished by providing a static image of the front panel of a VI that is now in memory on the server machine using the Snap feature in the Web Publishing Tool. All Virtual Instrumentation files and activities are accessible to all Web browsers. During calibration procedures, a client computer can monitor and manipulate the front panel remotely attributable to an embedded animated image of the front panel.

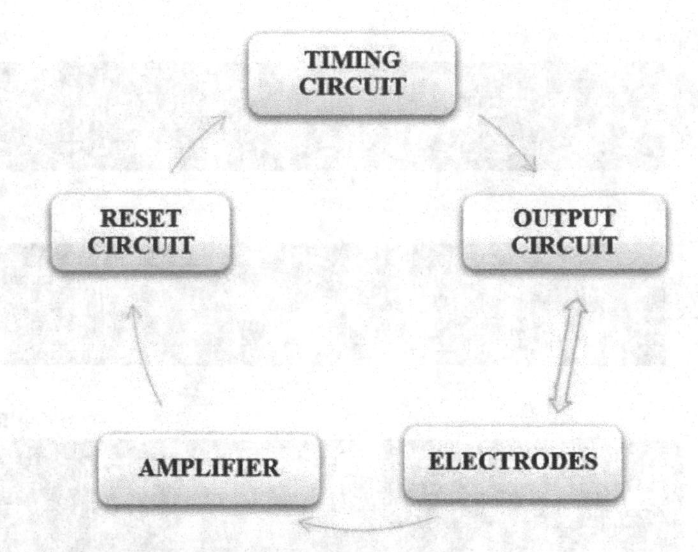

Fig. 8. Flow of the components in a Synchronous Pacemaker

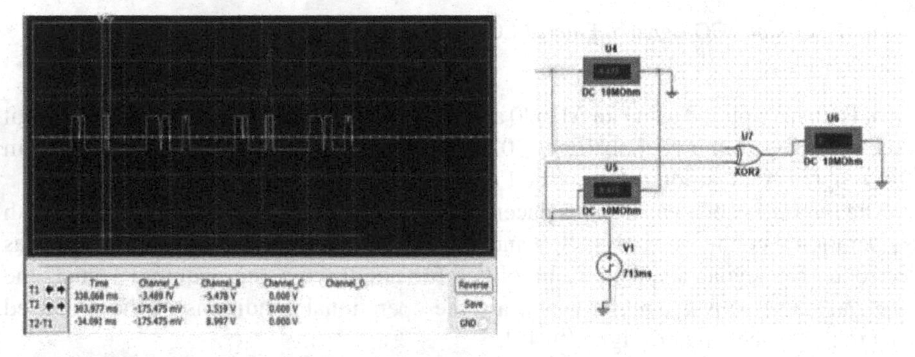

Fig. 9. Simulation result showing reset condition of the pacemaker with no voltage detected.

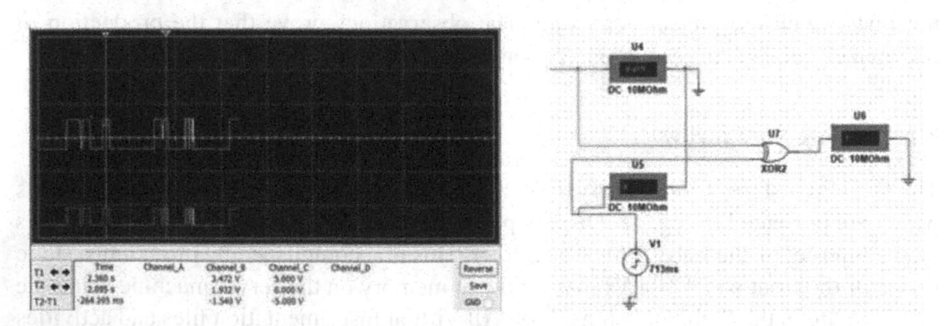

Fig. 10. Simulation result showing deliver condition of the pacemaker when then natural pulse is not detected.

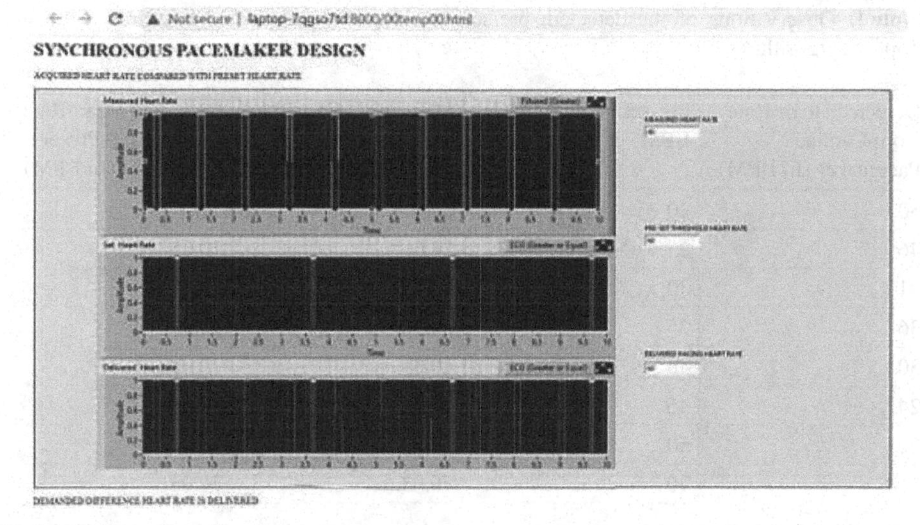

Fig. 11. a) Pre-determined Heart Rate b) Detected Heart Rate c) Paced Heart Rate using LabVIEW Web Publishing Tool

The intended pacemaker algorithm measures the subject's heart rate and compares it to the pacemaker's pre-set heart rate threshold in order to supply the extra pulse the subject demands. The design was evaluated without a pulse in order to calibrate the circuitry that produces the pulse so that it provides the complete pre-set heart rate. The observations made are recorded in Table 1. And the proposed design's delivered pacing pulses are indicated and presented schematically in Fig. 12 a) and b). This validates the study's theory that the pacemaker's design only transmits pulses to the patient when they are needed, without interfering the natural cardiac conduction pathway.

3 Results and Discussion

The suggested method thresholds the bandpass-filtered ECG signal appropriately. This led to a less complex algorithm than those currently in use that used many decomposition stages. [7] Owing to the complexity and non linearity of acquiring and extracting ECG information, [1] the proposed algorithm can automatically find the best pacing modes with the least amount of delay. This is a real-time application where the output pacemaker pulses can be interfaced with suitable NI-ELVIS boards to the electrodes [42–44]. Web publishing capabilities were simultaneously used to remotely monitor the ECG [47] with little time lag. For our investigation, the same outcomes were accomplished. LabVIEW was employed to implement the ECG pre-processing and feature extraction with minimal energy utilization [48] and without exhausting the pacemaker batteries. Statistics were used to validate the consistency of the test and the accuracy of the results.

In order to deliver the required pacing heartbeat to the patient and prevent clinical complications, the aforementioned illustration provides an analytic overview of the predefined threshold heart rate in the pacemaker circuitry and the measured heart rate recorded from the subject.

Table 1. Observations of the detected, pre-set, measured, and delivered heart rate from the proposed algorithm.

Detected Heart Rate from Cardiac Pacemaker (In BPM)	Pre-set Heart Rate (In BPM)	Measured Heart Rate from the proposed algorithm (In BPM)	Delivered Heart Rate [Detected + Preset Heart rate] (In BPM)
50	20	20.07	70.07
46	25	24.65	70.65
41	30	29.62	70.62
36	35	34.96	70.96
30	40	40.08	70.08
24	46	46.92	70.92
20	50	50.02	70.02
0	70	70.03	70.03

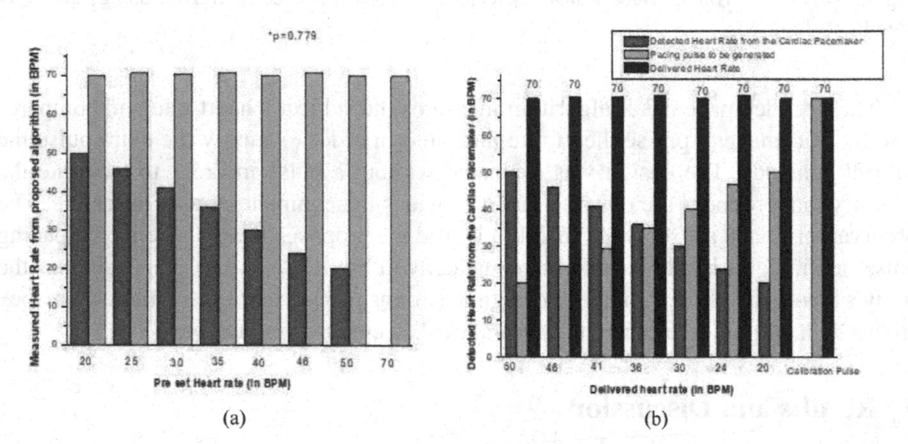

Fig. 12. a) Bar chart depicting the measured heart rate from the algorithm and pre-set rate.b) Bar diagram depicting the detected heart rate from the pacemaker and delivered heart rate.

The entire analysis was done using IBM SPSS Statistics 20. The proposed algorithm's recorded observations were contrasted to the pre-set heart rate values using the Wilcoxon Signed Rank test, a non-parametric evaluation metric to recognise significant differences between repeated measurements on a single sample to characterise whether the population mean ranks differ. The test findings indicate that there is a higher incidence of correlation between the two groups of observations than a significant variation.

3.1 Discussion

The appropriate thresholding is applied to the bandpass-filtered ECG signal in the proposed method. As a result, it produced an algorithm with less complexity than those using several decomposition levels in use today. [7] Acquiring and extracting ECG features, because of the intricacy and non-linearity, [1] the optimum pacing modes can automatically determine by the proposed algorithm with minimum delay. The ECG was simultaneously remotely monitored using web publishing tools [47] with minimal delay. The same results were obtained for our study. The ECG pre-processing and feature extraction were accomplished using LabVIEW with low energy consumption [48] without draining the pacemaker batteries. The test uniformity and the reliability of the findings were statistically validated.

4 Conclusion

The study concludes the proposed design's effective processing of the acquired ECG and generation of pacing pulse is based on the requirement. The data trend can be transmitted to the cardiologists for improving the well-being of the patients with cardiac pacemaker. The optimized proposed design will aid people with conduction problems in the heart with comprehensive extra safety features. This can be processed as a medical device which can further be extended for IoT applications with added security features as the software is vulnerable to cyber-attacks. This work can be further extended to remotely monitoring the patients with implanted pacemaker during arrhythmic actions and inform the nearby hospital to reduce fatalities.

References

1. Acharya, U.R., Fujita, H., Lih, O.S., Hagiwara, Y., Tan, J.H., Adam, M.: Automated detection of arrhythmias using different intervals of tachycardia ECG segments with convolutional neural network. Inf. Sci. **405**, 81–90 (2017)
2. Adnane, M., Jiang, Z., Choi, S.: Development of QRS detection algorithm designed for wearable cardiorespiratory system. Comput. Methods Programs Biomed. **93**(1), 20–31 (2009)
3. Afonso, V.X., Tompkins, W.J., Nguyen, T.Q., Luo, S.: ECG beat detection using filter banks. IEEE Trans. Biomed. Eng. **46**(2), 192–202 (1999)
4. Azmoudeh, B., Cvetkovic, D.: Wavelets in biomedical signal processing and analysis. Encycl. Biomed. Eng., 193–212 (2014)
5. Bahoura, M., Hassani, M., Hubin, M.: DSP implementation of wavelet transform for real time ECG wave forms detection and heartrate analysis. Comput. Methods Programs Biomed. **52**(1), 35–44 (1997)
6. Belkadi, M.A., Daamouche, A.: A robust QRS detection approach using stationary wavelet transform. Multimedia Tools Appl., 1–22 (2021).https://doi.org/10.1007/S11042-020-10500-9
7. Benmalek, M., Charef, A.: Digital fractional order operators for R-wave detection in electrocardiogram signal. IET Signal Proc. **3**(5), 381–391 (2009)
8. Berkaya, S.K., Uysal, A.K., Gunal, E.S., Ergin, S., Gunal, S., Gulmezoglu, M.B.: A survey on ECG analysis. Biomed. Signal Process. Control **43**, 216–235 (2018)
9. Bono, V., et al.: Development of an automated updated Selvester QRS scoring system using SWT-based QRS fractionation detection and classification. IEEE J. Biomed. Health Inform. **18**(1), 193–204 (2014)
10. Chen, S.W., Chen, H.C., Chan, H.L.: A real-time QRS detection method based on moving-averaging incorporating with wavelet denoising. Comput. Methods Programs Biomed. **82**(3), 187–195 (2006)

11. Chiarugi, F., Sakkalis, V., Emmanouilidou, D., Krontiris, T., Varanini, M., Tollis, I.: Adaptive threshold QRS detector with best channel selection based on a noise rating system, pp. 157–160. In: Computers in Cardiology. IEEE (2007)
12. Chouhan, V.S., Mehta, S.S.: Detection of QRS complexes in 12-lead ECG using adaptive quantized threshold. Int. J. Comput. Sci. Netw. Secur. **8**(1), 155–163 (2008)
13. Deepu, C.J., Lian, Y.: A joint QRS detection and data compression Technique for wearable sensors. IEEE Trans. Biomed. Eng. **62**(1), 165–175 (2015)
14. Dinh, H.A.N., Kumar, D.K., Pah, N.D., Burton, P.: Wavelets for QRS detection. Australas. Phys. Eng. Sci. Med. **24**(4), 207 (2001)
15. Donoho, D.L.: Denoising by soft-thresholding. IEEE Trans. Inf. Theory **41**(3), 613–627 (1995)
16. Elgendi, M., Eskofier, B., Dokos, S., Abbott, D.: Revisiting QRS detection methodologies for portable, wearable, battery-operated, and wireless ECG systems. PLoS ONE **9**(1), e84018 (2014)
17. Elgendi, M., Mohamed, A., Ward, R.: Efficient ECG compression and QRS detection for E-health applications. Sci. Rep. **7**(1), 459 (2017)
18. Faezipour, M., Saeed, A., Bulusu, S.C., Nourani, M., Minn, H., Tamil, L.: A patient-adaptive profiling technique for ECG beat classification. IEEE Trans. Inf. Technol. Biomed. **14**(5), 1153–1165 (2010)
19. Getreuer, P.: Filter coefficients to popular wavelets. MATLAB Central (2006)
20. Ghaffari, A., Golbayani, H., Ghasemi, M.: A new mathematical based QRS detector using continuous wavelet transform. Comput. Electr. Eng. **34**(2), 81–91 (2008)
21. Hou, Z., Dong, Y., Xiang, J., Li, X., Yang, B.: A real-time QRS detection method based on phase portraits and box-scoring calculation. IEEE Sens. J. **18**(9), 3694–3702 (2018)
22. https://www.who.int/news-room/fact-sheets/detail/cardiovascular-diseases-(cvds)
23. Jain, S., Kumar, A., Bajaj, V.: Technique for QRS complex detection using particle swarm optimisation. IET Sci. Meas. Technol. **10**(6), 626–636 (2016)
24. Junior, E.A., de Medeiros Valentim, R.A., Brandao, G.B.: Real time QRS detection based on redundant discrete wavelet transform. IEEE Latin Am. Trans. **14**(4), 1662–1668 (2016)
25. Kadambe, S., Murray, R., Boudreaux-Bartels, G.F.: Wavelet transform-based QRS complex detector. IEEE Trans. Biomed. Eng. **46**(7), 838–848 (1999)
26. Kohler, B.U., Hennig, C., Orglmeister, R.: The principles of software QRS detection. IEEE Eng. Med. Biol. Mag. **21**(1), 42–57 (2002)
27. Kumar, A., Berwal, D., Kumar, Y.: Design of high-performance ECG detector for implantable cardiac pacemaker systems using biorthogonal wavelet transform. Circuits Syst. Signal Process., 1–20 (2018a)
28. Kumar, A., Komaragiri, R., Kumar, M.: From pacemaker to wearable: techniques for ECG detection systems. J. Med. Syst. **42**(2), 34 (2018)
29. Kumaravel, N., Nithiyanandam, N.: Genetic-algorithm cancellation of sinusoidal powerline interference in electrocardiograms. Med. Biol. Eng. Comput. **36**(2), 191–196 (1998)
30. Kyrkos, A., Giakoumakis, E.A., Carayannis, G.: QRS detection through time recursive prediction techniques. Signal Process. **15**(4), 429–436 (1988)
31. Laguna, P., Mark, R.G., Goldberg, A,, Moody, G.B.: A database for evaluation of algorithms for measurement of QT and other waveform intervals in the ECG. In: Computers in Cardiology, pp. 673–676. IEEE (1997)
32. Lin, K.P., Chang, W.H.: QRS feature extraction using linear prediction. IEEE Trans. Biomed. Eng. **36**(10), 1050–1055 (1989)
33. Massagram, W., Hafner, N., Chen, M., Macchiarulo, L., Lubecke, V.M., Boric-Lubecke, O.: Digital heart-rate variability parameter monitoring and assessment ASIC. IEEE Trans. Biomed. Circuits Syst. **4**(1), 19–26 (2010)

34. Nayak, C., Saha, S.K., Kar, R., Mandal, D.: An efficient QRS complex detection using optimally designed digital differentiator. Circuits Syst. Signal Process., 1–34 (2018)
35. Pan, J., Tompkins, W.J.: A real-time QRS detection algorithm. IEEE Trans. Biomed. Eng. **32**(3), 230–236 (1985)
36. Qin, Q., Li, J., Yue, Y., Liu, C.: An adaptive and time-efficient ECG R-peak detection algorithm. J. Healthcare Eng. (2017)
37. Ramesh, M., Balasubramanian, S., Vijayan, V., Balasubramanian, G., Veezhinathan, M.: Design and development of a two channel telemedicine system for rural healthcare. Engineering **5**, 579–583 (2013). https://doi.org/10.4236/eng.2013.510B119
38. Rodrigues, J.N., Olsson, T., Sornmo, L., Owall, V.: Digital implementation of a wavelet-based event detector for cardiac pacemakers. IEEE Trans. Circuits Syst. I Regul. Pap. **52**(12), 2686–2698 (2005)
39. Sannino, G., De Pietro, G.: A deep learning approach for ECG-based heartbeat classification for arrhythmia detection. Futur. Gener. Comput. Syst. **86**, 446–455 (2018)
40. Satija, U., Ramkumar, B., Manikandan, M.S.: Automated ECG noise detection and classification system for unsupervised healthcare monitoring. IEEE J. Biomed. Health Inform. **22**(3), 722–732 (2018)
41. Sharma, L.D., Sunkaria, R.K.: Inferior myocardial infarction detection using stationary wavelet transform and machine learning approach. SIViP **12**(2), 199–206 (2018)
42. Suarez, K.V., Silva, J.C., Berthoumieu, Y., Gomis, P., Najim, M.: ECG beat detection using a geometrical matching approach. IEEE Trans. Biomed. Eng. **54**(4), 641–650 (2007)
43. Tang, X., Hu, Q., Tang, W.: A real-time QRS detection system with PR/RT interval and ST segment measurements for wearable ECG sensors using parallel delta modulators. IEEE Trans. Biomed. Circuits Syst. **99**, 1–11 (2018)
44. Thakor, N.V., Webster, J.G., Tompkins, W.J.: Optimal QRS detector. Med. Biol. Eng. Comput. **21**(3), 343–350 (1983)
45. Thungtong, A.: A robust algorithm for R peak detection based on optimal discrete wavelet transform. In: 2017 14th International Joint Conference on Computer Science and Software Engineering (JCSSE), pp. 1–6. IEEE (2017)
46. Tripathy, R.K., Dandapat, S.: Automated detection of heart ailments from 12-lead ECG using complex wavelet sub-band bi-spectrum features. Healthcare Technol. Lett. **4**(2), 57 (2017)
47. Shen, X.B., Zeng, J.P., Hin, T.D.: Remote healthcare monitoring system. In: International Conference on Biomedical Engineering and Informatics, vol. 3, pp. 1901–1905 (2010)
48. Zou, Y., Han, J., Weng, X., Zeng, X.: An ultra-low power QRS complex detection algorithm based on down-sampling wavelet transform. IEEE Signal Process. Lett. **20**(5), 515–518 (2013)

Common Smart Stick for Blind and Elderly People to Detect Environmental Factors and Free Navigation

Amit Kumar Singh[1,3](✉), P. Rajarajeshwari[1], R. Rathika[1], B. Shobiya[1], P. Keerthi[1], and M. Marieswaran[2]

[1] Department of Biomedical Engineering, VSB Engineering College, Karur, India
[2] Department of Biomedical Engineering, National Institute of Technology, Raipur, India
[3] Department of Biomedical Engineering, Vignan's Foundation for Science Technology & Research, Vadlamudi, Guntur, Andhra Pradesh, India
draks_bme@vignan.ac.in

Abstract. The ageing population effect is becoming predominant as the world population grows. Ageing leads to problems such as visual impairment, hearing impairment, arthritis, and others. As per the world health organization (WHO) report 2021, more than 25% of the world's population has distance vision impairment, of which 58% of the blind people were in the age group higher than 60 years. The blind and geriatric population require a stick for their free navigation. The existing passive blind sticks have the limitation of non-interacting with the nearby environment, for example, finding the location of the destination and detecting the fire nearby and water on the ground. Therefore, a smart stick is required that can be commonly used by blind and older people. In this paper, we have developed the advanced blind stick by integrating an obstacle detector in front of the user, a fire detector in the nearby environment, a water detector to detect water puddles on the ground, an emergency save our souls (SOS) button, a global positioning system (GPS) module for user location, and Internet of Things (IoT) integration for real-time data transfer to the website. The stick was tested by the ten older persons. The developed advanced stick uses some latest technologies that made the device affordable, durable, and easy to operate. We have also developed a website to be used by the user's caregiver and the hospital. With the help of a smart stick, an increase in travel speed, independence, confidence, safety, and reduction in collisions was found in target users.

Keywords: Blind Stick · environment parameter detection · obstacle detection · GPS navigation · geriatric population

1 Introduction

All the living beings of the universe should have the opportunity to live a long and healthy life. Environment plays a crucial role in providing a good healthy life. The world population is increasing daily, thus producing the ageing effect for human society.

B. K. Singh et al. (Eds.): ICBEST 2023, CCIS 2003, pp. 388–400, 2024.
https://doi.org/10.1007/978-3-031-54547-4_30

Ageing leads to problems such as visual impairment, hearing impairment, arthritis and other issues. Around one-eighth of the world population in 2019 was above the age of 60 years. This trend is growing, and the world's 2.1 billion population will be older by 2050. The 60 years plus population in the world will almost double from 2015 to 2050. In 2020 older people outnumbered children below 5 years. By 2050 around 80% of older people will live in low and middle-income countries [1]. So these countries face healthcare and social care for these older people [2]. Ageing is occurring at a record speed and will speed up in the coming decades, especially in developing countries like China and India. As per the world health organization (WHO) report 2021, more than 25% of the world's population has distance vision impairment [2]. Around 32% of the blind people were in the age group of 45–59 years, and 58% of the blind people were in the age group higher than 60 years [3]. So to help the mobility of the blind and the geriatric, we developed the common advanced blind stick.

The eye is one of the most crucial sensors of the human body. Due to various causes, there may be vision impairments in the eye. Vision impairment can be classified into different categories depending on sight loss. As per the WHO definition, blindness is the visual activity of less than 3/60 (sellers). Types of blindness are loss of central vision, peripheral (side) vision, blurred vision, generalized haze, extreme light sensitivity, and night blindness. In case loss of central vision causes the blind spot to leave the side vision normal. Hence mobility is unaffected. In a loss of peripheral vision/tunnel vision, the person can see in the front, but the sides can be affected. This makes the person comply with limited mobility and reading too. For blurred vision, the image formed is blurred, and the spectacle can correct the images formed. In a generalized haze, the glare occurs on the image formed and can affect the entire viewing angle of the person. The person affected with extreme light sensitivity is over- whelmed with the standard intensity levels of the light, and the image formed gets whitewashed. So, they suffer from discomfort even in standard lights. When people cannot see in dim light, it is known as night blindness. [4]

Most blindness can be taken care of with the help of advanced medical care, known as the rehabilitation engineering field. The four-fifths of blindness can be avoided. But the persons living in the third world don't have enough facilities and the budget to afford that healthcare facility, which can lead to a restricted or dependable life [4]. The risk factors associated with vision impairment are poor parental care, advancing age, poor nutrition, premature birth, not using safety glasses, poor hygiene, smoking, childhood blindness, and genetic-related cases. Medical conditions also lead to visual impairment, like diabetes mellitus, hypertension, cerebrovascular, trachoma, and cataract diseases [5]. The educated blind people who can still do many works cannot do their jobs due to the mobility problem and hence are financially dependent on others. Ageing is universal but is not uniformly distributed in the world. As age increases beyond a particular level, physiological changes occur, like reduced vision activity, hearing, degraded balance, and slow response. So, caregivers should be aware of these conditions to manage older people and help them make informed decisions, preventing falls and medication-related concerns. Functional deterioration can be avoided by being aware of depression, stroke and other disease symptoms. The increased use of vision and hearing sensors, socialization and activeness. The quadriceps muscle exercise knowledge helps to maximize older people's mobility. These muscles help them to maintain their body mass index and avoid the risk of obesity. Rehabilitation engineering will benefit more old-age people by providing home-based services, increasing mobility, and other applications to support caregivers [6].

Each country's health department and society require a culture to be sensitive and supportive of these subgroups of the human population. Some common disease problems related to ageing are falls, cardiovascular disease, hearing and vision loss, osteoporosis, and dementia are common chronic [2]. Hence the rehabilitation engineering field can help geriatric-blind people by providing them with a cane stick.

Rehabilitation engineering provides different solutions for blind-geriatric people for their various requirements. So, in this work, we are focused on developing an advanced stick that both geriatric and blind people can use. Based on the functionality and use, there are two types of sticks: passive and active. The passive blind stick is divided into four segments. These are white cane and long cane, symbol cane and guide cane. The white cane is the primary and simplest cane that allows the user to navigate the nearby known surroundings safely [7]. The type of white cane that reaches up to the user's sternum is used to avoid obstacles and by rolling or tapping by the low-vision user. The cane tip can be changed for different vision users [8]. The least known cane is the symbol cane. The low vision people use it. It just signifies to the nearby people that the user

Table 1. The literature review of different blind sticks developed recently and their advantages and disadvantages.

S No	Title of the paper	Advantages	Disadvantages
1	Third eye for the blind people using Arduino and ultra-sonic sensor [12]	Smart wristband for blind people with an ultrasonic sensor for detecting the obstacles	Does not detect the object nearer to the earth's surface
2	Smart stick for blind people [13]	Smart stick for blind people using microcontroller, different sensors and GPS features include	There was no mechanism for walking at night to indicate to other people about the blind person
3	Smart walking stick for visually impaired people [14]	Smart stick for blind people use different sensors like temperature and humidity for environment monitoring and detecting night conditions.	No mechanism to locate the stick if it is misplaced
4	Intelligent Walking Stick for Elderly and Blind people [15]	Developed blind stick detecting an obstacle, pits, water on the ground, location of the user, alert another person by sending the message	No mechanism for detect-ing the environmental fire nearby

is partially sighted. This cane is smaller in dimension than a standard long cane. The guide cane is used to find the immediate obstacle. It is also known as the shorter cane due to its size [9]. The user has to hold the cane diagonally across the user's body to find curbs and steps. It does not have the potent ial for long-range movement but still is helpful to many users. One more type is the tedstripped cane. It contains red stripes on the white cane to indicate to the nearby person that the stick user has both hearing and visual impairment [10, 11]

Based on the literature survey, we have found that other researchers have used ultrasonic sensors to detect the path's obstacle [16]; some have included the GPS to find the user's navigation path [17] and send the user's location to the concerned family member [16]. In addition, some researchers have used the light-dependent resistor to detect night or dark conditions [14], some have used a rechargeable battery for the digital system to function, some have used a global system for mobile communication for data transmission, a buzzer for alarming the user in uncomfortable situations, and a water sensor to detect the moisture on the walking floor. Some researchers have also used the passive infrared sensor to detect potholes [15]; some even use the pulse rate and the temperature sensor to detect the user's temperature. The literature review on some articles was done and shown in Table 1. The table shows the advantages and disadvantages of the developed blind stick. Different authors have demonstrated different features of the blind stick. For example, the work done by Shalini et al. shows that they have not included the nearby fire for blind people.

Some basic features of a smart stick taken from different research papers were:

(1) **Environmental Sensors:** Smart sticks are equipped with various sensors that detect the environment, including ultrasonic sensors, infrared sensors, and GPS. The- se sensors can detect obstacles, walls, curbs, and elevation changes, such as stairs and ramps.
(2) **Haptic Feedback:** Haptic feedback is a form of communication that uses vibration or tactile sensations to convey information to the user. Smart sticks use haptic feedback to alert the user to obstacles and changes in the environment, such as when they are approaching a curb or a set of stairs.
(3) **Audio Feedback:** Audio feedback is another form of communication used by smart sticks. The device can be programmed to provide voice prompts to the user, such as telling them the direction they are facing or the distance to an obstacle.
(4) **Lightweight and Portable:** Smart sticks are designed to be lightweight and portable, making them easy to carry and use. They are typically made of lightweight materials, such as aluminium or carbon fibre, and can be folded or collapsed for easy storage.

The above-discussed active/passive canes provide limited help to the blind-geriatric to navigate more freely and independently. So, to make the blind and geriatric navigate more freely, smart active sticks were developed. As time and technology progress, these active sticks include more features based on the different requirements of different types of blind-geriatric people, as classified earlier in the paper. Still, there are conditions where there could be a fire in the nearby place of the user. In their review article, Gaur et al. have shown different methods for detecting fire. [18]. So, in the developed smart stick, we tried to incorporate all the features in our work and to develop a cost-effective system for resource-poor countries with blind and geriatric conditions.

2 Methods

2.1 Development of Advanced Blind Stick

The block diagram for the electronic part of the smart stick consists of several integrated subsystems. The advanced stick has the following electronic components: the battery for the power supply circuit, an Arduino Uno R3 board having an ATMEGA328P Microcontroller to monitor, control and transmit the data in real-time, an emergency SOS button, an obstacle sensor (ultrasonic-based), a fire sensor (Infrared (IR)-based), the water moisture sensor (resistance change based), GPS for the live location of the user, Node microcontroller unit (MCU) for IoT to transfer real-time data [19], and buzzer for the emergency of the user. The functional blocks of the prototype system are shown in Fig. 1. The developed system uses the wooden stick for fitting all the electronic components. The advanced stick is used to detect and notify different parameters for the blind - geriatric to avoid bumps and to detect different parameters of the nearby surroundings so that the user would avoid bumping, fire, and water and navigate freely without any other person.

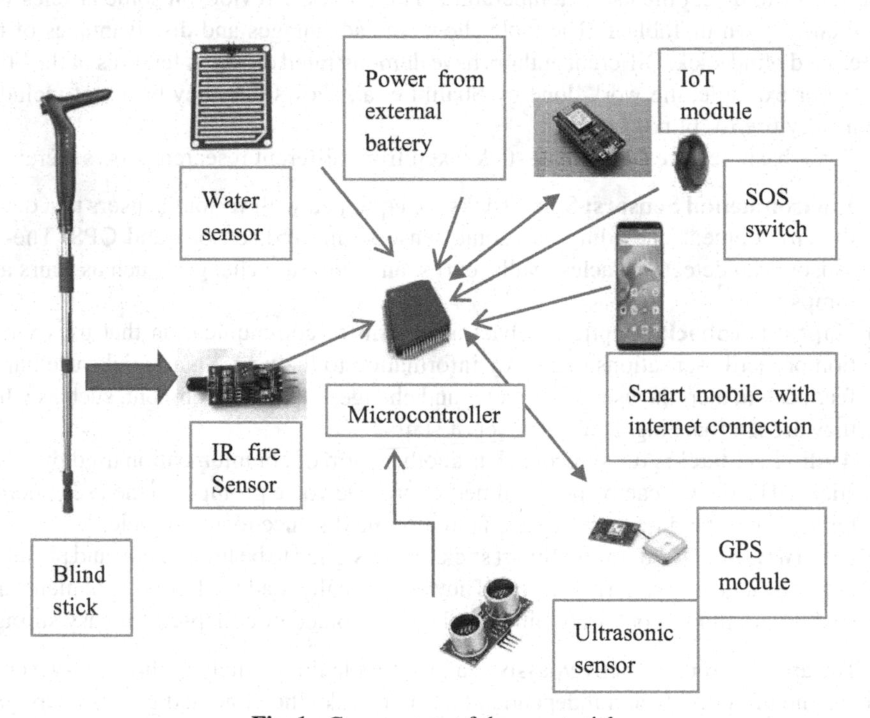

Fig. 1. Components of the smart stick

2.2 The Procedure of the Smart Blind Stick System

The procedure for using the advanced blind stick system is shown in Fig. 2. First, the user has to turn on the switch to power the electronic part of the blind stick. Then the sensors in the smart blind stick gather the environmental parameters like obstacle detection, fire detection, and water detection, and if any of the parameters are detected within the specified range, inform the user by sending the real-time data to the smart mobile. For example, the range for obstacle detection is 2m; for fire, it is 1m; for water/mud, it detects when the smart stick comes in contact with it, and the GPS location covers worldwide coverage as per the specification of the GPS module used.

The smart mobile sends the data to the user's home/hospital as per the requirements fed into the system's software. The stick also carries the SOS button if the user faces any difficulty or problem. If pressed by the user, it sends the GPS location of the user along with the other sensor parameters to the user's home/hospital. The buzzer also signals the users for any emergency.

A common smart stick for blind and geriatric people typically works by using a combination of sensors and feedback mechanisms to detect and communicate information about the environment. Here is a step-by-step breakdown of how a smart stick works:

(1) The user holds the smart stick in front of them and activates it.
(2) The smart stick's sensors detect the environment, including obstacles, walls, curbs, and elevation changes.
(3) The sensors send information to the device's processor, which analyzes the data and determines the appropriate feedback provided to the user using a buzzer and alarm sound in the smart mobile of the user.
(4) The device provides feedback to the user through haptic or audio cues, such as vibrations or voice prompts for GPS locations and turns.
(5) The user can use the feedback to navigate their environment safely and confidently.

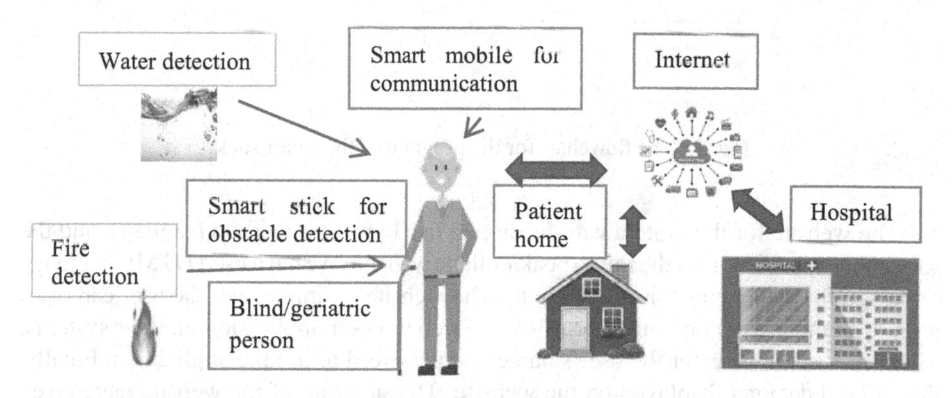

Fig. 2. The block diagram briefing the operation of the smart stick

2.3 Flowchart of the Program Used in the Blind Stick System

The flow chart for the electronic system was written in Arduino IDE using C language and is shown in Fig. 3. First, the microcontroller was initialized, and then it initialized all the sensors, GPS, and liquid crystal display (LCD) [15] attached to the system and the IoT module [20]. The water sensor senses the presence of water, and the ultrasonic sensor senses the obstacle, the IR sensor senses fire and all the sensors with proper signal conditioning circuits send the data to the microcontroller. The microcontroller converts the sensor data to the display format and sends it to the LCD for the user display. The system also reads the GPS data and sends it to the microcontroller. Finally, the microcontroller sends all the data to the IoT to update the website for real-time monitoring. The system also continuously monitors all the environmental parameters, and if any emergency occurs, it on the buzzer so the user can be informed about any emergency. The system also alerts the home and hospital of the user during emergencies by pressing the SOS button for help. The sensor data, along with GPS coordinates, goes to the user's home or hospital for any requirement of critical support.

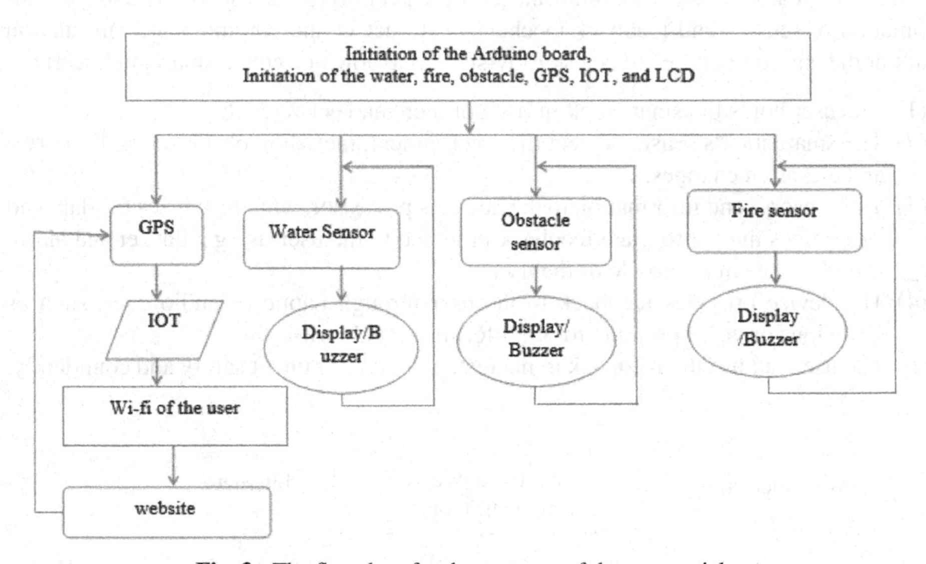

Fig. 3. The flowchart for the program of the smart stick.

The website for the system was developed in a hypertext markup language, and the server was provided from the thingspeak/college website. We have used HTML language to design the web pages for our program. The website is simple, and the user can open it on their mobile/laptop/computer using the web browser application on their systems. Then, the user must enter the user's name and password to use the application. Finally, the entered data get displayed on the website. The snapshot of the website page taken on the smart mobile is shown in Fig. 4. It shows the sensors' data along with the GPS location. After using it, the user closes the application. The website's data is refreshed every 10 s in real-time on the web browser and stored in the excel sheet.

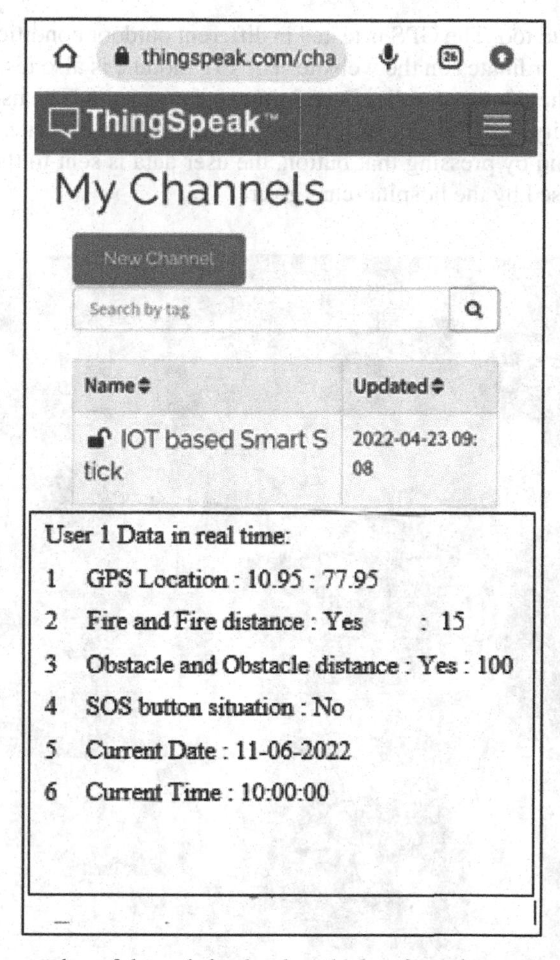

Fig. 4. The snapshot of the website developed taken from the user's smart mobile

3 Results

The work is divided into two parts. First, we developed the board containing all the electronic components to check all the sensor data in real-time. The electronic part of the advanced blind stick is shown in Fig. 5. The developed system contains all three sensors with their signal conditioning circuit, IoT module, GPS module, and power supply unit, including the battery. The IR sensor used for fire detection contains the potentiometer for calibrating the sensor. The detection range for the sensor is around 80 cm, and it was tested for 5 cm to the size of the fire caused by the match stick. The ultrasonic sensor for obstacle detection is a self-calibrated module that sends the obstacle's distance up to 2 m range from the sensor. The sensor is tested for the 1 m range and is helpful for blind-geriatric people. The third is the water sensor. It also comes with a signal conditioning circuit and contains a potentiometer for calibrating the sensor. The sensor is tested for different wet conditions as can be found, for example, dirty water and mud too. The GPS is tested in different outdoor conditions and sends the user's real-time coordinates on the website. The IoT module is also tested and sends the data received by the microcontroller in real time to the website. The user has to provide internet connectivity to the IoT module to send the data to the website. The SOS button was also tested, and by pressing that button, the user data is sent to the website in real time and can be used by the hospital-care giver.

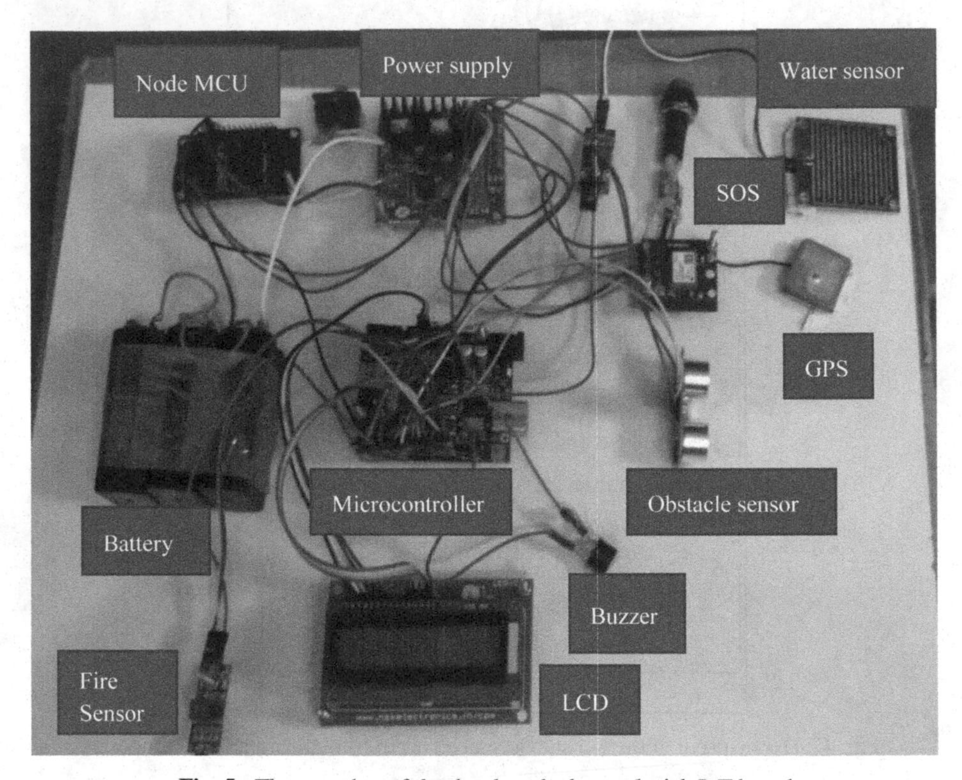

Fig. 5. The snapshot of the developed advanced stick IoT board

In the second part of the development of our work, we put all the electronic components on the wooden stick. The wooden stick was used as it is affordable, readily available in rural localities, and the weight was also less. The sensors were placed at different locations on the stick. The low-vision person experience was good and tested in our college's indoor and limited outdoor environments. The smart stick was also tested by one of the old age people aged 65 plus years, and he also said that the developed stick helps him provide good balance and increased mobility. They also say that they can feel better confidence while going outside to the park or the market because of the presence of the SOS emergency button on the stick, and also, the caregiver - hospital monitors them every time they require help in real-time. The photograph of the developed blind stick along with the student's photo, is shown in Fig. 6(a). Fig. 6(b) shows the developed blind stick with different sensors and electronics components attached in a specific arrangement as per the user's requirement. In the blind stick, we used the wood-based stick and placed all the sensors on it. The electronic components were placed on the stick from bottom to top in the sequence: water sensor, distance sensor, GPS module, fire sensor, Arduino board, IoT module, buzzer, 5 V battery, power switch and at the top, SOS button closer to the hand of the user.

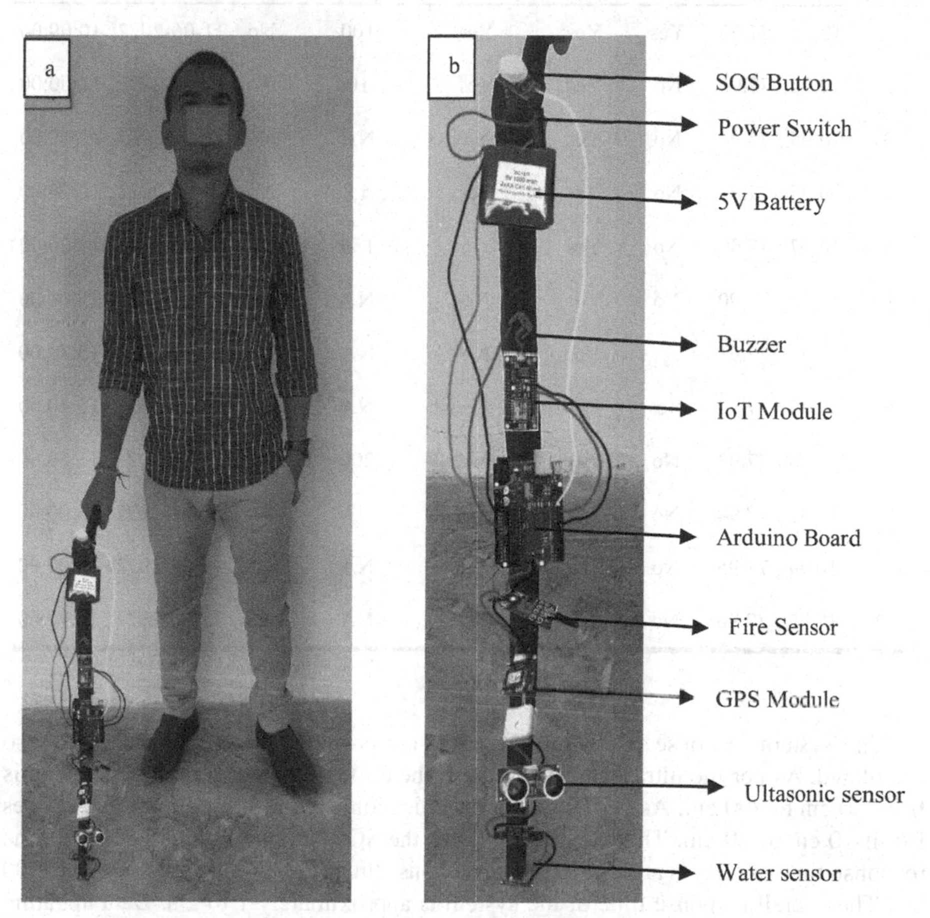

Fig. 6. **(a)** The snapshot of the student using the developed smart stick **(b)** The snapshot of the smart stick

The user was then asked to use the smart stick, and all the data was recorded on the developed website server. From there, it will automatically save all the data on the smart mobile of the user in excel format. The snapshot of the data saved in the excel sheet is shown in Table 2. The data contains the user's GPS longitude and latitude locations. Then the fire detection status and the distance are shown; if the fire is not present, the display shows it as NA (not applicable). After that, the obstacle distance is shown, including the obstacle presence and distance. Then the SOS button status was shown. Then, at last, the current date and time of the system were shown in real- time.

Table 2. Table of the sensors data stored by the web server in an excel sheet for the end user

S No	GPS Location (Lat : Long)	Fire	Water	Obstacle	Obstacle Distance (cm)	SOS	Date	Time
1	10.95 : 77.95	Yes	Yes	Yes	100	No	11-06-2022	10:00:00
2	10.96 : 77.96	No	Yes	Yes	10	No	11-06-2022	11:00:00
3	10.95 : 77.94	No	Yes	No	NA	No	11-06-2022	12:00:00
4	10.95 : 77.98	No	Yes	Yes	15	No	11-06-2022	13:00:00
5	10.97 : 77.99	No	Yes	Yes	150	No	11-06-2022	14:00:00
6	10.98 : 77.90	No	Yes	No	NA	No	11-06-2022	15:00:00
7	10.95 : 77.91	No	Yes	No	NA	No	11-06-2022	15:30:00
8	10.95 : 77.92	Yes	No	No	NA	Yes	11-06-2022	15:40:00
9	10.98 : 77.93	No	No	No	200	No	11-06-2022	13:50:00
10	10.97 : 77.94	No	No	No	NA	No	11-06-2022	16:00:00
11	10.99 : 77.95	No	No	No	NA	No	11-06-2022	16:00:00
12	10.95 : 77.96	No	No	No	NA	No	11-06-2022	16:10:00

The system response times and the hardware cost of the components were also calculated. As per the ultrasonic sensor used, the obstacle range: for the smart stick is from 30 cm to 200 cm. As per IR sensor specifications, the fire detection range ranges fro m 10 cm to 80 cm. The response time for the SOS button is around 500 ms. The response time of GPS is around 10 s. The response time of the water sensor used is 500 ms. The overall response time of the system is approximately 1 to 2 s. Data updating rate over the website: once every 5 s. The energy consumption of the board developed was about 1.5 Wh. The power consumption of the developed smart blind stick is around 400 mWh. The hardware price for the developed board is around 30$ as of date. The hardware price of the smart stick was around 25$ as of date. If the device was built in bulk quantities, the cost of the smart stick could be reduced, making it affordable to the end user.

4 Conclusion

The developed smart stick detected the obstacle efficiently. It helps the blind geriatric navigate freely in indoor and outdoor environmental conditions without any caregiver's help. The obstacle information was given through a buzzer and voice alert on on user's smartphone. The developed system is cost-effective and valuable for developing and resource-poor countries. The developed device is light in weight and portable too. The stick can be beneficial in avoiding accidents during blind-geriatric persons travel alone. The user of the advanced blind stick would always be in touch with the caregiver. In case of any emergency, the user can alert the hospital home by just pressing the SOS button attached to the stick. Therefore, the developed smart blind stick would be helpful to geriatric and blind people in terms of health monitoring management and the feature of safe and independent travel. The increased independent mobility also increases the chance of getting employment of the blind person as they would become more independent in terms of mobility and hence can become self-reliant too by using the developed advanced blind stick. We have also developed a website to be used by the user's caregiver and the hospital. The stick has common features (GPS, SOS, environment fire, water detection, obstacle detection, IoT) to be used by both geriatric and blind persons. It is cost-effective and valuable for developing and resource-poor countries.

Some benefits of the developed smart stick are as follows:

(1) **Increased Independence:** Smart sticks allow blind and geriatric people to navigate their environment more independently, reducing their reliance on others for assistance.
(2) **Improved Safety:** By detecting obstacles and changes in the environment, smart sticks can help prevent accidents and injuries, such as wire, water and collisions.
(3) **Greater Confidence:** Smart sticks provide feedback and information to the user, giving them greater confidence in their ability to navigate their environment.
(4) **Customizable Settings:** Smart sticks can be customized to meet the user's specific needs, such as adjusting the sensitivity of the sensors or the type of feedback provided.
(5) **Cost-Effective:** Smart sticks are a cost-effective solution compared to other mobility aids, such as guide dogs or mobility scooters.
(6) **Increased Employment:** Helps the blind person get a job, become self-reliant, and set a good example for society.

There are some limitations of the developed system observed during field trials. They cannot detect potholes, stairs and other hidden obstacles. The user has to be trained to use all the features of the smart blind stick. The user is still not 100% independent of the caregiver in many areas of his daily work. Our work can help future researchers develop more advanced smart sticks by incorporating other features and AI techniques. The future scope of the existing bright stick guides the visually impaired person in their navigation independently and efficiently, ensuring user safety.

Acknowledgements. We thanks the VSB Engineering College, Karur, for providing the infrastructure support for the making of the prototype.

References

1. Carlsen, A.: GH introduction. In: International Review of Education, pp. 311–318 (2013)
2. WHO Ageing and health, WHO Report 2021 (2021)

3. Kaphle, D., et al.: Vision impairment and ocular morbidity in a refugee population in Malawi. Optom. Vis. Sci. Off. Publ. Am. Acad. Optom. **93**, 188–193 (2016). https://doi.org/10.1097/OPX.0000000000000775
4. Coa, E.: Common types of visual impairment. Calif. Optom. Assocaition (2021). https://www.coavision.org/m/pages.cfm?pageid=3625. Accessed 21 Jul 2022
5. Dahl, A.A.: Blindness: causes, 3 types, treatment & symptoms. In: RxList (2022). https://www.rxlist.com/blindness/article.htm. Accessed 21 Jul 2022
6. Jaul, E., Barron, J.: Age-related diseases and clinical and public health implications for the 85 years old and over population. Front. Public Health **5**, 335 (2017). https://doi.org/10.3389/FPUBH.2017.00335/BIBTEX
7. Chinmayi, A.B., Lakshmi, H., Shivaranjini Rajashekarappa, D.: Smart blind stick. Int. J. Eng. Res. Technol. 7 (2019). https://doi.org/10.17577/IJERTCONV7IS10069
8. Arunbalaji, T.E., et al.: Smart blind walking stick. Int. J. Creat. Res. Thoughts **9**, 2320–2882 (2021)
9. Sathya, D., et al.: Smart walking stick for blind person. Int. J. Pure Appl. Math. **118**, 4531–4536 (2018)
10. Cabvi, E.: What are the different types of white canes? Cent. Assoc. Blind Vis. Impair (2020). https://www.cabvi.org/articles/what-are-the-different-types-of-white-canes/. Accessed 21 Jul 2022
11. Scotland, E.S.: Types of mobility canes for vision impairment. Sight. Scotl. (2022). https://sightscotland.org.uk/articles/information-and-advice/types-mobility-canes-vision-impairment. Accessed 21 Jul 2022
12. Narendran, M., Sarmistha Padhi, A.T.: Third eye for the blind using Arduino and ultra-sonic sensors. Natl. J. Multidiscip. Res. Dev. **3**, 752–756 (2018)
13. Loganathan, N., et al.: Smart stick for blind people. In: 2020 6th International Conference on Advanced Computing and Communication Systems (ICACCS). pp 65–67 (2020)
14. Kumar, A.R.V.: Smart blind stick for visually impaired people. In: IJRECE (2018)
15. Singh, S., Singh, B.: Intelligent walking stick for elderly and blind. Int. J. Eng. Res. Technol. (2020)
16. Vinutha, H., Sarwath, N., Lavanya, R., A literature survey on : walking aid stick for visually challenged people. Int. Res. J. Eng. Technol., 1073–1080 (2021)
17. Munirathnam, S,, Amruthavalli, S.: Assistive voice alert based smart stick for blind people with GPS **5**, 173–178 (2018)
18. Gaur, A., et al.: Video flame and smoke based fire detection algorithms: a literature review. Fire Technol. **56**, 1943–1980 (2020). https://doi.org/10.1007/s10694-020-00986-y
19. Ktron, E.: ESP8266 ESP-12E WiFi Module - 2.4 Ghz. In: KTRON India (2022). https://www.ktron.in/product/esp8266-esp-12e-wifi-module/. Accessed 7 Jul 2022
20. Apu, A., Nayan, A.-A., Ferdaous, J., Kibria, M.: IoT-based smart blind stick. In: Lecture Notes on Data Engineering and Communications Technologies, pp. 447–460 (2022)

Wearable Health Monitoring System for Sweat Analysis

Vandana Pagar[1(✉)], Pravin Bhadane[2], and Arvind Shaligram[3]

[1] MAAER's MIT Arts Commerce and Science College Alandi, Pune, Maharashtra, India
vnpagar@mitacsc.ac.in
[2] Nowrosjee Wadia College, Pune, Maharashtra, India
[3] Department of Electronics, Savitribai Phule Pune University, Pune, India

Abstract. Noninvasive diagnosis and continuous health monitoring are fast and convenient alternatives to blood-based diagnosis. This is made possible by the presence of biomarkers in bodily fluids that are released from the human body externally, such as sweat, saliva, and tears. Out of this sweat is the most easily accessible biofluid for continuous health monitoring. Recent developments in sweat analysis systems show that many researchers presently proposed wearable, miniaturized, integrated, and multiplexed sensor systems. The major challenge is to accurately analyze analytes by improving calibration in real-time analysis and simultaneous monitoring of multiple analytes. Current advances in wearable electronics have overcome many limitations of sweat sensing. Now, these sensors can offer a non-invasive way to gain molecular understanding of the dynamics of our bodies. Here we present a wearable health monitoring system based on sweat analysis using ion-selective electrodes of sodium and chloride. The proposed system measures the concentrations of sodium, chloride, and glucose in human sweat. This can be used as an indication of the dehydrated state of the human body. For diabetic patients, a glucose sensor's noninvasive monitoring of glucose levels is very beneficial. The proposed system is compact so can be used as a wearable device that is capable of continuously monitoring and wirelessly transmitting sensor data to the personal computer or smartphone.

Keywords: Sweat · Wearable · Health monitoring · Electrochemical sensors · Ion Selective Electrodes

1 Introduction

The ability of water to evaporate to cool the body is the primary cause of perspiration. The production of fluids produced by the sweat glands in the skin of mammals is known as perspiration, sometimes known as sweating or diaphoresis. Humans have eccrine glands and apocrine glands, two different types of sweat glands. There are numerous locations in the body where eccrine sweat glands are found. A typical person has more than 2.5 million sweat glands. The excretion of poisons and waste materials is greatly assisted by perspiration. Although the chemical composition of sweat and plasma are identical, some components are preferentially kept or expelled. By evaporating on the skin's surface,

B. K. Singh et al. (Eds.): ICBEST 2023, CCIS 2003, pp. 401–417, 2024.
https://doi.org/10.1007/978-3-031-54547-4_31

sweat maintains a balance between heat gains and losses. The major constituent of sweat is water (99%). But it also contains many different chemical and biochemical compounds [1]. Similar to sweat, urine is another important biofluid to which more attention is given for its pathological study. This study resulted in establishing a relationship between the presence of certain bodies and variations in the percentage of the normal constituent as a result of some pathological disorder [7]. It is revealed from the literature that very less attention has been given to the analysis of the important biofluid sweat. But nowadays sweat has grabbed the attention of researchers because it contains various biomarkers which can be used for the detection of certain diseases or health conditions. As one of the most readily accessible human biofluids, sweat can be used to provide information about one's electrolyte concentrations [3].

A number of sweat sensors have been developed in recent years to monitor this important biomarker which provides non-invasive and continuous health monitoring of athletes and patients with certain diseases [2–5, 40]. Recent research has focused heavily on the detection of analytes that are related to human diseases and disorders using wearable sweat-based sensors. A few studies have also designed wearable electronics with circuitry and multi-analyte sensing for in-situ calibration and analysis. Many researchers are working on how wearable technology can be used for disease diagnosis, exercise monitoring, drug metabolism monitoring [39]. Hypohydration and hyperhydration have received the greatest attention. To realize these applications a small miniaturized, integrated, wearable system should be developed. Recently developed microfabrication technologies and advances in the VLSI field have enabled the integration of sensors and controllers and analyzing acts into a small portable device. With the help of IoT and wireless collection of health monitoring data and its use for future advanced medical and healthcare applications is possible.

More disease-related, calibration-free sensors must be added to systems. Due to current limitations, wearable sweat devices have not yet been made clinically useful. However, it can be said that non-invasive health monitoring systems for sweat analysis have only just begun to tap into their potential for health monitoring, and further significant development in the medical sector is surely on the horizon based on recent, significant advancements. More research in this area is needed before this approach is widely used. Additionally, it's crucial to have simple, dependable design solutions that are affordable and simple to reproduce.

2 Constituents of Sweat

Sweat is chemically alike to the plasma, but specific components are selectively retained or excreted in sweat. By evaporating on the skin's surface, sweat maintains a balance between heat gains and losses.

The major constituent of sweat is water (99%). But it also contains many different chemical and biochemical compounds. Table 1 shows the standard contents of sweat and their relative concentrations observed by many researchers using different methods such as high-resolution NMR spectroscopy, ion chromatography, inductively coupled plasma sector field mass spectrometry, etc.

Table 1. Summary of standard sweat constituents and their concentrations found by researchers.

Sweat Constituent	Concentration	Method Used	Reference
Lactate	1.34 to 4.12 ppm	high-resolution NMR Spectroscopy	[15]
Na+	60 Mm	high-resolution NMR Spectroscopy	[15]
K+	40 Mm	high-resolution NMR Spectroscopy	[15]
Cl	70 Mm	high-resolution NMR Spectroscopy	[15]
Alcohol	5 mM	high-resolution NMR Spectroscopy	[15]
Sodium	35.2–81.0 mM	ion chromatography	[16]
Potassium	2.7–6.8 mM	ion chromatography	[16]
Calcium	0.0–2.9 mM	ion chromatography	[16]
Magnesium	0.0–1.5 mM	ion chromatography	[16]
Chloride	31.6–70.4 mM	ion chromatography	[16]
Arsenic	(Range 3.7–22) ug/L		
Cadmium	(0.36–36) ug/L,	inductively coupled plasma sector field mass spectrometry	[18]
Lead	(1.5–94) ug/L	inductively coupled plasma sector field mass spectrometry	[18]
Mercury	(0.48–1.5) ug/L	inductively coupled plasma sector field mass spectrometry	[18]
Sodium	(0.9 gm/l)	inductively coupled plasma sector field mass spectrometry	[18]
Calcium	(0.015 g/l)	inductively coupled plasma sector field mass spectrometry	[18]
Potassium	(0.2 g/l)	Inductively coupled plasma sector field mass spectrometry	[18]
Magnesium	(0.0013 g/l)	Inductively coupled plasma sector field mass spectrometry	[18]
Glucose	200 uM	Enzymatic amperometric	[4]
Lactate	30000 uM	Enzymatic amperometric	[4]
Potassium	1000–30000 uM	Potentiometric ISE	[4]
Sodium	10000–60000 uM	Potentiometric ISE	[4]
Calcium	125–2000 uM	Potentiometric ISE	[4]

3 Sweat Collection

Designing wearable sweat analysis devices requires careful consideration of sampling or sweat collection techniques. Use an appropriate sample procedure that detects very small amounts of an analyte and supplies a regulated amount of that analyte to a receptor without contaminating it. Sweat collection involved the use of numerous conventional methods. The whole-body washdown approach was the most traditional method. In this method, the person is weighed before and after exercise, and the amount of sweat on their entire body is calculated. For analysis, all of the fluid lost throughout the workout is kept. It has been found that the method produces findings with a large variance coefficient and that it cannot be used in wearable technology. Other sweat collection techniques that can be employed for continuous sweat analysis were developed as a result. Low or no power passive pumps based on the capillary force are utilized for sweat sampling and collection. One of the known approaches in these procedures is the employment of a macroduct sweat collector [25], which directs the sweat excreted by the sweat glands to the skin pores and a collection channel. This method involves making a tiny hole in a plastic object through which sweat is piped. The polydimethylsiloxane-based sweat collector created by G. Liu, C. Ho, et al. is an illustration of this kind of sweat collection system. Due to the pressure difference, they mounted this sweat collector so that it comes into contact with the user's skin [2]. Although capillary forces are also present in the tubing, sweat secretion from the sweat glands, which work as a hydraulic pump, is a major force. A capillary-based macroduct system employed by a research team for a study on cystic fibrosis is another illustration of this kind of sweat collector [13]. The wrist was subjected to iontophoresis before the macroduct system was installed. They employed capillary tubing that was coiled and put on a wrist strap [13]. Additionally, Benjamin Schazmann measured the salt levels in sweat using a macroduct system. It comprises coiled-length capillary tubing attached to a wrist strap [10]. Sweat enters and mixes inside the coil through a tiny opening on the underside that is maintained in place against the skin. Due to the narrow collection aperture, very little contamination happens [10]. Many scientists also choose to obtain samples from textiles. Without the use of external pumps, it can deliver controlled fluid flow. Deirdre Morrisa, Shirley Coyle, and others developed a textile-based fluid management platform as a recent example of the textile-based patch. Sweat has been moved through a pre-determined channel using a passive pump to collect it for analysis. This sweat collector's design is based on a specific fabric with moisture-wicking capabilities. It gathers sweat from the skin's surface and wicks the sample to the detecting area through a pre-established route [4]. One more example is a wearable, flexible, and non-toxic pH sensor fabricated by assembling a cotton fabric treated with an organically modified silicate (ORMOSIL). Vincenzo F. Curto et al. have developed a flexible and wearable microfluidic patch for the measurement of the pH value of sweat consisting of four independent reservoirs. The platform of this patch is constructed on poly(methylmethacrylate) (PMMA) and pressure-sensitive adhesive (PSA) [8]. A Wearable electronic-free microfluidic device for the continuous monitoring of pH in sweat is developed by them.

4 Wearable Systems for Sweat Analysis

The main requirement for a wearable sweat analysis system is that the sensors must be in close contact with the skin. The use of flexible electronics allows us to design flexible wearable sensors. Materials used for flexible patches are required to have properties such as porosity, conductivity, and flexibility. A mechanically flexible polyethylene terephthalate (PET) substrate is used to pattern the electrodes [20]. Polydimethylsiloxane (PDMS) [12, 13], Polyimide (PI) [14], Polyethylene naphthalate (PEN) [22], Polyethylene terephthalate (PET) [4], Poly (methyl methacrylate) (PMMA) [16], Poly(vinylidenefluoride-co-trifluoroethylene) (PVDF-TrFE) [26] and Parylene [29], Polyethylene tereftalate and a Polyurethane matrix [6] etc. To trigger the sweat secretions generally iontophoresis method is used. A lot of cholinergic substances (like pilocarpine) that stimulate sweat gland secretion are present in the hydrogels [21]. Different sweat secretion patterns can be achieved depending on the chemical formulation used [21].The amount of sweat collected in an ambient indoor workplace for half an hour is nearly 475 ul for males and 180 ul for females [11]. But still with some challenges, such as the convenience to wear it for a long time, the capability of enduring dynamic use, etc. Fast response times, excellent sensitivity and selectivity, stability in a variety of environmental conditions, and power-efficient operation are essential features for skin-interfaced sweat sensors. These wearable sensors are designed with integrated signal processing circuitry for real-time data analysis and wireless transmission of data to computing devices. Communication technologies such as Bluetooth, Wi-Fi, or Zigbee can be used (Fig. 1).

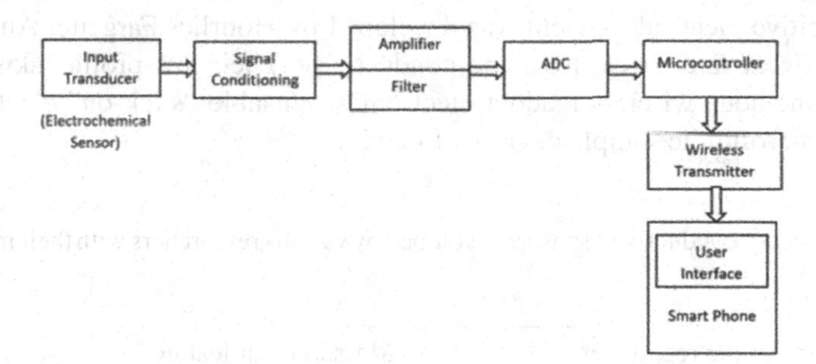

Fig. 1. General block Diagram of wearable sweat analysis system

5 Sweat Analysis Sensors

5.1 Conductivity Based

Applying an external voltage source to a liquid sample and measuring the current that flows through it allows for the measurement of conductivity [2]. The electrode polarisation, also known as Faraday's impedance, the double-layer effect, and the charge-transfer constraint all have a significant impact on how accurate the experiment is. Due to the interaction of ions with electrodes when an electrode emerges into an electrolyte solution, two layers with opposite polarity are created. A wearable conductivity sensor for wireless real-time sweat monitoring created by G. Liu, C. Ho, ET. Al. is an example of this type of measurement system. A wearable device was created by integrating a conductivity sensor, a sweat collection, and an interface circuit. A sweat collector made of polydimethylsiloxone was developed [2]. The design of this sweat collector was based on the hydraulic pumping action theory. The interface circuit and conductivity sensor were coupled to create a wearable device. The electrochemical dynamics of the conductivity measurement of sweat electrolytes are highlighted by their study. Gender, fitness level, food, environment, and the amount of activity or job engaged all have a significant impact on the concentration of sweat chemicals and the rate of sweating. [2]. Real-time sweat analysis by alternating current conductivity of synthetic and human sweat was developed in 2015 by Gengchen Liu, Mahmoud Alomar, et al. They have created a 3D printing-based device for effectively collecting and detecting sweat. They have done human studies as well. Using parallel-plate copper electrodes, the impedance of the solutions was measured. According to this technique, if no fluids are consumed while exercising, dehydration can occur as early as 40 min into the physical activity [5]. A novel sweat rate and conductivity sensor patch built with inexpensive manufacturing methods has been created by Steijlen, A. S. M., et al. A thin PCB with gold electrodes, a hydrophilic PET foil patch, and a double-sided adhesive is used to make the device. Sweat conductivity is measured by two electrodes that are constantly in contact with the inflowing fluid, and reservoir filling is monitored by a separate system of interdigitated electrodes. [6] They have found that the frequency sweep (5–100 kHz) results show that undesired capacitances do not influence the measurement above approximately 50 kHz for all solutions. For all measurements, the impedance's magnitude ranges between 3.6 KΩ and 250 KΩ [6]. A capacitive electrode system was developed by Hourlier-Fargette, Aurélie, et al. to evaluate sweat rate, sweat loss, and conductivity. Their low-profile, skininterfaced technology includes wireless readout electronics, reusable "stick-on" electrodes, and disposable microfluidic sample devices (Table 2).

Table 2. Summary conductivity sensors developed by various researchers with their measurement features

Sr. No	Name of the researcher	Measurement feature
1	G. liu, C. Ho. et al. [2]	EIS of electrodes studied to monitor hydration Status
2	Gengchen Liu, 1 Mahmoud Alomar et al. [5]	Impedance measurement using parallel plate copper electrodes
3	Steijlen, A. S. M., et al. [6]	Sweat rate and conductivity measurement using interdigitated electrodes
4	Hourlier-Fargette, Aurélie et al. [41]	frequency dependent capacitive behavior for fluid (sweat) conductivities to monitor electrolyte levels

5.2 Optical Detection

Optical detection technique is mostly used for pH measurement of sweat. In this technique, a pH sensitive dye is used which changes color according to the value of pH. In the optical detection technique, a pair of LED and a photodiode is used to detect the color changes of the dye.

Michele Caldara et al. [3] have constructed a bio-sensing textile-based patch with an integrated optical detection system for sweat monitoring. Sweat pH has been found to rise during excercise. The advantage of an optical pH sensor over an electrochemical pH sensor is that it does not suffer from electromagnetic interference and thus has a high SNR. It has a longer lifetime, it is reversible, and gives a fast response. But its drawback is that ambient light can interfere with their operation. Stability over a long time is not guaranteed [3]. Morris, D. et al. developed a bio-sensing textile-based patch with an integrated optical detection system. An optical pH sensor, which relies on the employment of a pH-sensitive dye and coupled emitter detector LEDs to measure colour changes, has been utilized. It has been created as a textile-based fluid handling platform that uses a passive pump to collect sweat and transfer it through a designated channel for analysis. The sensor has the ability to record real-time variations in sweat during exercise, according to in vitro and on-body tests [7]. The pH sensor developed in this work has demonstrated that sweat pH increases during exercise [7]. An autonomous wearable micro-fluidic system for in-sweat pH analysis was developed by Vincenzo F. Curto et al. Curto, Vincenzo F., et al. developed an autonomous wearable micro-fluidic platform for real time pH sweat analysis. Using a surface mount light emitting diode (SMD LED) and a photo sensor as a detector, they have implemented an optical detection system. They have observed a drastic increase in pH, from 2.5 to 6.2 within 10 min of physical activity such as cycling [8]. Xiao, G et al. Developed a low-cost, lightweight, and skin conformable microfluidic thread/paper-based multi-sensing system using a needle and a pair of scissors for rapid and accurate in situ monitoring of human sweat pH and lactate [9]. Their system consists of a liquid reservoir for sweat collection and storage as well as a colorimetric multi-sensing component for sensitive detection [9]. The pH sensor designed by them showed a linear detection range from pH 4.0 to 8.0 with a sensitivity of 10.43. The lactate sensor possesses a dynamic range from 0 to 25 mM and a sensitivity of $-$ 3.07 mM^{-1}. It has been found that both can be used as reliable devices for sweat analysis [9].

5.3 Electrochemical Sensing

Due to improvements in material sciences along with digital communication technologies and wireless sensor networks, the field of electrochemical and biochemical sensors has experienced enormous growth in recent years. Chemical sensors work in a complicated manner. The sample medium's unique molecular targets must interact with these sensors. A chemical sensor typically comprises of a transduction element joined to a recognition element (receptor). Chemical information, such as the concentration of a certain substance, is turned into an energy that may be measured by a transducer in the receptor component. The receptor engages in chemical equilibrium with the analyte, catalyzes the reaction, or interacts with the analyte molecules (Fig. 2).

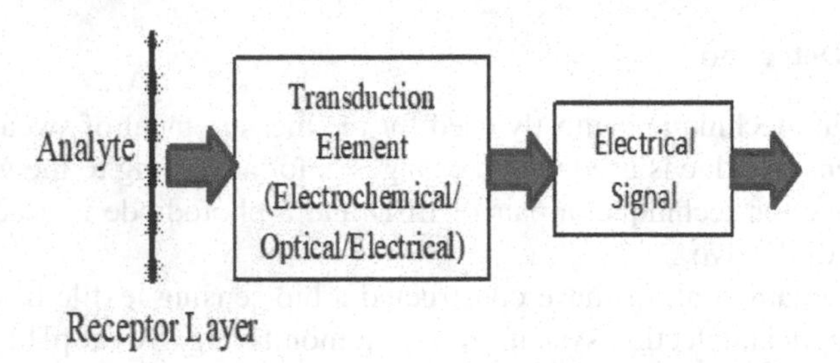

Fig. 2. Sensing Mechanism of Electrochemical sensors

Many challenges are faced by the researcher in designing of wearable chemical sensors which can be used in real life conditions. These chemical sensors get easily affected by many environmental effects which affect sensitivity, stability and reproducibility. Designing of low power miniaturized interfacing circuit is also a challenging task but with the remarkable development of nanoscience in last decade which helped researchers to develop miniaturized and low power circuits. Consequently, a number of chemical and electrochemical sensors have been developed recently. They incorporate sensors for measuring the pH of sweat as well as a number of electrolytes such lactates, sodium, and chloride. The selectivity of sweat sensors is vital because various electrolytes and metabolites in sweat influence the accuracy of the sensor readings. Presence of non-target electrolytes and metabolites in sweat can influence response of each sensor [4].

Various Methods such as Screen Printing, Stamp transferring, Photolithography, Electron beam evaporation used to deposit 150-nm Au [11], electrodeposition of Au nanodntrites by cyclic voltammetry to deposit layers of Prussian dlue mediator and nickel hexacyanoferrate (NiHCF) [11], Ag/AgCl reference electrodes fabricated by depositing Ag/AgCl conductive ink. Etc. Are reported for manufacturing of wearable sweat sensors.

An example of this method is a wearable electrochemical sensor developed by Kevin J. Fraser and Vincenzo F. Curtin et al. [11]. A sodium sensor belt was developed to monitor sodium concentration of sweat in real time. It consists of an integrated Ion selective electrode platform that can interface with the human body while exercising [11]. They have shown how an on-body sampling and sensing technology can analyse sweat electrolytes in real time. The findings found in this analysis were confirmed with an AAS reference method and are largely consistent with literature ranges. According to the several ionophores that can be used in ISEs, a huge number of cations and anions can theoretically be selectively evaluated. According to the scientists, they can either be utilized as a sensor array or to simply replace the present sodium selective ionophore. The delay might be decreased if these sensors were placed closer to the sweat sample inlet and the system was made smaller.

A wearable electrochemical sensor for the real-time monitoring of sweat sodium concentration was developed by Benjamin Schazmann et al. [23]. Capillary zone electrophoresis was used to quantify the amount of sodium in the sweat. They built ISE with PVC tubing acting as the barrels. 0.1 M KCL solution was used as a filling solution for reference electrodes which are connected to a potentiometer [23]. According to this paper there is no definite relation between sweat electrolyte concentrations and gender, maturation and aging and different body parts [23]. Evan K. Wujcik et. Al. Have developed an electrochemical Ion Sensor for the Quantification of Sodium in Sweat Samples [10]. They have used nylon 6/MWNT/calixarene nanocomposites and found it be effective for selective and sensitive sodium ion detection in liquids, particularly sweat samples.

One more example is a flexible chloride electrode designed by V.A.T. Dam, et al. For real-time monitoring of the chloride content in sweat, it is made up of a reference electrode and an array of chloride selective electrodes. AgCl paste has been printed using screen printing method on a polyethylene terephthalate (PET) substrate. During calibration, they achieved a chloride sensitivity of 56 mV/dec. A small amount of sweat that had been absorbed in a layer of gauze was able to be continually detected by the sweat patch for the chloride ions. The patch revealed that the test subject's sweat had a chloride concentration close to 55 mm, which is within the range noted in the literature [27]. A wearable multielectrode platform based on electrodeposited platinum nanostructures was developed by Francesca Criscuolo et. al. In artificial sweat, they were able to get a linear response with accuracy of 0.97 and 0.81 for Na+ and K+, respectively [38].

6 Detection of Diseases

Sweat consists of various disease-specific markers. Recent investigations have revealed the potential of sweat as a significant technique for detection of various diseases and health conditions. Recent advances in wearable electronics have made it possible to use sweat based wearable biosensors for point-of-care medical or continuous health monitoring. Various diseases and health conditions can be detected and monitored in more convenient manner and with a less invasive approach than conventional invasive monitoring methods. In the following examples, an overview of the current research and potential impact of sweat as a biofluid in clinical diagnostics is provided.

6.1 Cystic Fibrosis

One of the most common examples is for diagnosis of Cystic Fibrosis in children. Sodium and Chloride levels in sweat are measured for detection of sweat analysis.

6.2 Immunology

Different cytokine receptors which are hosted by human skin plays a vital role in the immune response of human body. Therefore, the presence of a broad range of cytokines in sweat samples is not unexpected. Katchman et al. [31] detected immunoglobulin A (IgA), IgD, IgG, and IgM, which suggests the possibility to profile antibodies as well as innate immune reactions in sweat [30]. Xiuju Dai et al. [32] have investigated that Sweat is secreted onto the skin's surface and does not come into contact with keratinocytes in normal skin. However, in skin with a defective cutaneous barrier, such as atopic dermatitis-affected skin, sweat cytokines can directly act on epidermal keratinocytes, resulting in their activation. In conclusion, eccrine sweat contains proinflammatory cytokines, IL-1 and IL-31, and activates epidermal keratinocytes as a danger signal [32].

6.3 Dehydration

Sweat can be used to provide information about one's electrolyte concentrations [3]. Hypohydration and hyper hydration have received the greatest attention. Dehydration leads to increase in sodium concentrations in sweat. Electrolyte levels in sweat including sodium (Na+) levels vary considerably. Concentration of these electrolytes also depends on the point of the body where they are measured, genetic predisposition, diet, and heat acclimatization rate. Conversely, gender and aging do not have a large effect on sweat electrolyte concentrations.

6.4 Diabetes

Sweat has also been considered for continuous monitoring of glucose levels of diabetes patients by noninvasive method. Blood glucose levels can be reliably evaluated through glucose detection in sweat [29]. Terri D. La Count et al. have studied glucose transport from systemic circulation to Sweat. They have prepared glucose model, calibrated under a variety of experimental conditions including electrical e nhancement, revealed a 10 min blood-to-sweat lag time and a sweat/blood glucose level ranging from 0.001 to 0.02, depending on the sweat rate [33]. Nyein, H.Y.Y et al. have developed high-throughput microfluidic sensing patches toward decoding sweat [34]. They have presented a device as a crucial tool for advancing sweat testing beyond the research stage for point-of-care medical and athletic applications [34].

7 Experimental Work

Here we have designed a wearable health monitoring system based on human sweat analysis. Ion-selective electrodes, for sodium, chloride, and glucose are used. For designing of the wearable health monitoring system, we are intended to monitor three significant constituents of sweat, Na+, Cl−, and Glucose. The concentration of sodium is essential for cystic fibrosis patients. Sodium and Chloride is a crucial diagnostic biomarker to monitor hydration status in athletes. Non-Invasive Continuous glucose monitoring can prevent hyperglycemia and its life-threatening effects. For designing this system, we have selected to use screen-printed electrodes which are best suitable for wearable applications due to their small size and ease of use. Three commercial screen-printed three-electrode cells from Zimmer Peacock are used. These Na+ and Cl− ion-selective electrodes have measurement range of 40 mM to 360 mM concentrations, which is adequate for measuring concentrations of Na+ and Cl− in human sweat. As depicted in Table 1, Na+ and Cl− median concentrations are 60 mM and 70 mM, respectively. A capillary filling technique is provided for all the sensors which makes it suitable for continuous sweat monitoring. For this system, a small sweat volume is sufficient to get the measurement of sweat constituents. All three sensors have three electrodes, a working electrode, a reference electrode, and a counter electrode as shown in Fig. 3 (Fig. 4).

Here we used Ecflex which is a stamp-sized, flexible electronics platform that supports the readout of electrochemical sensors wirelessly [35]. It can be deployed in the field in most applications due to its compact size and light weight (22 × 25 × 1 mm,

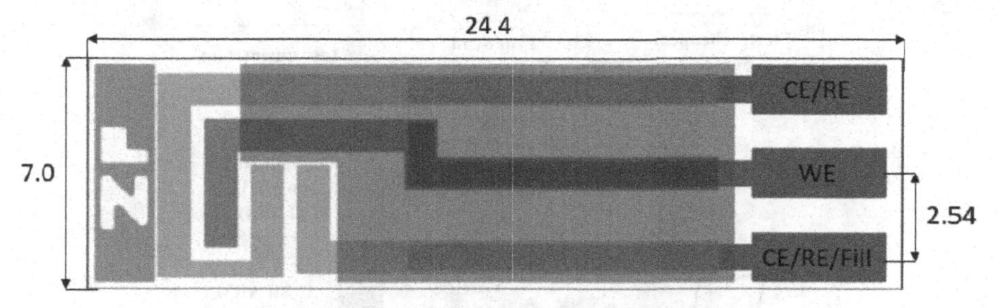

Fig. 3. Schematic of Sensitive Electrodes

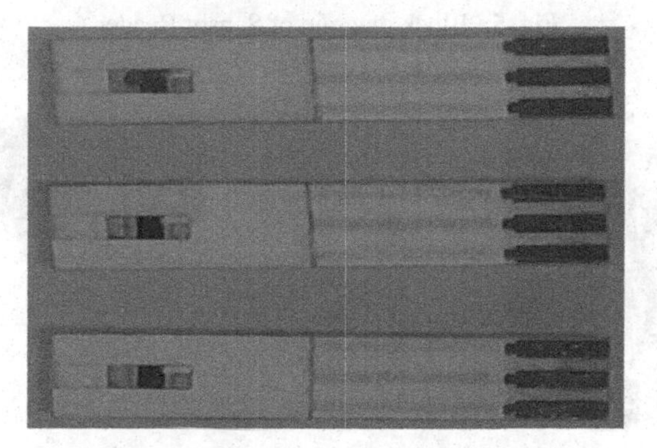

Fig. 4. Na+, Cl− & Glucose Sensors

21 g). Because of its flexibility, it may be fixed to curving or moving surfaces, such as the human body [35]. It can be modified to do more dynamic techniques like square-wave voltammetry, but its primary use is for reporting continuous, steady-state signals vs. time, such as in chronoamperometry and open-circuit chronopotentiometry [35]. The potentiostat is the most elementary piece of instrumentation used in electrochemistry in three electrode systems. Its objective is to keep the voltage between the reference and working electrodes constant. It accomplishes this by attempting to balance the potential difference between the working electrode and reference electrode due to any additional external potential at the input terminals. A potentiostat IC LMP9100 is used for measurement of output in three electrode system. Instrumentation amplifier chip MAX4461 is used for reading open circuit-potentials. Voltage regulator IC TPS 61220 is used for power management. 32-bit Arm CC2541 is used which supports Bluetooth Low Energy wireless communication it also contains integrated 12-bit ADC (Fig. 5).

For contact with human skin, the circuit and sensor assembly were bonded to the backside of the platform. It provides wireless communication with BLE protocol. The circuit is made fit on arm with the help of an adjustable sports band, common in sportswear. Sports band is wrapped on arm in such a way that sensor is in contact with the skin to sense the sweat (Figs. 6, 7 and 8).

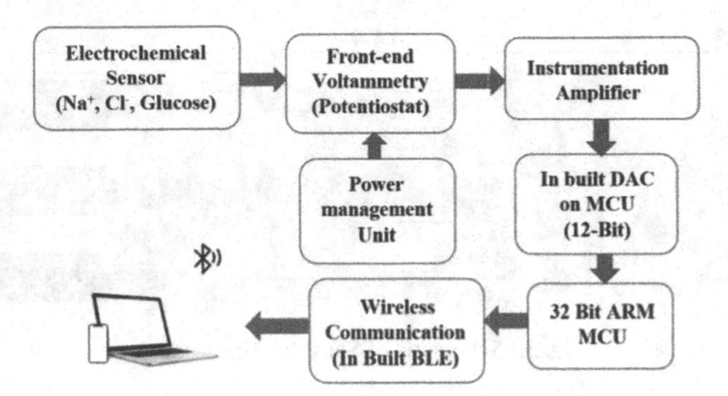

Fig. 5. Block diagram of Sensor Reader

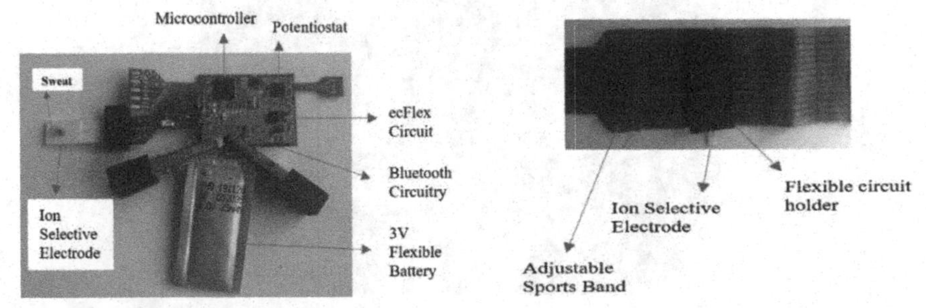

Fig. 6. Sensor and reader interfacing

Fig. 7. Circuit bonded on adjustable sports band

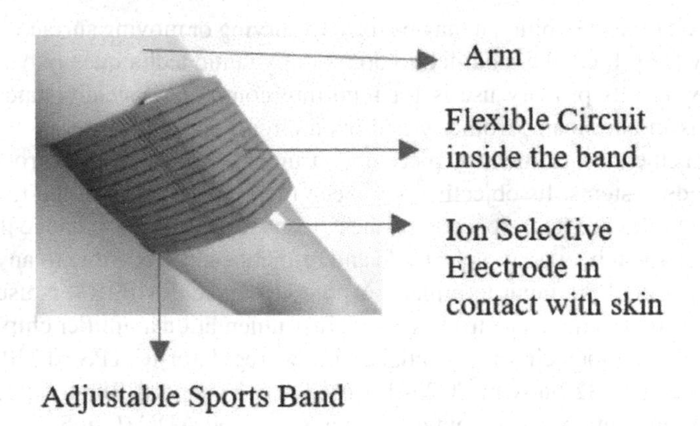

Fig. 8. Wearable sweat monitoring system on Arm

8 Results and Discussion

All the sensors were initially tested by preparing a solution of sodium chloride and glucose in deionized water. The range was based on standards of 10, 50 and 100 mM NaCl and was selected to encompass sodium concentrations in sweat that are frequently observed. Additionally, the sensor's correct operation was ensured by doing this. The sensor's output can be seen in graphical format on the ecflex BLE readout platform. Figure 9 shows the output of sodium sensor for different concentrations of NaCl solutions. It is depicted from graph that the output voltage level increases with increase in sodium ion concentration. Ecflex BLE readout platform is used for reading sensor data. Data is also available in graphical form as shown in Fig. 10. It shows Output of Na+ sensor for 10 mM, 50 mM and 100 mM solution. Sensor data downloaded in Microsoft excel format which can be used for further analysis. On body trail was taken on 38-year-old healthy male individual. Proper consent was obtained from the person. Figure 10, 11 & 12 shows the graphs plotted for Sodium, Chloride and Glucose sensor from the received data.

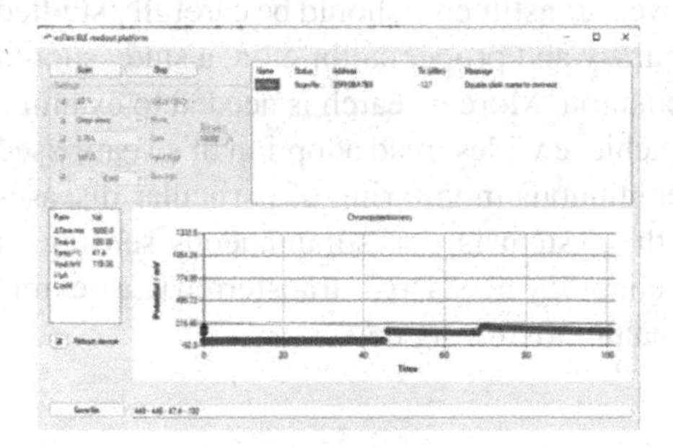

Fig. 9. Output of Na + sensor for 10mM, 50mM and 100 mM solution

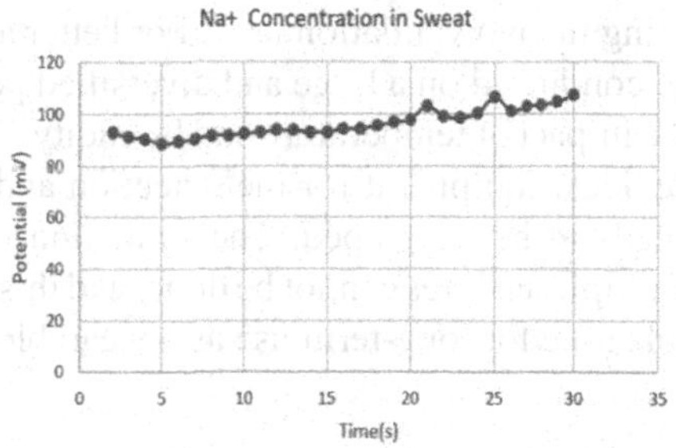

Fig. 10. Sodium sensor response for Sweat

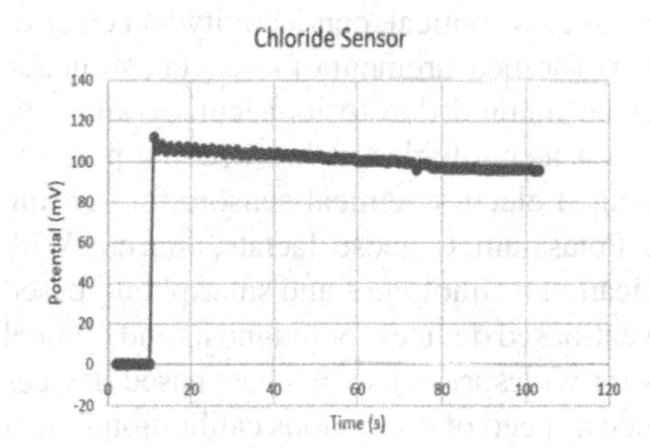

Fig.11. Chloride sensor response for sweat

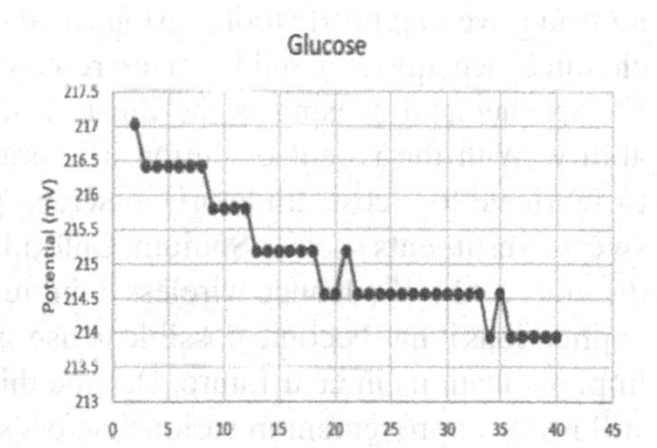

Fig. 12. Glucose sensor response for sweat

This Sweat sensor based electronic system is suitable for continuous health monitoring. The sensors are working in the prescribed range. This wearable health monitoring system provides concentration of sodium chloride and glucose levels in sweat, which can be used as an indicator of current hydration status and glucose levels in individuals. Further data is to be collect from different individuals with different health conditions and age groups and then proper calibration can be done to send alert signals. In this system, small size screen printed electrochemical sensors are used which are suitable for wearable applications and provides good accuracy as compare to optical sensors. Sensors comes with inbuilt capillary actions which eliminates the need of separate sweat collector. This system provides wireless data transmission using BLE protocol which makes it convenient for wearable health monitoring applications. The system can be easily reproduced at low cost for wide applications.

9 Challenges for Future Research

With recent advances in nanoelectronics and screen-printed technologies leading to make it possible to fabricate miniaturized wearable and flexible wearable sweat analysis systems with screen printed sensors and miniaturized low power signal conditioning circuits along with wireless communication technology. Furthermore, advances must be made in sensor manufacturing to improve sensitivity, selectivity, repeatability, and reliability. Correlation between blood and sweat constituents should be carefully studied using in vivo validation tests. For better accuracy and proper calibration, a study should be conducted on a large and diversified population. More research is needed to examine the impact of temperature and humidity. To achieve widespread adoption of sweat-based devices, equipment for the detection and continuous monitoring of particular diseases needs to be developed. The limitation of the system is that simultaneous sensing of multiple analytes cannot be done, and this technology needs to be transferred to a flexible substrate for long-term use as a wearable health monitoring device.

10 Conclusion

Sweat is a potential biomarker, can be used for continuous health monitoring and as a non-invasive diagnostic tool. Various methods such as optical, conductivity and electrochemical sensing are used by many researchers for measurement of sweat constituents. Electrochemical sensing is the most promising method due to its accuracy and reliability. With the recent evolution in fabrication technologies, it has become possible to fabricate low-cost miniaturized screen-printed electrochemical sensors for sensing sweat constituents such as Sodium, Chloride, Potassium, Glucose, lactate, urea etc. With the emergence of advance wireless communication technologies and smartphone-based applications it has become possible to use sweat-based devices for diagnosis and clinical implementations in near future. Despite this for widespread use of sweat based devices still needs improvement in various factors such as need of continuous calibration, effect of ambient temperature and humidity on concentration of sweat composites and correlation of blood and sweat components. The designed Sweat sensor based electronic system is suitable for continuous health monitoring. It can provide information about current hydration status and glucose levels.

The system is highly replicable and performing, which addresses some of the inherent problems with current wearable technology. A wearable configuration is used to demonstrate the system's reversibility and selectivity in artificial sweat.

Further work is to be done to monitor multiple analytes simultaneously. Further data is to be collect from different individuals with different health conditions and age groups and then proper calibration can be done to send alert signals.

References

1. Rizzo, D.C.: Fundamentals of anatomy and physiology. 4th edn.
2. Liua, G., et al.: A wearable conductivity sensor for wireless realtime sweat monitoring. Sens. Actuat. B: Chem. **227**, 35–42 (2016)
3. Caldara, M., Colleoni, C., Guido, E., Re, V., Rosace, G.: Optical monitoring of sweat pH by a textile fabric wearable sensor based on covalently bondedlitmus-3-glycidoxypropyltrimethoxysilane coating". Sens. Actuat. B Chem. **222**, 213–220 (2016)
4. Gao, W., et al.: Fully integrated wearable sensor arrays for multiplexed in situ perspiration analysis. Nature **529**(7587), 509–514 (2016). https://doi.org/10.1038/nature16521
5. Liu, G., et al.: Real-time sweat analysis via alternating current conductivity of artificial and human sweat. Appl. Phys. Lett. **106**(13), 133702 (2015)
6. Steijlen, A.S.M., et al.: A novel sweat rate and conductivity sensor patch made with low-cost fabrication techniques. In: 2020 IEEE Sensors. IEEE (2020)
7. Morris, D., Coyle, S., Wu, Y., Lau, K.T., Wallace, G., Diamond, D.: Bio-sensing textile based patch with integrated optical detection system for sweat monitoring. Sens. Actuat. B Chem. **139**(1), 231–236 (2009)
8. Curto, V.F., Coyle, S., Byrne, R., Angelov, N., Diamond, D., Benito-Lopez, F.: Concept and development of an autonomous wearable micro-fluidic platform for real time pH sweat analysis. Sens. Actuat. B Chem. **175**, 263–270 (2012)
9. Xiao, G., et al.: Facile and low-cost fabrication of a thread/paper-based wearable system for simultaneous detection of lactate and pH in human sweat. Adv. Fiber Mater. **2**(5), 265–278 (2020)
10. Wujcik, E.K., Blasdel, N.J., Trowbridge, D., Monty, C.N.: Ion sensor for the quantification of sodium in sweat samples. IEEE Sens. J. **13**(9), 3430–3436 (2013)
11. Fraser, K.J., et al.: Wearable electrochemical sensors for monitoring performance athletes. In: SPIE Photonic Devices+ Applications, pp. 81180C–81180C. International Society for Optics and Photonics (2011)
12. Lee, H., et al.: Wearable/disposable sweat-based glucose monitoring device with multistage transdermal drug delivery module. Sci. Adv. **3**, 1–9 (2017)
13. Chen, C.Y., Chang, C.L., Chien, T.F., Luo, C.H.: Flexible PDMS electrode for one-point wearable wireless bio-potential acquisition. Sens. Actuat. A Phys. **203**, 20–28 (2013). https://doi.org/10.1016/j.sna.2013.08.010
14. Roy, S., David-Pur, M., Hanein, Y.: Carbon nanotube-based ion selective sensors for wearable applications. ACS Appl. Mater. Interfaces **9**, 35169–35177 (2017). https://doi.org/10.1021/acsami.7b07346
15. Kutyshenko, V.P., Molchanov, M., Beskaravayny, P., Uversky, V.N., Timchenko, M.A.: Analyzing and mapping sweat metabolomics by high-resolution NMR spectroscopy. PLoS One (2011)
16. Shirreffs, S.M., Maughan, R.J.: Whole body sweat collection in humans: an improved method with preliminary data on electrolyte content. University Medical School, Foresterhill, Aberdeen AB25 2ZD, United Kingdom (1997)
17. Sears, M.E., Kerr, K.J., Bray, R.I.: Arsenic, cadmium, lead, and mercury in sweat: a systematic review. J. Environ. Public Health **2012**, 10 (2012). Article ID 184745
18. Genuis, S.J., Birkholz, D., Rodushkin, I., Beesoon, S.: Blood, urine, and sweat (BUS) study: monitoring and elimination of bioaccumulated toxic elements. Arch. Environ. Contamination Toxicol. **61**(2), 344–357 (2011)
19. Choi, D.H., et al.: Sweat test for cystic fibrosis: wearable sweat sensor vs. standard laboratory test. J. Cyst. Fibros. **17**, e35–e38 (2018). https://doi.org/10.1016/j.jcf.2018.03.005

20. Nyein, H.Y.Y., et al.: A wearable electrochemical platform for noninvasive simultaneous monitoring of Ca2+ and pH. ACS Nano **10**, 7216–7224 (2016). https://doi.org/10.1021/acsnano.6b04005
21. Autonomous sweat extraction and analysis applied to cystic fibrosis and glucose monitoring using a fully integrated wearable platform
22. Someya, T.: Bionic skins using flexible organic devices. In: 44th European Solid State Device Research Conference (ESSDERC), Venice, Italy, 22–26 September, pp. 68–71. IEEE, Piscataway (2014)
23. Schazmann, B., et al.: A wearable electrochemical sensor for the real-time measurement of sweat sodium concentration. Anal. Methods **2**(4), 342–348 (2010)
24. Klous, L., de Ruiter, C., Alkemade, P., Daanen, H., Gerrett, N.: Sweat rate and sweat composition following active or passive heat re-acclimation: a pilot study. Temperature **8**(1), 90–104 (2021)
25. Klous, L., De Ruiter, C.J., Scherrer, S., Gerrett, N., Daanen, H.A.M.: The (in) dependency of blood and sweat sodium, chloride, potassium, ammonia, lactate and glucose concentrations during submaximal exercise. Eur. J. Appl. Physiol. **121**(3), 803–816 (2021)
26. Fujita, T., Shiono, S., Kanda, K., Maenaka, K., Hamada, H., Higuchi, K.: Flexible sensor for human monitoring system by using P(VDF/TrFE) thin film. In: International Conference on Emerging Trends in Engineering and Technology, ICETET, Himeji, Japan, 5–7 November, pp. 75–79. IEEE, Piscataway (2012)
27. Dam, V.A.T., Zevenbergen, M.A.G., van Schaijk, R.: Flexible chloride sensor for sweat analysis. Procedia Eng. **120**, 237–240 (2015)
28. Ha, D., De Vries, W.N., John, S.W.M., Irazoqui, P.P., Chappell, W.J.: Polymer-based miniature flexible capacitive pressure sensor for intraocular pressure (IOP) monitoring inside a mouse eye. Biomed. Microdevices **14**, 207–215 (2012). https://doi.org/10.1007/s10544011-9598-3
29. Karyakin, A.A.: Glucose biosensors for clinical and personal use. Electrochem. Commun. **125**, 106973 (2021)
30. Brasier, N., Eckstein, J.: Sweat as a source of next-generation digital biomarkers. Digit. Biomarkers **3**(3), 155–165 (2019)
31. Katchman, B.A., Zhu, M., Blain Christen, J., Anderson, K.S.: Eccrine sweat as a biofluid for profiling immune biomarkers. PROTEOMICS–Clin. Appl. **12**(6), 1800010 (2018)
32. Dai, X., et al.: Eccrine sweat contains IL-1α, IL-1β and IL-31 and activates epidermal keratinocytes as a danger signal. PLoS ONE **8**(7), e67666 (2013)
33. La Count, T.D., Jajack, A., Heikenfeld, J., Kasting, G.B.: Modeling glucose transport from systemic circulation to sweat. J. Pharm. Sci. **108**(1), 364–371 (2019)
34. Nyein, H.Y.Y., et al.: Regional and correlative sweat analysis using high-throughput microfluidic sensing patches toward decoding sweat. Sci. Adv. **5**(8), eaaw9906 (2019)
35. ecFlex quick-start developer guide Rev: 0.3, Zimmer and Peacock AS (2020). https://s46c23316fd696b31.jimcontent.com/download/version/1596730497/module/11782510898/name/ecFlex%20quick-start%20v0.3.pdf
36. Ghaffari, R., Rogers, J.A., Ray, T.R.: Recent progress, challenges, and opportunities for wearable biochemical sensors for sweat analysis. Sens. Actuat. B Chem. **332**, 129447 (2021)
37. Gai, Y., et al.: A self-powered wearable sensor for continuous wireless sweat monitoring. Small Methods 2200653 (2022)
38. Criscuolo, F., et al.: Wearable multifunctional sweat-sensing system for efficient healthcare monitoring. Sens. Actuat. B Chem. **328**, 129017 (2021)
39. Yang, P., Wei, G., Liu, A., Huo, F., Zhang, Z.: A review of sampling, energy supply and intelligent monitoring for long-term sweat sensors. NPJ Flexible Electron. **6**(1), 33 (2022)

40. Heikenfeld, J., et al.: Accessing analytes in biofluids for peripheral biochemical monitoring. Nat. Biotechnol. **37**(4), 407–419 (2019)
41. Hourlier-Fargette, A., et al.: Skin-interfaced soft microfluidic systems with modular and reusable electronics for in situ capacitive sensing of sweat loss, rate and conductivity. Lab Chip **20**(23), 4391–4403 (2020)

FPGA Implementation of DNA Computing and Genetic Algorithm Based Image Encryption Technique

R. Rajashree[1] and S. Ananiah Durai[2](✉)

[1] SENSE, VIT, Chennai, India
[2] CNVD, VIT, Chennai, India
ananiahdurai.s@vit.ac.in

Abstract. In the digital age, an increasing dependency of humans on digital technologies is much prevalent. A high demand in securing the private and confidential data is essential, which would otherwise lead to mishandling and misuse of personal information for ulterior motives. To avoid such vulnerabilities many Cryptographic and stenographic techniques were developed over the past decades. Though such existing schemes provided satisfactory shielding, high computational cost, increased processing time, low embedding capacity, and low imperceptibility, renders it as unsuitable for applications requiring high security and high speed such as medical data security. A hybrid DNA computing technology combined with a genetic algorithm is developed in this work to cater for medical image encryption, which will address the aforementioned limitations. The suggested approach converts plaintext information into binary by first mapping it to the DNA sequence. The binary information is then embedded in the cover image using genetic algorithm to yield a stego image. Elliptic Curve Cryptographic hardware (ECC) encrypts the generated stego image, providing a more secure means of storage/communication. The suggested hardware system is implemented in the Zynq 7000 FPGA device, which occupied a comparatively low overall LUT and DSP slices of around 5796 and 19 respectively.

Keywords: DNA computation · Genetic Algorithm · Steganography · Image encryption · Binary data conversion · Elliptic Curve Cryptography

1 Introduction

The confidentiality of data is a crucial consideration for any database that is a compilation of an individual's personal, financial, and other confidential information. This information requires a stringent securing measures that provide increased shielding against various threats. Cryptography is one such technique which secures confidential data [1]. Legitimate access to such crucial messages are authenticated by cryptography, which also shields the system from severe attacks. As it implies a secret key to enable such access, a hacker might successfully break and retrieve the ciphered data. Alternatively, data hiding schemes that is not obvious to the third party will render unawareness of

B. K. Singh et al. (Eds.): ICBEST 2023, CCIS 2003, pp. 418–432, 2024.
https://doi.org/10.1007/978-3-031-54547-4_32

its existence [2]. Image, audio, and video steganography are examples of such digital hiding schemes. Image steganography that hides confidential information in plaintext within a cover image is preferable as techniques of hiding is efficient.

The encrypted or hidden data is eventually stored in the electronic repository popularly known as database. Protection of confidential data stored in such database is essential irrespective of whether it is encrypted or hidden [3]. The intrusion of third party into the database is a major threat or concern faced widely. The integrity of a database can be jeopardized by loss, destruction, or misuse, either unintentionally or purposefully. Securing the database through DNA computing and Genetic Algorithm (GA) based Image Encryption Technique if preferred, will offer high level of security [4]. The foundation of genetic algorithms is the idea of generational reproduction, fitness based selection, and replacement of inferior offspring. In other words, a Genetic Algorithm starts with a population of solutions rather than a single answer, which serves as the foundation upon which the system evolves. A Genetic Algorithm finds the best solution that optimizes the fitness function by repeatedly performing this process under specific conditions. As a result, this has become popular among the population based algorithms due to its high randomness thus yielding greater security. String of characters, integers, float, and bits are used to represent an individual. This string is analogous to the Chromosome. To further enhance the randomness, a cryptographic scheme can be employed as a second tire. ECC (Elliptic Curve Cryptography), a popular asymmetric encryption that utilizes the algebraic structure of an elliptic curve over finite fields is preferred here, as it involves comparatively smaller bit length that reduces latency. Further, high level of encryption strength due to the discrete logarithms developed by Koblitz and Miller, makes ECC the right candidate. The four steps involved in the encryption process of ECC are field operation, elliptic curve group operations, elliptic curve point multiplication, and ECC protocols.

1.1 Image Steganography

Steganography is an art of storing one image or text information on cover information. It can be said as an invisible communication. The process of hiding a plaintext (secret message) in an image is known as image steganography. The cover image is the picture that contains the hidden message, and the stego image is the picture that comes out of it. Image files comply with the requirement of redundancy. The bits that can be changed without being seen are known as redundant bits [9]. As a result, image steganography is favored over other types of steganography such as audio and video steganography. Larger photos are difficult to send over a regular internet connection. This leads to need for finding newer techniques to reduce the size of the image. For that, a process named compression is used which with the help of certain mathematical equations reduces the size of files.

There are two types of compression, viz. Lossy and lossless, which differ largely in implementation. Lossy compression reduces the size of files by removing image that are not sensitive to human eye. Every bit of data is analyzed and represented by mathematical formula in this lossless technique, thus the original image's integrity is preserved. Compressed image is recovered back bit by bit during decompression, which will be identical as that of the original image. Picture steganography encompasses data domain and image domain. The least significant bits in the image matrix represents the image domain. Least significant bit (LSB) insertion in this domain is widely used technique for hiding information in a cover image's LSB as the intensity of the colors is minimally altered. Because these alterations are not visible to the naked eye, the image is successfully disguised. In image steganography, invisibility, payload capacity, and other considerations must be addressed for better performance [9].

1.2 DNA Computing Techniques

The feasibility to map any amount of data through DNA (deoxyribonucleic acid) renders it as a popular choice as add-on in encryption. Further, DNA is increasingly preferred as it is affordable, simple to generate, and adaptable [4]. Due to the salient features of low bit mapping in DNA, low power dissipation is observed, also when more compounds are being added still the power performance is comparatively better. While traditionally used silicon transistors can conduct only one item at a time, the DNA structures can conduct as many operations at a time.

1.3 Elliptic Curve Cryptography

An Elliptic Curve Cryptography is derived from Elliptic Curve equation as shown in Eq. (1). The points generated from elliptic curve equation are used to perform Elliptic Curve Cryptosystem. Also parameters 'a' and 'b' helps to generate the point which satisfies the abelian group operation a given by;

$$y^2 = x^3 + ax + b \tag{1}$$

Encryption, pseudo-random generators, digital signatures etc., are the prominent applications of ECC [13]. Data is encrypted using the key that has been generated through the points over the prime field. Public and private keys are used for decryption and encryption. ECC is attractive due to the fact that it requires a lower key length than RSA to achieve the similar level of security. For example, ECC encryption keys of 160, 224, and 256 bits provide equal security to RSA encryption keys of 1024, 2048, and 3072 bits, respectively. The benefit of small key sizes forms the basics of ECC to best suitable for the high-velocity Cryptographic processors [14].

2 Related Works

The authors of [1] created a novel steganography and fuzzy DNA picture encryption technique that yields a very effective stego image. Hybrid DNA encoding and Choquet's Fuzzy Integral sequences are used. Choquet's Fuzzy Integral sequences are used to generate a pseudorandom sequence. The encrypted image is created using the wavelet fusion process. The main objective is to achieve enhanced security with highly efficient image encryption so that data won't be vulnerable to attack. A gray scale image is used to perform simulation model. The efficiency of the technique was analyzed using parameters like Differential attack, Correlation coefficient, Entropy information analysis,

Peak Signal to Noise Ratio (PSNR), Histogram Analysis, Key Sensitive analysis, Speed analysis. The results show that the new method is resistant to wide range of attacks. It is a highly sensitive stochastic method that is sensitive to small changes in the encryption key and has high entropy.

Kulkarni et al. [2], successfully designed a key Generating Algorithm based on DNA Cryptography. It aims on providing an extra layer of security for protecting the key. DNA string is generated as the key to provide enhanced security and to optimize the design. AES algorithm is used for key generation. The algorithm is developed using verilog coding, which is then simulated and synthesized in Vivado simulator. The technique proposed in this has been found to be highly robust against different attacks since a random key is produced at the sender every time and is used to decrypt the cipher text at the receiver.

Video steganography based on chaos theory was created by Nirmalya Kar et al. Information is concealed within a video file making use of DNA's characteristics. A chaotic map is used to pinpoint the precise pixel location of video frames. A linear congruential generator creates the pseudorandom sequence. Quantitative metrics like PSNR and MSE are used to do qualitative analysis. Even though the stego movie is nearly impossible to tell apart from the original, discrepancies in the histograms and pixel correlations between the two videos suggest that it was created to a high standard and with a decent visual quality [3].

To ensure the security using RSA algorithm, DNA cryptography, and DNA steganography were compared with a Hyperelliptic Curve Cryptography (HECC)-based DNA steganography by the authors [4]. Aside from improving the safety of digital material, HECC also improves the safety of image files. Improved image quality is achieved without the drawbacks of conventional methods, such as larger keys and higher processing costs, even when the volume of secret data to be protected is substantial. The XOR operation is used to encrypt the cover image with the secret image. To convert decimal values to HEC points, the Koblitz algorithm is utilized. Qualitative analysis is done based on the parameters like Embedding capacity, PSNR, MSE and all proves that proposed steganographic technique is highly efficient when co mpared to existing steganographic technique. It was obvious from the simulation results that the suggested work produces an efficient result.

Abdelhamid et al. [5] discussed on the techniques to increase the resistance to steganalysis. That is to reduce the chances of finding out whether a message is hidden or not and hence changes in the cover message have to be minimized. Using the DCTM3 technique, the hidden message is first compressed twice, and then inserted into the cover image. Lastly, it is demonstrated that the method is more effective than the LSB steganography method.

Faiza Al-Shaarani, Adnan Gutub, introduced a matrix based secret sharing technique. Secret sharing involves splitting of the secret message, and handing over to different people. The message cannot be retrieved until all the parts are received and combined. As a result, while dealing with sensitive data, an extra layer of security is applied. The two main types of steganography are linear space-based and discrete wavelet transform. The XOR operation is used for both encrypting and decrypting data. The cypher key is created by XORing the plaintext with the secret text [6].

In [7], the authors main objective was to increase the storage capacity by co mpressing the image. Encryption is performed using RSA, and decryption is performed using Huffman coding and DWT. The compression of data leads to much easier handling of data and making it more secure. By reducing the size, it ensures that the message is send fast, thereby reducing the time that the intruders might get. LSB method is used for embedding. [7] Proved to have better capacity, average PSNR and average SSIM.

Subramanian et al. [8] explained and analyzed different techniques of image steganography. Though cryptography was widely used, steganography is found to be much better. In steganography the secret message is hidden inside a cover whereas in cryptography the secret message itself is altered. Image steganography is divided into three: traditional based stenographic methods (LSB method), CNN based stenographic method and GAN (General Adversarial method). In terms of performance, the GAN approach is believed to be superior. This is so because it uses a comparative study within its models itself, thereby creating a more realistic image. Data availability, convergence of GAN and real time steganography are some of the challenges discussed.

In [9] the selection of a suitable carrier image was made possible by combining image steganography with genetic algorithm. Since the embedding procedure can cause subtle modifications to the stego picture, determining whether or not a given image is a carrier is dependent on whether or not its secret data is compatible with the LSBs of the carrier image. The alterations to the stego image can be kept to a minimum with the assistance of GA. The cover picture is chosen via fitness proportionate selection, often known as roulette wheel selection, which finds appropriate images for embedding. Both the cover and payload images are combined using the LSB technique. MATLAB 2017a is used to run the suggested technique on the USC-SIPI image database.

Ranyiah Wazirali et al. [10] proposed an innovative steganography method using genetic algorithms (GAs). It uses genetic algorithm for LSB matching of secret data with carrier image so that there will not be a degradation in stego image and thereby increasing embedding capacity and reduce distortion. Steganography in the spatial realm, in which hidden information is encoded within the pixels of a carrier image. With the help of the given strategy, we can find the best places to conceal LSB pixels. The mapping process entails performing a number of procedures on the carrier picture to pinpoint the hidden data's correct locations. MATLAB was used to realise [10]. Parameters such as PSNR are used in qualitative analysis. Despite the obvious statistical loss, the quality of the stego-images is not discernible to the human eye. A cover image's data can be concealed using the XOR method.

In [11], the authors proposed a new method in steganography which is genetic algorithm. This method helps in identifying the best position in the DNA in order to hide the message. RSA algorithm is also used for encryption before embedding the message which adds more security. By this approach it is seen that the robustness, capacity and imperceptibility is increased, thereby making it an efficient method. Ginu Alexander et al. [12] used use the Least Significant Bit (LSB) technique. If you know the reference sequence, you can decode the hidden message by looking at its least significant bit. Here, instead of the normal pairing of 2 bits to 2 nucleotides, 4 bits are assigned to those bases. More safety is afforded by this modified binary e ncoding rule. As just substitution occurs, the procedure has no cost and the length of the original DNA sequence is preserved.

3 Methodology

3.1 Proposed Hybrid DNA Computing and Genetic Algorithm Based Image Encryption Technique

The secret message and the cover image to which the plaintext is going to be embedded are converted to DNA nucleotide triplet using DNA conversion from Table 1. The obtained result is converted into binary code using DNA to binary conversion from Table 2. The idea of a genetic algorithm is utilized to find the best answer to the problem of how to locate the LSB places in a cover image and a plaintext that have the most in common. Methods such as horizontal scanning, vertical scanning and mirror shifting are performed in order to achieve the above objective. The most effective approach is determined by the fitness function, which is efficiency, given by the (2);

$$\text{Efficiency} = \frac{\text{Number of matching bits}}{\text{Total number of secret bits}} \tag{2}$$

Any given approach is selected if its efficiency is better than 50%; else, the next best approach is used. First, the cover image is scanned vertically, starting with the least significant bit (LSB). As a second step, we employ vertical scanning. Furthermore, after horizontal and vertical scanning, the following step is mirror shifting the binary code of plain text. Efficiency is calculated when each stage is completed. If the fitness function is satisfied, the hidden message will be found in the least significant bit (LSB) of the cover photo. When that option fails, the following one is tried. When the best answer is found, the binary representation of the plain text is inserted into the least significant bit (LSB) of the binary representation of the cover art. The generated steganographic image is encrypted using Elliptic Curve Cryptography (ECC). The encrypted stego image is the end product, as illustrated in Fig. 1.

Table 1. Conversion of DNA nucleotide into binary code

Nucleotide	Binary equivalent
A	00
C	01
G	10
T	11

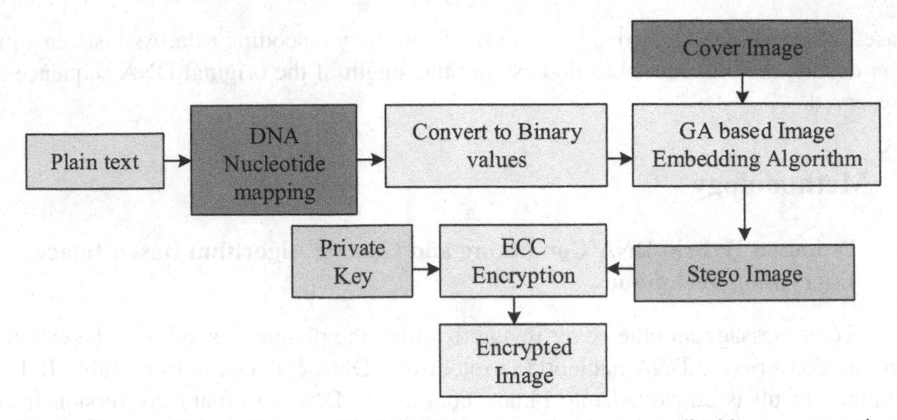

Fig. 1. Proposed block diagram of hybrid DNA computing based medical image encryption

3.2 Proposed DNA Computing Algorithm

The following steps are followed to perform the medical image encryption algorithm by implementing DNA computing techniques, genetic algorithm and ECC;

Step 1: Choose the plaintext that has to be kept confidential.
Step 2: Pick an appropriate cover photo to disguise the text inside.
Step 3: Apply Table 1 [4] to transform both the text and the pixel values into a DNA nucleotide triplet, as demonstrated in Fig. 3.
Step 4: Use Table 2, as depicted in Fig. 4, to translate DNA nucleotide into binary.
Step 5: Both horizontal and vertical scanning are applied to the carrier image, as demonstrated in Fig. 5 and 6 and 7.
Step 6: The least significant bit (LSB) of the carrier bits and the secret bits undergo an exclusive OR operation.
Step 7: If the fitness function is not met, then mirror shifting of the secret code followed by horizontal and vertical scanning is performed which is illustrated by Fig. 8, 9 and 10.
Step 8: Step 6 is repeated.
Step 9: Using the best approach discovered from the preceding procedures, the secret code is implanted into the cover artwork.
Step 10: The resultant stego image is encrypted using elliptic curve cryptography to obtain the encrypted stego image (Fig. 2).

Table 2. Mapped DNA table

A	AAT	M	ATA	Y	GTA
B	ATG	N	TGC	Z	TAA
C	CGA	O	CTC	0	CTC
D	TGA	P	GCC	1	TTC
E	TTC	Q	GAG	2	GGT
F	CTG	R	CTA	3	AGC
G	CCA	S	CCC	4	GAC
H	TTA	T	GGG	5	GAA
I	TAT	U	CTA	6	TTG
J	AGC	V	CTA	7	CTA
K	ATT	W	ATC	8	TAG
L	GTC	X	GAC	9	GAT

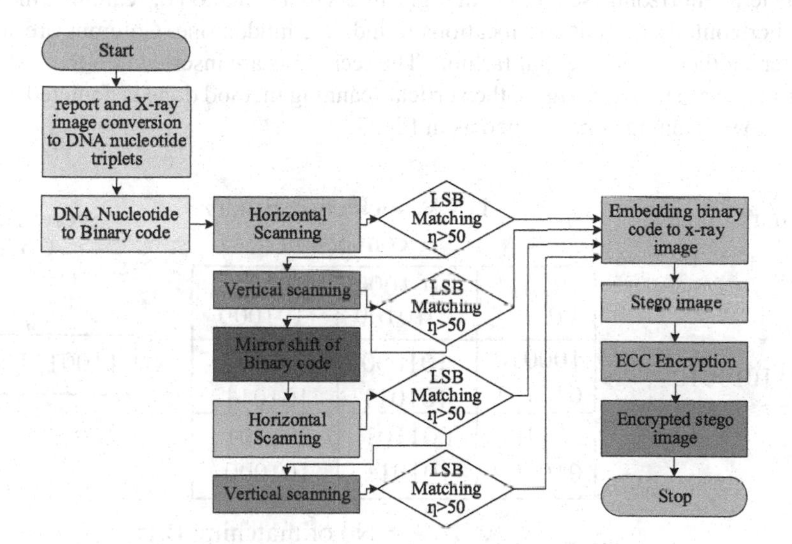

Fig. 2. Process of Genetic Algorithm

3.3 Typical Example

The above mentioned algorithm is shown with some typical values as given below:

- The plaintext chosen is Hi. Step 3 - Step 9 of the algorithm is shown with the help of an example in Fig. 3:

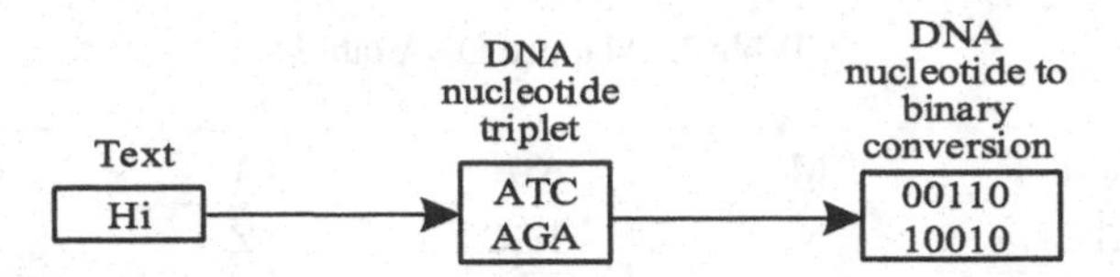

Fig. 3. DNA nucleotide conversion of Plain text with cover image

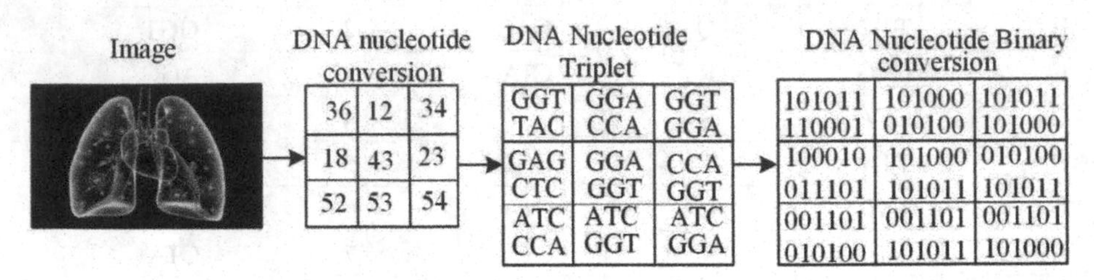

Fig. 4. DNA nucleotide conversion of cover image

The term "horizontal scanning" in Fig. 6 refers to the method of searching the carrier image horizontally for suitable locations to hide the hidden one. Columns are scanned one after another in a sequential fashion. The secret bits are inserted into the LSB of the carrier image as shown in Fig. 4, the vertical scanning method done is depicted in Fig.7. Row by row scanning is performed as in Fig. 5.

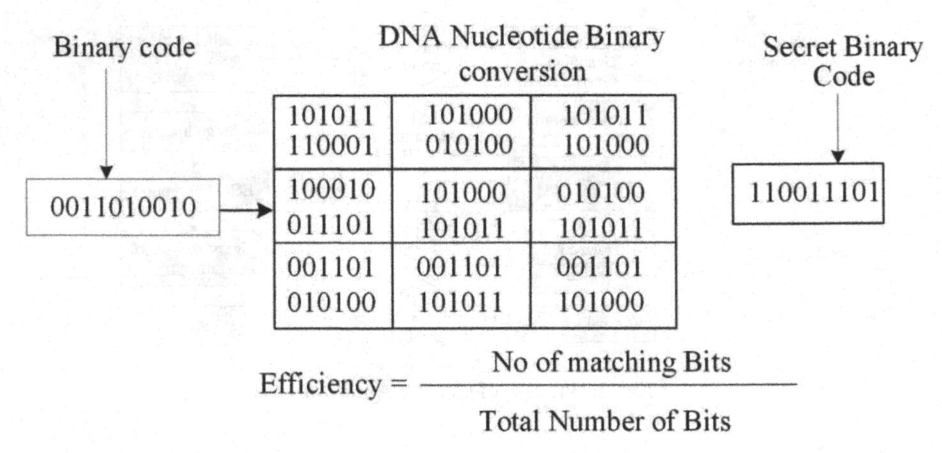

Fig. 5. Binary to DNA Nucleotide Binary conversion

The bits of secret image is flipped to obtain mirrored version of the plaintext Fig. 8. This increases the chances of matching between the plaintext and the carrier image in both Horizontal and vertical scanning in Fig. 9 and Fig. 10.

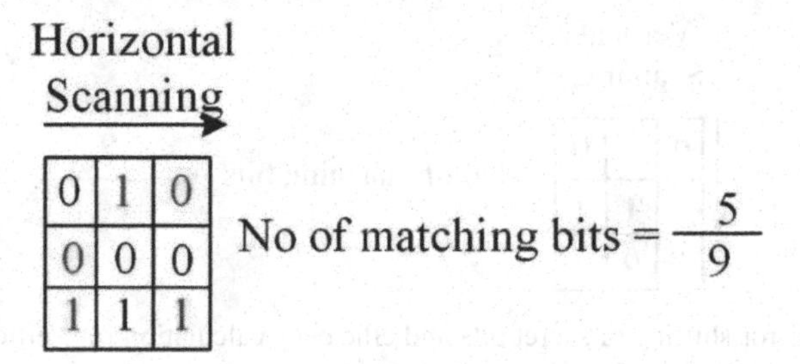

Fig. 6. Horizontal Scanning

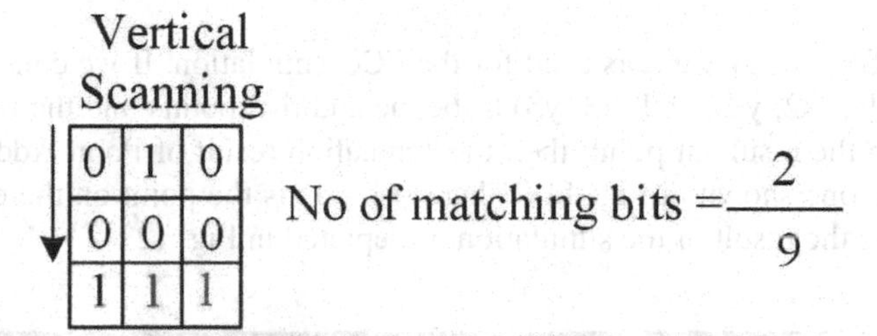

Fig. 7. Vertical Scanning

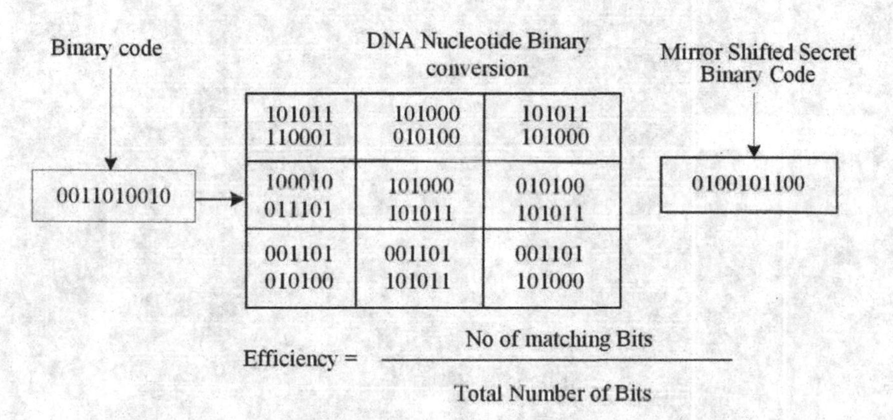

Fig. 8. Binary to DNA Nucleotide Binary conversion for Mirrored Binary Code

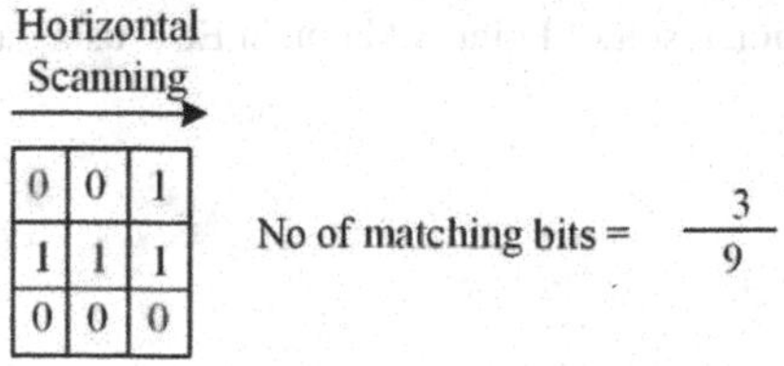

Horizontal Scanning

0	0	1
1	1	1
0	0	0

No of matching bits = $\frac{3}{9}$

Fig. 9. Mirror shifting of secret bits and efficiency calculation for Horizontal Scanning

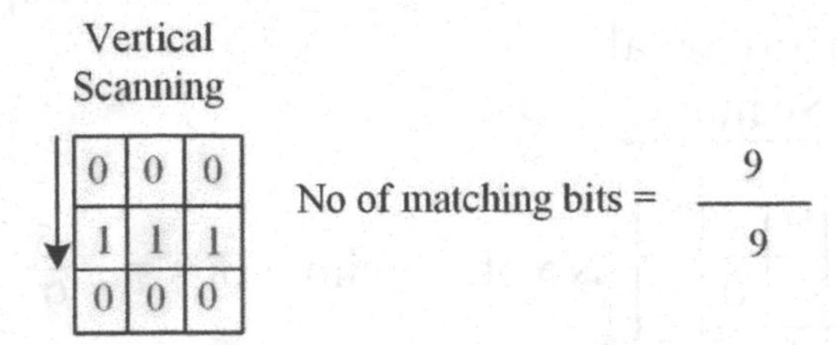

Fig. 10. Mirror shifting of secret bits and efficiency calculation for Vertical Scanning

4 Simulation Result

Xilinx 14.1 ISE Design Suite is used for the ECC simulation. If we consider the three points (x1, y1), (x2, y2), and (x3, y3) to be the additive points and the resulting point (x3, y3) to be the resultant point, then the simulation result of Point Addition in ECC looks like the one shown in Fig. 11. Where (x1, y1) is the point on the elliptic curve, and (x2, y2) is the result of the simulation, is depicted in Fig. 12.

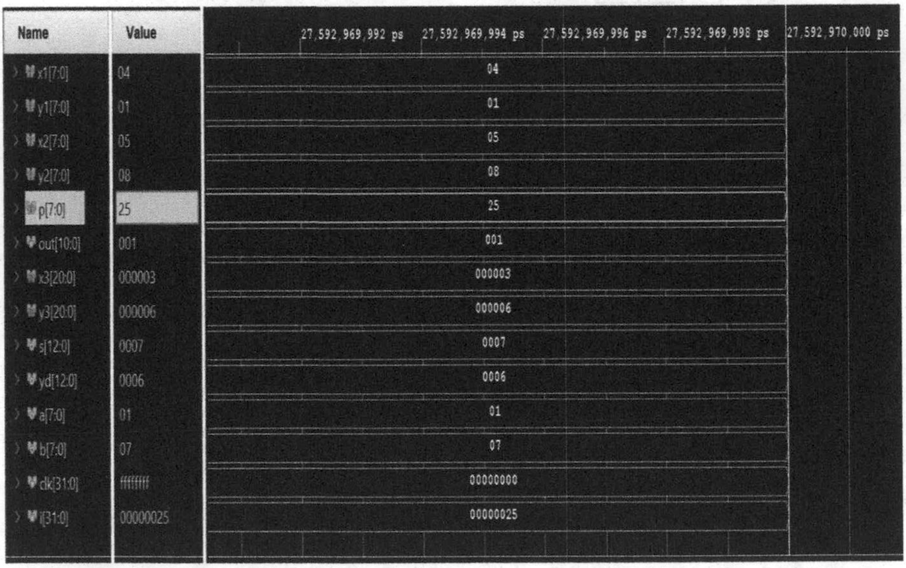

Name	Value	27,592,969,992 ps – 27,592,970,000 ps
x1[7:0]	04	04
y1[7:0]	01	01
x2[7:0]	05	05
y2[7:0]	08	08
p[7:0]	25	25
out[10:0]	001	001
x3[20:0]	000003	000003
y3[20:0]	000006	000006
s[12:0]	0007	0007
yd[12:0]	0006	0006
a[7:0]	01	01
b[7:0]	07	07
clk[31:0]	ffffffff	00000000
i[31:0]	00000025	00000025

Fig. 11. Simulation result of Point Addition in ECC using affine Coordinates

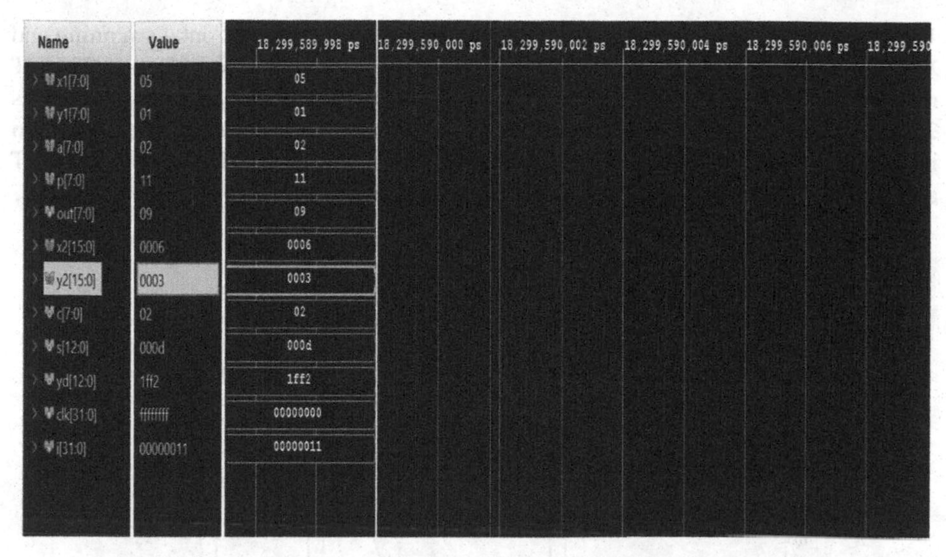

Fig. 12. Simulation result of Point Doubling in ECC using Affine Coordinates

4.1 ECC Based Arithmetic Operation

In this section, as shown in Fig. 9 and Fig. 10, the arithmetic Operations of ECC like Point Doubling,Point Doubling,DNA Function,Horizontal Scanning,Vertical scanning,Mirror shifted horizontal scanning and Mirror shifted vertical scanning were summarized and comparative analysis is given. The values like CPU Timing, Elapsed Timing, LUT and DSP are tabulated in Table 3.

Table 3. Comparison of Different ECC based Arithmetic Operation Metrics

ECC Operation	CPU Timing (s)	Elapsed Timing (s)	LUT	DSP
Point Addition	2	7	3450	7
Point Doubling	2	8	2218	4
DNA Function	19	116	24	–
Horizontal Scanning	3	65	26	1
Vertical Scanning	3	65	27	2
Mirror shifted Horizontal Scanning	23	181	23	2
Mirror shifted Vertical Scanning	23	165	28	3

From the Fig. 11 & Fig. 12, it is inferred that Point addition operations use 2s of CPU timing, have consumed 3450 LUT and 7 DSP slices 7s of point doubling timing, it has consumed 2218 LUT and 4 DSP slices. DNA function uses 19s of CPU timing, and

116s of elapsed timing, also it has 24 LUT and no DSP slices. Horizontal sca nning and Vertical scanning have 3s of CPU timing, 65s of elapsed timing, 26 LUT, and 27 LUT respectively with 1 and 2 DSP slices. Mirror-shifted horizontal and vertical scanning has 23 LUT and 28 LUT without DSP slices for 32-bit size. Their graphical representation is given below in Fig 13 & Fig. 14 and Fig. 15 for CPU Timing, Elapsed Timing, LUT and DSP Slices using the graph which is shown below.

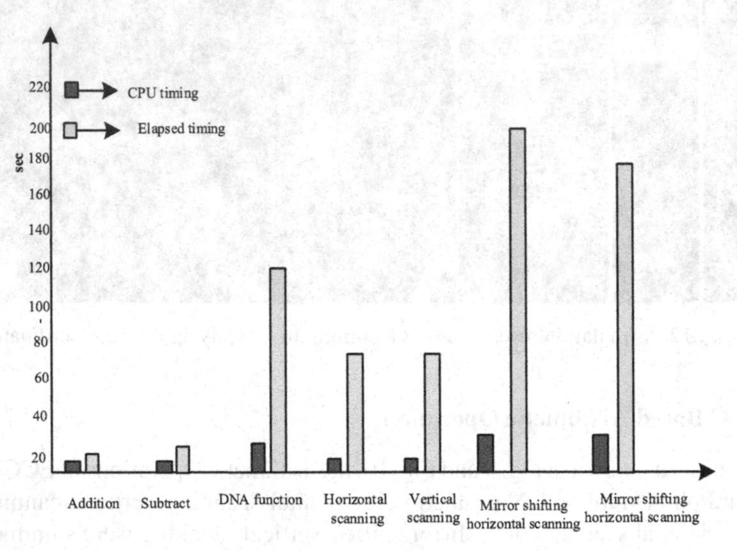

Fig. 13. Comparison of CPU and Elapsed timing Operations

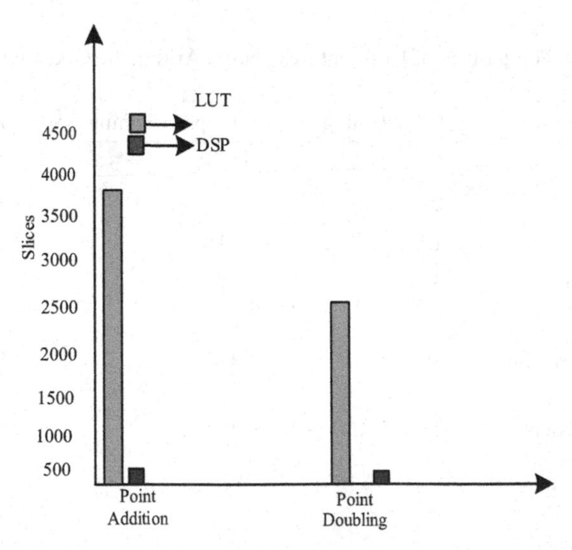

Fig. 14. Comparison of LUT's and DSP for Point addition and point doubling

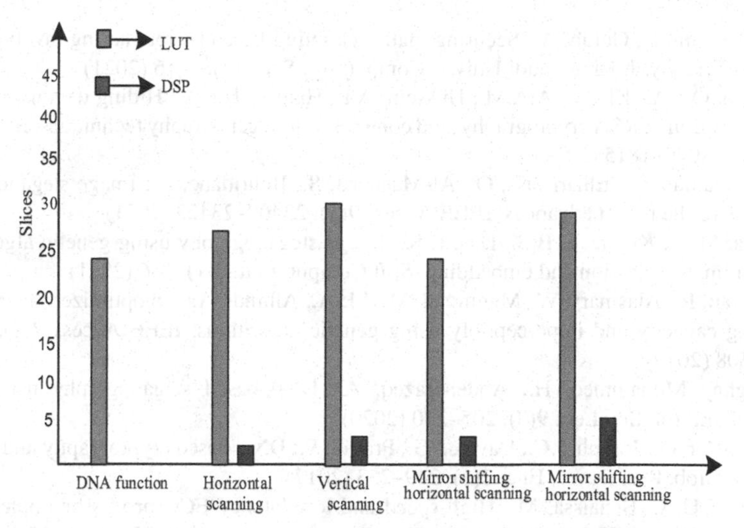

Fig. 15. Comparison of LUT's and DSP for DNA function for horizontal and vertical scanning method

5 Conclusion

In the proposed DNA computing based Medical image encryption techniques, genetic algorithm and Elliptic Curve Cryptography technique plays an important role to embed the plaintext into cover image and encrypt the stego image. Also, it proves that a proposed image encryption technique provides increased level of security with less computational overhead. Finally, the implemented results shows that proposed algorithm requires low CPU time, utilizes lesser number of gates and LUT's, and also occupies low memory space in Mirror shifted horizontal scanning method. In future, it will be improved by involving image steganography method to provide the two level of security to two different medical images.

References

1. El-Khamy, E., Korany, N.O., Mohamed, A.G.: A new fuzzy-DNA image encryption and steganography technique. IEEE Access **8**(1), 148935–148951 (2020)
2. Kulkarni, R., Manjunath, H.S., Darshan, T., Sudha, H.: VLSI design of DNA key generating algorithm for symmetric cryptographic systems. Int. Res. J. Eng. Technol. **8**(7), 1708–1711 (2021)
3. Kar, N., Mandal, K., Bhattacharya, B.: Improved chaos-based video steganography using DNA alphabets. ICT Express **4**(1), 6–13 (2018)
4. Vijayakumar, P., Vijayalakshmi, V., Zayaraz, G.: An improved level of security for DNA steganography using hyper elliptic curve cryptography. Wireless Pers. Commun. **89**(4), 1–22 (2016)
5. Attaby, A.A., Ahmed, M.M.F.M., Alsammak, A.K.: Data hiding inside JPEG images with high resistance to steganalysis using a novel technique: DCT-M3. Ain Shams Eng. J. **9**(4), 1965–1974 (2018)

6. Al-Shaarani, F., Gutub, A.: Securing matrix counting-based secret-sharing involving crypto steganography. J. King Saudi Univ. – Comput. Inf. Sci. **1**(1), 1–16 (2021)
7. Wahab, O.F.A., Khalaf, A.A.M., Hussein, A.I., Hamed, H.F.A.: Hiding data using efficient combination of RSA cryptography, and compression steganography techniques. IEEE Access **9**(1), 31805–31815 (2021)
8. Subramanian, N., Elharrouss, O., Al-Maadeed, S., Bouridane, A.: Image steganography: a review of the recent advances. IEEE Access **9**(1), 23409–23423 (2021)
9. Shyla, M.K., Kumar, K.B.S., Das, R.K.: Image steganography using genetic algorithm for cover image selection and embedding. Soft Comput. Lett. **3**(1), 1–6 (2021)
10. Wazirali, R., Alasmary, W., Mahmoud, M.M.E.A., Alhindi, A.: An optimized steganography hiding capacity and imperceptibly using genetic algorithms. IEEE Access **7**(1), 133496–133508 (2019)
11. Marghny Mohammed, H., Abdel-Razeq, A.: DNA-based steganography using genetic algorithm. Inf. Sci. Lett. **9**(3), 205–210 (2020)
12. Alexander, G., Joseph, J.C., David, J.G., Prasad, V.: DNA based cryptography and steganography. Glob. Res. Dev. J. Eng. **2**(6), 249–253 (2017)
13. Khan, Z.U.A., Benaissa, M.: High-speed and low latency ECC processor implementation over GF(2^m) on FPGA. IEEE Trans. Very Large Scale Integr. (VLSI) Syst. **25**(1), 165–176 (2017)
14. Choi, P., Lee, M., Kim, K., Kim, D.K.: Low-complexity elliptic curve cryptography processor based on configurable partial modular reduction over NIST prime fields. IEEE Trans. Circuits Syst. II: Express Briefs **65**(11), 1703–1707 (2018)

Author Index

B. K. Singh et al. (Eds.): ICBEST 2023, CCIS 2003, pp. 433–434, 2024.
https://doi.org/10.1007/978-3-031-54547-4